Perspectives on Process Safety from Around the World 2015

Topical Conference at the 2015 AIChE Spring Meeting and 11th Global Congress on Process Safety

Austin, Texas, USA
26-30 April 2015

ISBN: 978-1-5108-0687-0

Printed from e-media with permission by:

Curran Associates, Inc.
57 Morehouse Lane
Red Hook, NY 12571

CURRAN ASSOCIATES INC.
proceedings.com

Some format issues inherent in the e-media version may also appear in this print version.

Printed by Curran Associates, Inc. (2015)

AIChE
3 Park Avenue
New York, NY 10016-5991

Phone: (203) 702-7660
Fax: (203) 775-5177

www.aiche.org

Additional copies of this publication are available from:

Curran Associates, Inc.
57 Morehouse Lane
Red Hook, NY 12571 USA
Phone: 845-758-0400
Fax: 845-758-2634
Email: curran@proceedings.com
Web: www.proceedings.com

TABLE OF CONTENTS

409505 Chemical Engineering: Is This the Golden Age?

Abstract:

Schlanger will offer his insights on the current opportunities in the chemical process industries in his talk entitled "Chemical Engineering: Is This the Golden Age?" on Monday morning, April 27. At the conclusion of his lecture, Schlanger will receive the AGILE Award in recognition of his leadership at CEVA Group, a $9 billion logistics supplier to industry worldwide; LyondellBasell, a major player in the plastics, chemicals, and refining industries; Arco Chemical; and at the other companies where he has held leadership positions.

Development and Research Activities of Advanced System Safety Laboratory for Safety Education/Culture

Atsuko Nakai
Center for Safe and Disaster-Resistant Society
Okayama University,
Tsushima-naka 3-1-1, Kita-ku, Okayama, 700-8530, JAPAN
fumoto@safelab.sys.okayama-u.ac.jp

Kazuhiko Suzuki
Division of Industrial Innovation Sciences,
Graduate School of Natural Science & Technology,
Okayama University,
Tsushima-naka 3-1-1, Kita-ku, Okayama, 700-8530, JAPAN
kazu@sys.okayama-u.ac.jp

Prepared for Presentation at
American Institute of Chemical Engineers
2015 Spring Meeting
11th Global Congress on Process Safety
Austin, Texas
April 27-29, 2015

UNPUBLISHED

Development and Research Activities of Advanced System Safety Laboratory for Safety Education/Culture

Atsuko Nakai
Center for Safe and Disaster-Resistant Society
Okayama University,
Tsushima-naka 3-1-1, Kita-ku, Okayama, 700-8530, JAPAN
fumoto@safelab.sys.okayama-u.ac.jp

Kazuhiko Suzuki
Division of Industrial Innovation Sciences,
Graduate School of Natural Science & Technology,
Okayama University,
Tsushima-naka 3-1-1, Kita-ku, Okayama, 700-8530, JAPAN
kazu@sys.okayama-u.ac.jp

Keywords: Safety culture, Operation support, dynamic simulator

Abstract

In recent years, serious accidents have been continually occurring in large industrial facilities in Japan. We have a responsibility to convey what we have learned from experiences to the younger generation to provide them with safe working environments and protect people's lives. Our society needs to be assured about the safety and reliability of chemical plants to prevent accidents. Safety education is especially important because safe practices are based on a strong foundation of safety experience and safety culture. This paper presents our research activities concerning the Advanced System Safety Laboratory. We have developed a manual operations support system using augmented reality (AR) technology and an operator training system using virtual reality (VR) technology. Both systems will improve safety awareness.

1. Introduction

As is well-known, many kinds of hazardous materials are being controlled in chemical facilities. If a severe accident occurs, there is the potential for serious damage to employees and the residents of local communities. Our society needs to be assured about the safety and reliability of chemical plants to prevent accidents. Safety education is important because safe practices are based on a strong foundation of safety experience and safety culture. Safety culture exists in people's minds. For our future generations, we need a steady and sustainable culture of safety in the field of chemical engineering. To construct a safety culture for all persons involved with safety engineering, the following aims are particularly important:

1. Involving individuals and organizations for safety responsibilities.
2. Having a mutual understanding and willingness to share information about risks and safety.

3. Promoting the inheritance of technology and safety knowledge.

A safety culture concerns the comprehensive attitudes and characteristics of both individuals and organizations. Every person may have a different sense of values, which ultimately lowers social awareness and creates difficulties in partnership. In this paper, we aim to present our research activities concerning the Advanced System Safety Laboratory to promote safety education and culture.

2. Development and Research Activities

2.1 Advanced System Safety Laboratory

The Advanced System Laboratory is one of the leading laboratories in Okayama University. As researchers, we have run a research and education project related to safety engineering and risk management. Its main aim is to address the urgent need for highly-skilled and knowledgeable human resources in Japan. We established the Center for Safe and Disaster-Resistant Societies in January of 2014. The key topics include safety design features for disaster-resistant societies: safety culture, risk management, security management, and public communication. We will provide graduate programs, including Master's and Doctoral courses. Intensive courses will be available to post-graduates as well as management and technical staff working in relevant industries.

2.2 Research Challenges

Improving the safety and reliability of plants is extremely important for workers' safety. Consequently, a number of activities, countermeasures, and assessment techniques have been proposed. We are required to predict potential risks before any accident or disasters occurs, take preventative/reactionary measures and organize a management system that is compatible with the risks. In addition, a safe process design is important for preventing human error. Control problems are increasingly significant in both software and hardware as the economic situation changes over time. Our main research challenges concern the following areas: advanced operating support (OS) and safety management; safety education and training using virtual reality (VR) and augmented reality (AR); and hazard identification based on dynamic simulation.

2.3 Development Systems

We have developed the manual operations support system using AR technology and an operator training system using VR technology. Both systems are based on plant information and use a dynamic simulator. The manual operation support system uses AR technology to help operators to confirm quantitative data in the field. The operator training system can offer safety education and training for chemical plant workers. A trainee can learn the correct operations through simulated experiences. In Japan, there have been rapid generational changes in operators. The inheritance of technology and accomplishments is essential to ensuring the safety of environments. When responding to an accident, quick decisions are required. However, at present, there has been insufficient training due to safety and cost issues. This operator training system will lead to improved safety awareness.

3. Manual operation support system using AR

This system can display changing sensor data on a device for field workers. The field worker operates the instruction device like tablet pc, and the sensor data are updated in real time. Using AR technology, field workers can visually receive information about the instruments. Also, the instructions take the form of quantitative information from dynamic simulator instead of spoken dialogue. This system can reduce transmission errors from operators to field workers.

3.1 Outline of Manual operations support system

This system can provide quantitative instructions for the operation of field workers. After setting the sensors to be displayed, the target value varies for each sensor and is displayed on field workers' tablet PCs. When field workers operate the devices, the fluctuant sensor data are updated in real time. The system can display different colors that represent differences between the target values. A field worker can intuitively ascertained sensor data using the different colors between the set target values. Figure 1 shows an outline of the system.

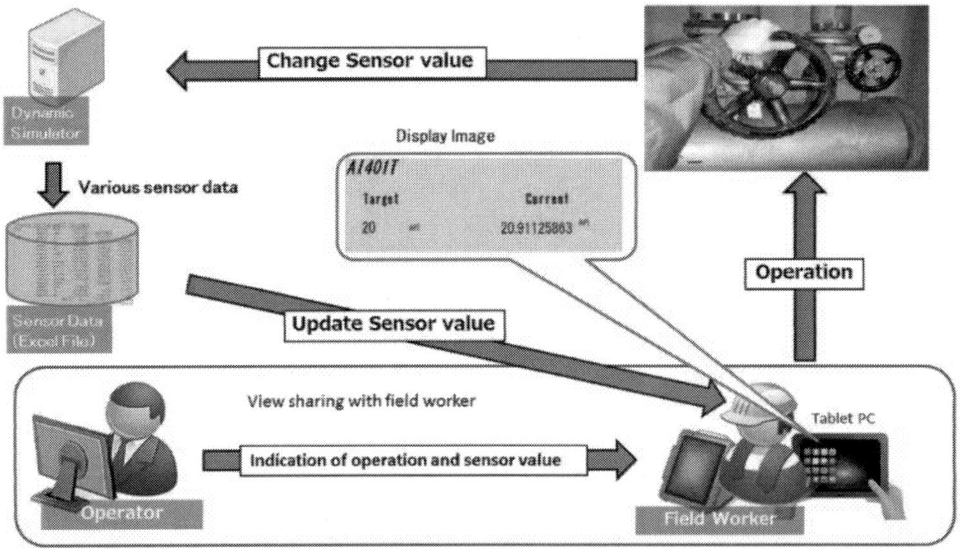

Figure 1 Outline of manual operation support system

By capturing the screen of the field workers' devices, the DCS operator and field worker can share the field of view. Thus, the DCS operator can also check the operation and prevent any errors from being made by field workers.

3.2 Capture the sensor data

This system is linked with the dynamic plant simulator (DS). The field worker can confirm current sensor data using the device. When the equipment is operating, the system indicates the optimum operations for the field worker through color changes. The display shows the different colors via the operation of the field worker, as shown in Figure 2. Green color means that the operation is safe and red color means that the operation is dangerous.

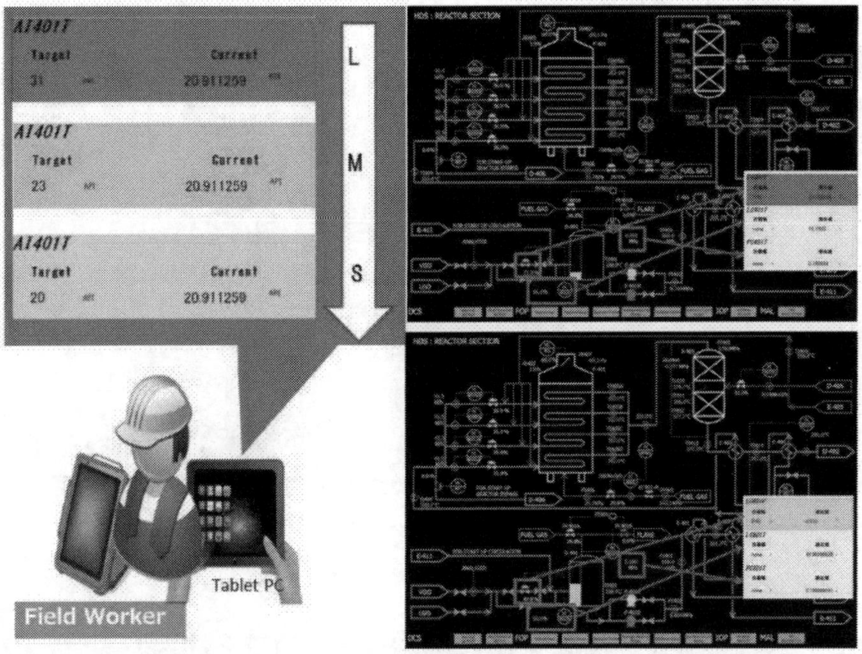

Figure2 Display color linked with dynamic simulator

Using this system, a field operator can visually confirm real time data. In addition, the instructions from the DCS operator to the field worker can include quantitative instructions from qualitative information, such as an individual's voice.

4. Safety Training System Using VR Technology

Workers' safety awareness can be improved by experiential learning. However, reproducing the non-stationary conditions of actual equipment or mock-up equipment is not possible because it is too dangerous. Using VR technology, we can build a virtual chemical plant with lower costs as compared to a real plant. The operator can experience an accident involving fire and explosions in a virtual space. This system is linked with plant models within the VR environment using the process values of the DS. To safely operate the device in various chemical plants, operators need to be responsible for their own areas. Plant operators are responsible for filling two positions: field men and board men. Field men have the role of operating equipment in the field. Board men have the role of controlling refineries in the control center or control room. A trainee in this system can experience work as a field man or a board man in an environment that is similar to actual chemical plants.

4.1 Outline of the training system

In an emergency, an operator is required to make quick decisions to prevent the expansion of an accident. An untrained operator cannot properly respond to emergencies. Education and training of non-stationary operations is required for workers. This paper will present a training system for chemical plants using the VR technology that we have developed. Figure 3 shows an outline of the training system. Accidents do not always happen the way they are described in manuals. Individuals should be made aware of what to do in the case of an emergency.

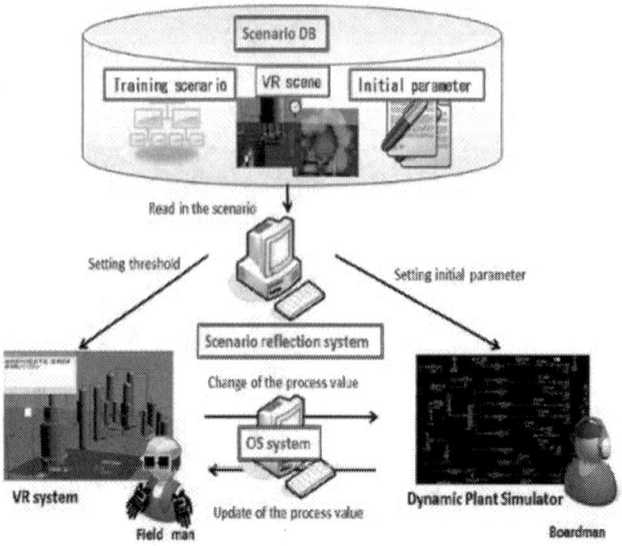

Figure3 Outline of the training system

The system was developed using VR as a new training method. This is a technology that artificially creates a sense of reality. Our system is composed of a combination of computer graphics and DS information. The trainee can feel as if they are actually experiencing things in the virtual plant. The DS integrates dynamic simulation environments that reproduce the behaviors of actual plant operations; it provides a sense of realism. Therefore, the DS has been used for operator training systems. This proposed system is composed of a virtual plant that works in concurrence with the DS in VR environments. As a result, it became possible to reflect on changes in process values corresponding to the passage of time in a VR environment. Operators and workers can learn the correct operations through these simulated experiences to prevent accidents from occurring. In this study, the system was developed to consider both field men and board men. The board men input the process values in the DS display. The result is then reflected in the virtual plant. The field men operate the equipment using an avatar in the virtual plant through the virtual plant screen. An event qualitatively progresses by using animations in the virtual plant and using the process values corresponding to the DS, according to the amount of operation change. The training system is constructed using three subsystems: a scenario database (DB), Operations Support (OS) system, and scenario reflection system. It also consists of two parts: a virtual plant and dynamic simulator. The biggest difference between web learning and VR environments is that a trainee can reach out into the virtual world and use the VR technology.

4.2 The OS system

This system reflects changes in process values as time progresses in the VR environment by synchronizing the plant model in the VR environment with the process values of the plant simulator. Each process value from the plant simulator is stored in Excel. The system outputs process values as a CSV format. The VR environment reflects behavior as a change in the process value of the equipment by storing the value in the array in the VR environment. Moreover, the equipment operation is synchronized using the simulator. When the equipment operation in the VR environment is complete, the OS system writes the data in the CSV for the dynamic simulators' operation through Excel. Figure 4 shows the OS system.

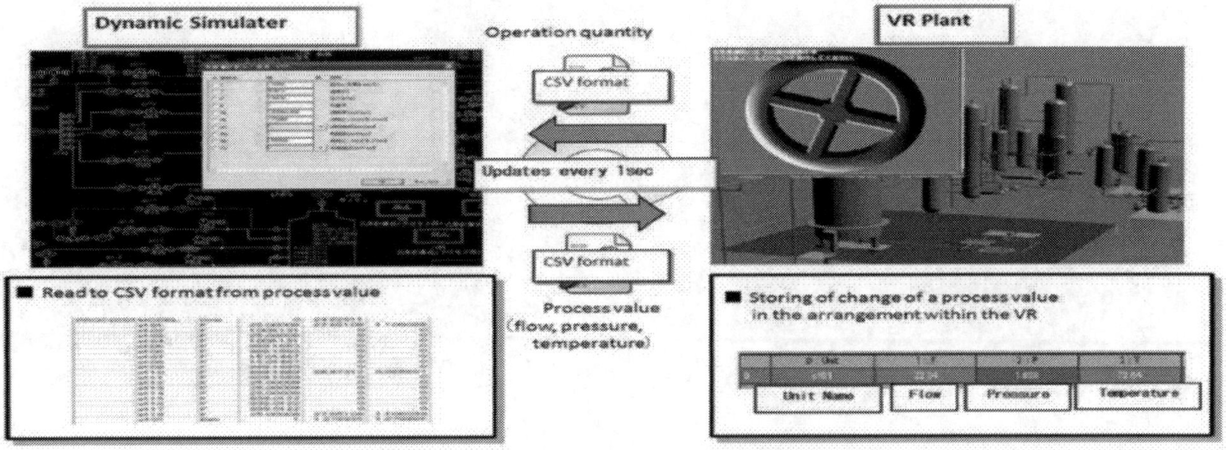

Figure4 Composition of OS system

4.3 Output of VR scene linked DS

In this system, the process value is linked with the plant model in the VR environment using the OS system between the DS. The scenario reflection system compares the virtual plant conditions in the scenario using the process values stored in the array in the VR environment and outputs the corresponding scene. Figure 5 indicates that the system can change the output based on the process values.

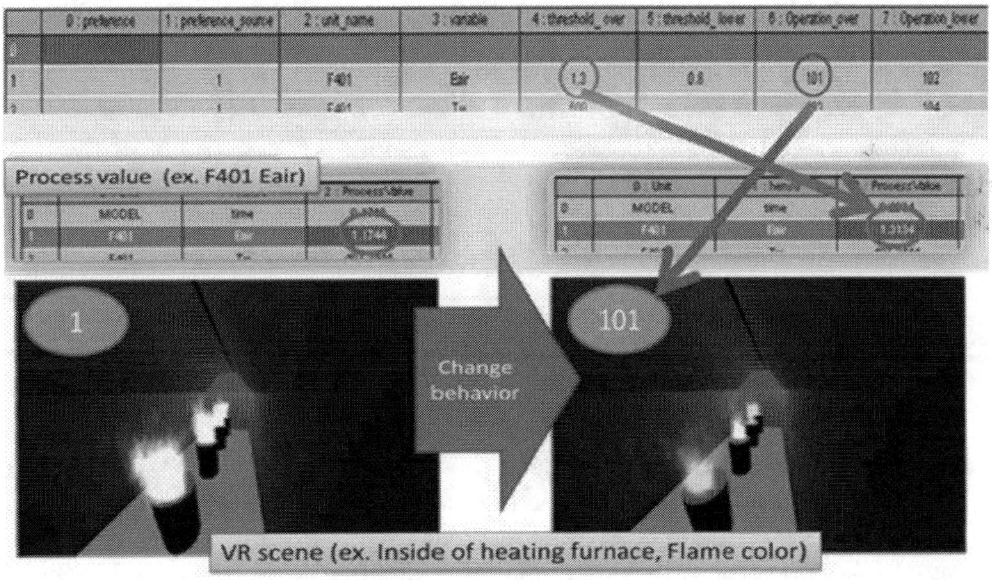

Figure 5 Output of VR scene linked DS

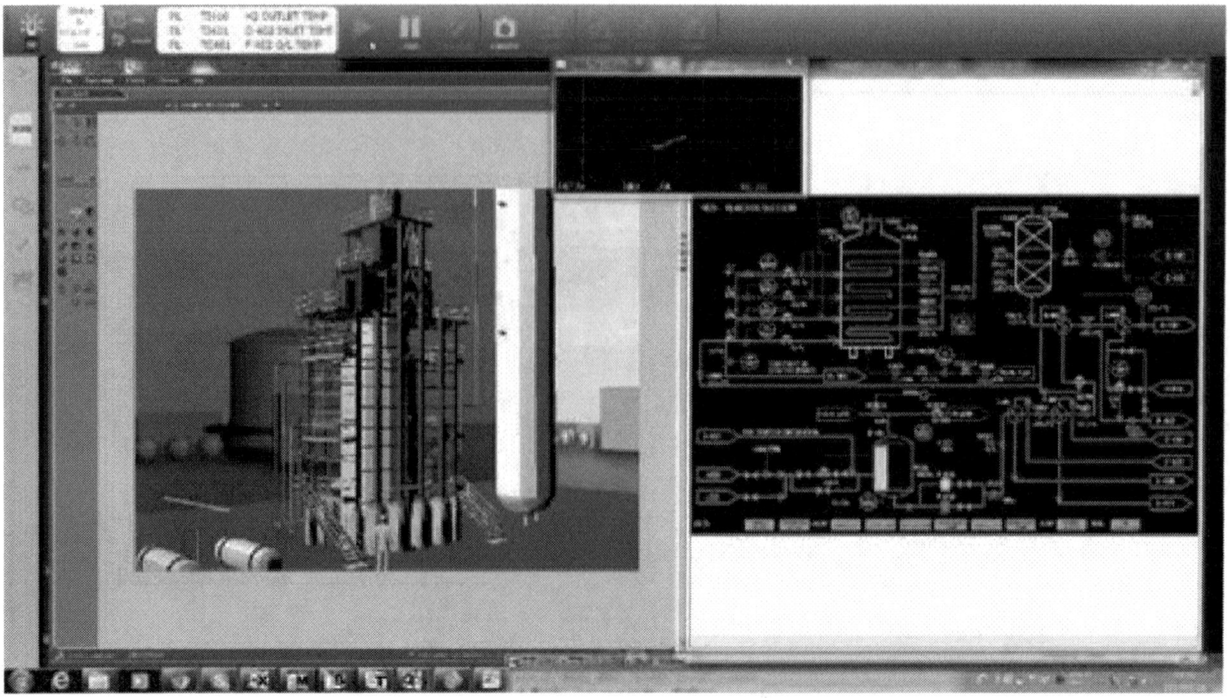

Figure 6 The training system of the hydrodesulfurization process

Figure 6 shows the screen of a training system concerning the hydrodesulphurization process. Board men and field men can learn to operate in another role in this training system. The trainee can safely perform the training of the corresponding operations in a non-steady situation. In this system, VR environments and the DS are integrated. A trainee in the system can experience cooperative work between field men and board men in an environment that is similar to a real chemical plant. Through training in a virtual plant, a trainee can experience hazardous conditions that cannot be experienced in real life. This training is useful for workers' safety education and training, especially when they have no experience. After engaging in experience-based learning in a virtual plant, the operator should review the corresponding operations of the accident. We will also add a re-learning part that a trainee can find the cause of the accident and safety operation.

5 Conclusion

In chemical plants in Japan, there have been rapid generational changes in the workers. Inheriting technology and past accomplishments in a chemical plant is essential for enhancing competitive production. When an accident or disaster happens, the correct operations must be taken without human error. However, at present, sufficient training is not being done due to safety and cost issues. Our laboratory developed a manual operations support system that uses AR as well as an experience-based safety training system that uses VR. The proposed system will lead to improvements in safety awareness during day-to-day operations due to increases in skill, sensibility, knowledge, and overall competence.

6 References

[1] S. Jürgen, "Process and Plant Safety – Research & Education Strategy to Keep Long Term Competences", Chemical Engineering Transactions, vol.31, 2013.

[2] T. Ogawa, "The security measure subject of the latest chemical factory", SCAS NEWS 2003 pp.1-2.

[3] IAEA Safety Series No.75-INSAG-1 "Summary Report on the Post-Accident Review Meeting on the Chernobyl Accident", Vienna, 1986

[4] D. Schofield, "Mass Effect: A Chemical Engineering Education Application of Virtual Reality Simulator Technology", MERLOT Journal of Online Learning and Teaching vol. 8, No. 1, March 2012.

[5] Investigation Committee on the Accident at Fukushima Nuclear Power Stations of Tokyo Electric Power Company, Executive Summary of the Final Report, 23 July 2012.

[6] G. Fukano, K. Yokoyama, Y. Yahata, "MIRROR PLANT On-line Plant Simulator and its Applications", Yokogawa Technical Report English Edition ,vol.56 No.1, 2013 pp. 11-14.

[7] S. Nazir, S. Colombo, D. Manca, "Minimizing the Risk in the Process Industry by Using a Plant Simulator: a Novel Approach", Chemical Engineering Transactions, vol.32, 2013 pp.109-114.

[8] K. Yamamoto, A. Nakai, K. Suzuki, "Development of Experienced-based Training System combined with Process Dynamic Simulation", Asia Pacific Symposium on Safety 2013, October 2013.

[9] C. Norton, I. Cameron, C. Crosthwaite, N. Balliu, M. Tade, D. Shallcross, A. Hoadley , G. Barton, J. Kavanagh, "Development and deployment of an immersive learning environment for enhancing process systems engineering concepts" , Education for Chemical Engineers,vol. 3, 2, Decwmber 2008 pp.75-83.U.S.

11TH GLOBAL CONGRESS ON PROCESS SAFETY

Development of Quantitative Hazard Analysis Method for Inherently Safer Chemical Processes

Yuto Mizuta
Safety Technology Laboratory, MCHC R&D Synergy Center
1000, Kamoshida-cho Aoba-ku, Yokohama 227-8502 JAPAN
mizuta.yuuto@mf.mchc-rdsc.co.jp

Prepared for Presentation at
American Institute of Chemical Engineers
2015 Spring Meeting
11th Global Congress on Process Safety
Austin, Texas
April 27-29, 2015

UNPUBLISHED

Development of Quantitative Hazard Analysis Method for Inherently Safer Chemical Processes

Yuto Mizuta
Safety Technology Laboratory, MCHC R&D Synergy Center
1000, Kamoshida-cho Aoba-ku, Yokohama 227-8502 JAPAN
mizuta.yuuto@mf.mchc-rdsc.co.jp

Keywords: inherently safer chemical processes, quantitative hazard analysis, worst case scenario, abnormal scenario

Abstract

Chemical plants have many kinds of chemical materials and operating conditions are also various. If the accident occurred, it has a possibility to give severe consequence. Therefore it is important to analyze the hazard potentials in chemical plants. In this work, we reconsider the worst case scenario which is widely used in the world, and in addition we defined the abnormal scenario to become able to analyze the hazard potentials more accurately. Hazard potentials are plotted on the graph which takes fatalities and distance for each axis, and they are ranked to 3 categories according to their consequences. Furthermore, the facilities which have high hazard potentials should be practiced safety measures to reduce their hazard potentials, and we show the example the way to reduce hazard potentials by using inherently safety measures, and it clarifies which safety measure is how effective, and also able to select more effective safety measure.

1. Introduction

There are many accidents of chemical processes in Japan recently such as the explosion accident of the oxidation reactor caused by reaction runaway and the fire accident of the monomer tank caused by polymerization. Therefore it is re-realized the importance of review about the safety measures in upset condition of chemical processes. US.EPA established RMP regulation in 1996 and API has the guideline RBI for instance, and chemical companies implement consequence analysis to identify their hazard potentials and design prevention and mitigation programs. In Europe, in the same way, the Seveso Directive was established in 1982, and chemical companies in each country implement consequence analysis such as TNO full-QRA. In the case of Japan, FDMA(Fire and Disaster Management Agency) of the ministry of internal affairs and communications established the Disaster Prevention Plan for Petrochemical Complexes in 1994, and it was revised in the wake of 2 big earthquakes, the great Hanshin-Awaji earthquake on 1995 and the great east Japan earthquake on 2011. The consequence calculation method described in the plan of FDMA is the same one used in the world. Such methods would give accurate results because they are verified with many experimental data. However, because there are not detail description in the plan which equipment we should select and the scenario what will occur, we cannot understand whether our analysis is right or not. Even in the guideline in the world, it is necessary to decide what will occur by considering the relation between process conditions and material characteristics. In this work, for the quantitative identification of hazard potentials, we reconsider the assumptions of the worst case scenario (WCS) everyone obtains the same and accurate results. In addition, we defined abnormal scenario (ANS), because there are some cases it is not possible to identify hazard potentials by only WCS. Analysis of both scenarios makes it possible that we can comprehend the hazard potentials more accurately.

2. Worst Case Scenario (WCS)

There is the description about the accident events we should assume in the plan of FDMA, and that includes not only the accident events caused by release, fire and explosion in the normal process condition but also earthquake and tsunami. Therefore we should analyze the consequence in case of release of whole amount of chemical substances from the process equipment as WCS, because earthquake could make it real. We reviewed WCS of US.EPA and release scenario of API and TNO, and we redefined WCS so that everyone obtained the same results. We redefined the selection definition of the equipment, the release amount from the process equipment and the accident events, and they are introduced below.

2.1 Preparation for the calculation of hazard potentials

It is necessary to gather information shown in Table.1 to analyze the hazard potentials by WCS, ANS is also the same, for selection of equipment, identification of accident events and release amount from process equipment.

Table 1. Plant information for consequence analysis

Process conditions	Process equipment	Characteristic of materials	Other plant information
Temperature (deg.C)	Volume of equipment (m^3)	Formula of material	Flow sheet
Pressure (kPa)	Internal length (m)	Molecule weight (g/mol)	Material balance sheet
Composition (wt.%)	Hold up volume (m^3)	Boiling point (deg.C)	Equipment list
Flow rate (kg/hr)	Design temperature (deg.C)	Flash point (deg.C)	Information of dike
Phase of material	Design pressure (kPa)		

2.2 Selection of the equipment

In this paper, the analysis method is explained by using ammonia production plant shown in Fugure.1. The first step for the analysis is the selection of the equipment. In the plan of FDMA, the selection of the equipment for the analysis is that basically there are in the chemical plants in the petrochemical complex and the equipment deal with large amount of chemical substances such as production facilities, storage installations, shipping facilities and transport facilities. However, in case of highly toxic substances, even though the amount of substances in the equipment is not so large, it seems that the hazard potentials are so high. The selection of equipment is all ones having the large inventories described in PFD or P&ID, and it except pumps, short and small pipes and small heat exchangers. In Figure.1, the inventories of the tank, column or reactor are larger than pumps or pipes, and the hazard potentials of the equipment having small inventories are assumed to be represented by the nearest and larger equipment. In the ammonia production plant shown in Figure.1, encircled 12 equipment are selected for the analysis.

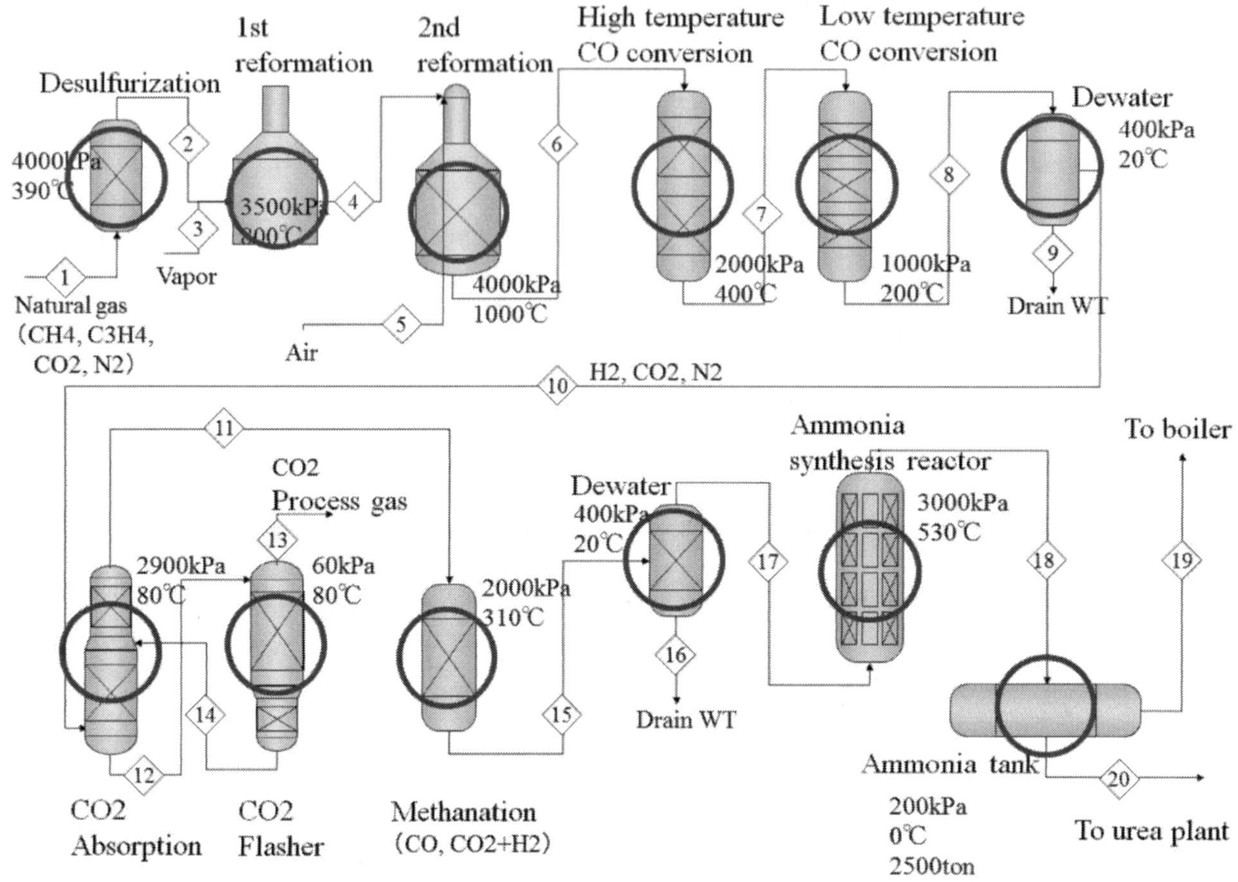

Figure 1. Process flow diagram of ammonia production plant

2.3 Definition of the accident events and the amount of release substances

It is necessary to clarify the assumption to obtain the same results among any person who analyze the hazard potentials. The accident events, what will occur after release from the equipment, are defined by considering the correlation among the boiling point, the flash point and ignition point of the released substances, process temperature and ambient temperature. The correlation of those parameters gives different release condition, i.e. gas phase release or liquid phase release or both, and gives different events after release. The flowchart of the accident events is shown in Figure.2. Here, the ambient temperature is defined as 40 degrees C which is the highest temperature in Japan. High ambient temperature urges gasification at the release from the equipment and at the evaporation from pool, and also gives downward movement of released cloud toward near the ground level which is the region we focus in the aspect of harmful for receptors. Furthermore, regarding wind speed and the atmospheric condition for toxic gas dispersion, they are defined as 1.5 meter per second and stability class F, which are same as the condition of WCS of EPA. The amount of release substances is shown in Figure.3. The amount of release substances is defined as the total amount of the inventory in the equipment and the amount of feed to the equipment in certain time period. The certain time period of feed is defined as 3 minutes in case of explosion and fireball, and in case of others defined as 10 minutes. The time period of 3 minutes is referred by API-RBI, which is based on the experience that the explosion would not occur after 3 minutes or later after release from the equipment because the concentration of the cloud would decrease until lower flammable limit by dispersion.

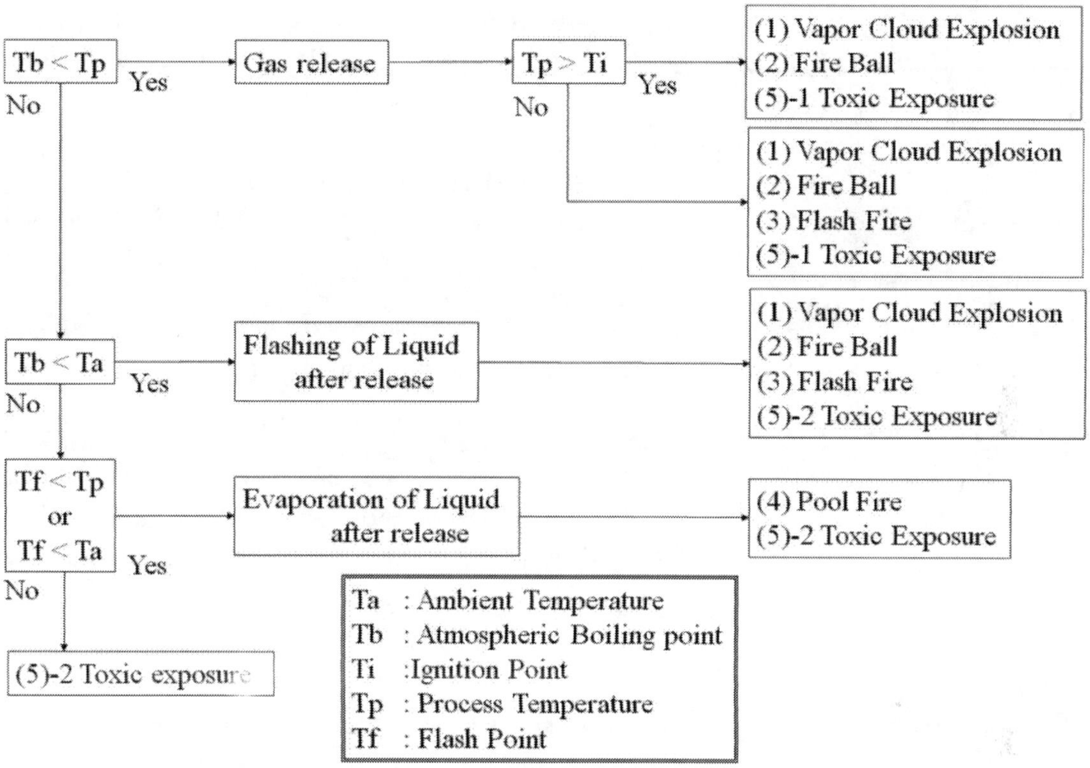

Figure 2. Release scenarios and event types

3. Abnormal Scenario (ANS)

The analysis method of the hazard in the chemical plants is often implemented by WCS, but traditional WCS is assumed the release from the equipment in normal operation condition. However, if some failure in the chemical process causes deviation of process condition from normal condition, there are possibilities that more severe accident events will occur. For example, even if it does not give so severe consequence in case of the release accident from the reactor controlled temperature in the normal condition, it would give more severe consequence if the runaway reaction occurs because of temperature rise of hot oil flowing into reactor jacket. Therefore, there is possibility to underestimate the hazard potentials if only WCS is assumed for the analysis, and it is important to assume ANS and add it to the analysis.

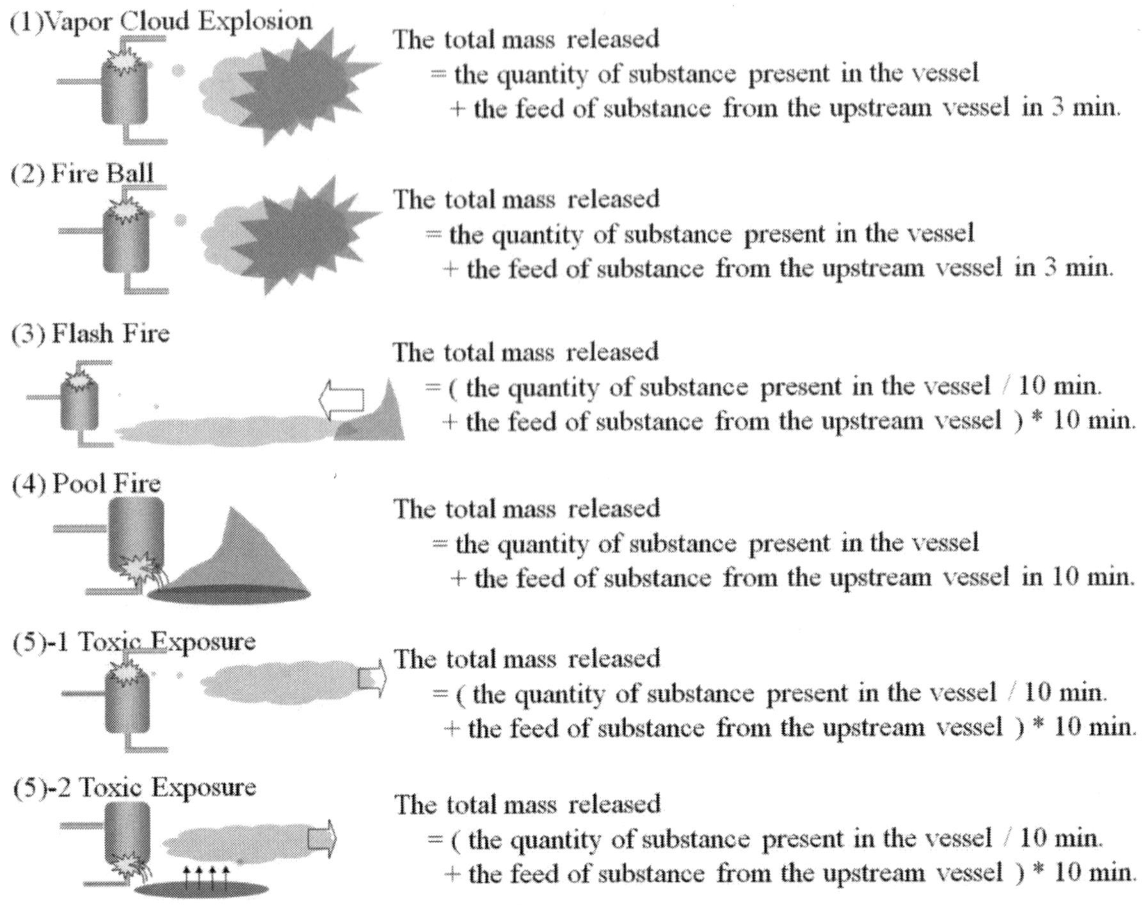

Figure 3. Release amounts defined by each event

3.1 Example of the analysis of ANS

Regarding ANS, it is necessary to make accurate scenarios. At first, some failure such as shutoff of valve or pump happen in the chemical process, and some deviation from normal condition occurs, and deviation propagate to around process, and finally explosion or release from equipment occur. Making scenarios of ANS are implemented by hazard identification method, e.g. HAZOP analysis or FMEA analysis. The example of ANS of ammonia production plant is shown below. Here, it assumes that TCV2021 shutoff occurs suddenly. TCV2021 is the valve to control of flow rate of cooling medium in the heat exchanger. If cooling medium does not flow into the heat exchanger, the temperature of ammonia tank gradually rises and at the same time internal pressure rises by ammonia vapor pressure. And then, the temperature alarm and the pressure alarm are alarmed and operators try to do recovery work. But if it is impossible to recovery, pressure in ammonia tank rises continuously. Safety valve is installed on ammonia tank so that pressure is released outside from ammonia tank at set-point below tank design pressure. However, if safety valve is not activated, tank will be ruptured and ammonia will be released at higher temperature than normal operation. According to the probability calculation by FTA analysis, the probability of tank rupture is not so high and it is within acceptance level. Because there is enough time to do recovery work during pressure-rise up to design pressure and safety valve is installed on ammonia tank. But if ammonia tank rupture occur, ammonia is released at higher temperature from tank and it causes flash of larger amount of ammonia, and large amount of ammonia gas will give more sever consequence.

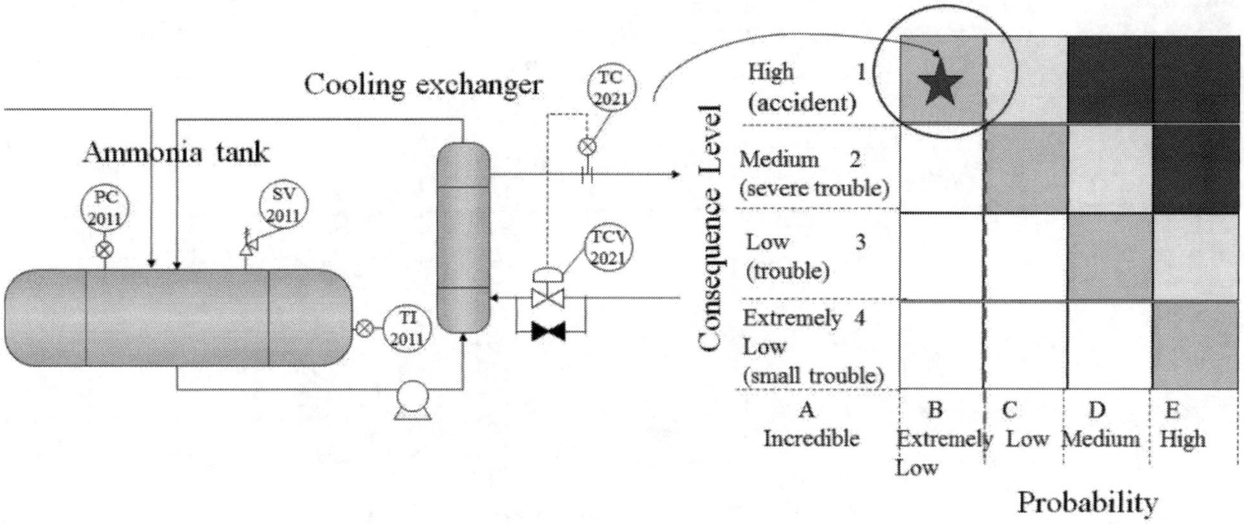

Figure 4. The example of ANS around ammonia tank

3.2 Definition of the accident events

If the release from the equipment is assumed, it is possible to use the accident events of WCS defined as Figure.2 for ANS. In addition to WCS, it is necessary to consider about the internal explosion of the equipment, e.g. reaction runaway, dust explosion. It has possibility that the internal explosion gives more severe consequences than semi-confined explosion in the chemical plant.

4. The presentation of the results of WCS and ANS

The results analyzed by WCS and ANS are expressed on the graph such as Figure.5. The vertical axis of the graph takes fatality number caused by accident events such as explosion, fireball, etc. The horizontal axis of the graph takes 10% lethality distance which consequence calculation gives 10% death probability to receptors. Furthermore, plant borderline and 1 fatality are taken as each criterion for severity, and the graph is divided 3 color categories. The region that fatality number is more than 1 fatality and 10% lethality distance is further than plant borderline is defined as the red color category. The region that fatality number is less than 1 fatality and 10% lethality distance is not further than plant borderline is defined as the green color category, and other 2 regions, right bottom and left top, is defined as the yellow color category. If process equipment plotted in the red area, that accident events affect over plant borderline and cause more than 1 fatality, safety practice should be considered immediately. Yellow color category means ARALP region and green color category means acceptance region respectively. According to this work, ammonia tank and ammonia synthesis reactor have high hazard potential in ammonia production plant. Furthermore, plotting past accidents on the graph and comparing to the calculation results, it is possible to imagine how large impact of the accident events would be. Table.2 shows the past accident examples, and fatality numbers and 10% lethality distances are calculated by using simulation software TRACE. Past accident examples are plotted on the graph in Figure.6 in addition to calculation results of ammonia production plant. According to comparison, the toxic consequence of ammonia release from 2500 ton tank implies higher hazard than Bhopal accident.

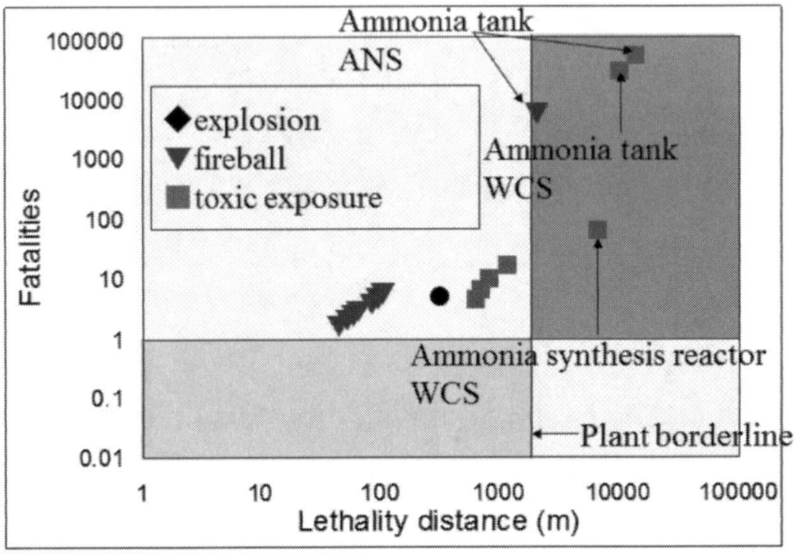

Figure 5. Result of consequence analysis of ammonia production plant

Table 2. Results of past accident example

Places of past accidents	Substances	Events	Release amount [ton]	Fatalities (Actual) [person]	Fatalities (calculation) [person]	Lethality distances [m]
Flixborough	Cyclohexane	VCE, fireball	13	28	50	227
Bhopal	MIC	Toxic gas dispersion	57	6,000	3,413	6,013
Mexico	LPG	VCE, fireball	800	500	338	1,476
Senegal	Ammonia	Toxic gas dispersion	22	129	163	1,052

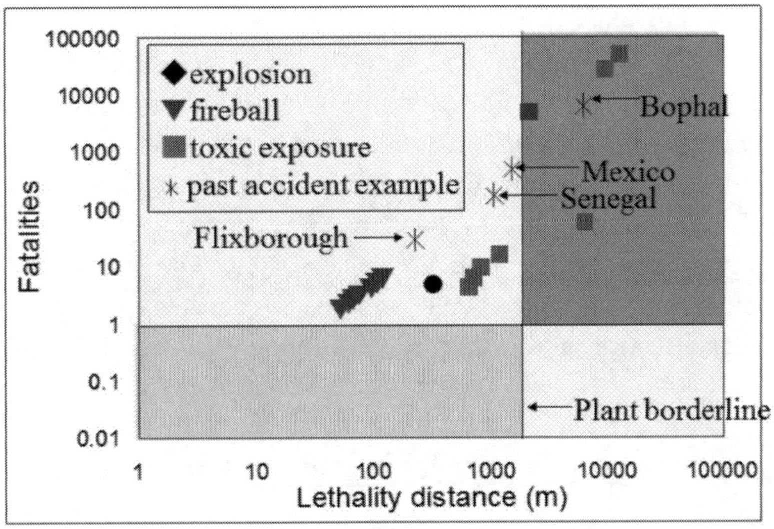

Figure 6. Comparison consequences to the past accidents

5. Applying inherently safety measures

In Figure.5, the hazard potentials of ammonia tank and ammonia synthesis reactor are clarified. Though their potentials are severe ones, it is possible to decrease hazard potentials by applying inherently safety measures. Regarding applying inherently safety measures, it is important when it should be done or when it is able to apply efficiently. Focusing the ammonia tank in this report, the applying inherently safety measures are introduced.

5.1 Inhelentry safety measures on each life cycle stage

The chemical plant goes through various life cycle stage such as research, process development, design and construction, operation, maintenance and modification, and finally, decommissioning. However, actually when we implement risk analysis is almost in process development stage and design and construction stage. It introduces how inherently safety measures could be applied for WCS and ANS on development stage and design and construction stage.

5.2 Inhelentry safety measures on process development stage

The process development stage is important stage to decide how process operation is done with what kinds of equipment. It is easier to consider various processes than after stages, because there are still many options to decide the processes. The detail process is not decided yet but it is possible to implement WCS for the hazard potential analysis. It is important opportunity to reduce the hazard potential based on WCS by trying to analyze many process options. As the inherently safety measure, "moderate" is efficient measure in this stage. For example, "moderate" reduces the quantity of gasification or release rate from the process equipment by changing temperature and pressure to more moderate condition. In addition, "substitute" is also efficient measure in this stage. For example, "substitute" reduces or eliminates the reactivity, flammability or toxicity of the substances in the equipment by changing substances to more unreactive substances. It introduces how the inherently safety measures should be applied to ammonia tank. There are 2 measures as "moderate", one is diluting ammonia by water and another one is cooling temperature of the ammonia tank. Regarding dilution, it is possible to reduce gasification and volatility of ammonia, and when accidental release occur the amount of flash is decreased and evaporation rate from pool in dike is also decreased. However, it is necessary to have distillation column to condense the concentration of ammonia if high concentration of ammonia is necessary. And it is necessary to confirm whether to generate new hazards, if the distillation column would be constructed. Regarding cooling tank temperature, it is possible to reduce gasification and volatility of ammonia by cooling down temperature to under boiling point.

5.3 Inhelentry safety measures on process design and construction stage

The process condition and substances are already decided by former stages, and the accident events are also decided. In this stage, the main strategy of the inherently safety measure is "minimize" and "simplify". To "minimize" the quantity of the inventory of the equipment is possible to reduce the consequence after release, and to "simplify" the process is possible to eliminate upset condition by improving spec or the layout of equipment. Furthermore, it is possible to analyze based on ANS, because detail design of process is decided in this stage. It is important opportunity to be able to review comprehensively how process is the inherently safer process, in addition to the results of WCS analyzed in previous stage. In this stage, regarding WCS, there are 3 ways to decrease the hazard potentials. At first, there is the way as "minimize" to divide tank to 2 or 3 tanks and reduce the inventory of each tank. Second, there is the way to change the shape of tank to vertical or spherical tank and reduce the surface area of the dike to reduce evaporation rate from pool in dike. Third, there is the way to eliminate the tank and change to pipeline. Regarding ANS, if the ammonia tank is implemented cooling tank temperature, when cooling medium does not flow into the heat exchanger, the temperature of ammonia tank gradually rises and at the same time internal pressure rises by ammonia vapor pressure and at last ammonia tank will rupture. Of course, the time of the temperature and pressure rise up to tank design pressure is extended, but the accident event is same before implementation.

5.4 Results of applying inherently safety measures

The results of WCS in process development stage and the results of after applying inherently safety measures are shown in Figure.7. The results of ANS in design and construction stage and the results of after applying inherently safety measures are shown in Figure.8. In process development stage in Figure.7, cooling down tank temperature under the boiling point of ammonia shows very effective safety measure. However, in design and construction stage in Figure.8, in case of ANS the cooling measure shows that it is not-effective and meaningless measure. In the aspect of the hazard potential, the most effective safety measure is eliminating the tank and changing to pipeline.

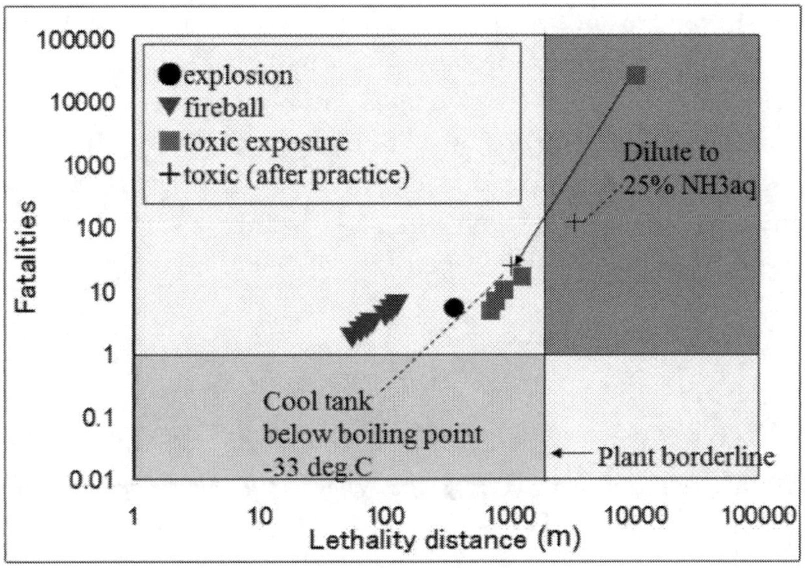

Figure 7. Applying inherently safety measures to WCSs

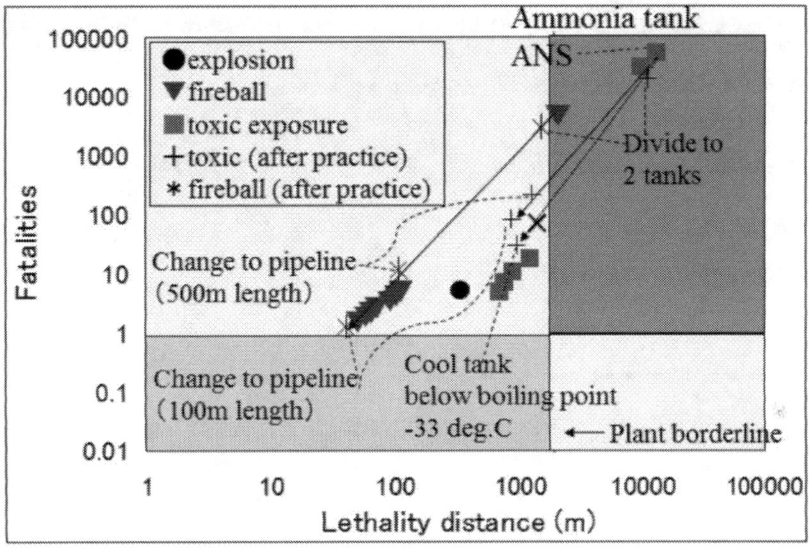

Figure 8. Applying inherently safety measures to ANSs

6. Conclusion

It is shown in this report that the analysis based on WCS and ANS make the results more accurate and effective, and that it is necessary to analyze ANS additionally to help identifying whether other hazards are hiding in the blind spot. Regarding WCS, WCS is redefined about the selection definition of the equipment, the release amount from the process equipment and the accident events so that everyone obtained the same results. Those analyses become more effective by implementing on each life cycle stage because of different perspectives. The results of hazard potentials are plotted on the graph and ranked by their severity, and also to express how large accidents would occur by comparing to the past accident examples. If the equipment has high hazard potential, it is necessary to consider applying inherently safety measures. The inherently safety measures are one of the thinking for safety options, and when the implementation of safety measures is considered, it is also necessary to consider whether safety measures are effective and whether they generate new hazards. By using this analysis method, it is able to confirm them, and it would give effective safety measure options.

7. References

[1] FDMA, the Disaster Prevention Plan for Petrochemical Complexes, 2013

[2] FDMA, the investigation report about revision of the Disaster Prevention Plan for Petrochemical Complexes, 2012

[3] U.S. EPA publication, Risk management program guidance for offsite consequence analysis, EPA 550-B-99-009, 1999

[4] API publication 581, Risk based inspection base resource document, 1st edition, 2000

[5] TNO purple book, Guidelines for quantitative risk assessment, CPR 18E, 1st edition, 1999

[6] Christian Kirchsteiger, Risk assessment and management in the context of the Seveso II Directive, 1998

[7] CCPS, Inherently safer chemical processes, 2nd edition, 2009

Challenge of Mitsubishi Chemical's Process Safety

Masaki Nakagawa
Process Safety Technology Department
Environment, Safety and Quality Department

Mitsubishi Chemical Corporation
1-1, Marunouchi 1-Chome
Chiyoda-ku, Tokyo 100-8251, Japan
2109797@cc.m-kagaku.co.jp

Prepared for Presentation at
American Institute of Chemical Engineers
2015 Spring Meeting
11th Global Congress on Process Safety
Austin, Texas
April 27-29, 2015

UNPUBLISHED

Challenge of Mitsubishi Chemical's Process Safety

Masaki Nakagawa
Process Safety Technology Department
Environment, Safety and Quality Department

Mitsubishi Chemical Corporation
1-1, Marunouchi 1-Chome, Chiyoda-ku
Tokyo 100-8251, Japan
2109797@cc.m-kagaku.co.jp

Keywords: HAZchart analysis, PHA_Organizer, Fault Tree

Abstract

The Mitsubishi Chemical Group has been working to improve process safety by implementing safety review (SR) activities. The aim of the SR activity is to improve safety of processes and operations by sharing information from a broad range of perspectives. The SR activities are joined not only by staffs who are familiar with routine operation, but also by staffs who belong to other departments. To strengthen the SR activity, SR Instructor Team was formed. The member of SR instructor team is composed of retired and senior engineers who have abundant knowledge and experience about process safety. The SR activity becomes a field where experiences, perspectives and knowledge are shared by participants.

One of risk evaluation methods used in the SR activity is HAZchart analysis. HAZchart analysis is a risk evaluation method which Mitsubishi Chemical Corporation and Mitsubishi Research Institute developed. It can carry out easily from development of an accident scenario to quantitative evaluation. HAZchart analysis method is packaged as a computer program called PHA_Organizer. PHA_Organizer has some risk analysis assistance functions which are minimal cut set display function, scenario modeling function and fault tree input assistance function. PHA_Organizer promotes sharing of the way of thinking at the SR activity.

1. Introduction

Recently, the number of accidents which occur owing to plant staff's deficient knowledge is increasing in Japan, because plant staff's opportunity to experience troubles is decreasing. Steady operation in a plant makes plant staffs unable to imagine abnormal situations of a plant. When performing risk evaluation, plant staffs do not regard the result of risk evaluation as a realistic situation due to poor understanding of their plant. Therefore, tools which improve plant staff's understanding and awareness for chemical plant safety are called for.

2. Safety activity

2.1 Safety Assessment and Safety Review Activity

The Mitsubishi Chemical Group identifies potential risk factors in the manufacturing process and takes steps to prevent their occurrence by manufacturing process risk assessment. The risk assessment is broadly divided into Safety Assessment (SA), which is implemented when beginning the manufacture of new products and when improving and upgrading existing manufacturing processes, and Safety Review (SR) activity, which is the full inspection of existing manufacturing processes, safety reassessment, and confirmation of countermeasure effectiveness.

The SR activity is an activity of identification of latent hazards, re-recognition of hazards and education of safety technologies to next generations. The SR activities are joined not only by staffs who are familiar with routine operation, but also by staffs who belong to other departments. The staffs exchange views from a wide range of perspectives at a meeting.

To strengthen the SR activity, SR instructor team was formed in 2003. The SR instructor team is composed of retired and senior engineers. They have a lot of knowledge and experience about process safety. When the SA and the SR activity meetings are held at a section in a plant site, the SR instructor participates together and makes comments and advices. The SR instructor reports contents of duties to a plant manager periodically. The SR instructors are exchanging information by gathering periodically.

2.2 Education

In 2009, we began offering process safety education to middle-tier engineers at each of our plant. The contents of this course are material hazard, explosion hazard, reactive hazard, risk management etc. We intend to continue offering the program with the aim of building on the capabilities of our workforce. Thus far, about 750 employees have received this training.

We also have been taking steps to educate engineers as a chemical process safety engineer (CPSE), with the aim of building a workforce that understands rules and principles of process safety to conduct the SA and the SR activity. With the aim of better understanding process safety, we have them take part in small-group seminars.

3. Risk evaluation method

3.1 HAZchart analysis

One of the risk evaluation methods used for the SA and the SR activity in the Mitsubishi Chemical Groups is HAZchart analysis. HAZchart analysis was developed by Mitsubishi Chemical Corporation and Mitsubishi Research Institute in 1996. It can carry out easily from development of an accident scenario to quantitative result.

HAZchart analysis standardizes a procedure of accident scenario modeling. An accident scenario is modeled by querying three situations repeatedly. After that, a HAZchart is transferred to a Fault Tree (FT) and the probability of accident can be calculated easily.

3.2 Procedure for HAZchart analysis

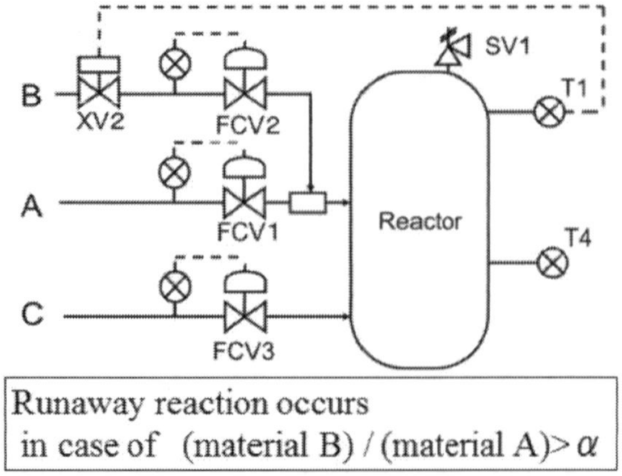

Figure 1. Example Flow

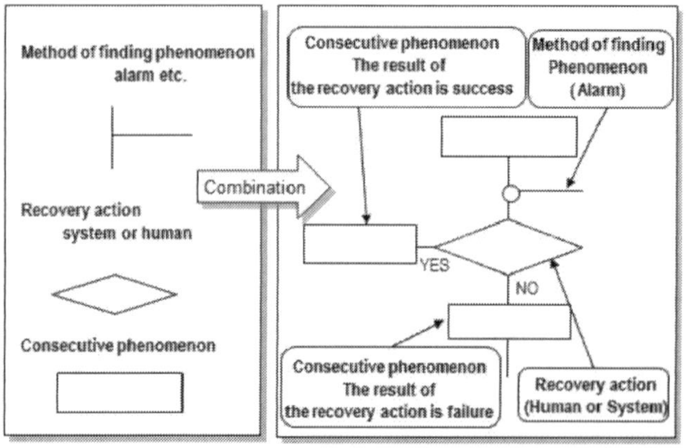

Figure 2. Three queries

The procedure for HAZchart analysis is described in the following example (Figure 1). Material A, B and catalyst C are fed in a certain proportion into the reactor. If the ration of the materials were changed, runaway reaction would occur.

HAZchart analysis consists of three steps.

1) Selection of the equipment and setting the fault mode
2) HAZchart modeling
3) Transformation from a HAZchart to a Fault tree

3.1.1 Selection of the fault mode of each equipment

One of the equipment which can potentially influence the process fluid is selected as a starting point for HAZchart. Then the fault modes are assumed to them. In this case, FCV2 which controls the mass flow of material B is chosen. Then the fault mode is assumed as control valve excessive opening.

3.1.2 Modeling HAZchart

After setting the fault mode, HAZchart is modeled by querying three situations which are "what is the method of the finding phenomena?", "what is the recovery action?" and "what is the consecutive phenomena ?". (Figure 2)

Each query corresponds to each symbol. The method of the finding phenomenon that indicates alarm or detector etc. corresponds to a straight line, the recovery action that indicates system action or operator action corresponds to a diamond and the consecutive phenomena that indicates the result of the recovery action correspond to a rectangle. The HAZchart that forms the scenario of an accident is modeled by repeatedly applying these three queries. Figure 3 shows HAZchart of FCV2 excessive opening.

3.1.3 Transformation from a HAZchart to a Fault tree

Next step is to transfer HAZchart to FT. "Methods of finding phenomenon", "recovery actions" and "consecutive phenomena" in the HAZchart correspond to FT logic. Thus, it is easy to transfer information from a HAZchart to a FT.

The FT corresponding to the HAZchart can be obtained by the following five steps (Figure 4).

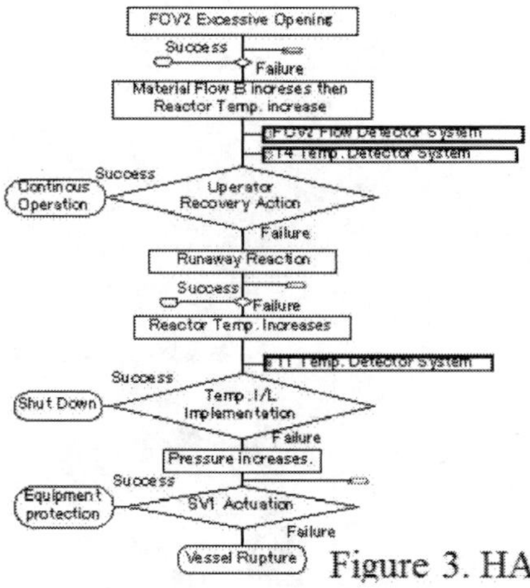

Figure 3. HAZchart

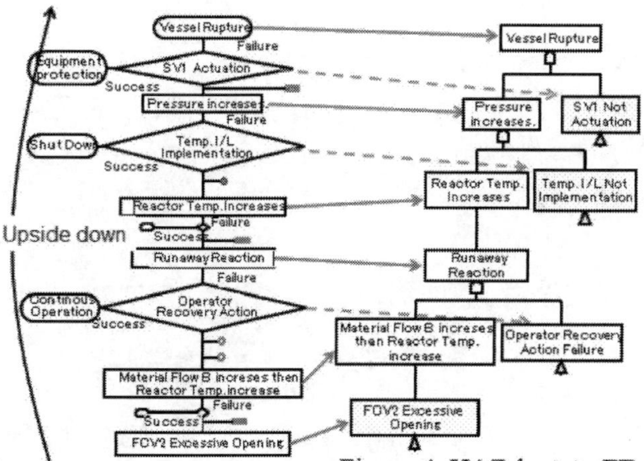

Figure 4. HAZchart to FT

1) HAZchart is turned upside down.

2) Rectangular parts which indicate "consecutive phenomena" are arranged on the left side of the FT. Diamond parts which indicate "recovery actions" are arranged on the right side of the FT.

3) The rectangular parts are combined with the diamond parts by " AND" symbol.

4) After the transformation, branches of FT are divided up to a level that has equipment reliability data or human error data.

5) After the equipment reliability data are set up in the FT, the probability of top event can be calculated (Figure 5).

3.3 Risk Matrix

The result of the HAZchart analysis is plotted in the risk matrix. Horizontal line means probability of accident and vertical line means accident levels. Red area and yellow area indicate unacceptable levels and White area and green area indicate acceptable levels. Figure 6 shows correspondence between the result of HAZchart analysis in the risk matrix and HAZchart and FT.

3.4 Correspondence between HAZchart and protection layers

The safety system is designed as a series of protection layers. So, it is natural for plant staffs to

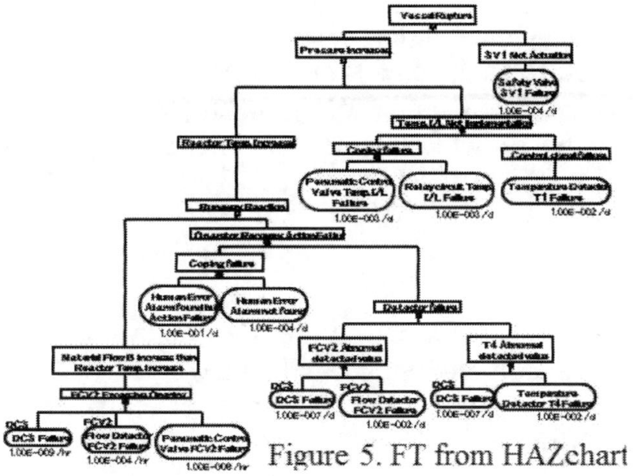

Figure 5. FT from HAZchart

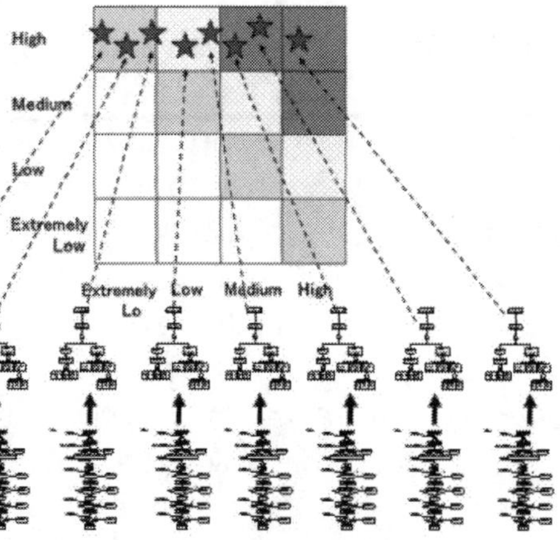

Figure 6. Risk Matrix

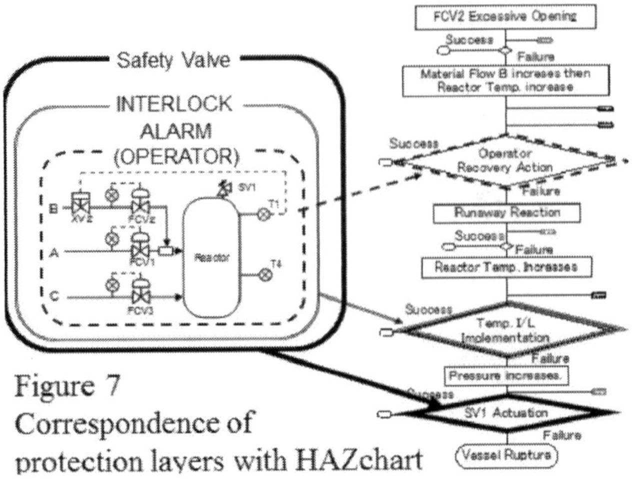

Figure 7
Correspondence of protection layers with HAZchart

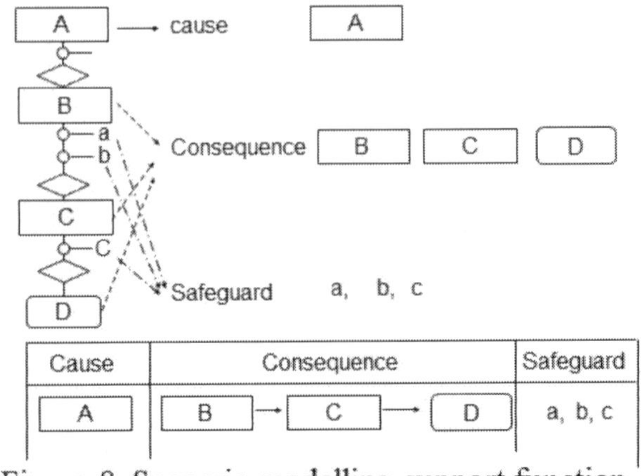

Figure 8. Scenario modelling support function

think that if a malfunction on a protection layer occurs the situation becomes worse. HAZchart was developed along the same concept of layers of protection. Each protection layer corresponds to each recovery action of the diamond. Figure 7 shows relationship between protection layers and HAZchart

4. PHA_Organizer

HAZchart analysis method was packaged as a computer program which was called PHA_Organizer. PHA_Organizer has a database which has equipment reliability data and branches of FT. Therefore, after modeling HAZchart, the probability of accident scenario can be calculated automatically. This software was privately used at first, but it became clear that PHA_Organizer could facilitate sharing a mutual thinking process and promoted discussion at a meeting. Consequently PHA_Organizer was significantly improved to make the tool even easier to use and anyone could easily conduct risk evaluations at a meeting.

4.1 Risk analysis supporting function of PHA_Organizer

4.1.1 Minimal cut set display function

A cut set is a combination of basic phenomena which generates the top phenomenon of FT. A minimal cut set is made required to generate a top phenomenon of a system, and is sufficient cut set. The minimal cut set was often expressed as a set. But this expression form is not so easy to understand intuitively. So the function of displaying minimal cut set relationship on the FT has been developed. By this function, it is easy to understand which basic phenomenon lowers the probability of the top phenomenon occurrence intuitively.

4.1.2 Scenario modelling support function

PHA_Organizer displays the results of HAZchart analysis automatically on HAZOP work sheet or what-if work sheet, where the detailed accident scenario and related safeguard are displayed. Figure 8 shows this mechanism. The scenario in the consequence column is constructed by connecting of " Consecutive phenomenon" in the HAZchart. Safeguards in the safeguard column are "Methods of finding phenomenon" in the HAZchart. This function assists to model more detailed and more intelligible accident scenario.

4.1.3 Redundant system support function

A redundant system is adopted as one of the means of increasing reliability of an interlock system. It is troublesome to describe a redundant system in a FT. Therefore, the support system that can describe a redundant system easily has been developed.

4.2 Case study by PHA_Organizer

4.2.1 Original Flow

Figure 9 shows an original flow of case study. Mixture of liquid and gas is fed into a separator and separated into liquid and gas. The LCV1 controls to maintain liquid level for preventing gas leakage into the liquid line. The starting point is LCV1 excessive opening. The accident scenario would occur in order of 1 to 4 in Figure 9. The diamond in the HAZchart shows there is one protection layer in this accident scenario.
Figure 10 shows FT of the original flow. The result of the HAZchart analysis is plotted in the risk matrix (Figure 8). The result of LCV1 excessive opening is in the red area. That means leakage to the liquid line due to the failure of LCV1 is unacceptable.

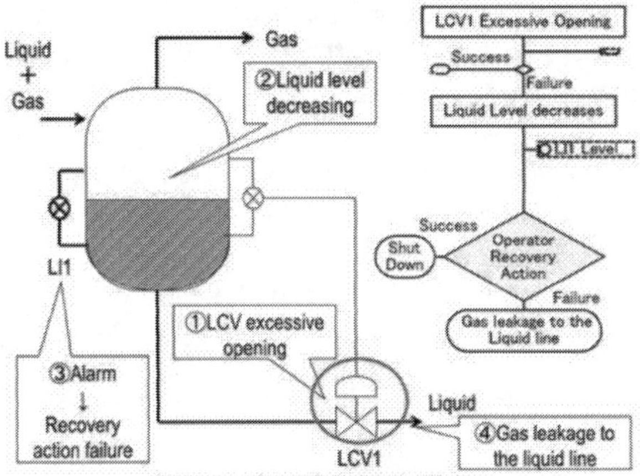

Figure 9. Original Flow

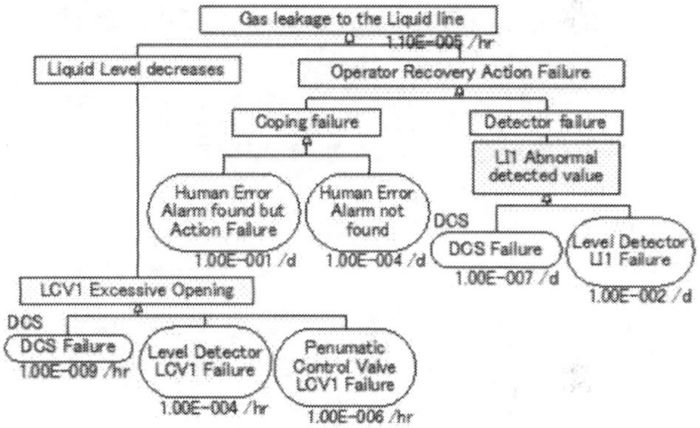

Figure 10. FT of original flow

4.2.2 Modified flow 1 (Case study by Minimal cut set display function)

Figure 11 shows modified flow 1. Two level detectors are added to this flow. In this case, if the liquid level decreased, operator would take a recovery action to prevent from gas leakage by three level alarms. The probability of gas leakage in the case of modified flow 1 is the same as in the original flow case. This result shows the countermeasures of three level detectors are not effective for decreasing the probability of gas leakage.
Figure 12 shows FT of the modified flow 1 on the PHA_Organizer screen, in which minimal cut set display function is operating. The combination of minimal cut set can be displayed on the FT by changing related line's color and thickness in PHA_Organizer. The highest demand rate of the basic event in the branch of "operator recovery action failure" is "Human error alarm found but

action failure". So, the minimal cut set of the highest probability is the combination of "Level Detector LCV1 Failure" and "Alarm found but Action Failure". That means that even if the number of alarms were increased, the probability of top event could not be reduced by them. The demand rate of "Alarm found but action failure" cannot be changed in this analysis. Therefore, the probability of top event can be reduced by adding other safety systems.

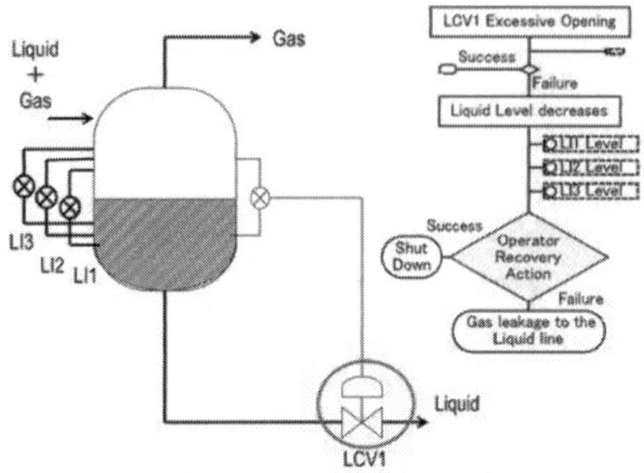

Figure 11. Modified flow 1

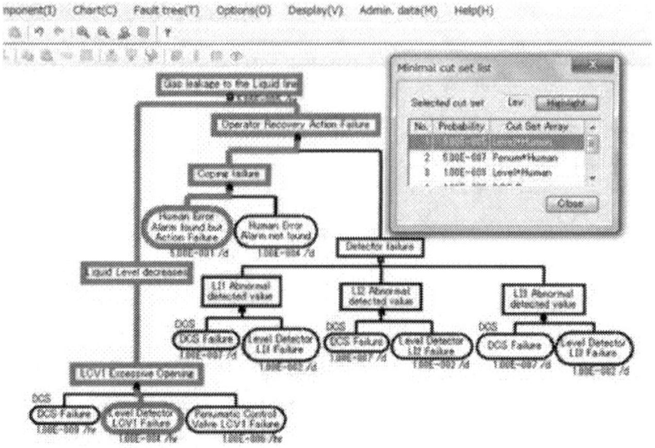

Figure 12. Minimal cut set display function

4.2.3 Modified flow 2 (Case study – common cause)

Figure 13 shows case study of modified flow 2. The interlock system is added to the flow. In this case, if the liquid level decreased, the interlock system would operate to prevent from gas leakage. The alarm and the interlock system are protection layers. But the level detector of LCV1 is shared with the interlock system and the level alarm.

Figure 14 shows the FT of modified flow 2. The FT has branches of operator recovery action failure and Interlock system failure. There are the same equipment in the left side and right side in the FT, which are level detector and DCS. They are common cause in the FT.

The minimal cut set with the highest probability is the combination of "Level Detector LCV1 Failure" and "Alarm found but Action Failure". The result of gas leakage of modified flow 2 is the same as the original flow and the modified flow 1 (Figure 18).

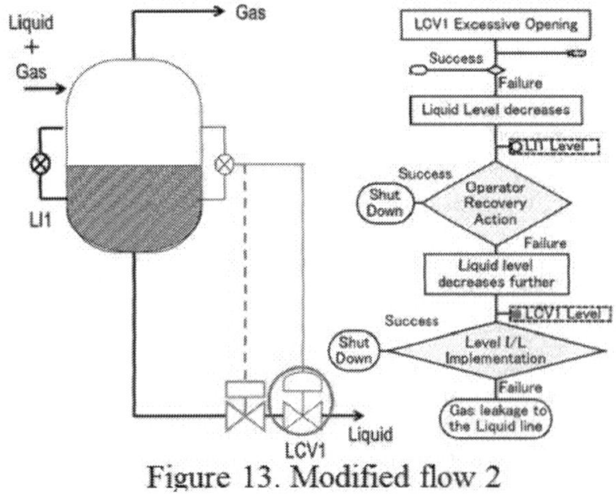

Figure 13. Modified flow 2

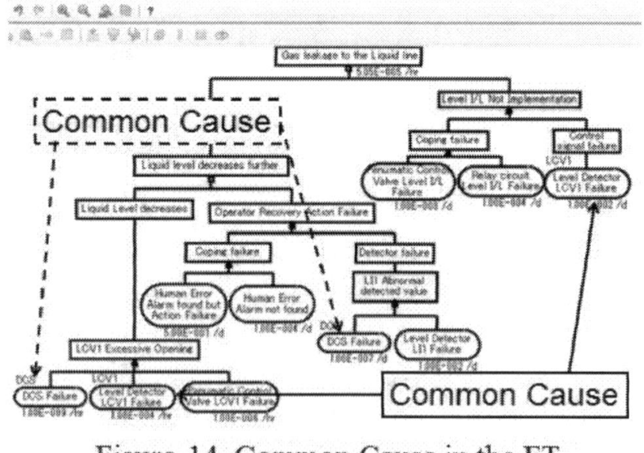

Figure 14. Common Cause in the FT

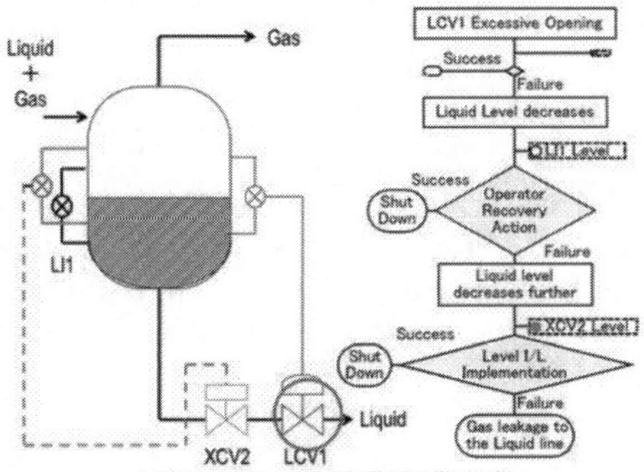

Figure 15. Modified flow 3

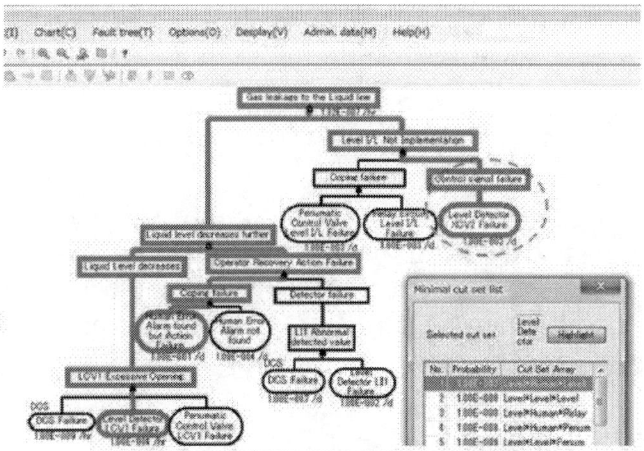

Figure 16. FT of modified case 3

4.2.4 Modified flow 3 (Case study by Redundant system support function)

Figure 15 shows modified flow 3. An independent interlock system and an alarm are added to this flow. If the liquid level decreased, the alarm and the interlock system would operate to prevent from gas leakage. This time, the interlock system and the alarm has different level detector. In the event that the level detector failed and the operators did not take an appropriate recovery action, the interlock system would prevent the liquid level from decreasing. Figure 18 shows the probability of gas leakage in this case can be reduced because the interlock system does not share common equipment. Figure 16 shows the FT of modified flow 3 displayed by minimal cut set function. The highest probability of the minimal cut set is the combination of "Level Detector LCV1 Failure", "Alarm found but action Failure" and "Level detector XCV2 failure". The demand rate of "Alarm found but action failure" cannot be changed in this analysis. Therefore, the probability of top event can be reduced by improvement of "Level detector XCV2".

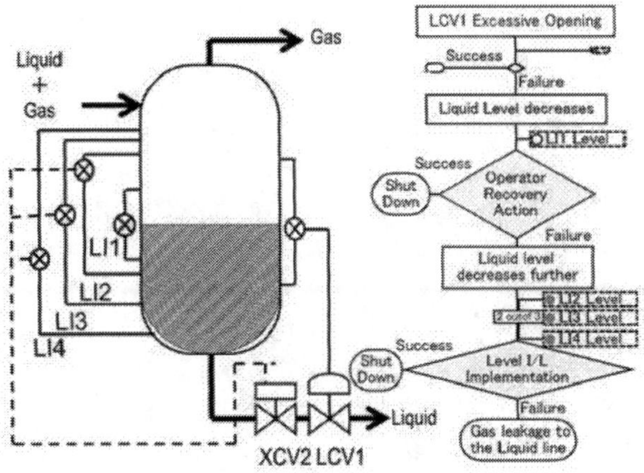

Figure 17. Modified flow 4

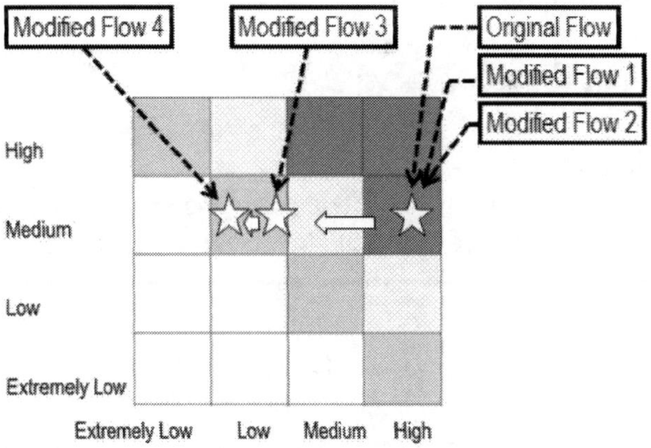

Figure 18. The result of case study

Redundant system is adopted as one of the means of increasing reliability of an interlock system. Figure 17 shows independent interlock system with 2 out of 3 redundant system and alarm are added to this flow. If the liquid level decreased, the alarm and the redundant interlock system would operate to prevent from gas leakage. By setting 2 out of 3 redundant system to the

HAZchart (Figure 17), redundant system of FT is modeled automatically by Redundant system support function.

4.3 Application of PHA_Organizer

PHA_Organizer shows thinking process that can facilitate sharing a mutual thinking process. Showing modeling process of accident scenario by PHA_Organizer induces various questions. Plant staff can understand their plant better by answering questions quantitatively through HAZchart analysis. PHA_Organizer is expected as a tool for providing a field which promotes discussion.

6. Conclusion

The SR activity is an activity of identification of latent hazards, re-recognition of hazards and education of safety technologies to next generations.

To facilitate the SR activity, HAZchart analysis is used for risk evaluation method. HAZchart analysis is a tool for identification of latent hazards. It can carry out easily from development of an accident scenario to quantitative evaluation. HAZchart analysis can facilitate sharing a mutual thinking process and promote discussion at a meeting.

PHA_Organizer has specific functions which are minimal cut set display function, scenario modelling function and redundant system support function. It is expected that these features contribute to improvement of plant staff's understanding and awareness for chemical plant safety.

REFERENCES

[1] M. Nakagawa., Y. Iizuka, Introduction and application of quantitative process hazard analysis – HAZchart, 11th Intl. Symposium of Loss Prevention and Safety Promotion in the Process Industries, 31 May – 3 June 2004, Praha, Czech Republic, 4388-4394.

[2] M. Nakagawa, Y. Mizuta, Development of a worst case scenario, 13th Intl. Symposium of Loss Prevention and Safety Promotion in the Process Industries, 6 - 9 June 2010, Brugge, Belgium, 455—458.

[3] M.Nakagawa, Promoting discussion in Safety Review Activity, Chemical Engineering Transactions, Vol. 31, 2013

[4] Lees F.P., 1996, Loss prevention in the process industries second edition, Butterworth-Heinemann, Oxford, UK, Vol.1, 9/18.

[5] Center for Chemical Process Safety (CCPS), 2000, Guidelines for Chemical Process Quantitative Risk Analysis Second Edition. American Institute of Chemical Engineers, New York, NY.

[6] Center for Chemical Process Safety (CCPS), 1994, Guidelines for Hazard Evaluation Procedures. American Institute of Chemical Engineers, New York, NY.

[7] Center for Chemical Process Safety (CCPS), 1993, Guidelines for Safe Automation of Chemical Processes. American Institute of Chemical Engineers, NewYork.

Aplicación de Análisis de Causa Raíz y Acciones – Basadas en Análisis de Pareto, como herramientas para Reducir LOPC Ocotlan – MEXICO.

Hugo Hernández

Process Safety Leader

Grupo Celanese

Km. 78.5 Carret. Guad. La Barca

Ocotlan, Jalisco, MEXICO 45965

52- (392) 925-9010

Hugo.Hernandez@celanese.com

Prepared for Presentation at
American Institute of Chemical Engineers
2015 Spring Meeting
11th Global Congress on Process Safety
Austin, Texas
April 26 – 29, 5

UNPUBLISHED

Aplicación de Análisis de Causa Raíz y Acciones – Basadas en Análisis de Pareto, como herramientas para Reducir LOPC Ocotlan – MEXICO..

Hugo Hernández

Process Safety Leader

Grupo Celanese

Km. 78.5 Carret. Guad. La Barca

Ocotlan, Jalisco, MEXICO 45965

52- (392) 925-9010

Hugo.Hernandez@celanese.com

Palabras Clave: RCA= Análisis de Causa Raíz, Análisis de Pareto, LOPC= Perdida de Contención Primaria, PSQ= Cantidad de Seguridad de Procesos, MOC = Administración de Cambio, RAGAGEP = Mejores Prácticas de Ingeniería Generalmente Aceptadas y Reconocidas. SOP= Procedimiento de Operación

Resumen

En 2011 Grupo Celanese estableció la Seguridad de Procesos en un Sistema Global de Administración de Seguridad de Procesos , conocido como LOPC (Perdida de Contención Primaria) para registrar estos incidentes ocurridos en el grupo y evaluar el éxito de los pasos introducidos para reducir incidentes LOPCs mediante el análisis de la tendencia a lo largo de los años.

Este trabajo muestra el proceso seguido en el Sitio Ocotlán con el fin de reducir los acontecimientos de estos incidentes LOPC y los resultados obtenidos hasta el año 2014, destacando el plan de medición, análisis, mejora y control establecidos hasta la fecha.

El objetivo de este artículo es compartir la mejor aplicación de prácticas para reducir los eventos LOPC que establezcan sistemas para la conducción de las operaciones y la integridad mecánica. El objetivo principal es reducir la probabilidad de fugas de materiales tóxicos y / o inflamables dentro de las áreas de proceso.

1. Introducción

En cualquier negocio la mejora continua es uno de los objetivos primarios. La Seguridad de Procesos es un factor clave.

Los eventos donde no se cumple efectivamente la Seguridad de Procesos son eventos que nadie quiere en su planta, pero cuando ocurren, tenemos que aprender de los análisis de las causas de raíz (RCA) que han llevado a este evento. A partir de este análisis, se llevan a cabo las acciones pertinentes con el fin de evitar la repetición.

Siguiendo la filosofía de mejora continua, Celanese declaró la reducción de eventos LOPC el principal objetivo año tras año. Para este fin, las causas de los acontecimientos que causó un LOPC en las unidades de producción y sistemas de gestión de la seguridad requerida a través de buenas prácticas e iniciativas mundiales fueron fuertemente analizadas.

Las unidades de producción adoptaron nuevas iniciativas para reducir los eventos LOPCs y mejorar los resultados obtenidos en años anteriores.

Junto con el campo de Integridad Mecánica, uno de los campos se reforzó fue la parte operativa de Celanese. Mediante el análisis de los LOPCs obtenidos, se observó que la Conducción de Operaciones representa un alto porcentaje del número total de eventos.

Como se puede ver en el Gráfico 1, el número total de LOPCs se han reducido en los últimos años a partir de 2012 a la fecha actual.

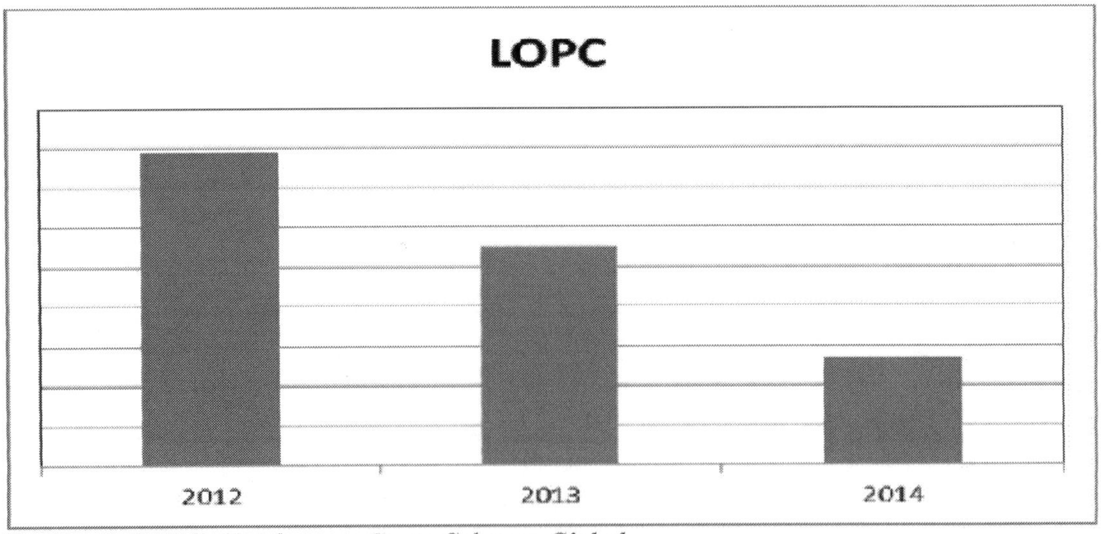

Grafica 1: LOPCs Totales por año en Celanese Global

Uno de los factores clave en la reducción de estos incidentes fue de localizar cuáles fueron las principales causas dentro de la Conducción de Operaciones e intensificar su reducción.

El siguiente gráfico 2 muestra los resultados obtenidos en el estudio de estos eventos y los incidentes generados a la fecha actual.

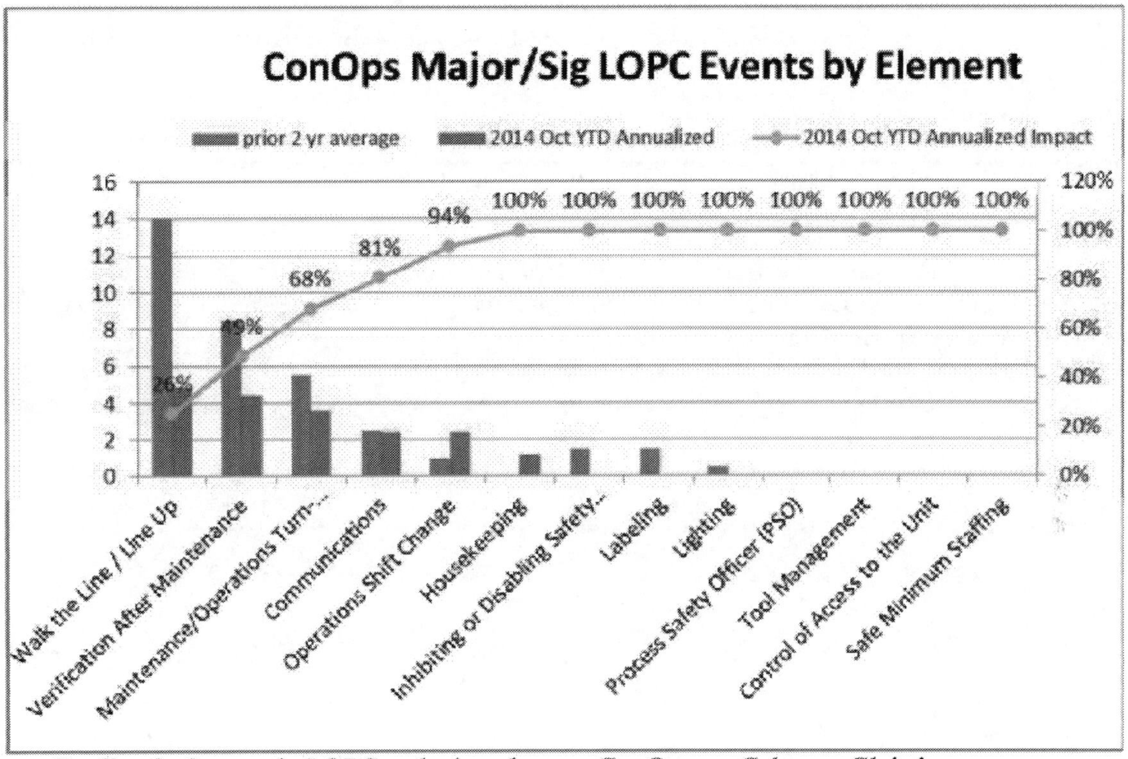

Grafica 2: Causas de LOPCs relacionadas con ConOps en Celanese Global

Mención especial de los esfuerzos para reducir los eventos LOPC debido a errores en la alineación de equipos de control y transferencia de energía del producto a partir de un punto de partida hasta el fin (camina la línea / Alinear). Estos eventos se consideran totalmente evitables y, aunque es difícil de completar la eliminación de cualquier evento, la reducción obtenida muestra que las prácticas e iniciativas emprendidas han tenido efecto.

A nivel global, podemos ver que las prácticas llevadas a cabo en todas las unidades de producción de Celanese han dado sus frutos y la reducción de eventos LOPCs entre los años 2011-2014 es considerable como se puede ver en el Gráfico 2.

Incluso habiendo reducido significativamente el porcentaje de eventos LOPC, y el espíritu de mejora continua para reducir los eventos de seguridad de procesos como sea posible, todavía hay este tipo de incidentes en la producción. Es por ello que es de suma importancia objeto de Celanese mantener sistemas de seguridad robustos y mejora continua para alcanzar un punto de excelencia en el campo de la seguridad industrial.

Principios Teóricos

2.3 Que es un RCA?

Es un método de resolución de problemas mediante la identificación de las causas o hechos que lo originan.

La práctica de la RCA se basa en la suposición de que los problemas se resuelven mejor tratando de corregir o eliminar las causas de raíz, en lugar de simplemente tratar los síntomas evidentes de inmediato.

Al dirigir las medidas correctivas a causas profundas, se espera que la probabilidad de recurrencia del problema se reduzca al mínimo. Sin embargo, se reconoce que la prevención total de la ocurrencia de un problema no es posible.

Por lo tanto, el RCA es a menudo visto como un proceso iterativo y se utiliza a menudo como una herramienta para la mejora continua.

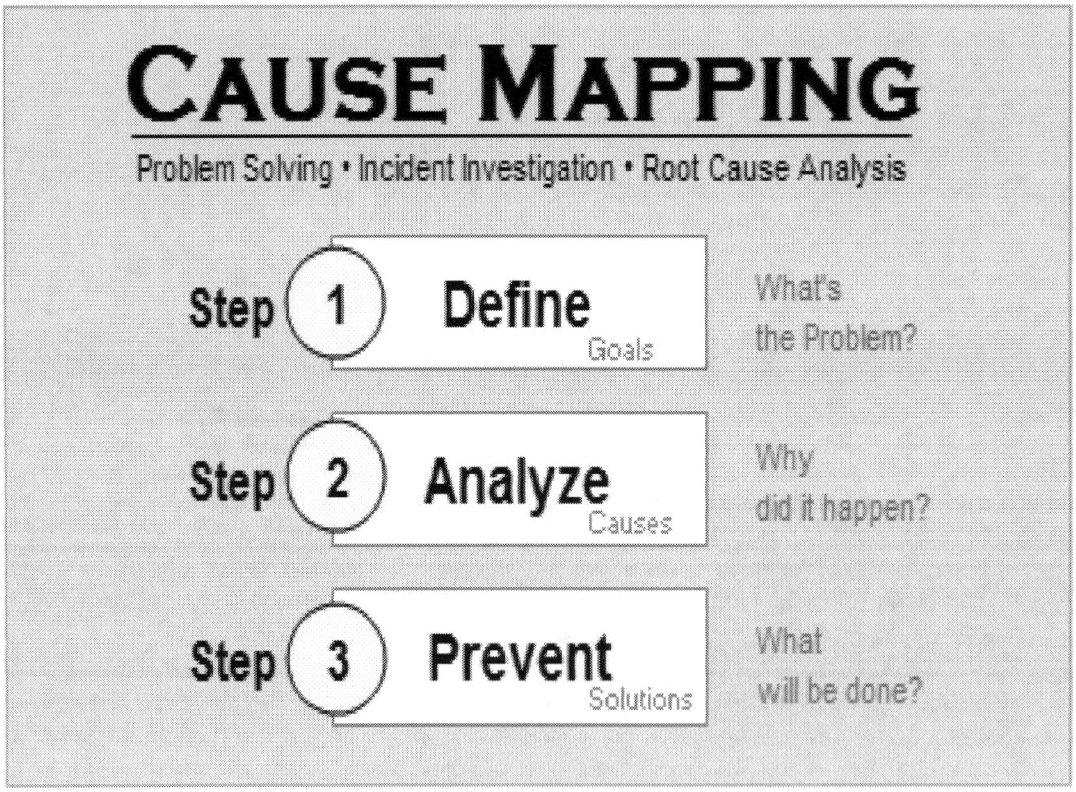

Figura 1: Mapeo de Causas (Think Reliability Copyright)

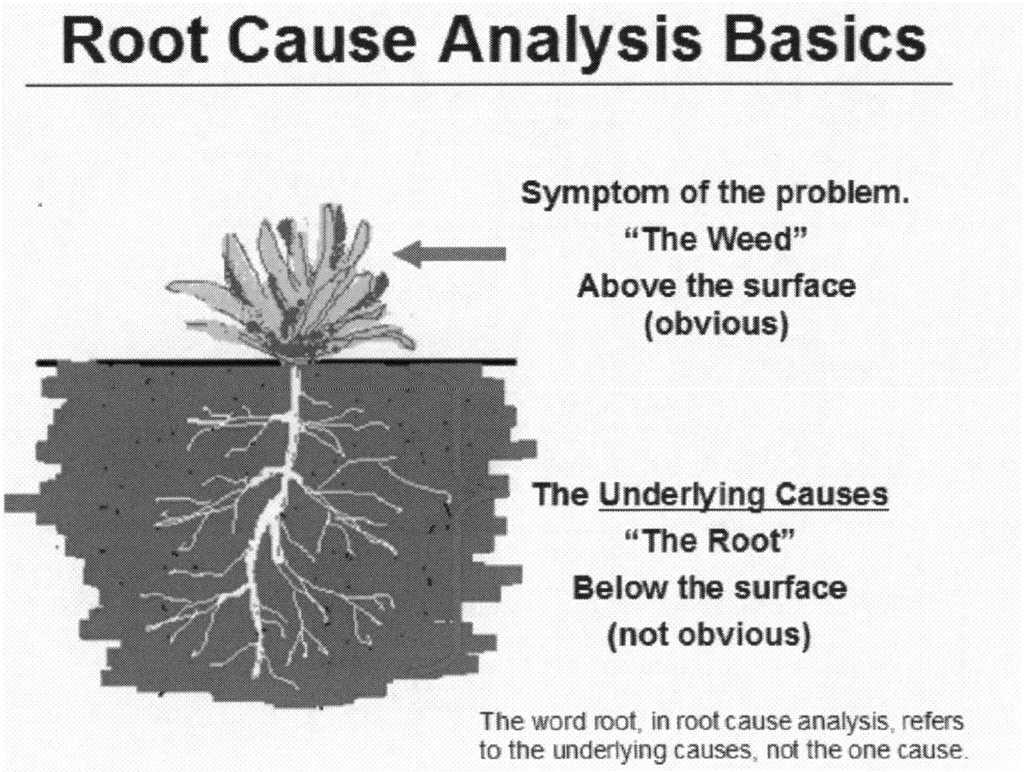

Figure 2: Muestra la filosofía de la causa raíz. .

2.4 Que es un Análisis de Pareto?

El análisis de Pareto es una serie de elementos cuantitativos y ordenados o factores comparados con su contribución a un efecto particular.

El propósito de esta comparación es clasificar estos elementos o factores en dos categorías: los " Vitales-Pocos" (los elementos muy importantes en su contribución) y (elementos sin importancia en ella) "Triviales-Varios".

Categoría de Factores "Pocos Vitales" representan aproximadamente el 80% causas de los hechos bajo estudio. Por lo tanto debemos estar enfocados en la principal causa que debemos trabajar para reducir el mayor número posible de eventos.

Este análisis se puede realizar de acuerdo con el siguiente diagrama de flujo en la Figura 3.

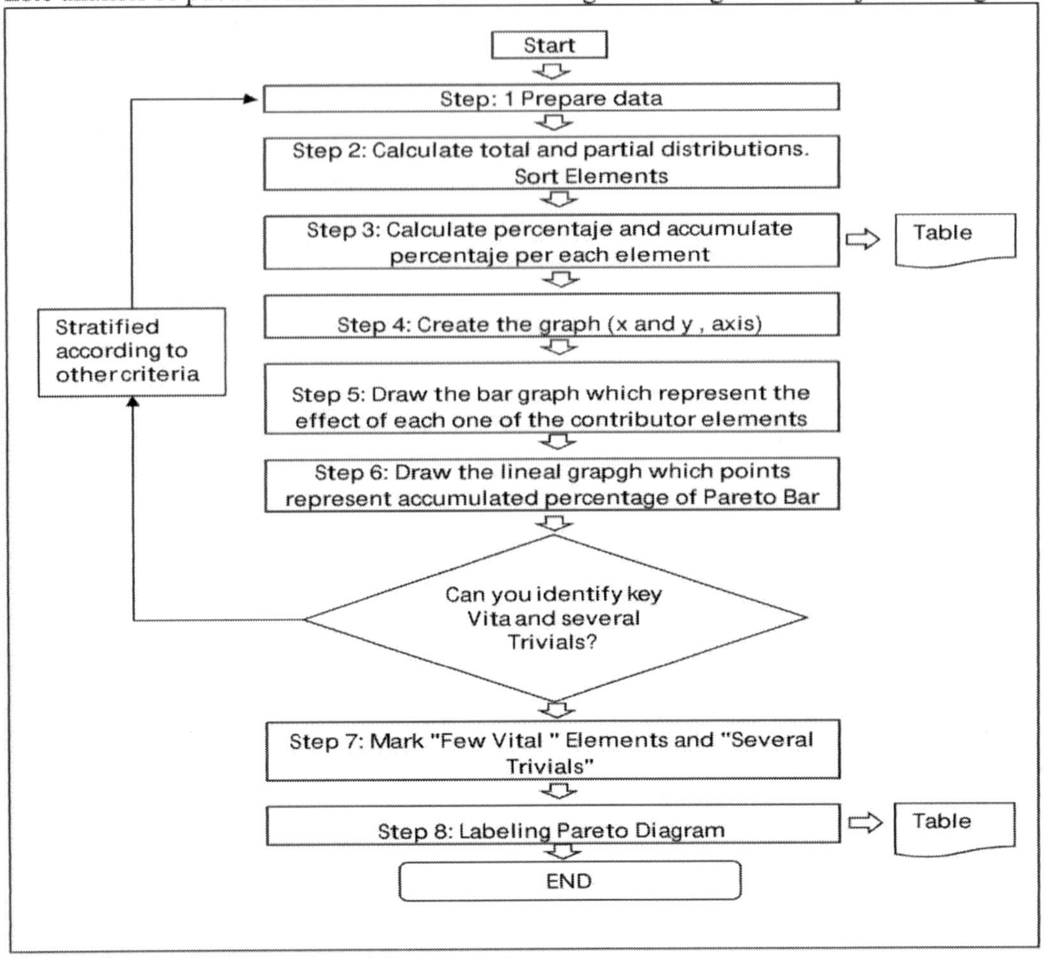

Figura 3: Secuencia para generar el diagrama de Pareto .

Este tipo de análisis es muy importante para dar prioridad a las acciones a tomar, ya que tanto los recursos económicos y humanos son limitados.

Así que detectando y solucionando las principales causas se puede obtener resultados en menos tiempo, logrando la eficiencia en el proceso de las acciones ya su vez alentar al grupo a continuar con las acciones, porque cuando la gente ve los resultados siguen participando logrando establecer una mejora de los procesos donde se les informó, analizó, las acciones tomadas y mejora.

Este proceso de mejora de la Seguridad de Procesos representa salvar vidas y evitar lesiones graves al personal, equipos o instalaciones de la industria

3. Aplicación del Análisis de Causa Raíz y Acciones Basadas en Pareto en el Sitio de Producción Ocotlan

3.1 Medición.

Globalmente, se solicitó a las áreas de proceso, proceder a reportar los incidentes que les sucedieron a ellos relacionados con LOPC. Estos incidentes se determinan como significativos cuando el 10% del valor de PSQ se alcanza y Mayor cuando se alcanza el 100% del valor de PSQ de la sustancia derramada. La metodología utilizada está alineada al RAGAGEP API 754: Indicadores de Desempeño de Seguridad de Procesos de Refinación e Industria Petroquímica. Globalmente, fue solicitado a las áreas de proceso, proceder a reportar los incidentes que les sucedieran

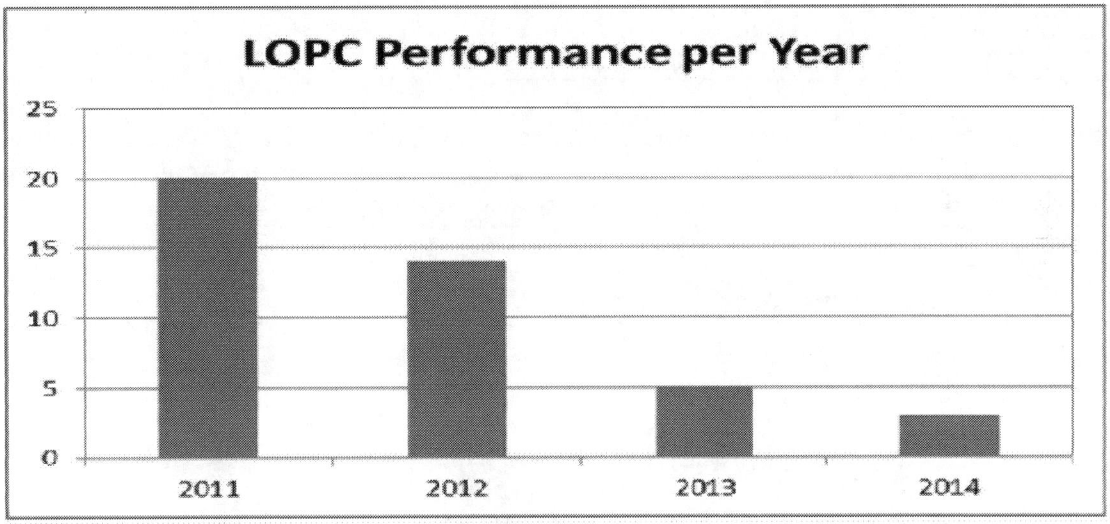

Grafica 3: Numero de LOPCs por año en el Sitio Ocotlan

El sitio procedió a clasificar los incidentes desde 2011, el gráfico muestra cómo los incidentes LOPC han disminuido de un máximo de 20 en 2011 a un mínimo de 3 en 2014.

3.2 Análisis

El rango de años que se ha analizado es 2011-2014, los incidentes de acuerdo con los criterios establecidos en el año 2012 fueron recalculados.

La información fue clasificada por área de proceso, encontrando por el análisis de Pareto los principales contribuyentes.

Encontrará en el Gráfico 4 a continuación, que el área con mayor número de incidentes es Área A, seguido de Área B, Área C y Área D como las cuatro primeras, representan más del 90%

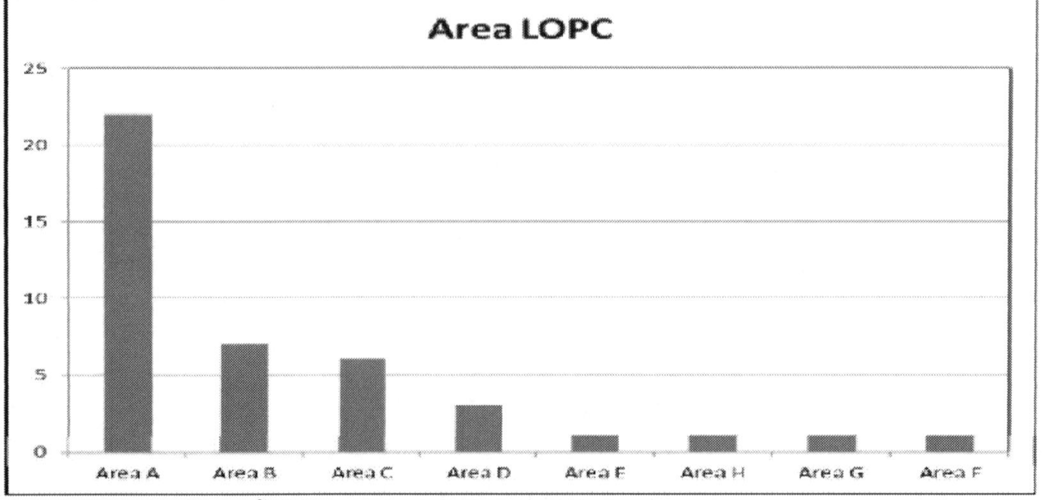

Grafica 4: LOPCs por Área de Proceso en Ocotlan del 2011 to 2014.

Análisis de Causa Raíz fue realizado para determinar el equipo involucrado. Encontrando que el equipo con el mayor número de LOPCs son Tanques, Bombas, Válvulas, Filtros Prensa y Tuberías, mencionando que los primeros 5 representan el 81% de las Causas.

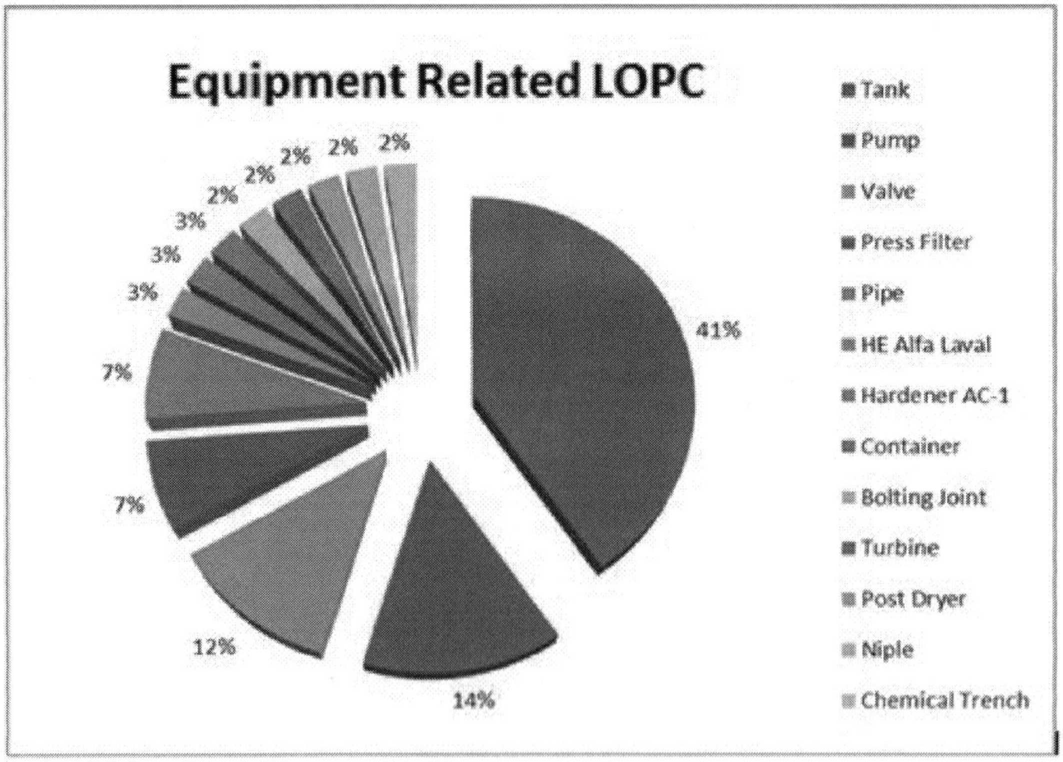

Grafica 5: Equipo donde sucedió un LOPC en el sitio Ocotlan del 2011al 2014.

Con información del RCA, se determinaron las causas raíces de los errores humanos, tratando de ser más específico y se encontró que los procedimientos (SOP), Capacitación, verificación después de mantenimiento, representan más del 90% de las causas.

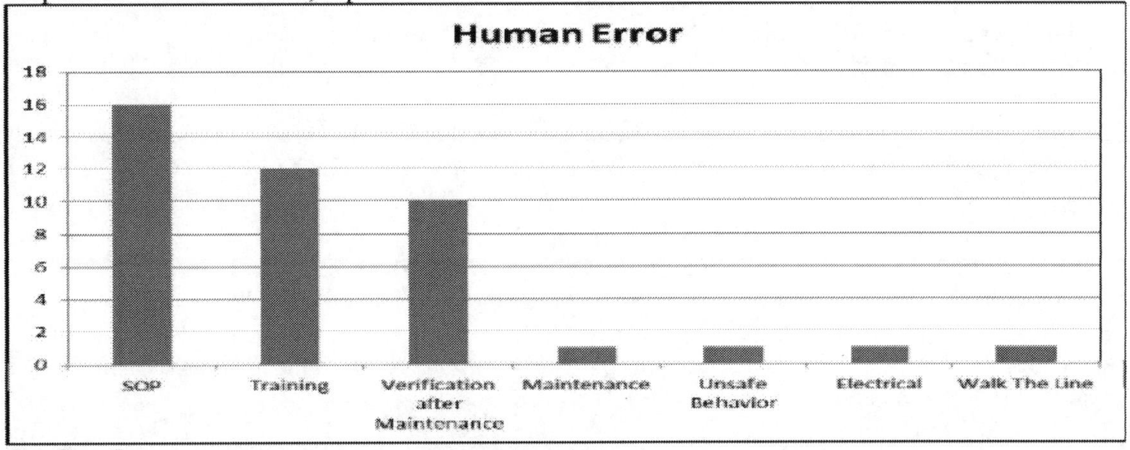

Grafica 6: Error Humano involucrado en LOPCs del Sitio Ocotlan

3.3 Acciones:

Después de la identificación de las causas principales, la administración determinó un plan de acción en el sitio Ocotlan, para hacer frente a estas causas.

Las principales causas se resumen en lo siguiente:

1. **Control de Nivel Tanques -** causa del evento Tanques de Almacenamiento
2. **Relocalización de Switch de descarga Tanques -** causa del evento: Tanques de Almacenamiento
3. **Sistema de Protección para derrame de Tanques**- causa del evento: Tanques de Almacenamiento
4. **Control de Presión en Bombas de Celusol –** causa del evento : Bombas
5. **Reapriete de Filtros Prensa -** causa del evento: Filtros Prensa
6. **Reducción en Temperatura de Mezclador -** causa del evento: Tanques de Almacenamiento
7. **Estandarizar procedimiento de muestreo en el campo -** causa del evento Tubería.
8. **Mejorar Capacitación en la Práctica de Muestreo** - causa del evento: Tubería
9. **Verificación posterior a Mantenimiento en forma Efectiva y Eficiente** - causa del evento: Turbina, Intercambiador Alfa Laval, Post secador, Niple.
10. **Actualizar Procedimiento de Apriete de Bridas capacitación práctica y certificación**-causa del evento: Intercambiador Alfa Laval, Ensamble de Bridas.
11. **Mejorar la Ejecución de MOC**- causa del evento: Tanques de Almacenamiento
12. **SOP para apriete de Intercambiadores de Espiral** - causa del evento: Intercambiador Alfa Laval, Ensamble de Bridas.
13. **Integridad Mecánica en Tuberías Soldadas** - causa del evento: tuberías.
14. **Mantenimiento de Equipo Rotativo -** causa del evento: Bombas
15. **Caminar la Línea -** causa del evento: Contenedor

16. **Evitar Comportamiento inseguro cuando se transfiere material**- causa del evento: Comportamiento Inseguro.

4. Experiencias de aprendizaje y mejoras después de Reducción LOPC

La comunicación es muy importante y el trabajo en equipo, ya que es importante que los dueños de los procesos conozcan cómo sus áreas están trabajando en los LOPCs generados.

Al principio sólo dijimos que Integridad Mecánica y Conducción de Operaciones eran las causas de LOPCS, pero le faltaba ser más específicos para facilitar la acción de los responsables de las áreas.

La participación de los Gerentes, Superintendentes, Supervisores y Operadores, todos los días para evitar los LOPCs es una actividad fundamental para la prevención

En la gráfica 7 muestra la reducción del 2011-2014 por área, encontrando áreas sin LOPCs durante tres, dos y un año.

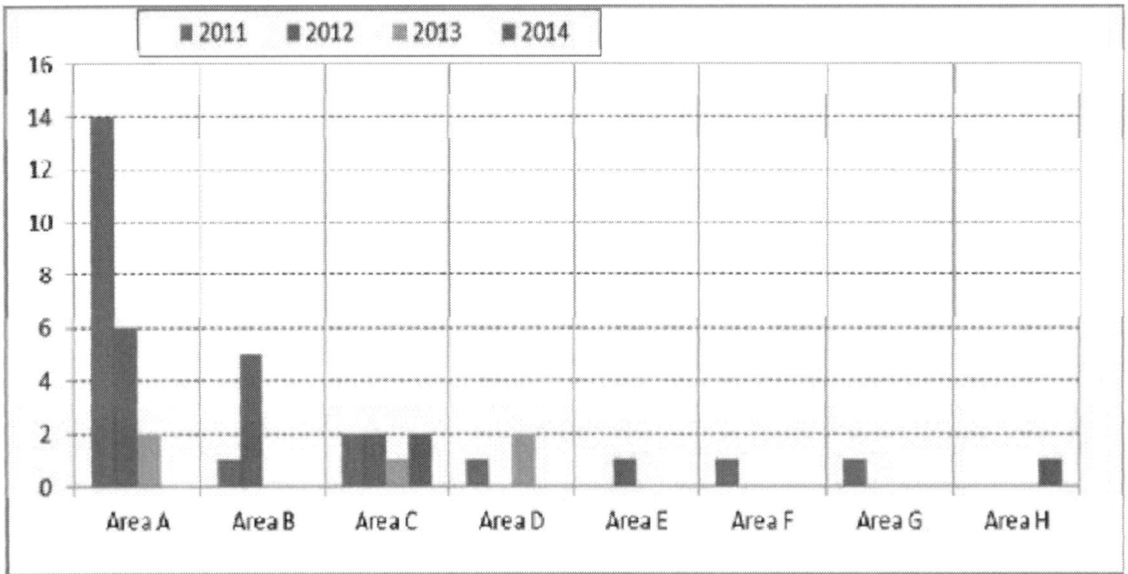

Area	2011	2012	2013	2014
Area A	14	6	2	0
Area B	1	5	0	0
Area C	2	2	1	2
Area D	1	0	2	0
Area E	0	1	0	0
Area F	1	0	0	0
Area G	1	0	0	0
Area H	0	0	0	1

Ocotlan LOPCs performance per Area until Dec-2014	
1 Year without LOPC	Area A, Area D
2 Years without LOPC	AreaB, Area E
3 Years without LOPC	Area F, Area G

Grafica 7: Reducción de LOPC por Área en el Sitio Ocotlan

Como se muestra en la Figura 8, el tiempo medio entre fallos aumentó de un LOPC incidente cada 18.6 días a 83.1 días. Ya tiene 8 puntos por encima de la media de 2013 que muestra que el cambio es estadísticamente sostenible.

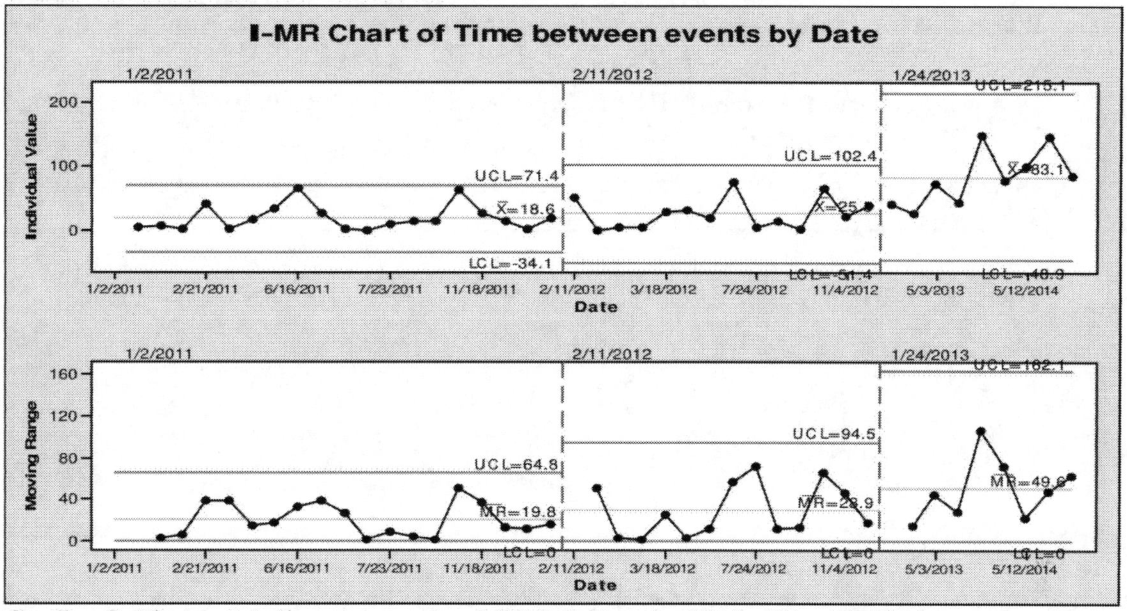

Grafica 8: Tiempo Medio entre eventos LOPC de Ene. 1, 2011 a Dic. 17, 2014

6. Conclusiones

Con la aplicación de análisis RCA y Pareto y la ejecución de acciones genera resultados, que en el caso de Ocotlán se están reflejando en cuatro años, logrando una reducción de 20 LOPCs en 2011 a 3 LOPCs en 2014. Esto da una reducción total de 85% LOPCs.

La participación de los líderes de las áreas es muy importante ya que los sistemas de seguridad de procesos se basan en la gente, si la gente Reconoce, se Compromete y Aprende, la mejora se sostiene, y esto debe hacerse todos los días. Se tiene revisión del Comité Director de Seguridad Mensual obligatoria desde el 2005

Culturalmente se refuerza el hecho de la necesidad de reportar todos los eventos debido a que al reconocer un problema se puede planear su solución y cuanto más sabemos lo que gastamos, más aceleramos el proceso de mejora.

También en el ámbito cultural, se crea una "Sentido de urgencia" sabiendo que son suficientes 51 kg de material para tener un evento LOPC significativo. Esto nos traslada a la acción.

El proceso se ha acelerado, con la comunicación de LOPCs ocurridos en plantas hermanas como Celco (Narrows, VA), se aprende de problemas antes de que sucedan en Ocotlan.

7. Referencias

[1] Process Safety Leading and Lagging Metrics (CCPS initial publication 2007)

[2] Process Safety Performance Indicators for the Refining and Petrochemical Industries (ANSI/API RECOMMENDED PRACTICE 754FIRST EDITION, APRIL 2010)

[3] M. Vela, Conduct of Operations Best Practice Networks (CCPS publication 2013)

[4] J. Forest. Improving conduct of operations using best practice tools from operating, engineering, and management discipline. Presented at the 8th global congress on process safety, April 1-4, 2012

[5] J. Forest. Management Discipline: Defining a Process Safety Strategy, Process Safety Progress 33 (2014), 162-165.

Paper Title

Observaciones Preventivas de Seguridad e indicadores como herramientas de Seguridad de Procesos

Presenter/Principal Author: Ing. Ricardo Ceskiavikus / Ing. Luis Landa
Affiliation: YPF S.A.
Address: Macacha Güemes 515 – Capital Federal, Argentina
Email address: ricardo.ceskiavikus@ypf.com / luis.landa@ypf.com

Prepared for Presentation at the
American Institute of Chemical Engineers
2015 Spring Meeting and 11th Global Congress on Process Safety

UNPUBLISHED

Paper Title

Observaciones Preventivas de Seguridad e indicadores como herramientas de Seguridad de Procesos

Presenter/Principal Author: Ing. Ricardo Ceskiavikus / Ing. Luis Landa
Affiliation: YPF S.A.
Address: Macacha Güemes 515 – Capital Federal, Argentina
Email address: ricardo.ceskiavikus@ypf.com / luis.landa@ypf.com

Keywords: OPS, Indicadores, Seguridad de Procesos.

Abstract

Son conocidas las múltiples herramientas o check list dedicadas a auditar o controlar las condiciones de seguridad de las instalaciones de las diferentes industrias. En este caso particular se explica una herramienta denominada Observaciones Planificadas de Seguridad (OPS) [1] cuyos objetivos son: involucrar directamente a las líneas Gerenciales y la Dirección, lograr objetividad y trazabilidad en los resultados mediante índices de desempeño de las Unidades de Procesos y facilitar la implementación de acciones correctivas monitoreadas a través de los Comité de Seguridad.

A su vez, la herramienta se ve complementada con indicadores de seguridad de procesos (predictivas y retrospectivas) orientados a mejorar la disciplina operacional y crear una cultura en la materia, para optimizar en forma continua el rendimiento de la seguridad de procesos.

1. Introducción

Implementar y mantener un programa integral de seguridad de los procesos es un elemento esencial que toda empresa Química y Petroquímica debe implementar y llevar adelante para optimizar el rendimiento de los procesos y mantener bajo niveles o estándar de seguridad las distintas variables productivas con el fin de evitar accidentes industriales.

El objetivo del presente Paper es mostrar una herramienta de control de procesos que es objetiva y llevada a cabo por los altos mando de una compañía. A su vez, la misma se ve complementada por indicadores de seguimiento mensual con el objetivo de mantener controladas las variables que pueden desencadenar en un accidente o incidente industrial.

2. Metodología de Observaciones Preventivas de Seguridad (OPS)

Una industria Química o Petroquímica se encuentra inmersa en distintos sistemas de control, planes preventivos y de gestión tendientes a mantener segura sus instalaciones y cuidar la vida de sus trabajadores. Dentro de estas herramientas pueden encontrarse:

- Seguridad en los diseños
- Planes estratégicos.
- Control de procesos.
- Instalaciones seguras.
- Procedimientos
- Análisis de incidentes y accidentes
- Formación y desarrollo
- Auditorias de frente de trabajo
- Comités de Seguridad, entre otros.

La metodología de OPS, es una herramienta de gestión que permite monitorear el funcionamiento del Sistema Integral de Seguridad con los siguientes beneficios esperados:

- Fortalecer la conducta del personal como medio para incrementar actos seguros.
- Medición del grado de cumplimiento de las políticas de seguridad con fácil procesamiento de la información.
- Retroalimentación positiva a las personas y al sistema, no agotándose en la observación y registro del acto inseguro.
- Proceso participativo que involucra a los altos mandos.

Dentro de sus características principales del sistema OPS se encuentran:

- Se realiza en forma bimestral a todas las Unidades de Procesos.
- Posee un sistema de evaluación objetiva valorando las observaciones a través de índices de desempeño.
- Los observadores son designados al azar, y los mismos son mandos medios o altos de la Compañía.
- Permite la acumulación de datos estadísticos del desempeño en seguridad de los distintos Aspectos observados.
- Calcula un índice de desempeño por Unidad de Procesos.
- Involucra a los observadores en el modelo de gestión de la Compañía.
- Carga de datos al sistema sin conocer los resultados finales.
- Fácil seguimiento de los resultado a través de los Comités de Seguridad.

2.1 Composición del Sistema OPS

Para entender el Sistema OPS debe tenerse en cuenta dos fases del proceso:

a) Proceso de realización de las auditorias.
b) Soporte informático e índice de desempeño.

a) Proceso de realización de las auditorias:

El proceso consiste en designar en forma semestral y al azar, dentro de una base de datos de un soporte informático, los observadores para cada Unidad de Procesos. Una vez concluida dicha etapa, el sistema emite correos electrónicos anunciando el comienzo del período para efectuar la OPS y la asignación de las personas a cada Unidad de Procesos. Como Aspecto relevante, es importante que las personas observadoras posean un nivel jerárquico con toma de decisión dentro de la estructura a fin de involucrarlos en el modelo de gestión de seguridad de la Compañía.

El proceso continúa con las auditorias en campo donde los Aspectos a observar son:

- *Instalaciones*: Estado de conservación de las mismas.
- *Conductas de las personas*: Comportamiento de las personas para evaluar una conducta segura.
- *Procedimientos*: Grado de cumplimiento de procedimientos internos y corporativos
- ***Estabilidad de los procesos*: conducción segura de la operación de las unidades de procesos, medidos a través de las alarmas y disponibilidad de los sistemas de paro de planta (enclavamientos).**

Para brindar mayor detalle de lo auditado relacionado a Seguridad de los Procesos se puede mencionar como ejemplo los siguientes puntos auditados:

Lista de control de sistemas de shutdown - Disponibilidad:

- Cantidad de sistemas de Shutdown que tiene la unidad de procesos auditada.
- Cantidad de sistemas de shutdown fuera de servicio o parcialmente fuera de servicio.
- Cumplimiento de los preventivos de los sistema mencionados.
- Si se cumplió con el procedimiento de by pass de enclavamiento de equipos.

Lista de control de sistemas de alarmas

- Número de alarmas de emergencias activadas.
- Motivo de las alarmas:
    - Instrumentes en falla.
    - Equipo fuera de servicio.
    - Mala configuración de alarmas.
    - Estabilidad de los procesos.

Números de alarmas de alta prioridad activadas

- Motivo de las alarmas:
  - Instrumento en falla.
  - Equipo fuera de servicio.
  - Mala configuración de alarma.
  - Estabilidad de los procesos.

Número de alarmas repetitivas (durante el tiempo que dura la auditoria)

- Motivo de las alarmas:
  - Instrumento en falla.
  - Equipo fuera de servicio.
  - Mala configuración de alarmas.
  - Estabilidad de procesos.

Drenajes de planta – Estado de conservación de las tapas rejillas de los sumideros, las tapas de cámaras se encuentran selladas correctamente, los drenajes no están obstruídos, etc.

Evaluación de pérdidas y fugas – Evaluar en os equipos las pérdidas de: hidrocarburo, agua, gases tóxicos, etc. Descripción de lo observado indicando la ubicación física o el nombre del equipo de referencia.

Soportación de equipos y cañerías – Verificar estado general de las soportaciones de máquinas, equipos y cañerías. Verificar la bulonería, estado de las plantas de los equipos, patines de cañerías, etc.

Purga y venteos de equipos – Las purgas de los equipos se encuentran con sus tapones colocados, es adecuado el lugar donde estos descarga, etc.

Las observaciones relevadas son cargadas en el soporte informático donde através de algoritmos matemáticos se establece como resultado índices de desempeño por Aspecto y por Unidad de Procesos. Como ejemplo, a continuación se detallan las variables que se tiene en cuenta para el cálculo del Aspecto de Estabilidad de los Procesos:

Shutdown de planta:

- Disponibilidad operativa de los shutdown de planta.
- Cumplimiento del preventivo.
- Cumplimiento del procedimiento de by pass de enclavamiento.

Alarmas:

- Si se registran o no alarmas de emergencias activas durante el transcurso de la auditoria.

Una vez efectuada la auditoria, los resultados son cargados por los mismos auditores en un soporte informático donde en forma automática y mediante el algoritmo matemático mencionado, se establecen índices de desempeño en seguridad en los distintos Aspectos observados.

La conformación de índices de desempeño, permite un seguimiento y tendencia de la performance en seguridad de la Unidad de Procesos a lo largo del tiempo, favoreciendo a su vez la implementación de planes de adecuación que son monitoreados a través de los Comité de Seguridad locales y corporativos.

En el diagrama siguiente se muestra en forma sintética el proceso de OPS:

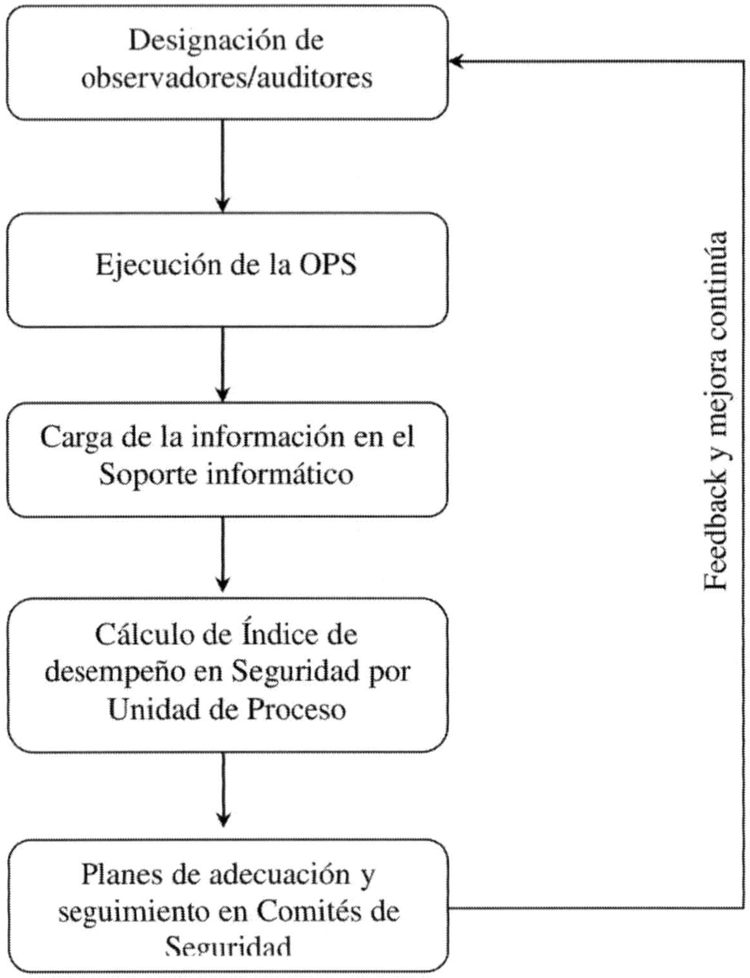

b) Soporte informático e índices de desempeños:

En el punto anterior se desarrolló el detalle del proceso de OPS, junto a los objetivos que persigue el mismo. En el presente apartado se mostrará un ejemplo de las características del soporte informático que es utilizado y los reportes de seguimiento de los índices de desempeño por Unidad de Procesos.

Reporte de campañas de auditorías:

En la siguiente figura puede observarse una síntesis de las distintas campañas realizadas en los últimos años en una Unidad de Procesos, y los resultados del Incide Global de desempeño de esa planta para cada OPS semestral realizada. A su vez, puede verse para un semestre en particular, el detalle de los índices por cada Aspecto evaluado. A la izquierda de la pantalla se observa el menú que puede contener el soporte.

Figura N° 1: Pantalla ejemplo del Soporte Informático OPS.

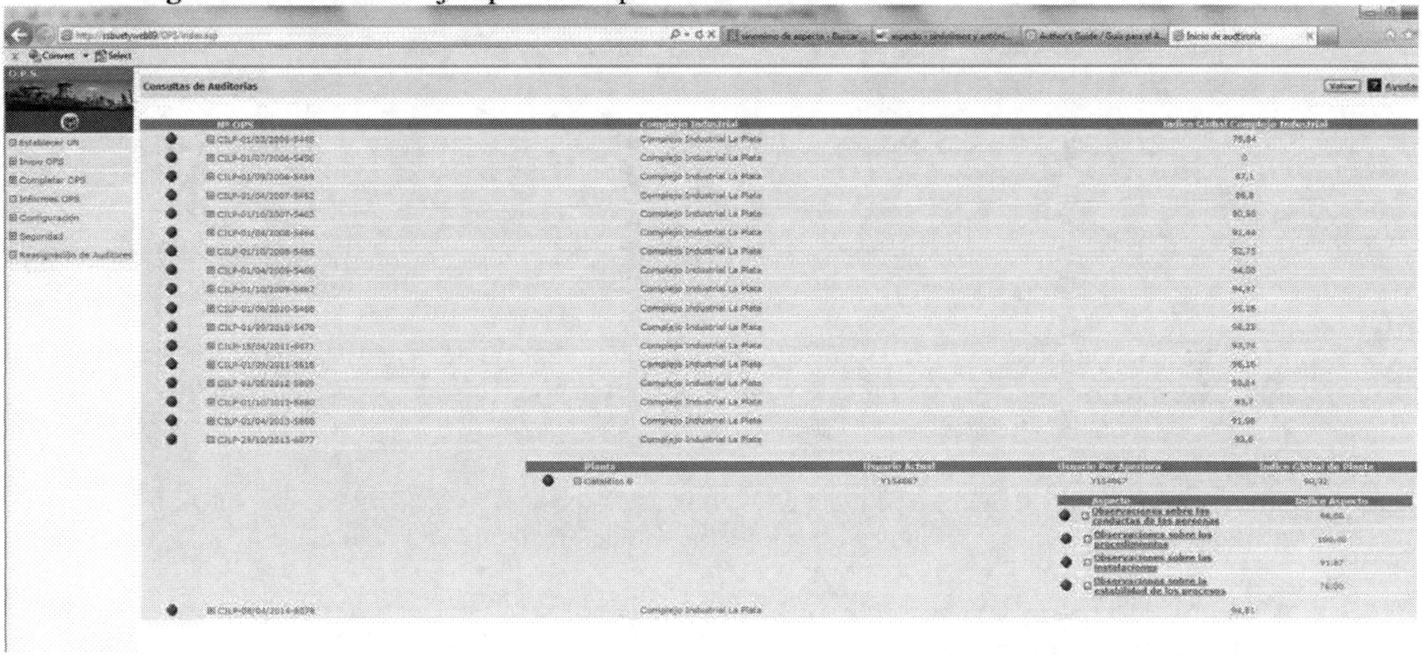

A continuación se observa la planilla donde se deben cargar los datos en el Aspecto relacionado en éste caso a "Observaciones sobre la estabilidad de los procesos", para una Unidad de Proceso dado. Aquí es donde el auditor carga las observaciones halladas durante la auditoría.

Figura N° 2: Ejemplo de Aspecto Estabilidad de los procesos.

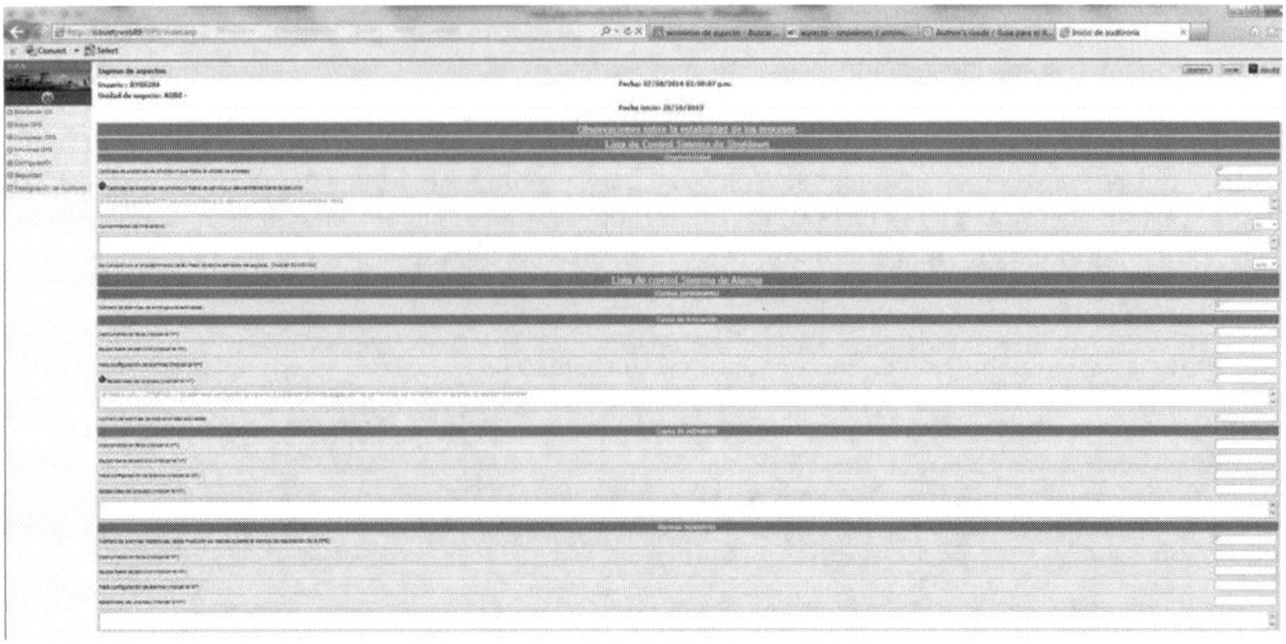

Por último, en la tabla y gráfico siguiente se observa un reporte de seguimiento de los cuatro Aspectos evaluados en la OPS. Puede generarse reportes por Aspecto evaluado, o un reporte de índice global de una Unidad de Procesos.

Un índice o resultado de 100% indica que la planta no posee observaciones sobre el Aspecto evaluado. Valores menores, indica que se realizaron hallazgos sobre lo evaluado.

Tabla N° 1: Reporte de índices por aspecto evaluado.

PLANTA	SEMESTRE	ESTABILIDAD DE LOS PROCESOS	INSTALACIONES	PROCEDIMIENTOS	CONDUCTAS DE LAS PERSONAS	INDICE GLOBAL DE LA PLANTA
Planta X	1° Semestre	66.80 %	91.67 %	100.00 %	100.00%	**89.62**
	2° Semestre	51.60 %	62.50 %	100.00 %	100.00%	**78.53**

A su vez, para cada Aspecto, puede graficarse su evolución o tendencia a los largo del tiempo.

Figura N° 3: Evolución de aspecto de Estabilidad de los procesos.

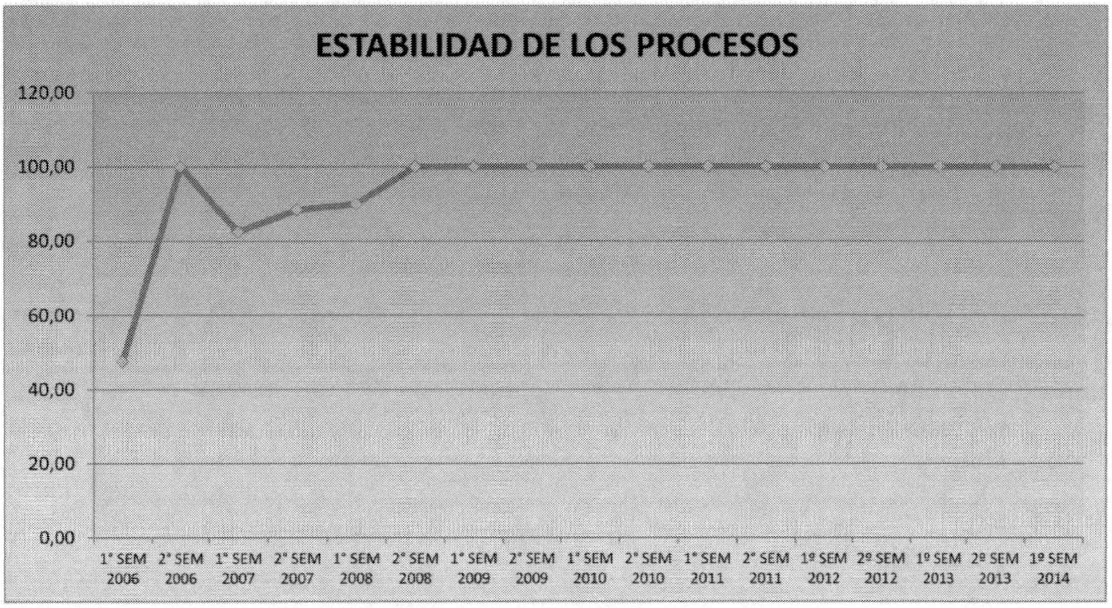

3. Indicadores de Seguridad de los Procesos

A nivel nacional existen indicadores de seguridad (legislados o no), orientados a mantener bajo control los aspectos de seguridad, principalmente los relacionados a puestos laborales. En la industria Química y Petroquímica, además del proceso de OPS, es de suma importancia en forma complementaria realizar gestión sobre la seguridad de los procesos mediante otros tipos de indicadores.

Hemos visto una herramienta dirigida a controlar y mejorar las variables que pueden desencadenar en un accidente o incidente industrial que indefectiblemente impactan sobre las instalaciones y las personas.

En consecuencia, y bajo los lineamientos de la Norma API 754 [2], como práctica recomendada es importante además establecer indicadores de los rendimientos de seguridad de procesos basados en los siguientes principios:

- Los indicadores deben impulsar la mejora de rendimiento de seguridad de proceso y el aprendizaje.

- Los indicadores deben ser relativamente fáciles de implementar y de fácil comprensión para todas las partes interesadas (por ejemplo, los trabajadores y el público).

- Los indicadores deben ser estadísticamente válidos en uno o más de los siguientes niveles: la industria, la empresa y el sitio.

- Los indicadores deben ser apropiados para la industria, la empresa, o la evaluación comparativa a nivel de sitio.

En 1931, H. W. Heinrich [3] introdujo el concepto de pirámide de accidente, que se divide en cuatro niveles diferentes:

- Los accidentes de seguridad de procesos.
- Otros incidentes.
- Cuasi accidentes
- Comportamientos inseguros / disciplina operativa insuficiente.

A su vez estos niveles llevan tres tipos de indicadores:

- *Reactivos*: Se basan en "hechos ya ocurridos" que alcanzan cierto umbral de gravedad y que deben ser reportados.

- *Proactivos*: Métricas prospectivas que indican el rendimiento de los procesos clave de trabajo, disciplina operativa, o capas de protección que previenen incidentes.

- *Cuasi - accidentes*: Incidentes menos graves o condiciones inseguras que activan una o más capas de protección. Aunque estos eventos son sucesos reales (es decir, "ya ocurridos"), generalmente se considera un buen indicador de las condiciones que en última instancia podría llevar a un incidente más grave.

Figura N° 4: Pirámide de accidentes

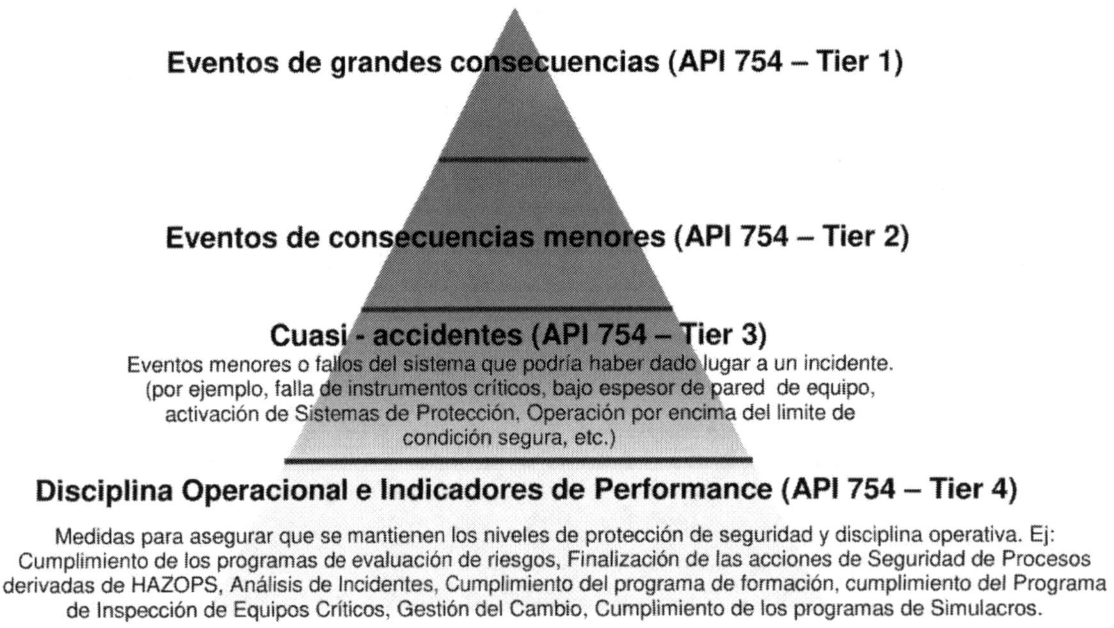

3.1. Indicador de desempeño Tier 1.

Un Incidente de Seguridad de Procesos (ISP) Tier 1 es toda Pérdida de Contención Primaria (PCP – toda liberación no planificada o descontrolada de cualquier material, incluyendo los no tóxicos o no inflamables, proveniente de un Proceso) que resulta en una de las siguientes consecuencias:

- Días perdidos o muerte de un empleado propio o contratista.
- Hospitalización y/o muerte de un tercero.
- Una evacuación de la comunidad por orden oficial.
- Fuego o explosión con un costo directo de más de U$S25.000
- Fugas incluyendo descargas de dispositivos de alivio de presión, ya sea directamente a la atmósfera o a través de un dispositivo de destrucción aguas abajo (ej. sistema de antorchas, lavador, etc.) y que resulta en:
    - Arrastre de líquido.
    - Descarga sobre un área potencialmente insegura.
    - Evacuación, refugio o medidas de protección públicas (ej: cierre de caminos) por orden oficial.
- Una liberación de material mayor a ciertas cantidades establecidas en API PR 754

Tier 1 ISP = (N° total de ISP / HH Trabajadas) x 200.000

3.2. Indicador de desempeño Tier 2.

Un Evento de Seguridad de Procesos (ESP) Tier 2 es toda Pérdida de Contención Primaria (PCP) con consecuencias menores. Un ESP Tier 2 es toda liberación no planificada o descontrolada de cualquier material, incluyendo los no tóxicos o no inflamables, proveniente de un Proceso) que resulta en una de las siguientes consecuencias y que no clasifican como ISP Tier 1:

- Primer Auxilio de un empleado propio o contratista.
- Fuego o explosión con un costo directo de mas de U$S2.500
- Accionamiento de un sistema de liberación de presión a la atmósfera o al sistema de antorchas y que tenga las siguientes consecuencias:
    - Arrastre de líquido.
    - Descarga sobre un área potencialmente insegura.
    - Evacuación o refugio.
    - Medidas de protección públicas (ej: cierre de caminos)
    - Y que no sobrepasa los valores establecidos por la API RP 754 en cualquier periodo de 2 horas.
- Una liberación de material mayor a ciertas cantidades establecidas en API PR 754 en cualquier periodo de 1 hora

Tier 2 ESP = (N° total de ESP / HH Trabajadas) x 200.000

3.3. Indicador de desempeño Tier 3.

Los Cuasi-Incidentes Tier 3 típicamente representan aquellas situaciones en las cuales las barreras de protección existentes actúan evitando el desarrollo de ESP Tier 2 o ISP Tier 1.

Los indicadores en este nivel proveen oportunidades adicionales para identificar y corregir debilidades en las barreras de protección existentes.

Algunos ejemplos de estos indicadores son:

- Operación por encima de los límites de operación segura de un proceso.
- Inspecciones o pruebas de recipientes o líneas que arrojan valores por debajo de los admisibles.
- Activación de las Lógicas de Seguridad.
- Pérdidas de Contención Primaria que no llegar a calificar como ISP o ESP.

3.4. Indicador de desempeño Tier 4.

Los Indicadores de Desempeño Tier 4 son los utilizados para medir la performance de los Sistemas de Gestión y la Disciplina Operativa.

Los indicadores Tier 4 representan el desempeño particular de los diferentes elementos que componen las barreras de protección del Sistema de Gestión de Seguridad de Procesos, evaluando también el impacto de la Disciplina Operativa sobre los mismos.

Los indicadores en este nivel proveen oportunidades adicionales para identificar y corregir debilidades en los elementos individuales que componen las barreras de protección existentes.

El aspecto más relevante de los indicadores Tier 4 es que proveen de información valiosa para evitar futuros ISP Tier 1 o ESP Tier 2.

Algunos ejemplos de estos indicadores son:

- Cumplimiento de los programas de evaluación de riesgos (HAZOPS).
- Finalización de las acciones de Seguridad de Procesos derivadas de HAZOPS, Análisis de Incidentes, etc.
- Cumplimiento del programa de formación.
- Cumplimiento del Procedimiento de Permisos de Trabajo.
- Cumplimiento del Programa de Inspección de Equipos Críticos.
- Gerenciamiento de Deficiencias de Equipos Críticos.
- Gestión del Cambio.
- Cumplimiento de los programas de Simulacros.

3.5. Otros Indicador proactivos.

3.5.1 Número de bypass activos de las lógicas de seguridad de procesos

Esta métrica identifica el estado de las barreras de protección lógicas existentes y que potencialmente pueden permitir la generación de un evento o incidente de Seguridad de los Procesos.

3.5.2 Porcentaje de Válvulas de Seguridad con desvío en el Pre Post Test

Esta métrica identifica aquellos Incidentes que no tiene la entidad para ser clasificados como ISP o ESP y permite generar una base de comparación para identificar comportamientos que puedan prever la ocurrencia de un ISP o ESP.

3.5.3 Operación por encima de los límites Seguro del Proceso

Límite seguro de proceso: representa el punto extremo o último antes del cual una acción predeterminada es capaz de retornar al proceso a un estado seguro. Esta acción predeterminada puede ser desde un procedimiento operativo ejecutado manualmente hasta un sistema totalmente automatizado como un SIS.

3.5.4 Cumplimiento del programa de inspección

Este indicador es una medida de la eficacia del sistema de gestión de seguridad de procesos para asegurar que la seguridad de la planta y equipos críticos es adecuada.

El cálculo de la métrica implica:

- Definir el período de medición de la actividad de inspección.
- Determinar el número de inspecciones de seguridad de la planta y el equipo crítico previsto para el período de medición.
- Determinar el número de inspecciones de seguridad de la planta y equipos críticos completado durante el período de medición.

3.5.5 Gestión del cambio debidamente gestionada

Esta métrica mide si la Instalación cumple con su Procedimiento de Gestión del Cambio. Implica una auditoría periódica de la documentación de los Cambios realizados.

La auditoría de cumplir los siguientes pasos:

- Definir el alcance de la auditoría: marco de tiempo, la frecuencia y área a auditar.
- Determinar el tamaño de muestra deseado y estadísticamente significativa. Para esto se puede hacer uso de tablas ampliamente disponibles, basado en el número total de Cambios.

- Revisar la documentación de los Cambios completados, incluyendo la documentación de respaldo, tales como la revisión de riesgos e información actualizada sobre seguridad de procesos (Instructivos de Operación y P&ID).

3.5.6 Número de alarmas por operador de consola

Esta métrica identifica aquellos Sistemas de Control que por diferentes motivos sobrecargan al Operador de Consola con un gran número de alarmas.

4. Conclusión

Con el fin de evitar accidentes industriales, las Observaciones Preventivas de Seguridad como los indicadores de seguridad de procesos son herramientas de gestión, que deben ser implementadas por toda industria Química y Petroquíca como practica recomendada para mantener bajo control los niveles o estándar de seguridad de las variables productivas..

Las OPS al ser ejecutadas por niveles organizacionales altos de la Compañía, le otorgan a ésta herramienta una importante jerarquía para obtener resultados concretos y objetivos a través de sus índices de desempeño por Aspecto evaluado. A su vez, facilita la elaboración de planes de adecuación con seguimiento desde los Comités de Seguridad orientado a la mejora de la disciplina operativa.

Por último, para que una organización adopte una cultura entre sus sistemas y trabajadores orientada a la seguridad de los procesos, es indispensable además implementar y tener bajo control indicadores reactivos y proactivos que detecten en forma temprana desvíos en los procesos que puedan generar consecuencias de riesgo alto o crítico hacia las personas como las instalaciones.

Referencias

[1] Sistema de Observaciones Preventivas de Seguridad utilizado en YPF S.A.

[2] API 754 RP Process Safety Performance Indicators for the Refining *and Petrochemical Industries*

[3] Heinrich, H.W., *Industrial Accident Prevention*, New York, McGraw-Hill, 1931.

Incidentes de Seguridad en la Industria del GNL

Ing. Juan Gabriel Haitzaguerre
Juan.haitzaguerre@ypf.com

Prepared for Presentation at
American Institute of Chemical Engineers
2015 Spring Meeting
11th Global Congress on Process Safety
Austin, Texas
April 27-29, 2015

UNPUBLISHED

Incidentes de Seguridad en la Industria del GNL

Ing. Juan Gabriel Haitzaguerre
Juan.haitzaguerre@ypf.com

Keywords: LNG, Incidentes, Capas de protección

Abstract

En este documento se hace una recopilación de los incidentes más importantes ocurridos a lo largo de la historia de la industria de GNL (Gas Natural Licuado), mediante la búsqueda de información provista por reportes e informes publicados en diferentes medios (páginas web, reportes de compañías, foros, publicaciones, etc.) con el objetivo de tomar real conocimiento de los peligros a los que se enfrenta la industria y estar preparados ante emergencias similares en las plantas que se encuentran en Argentina.

En este trabajo se hace una división de los incidentes según hayan ocurridos en terminales marítimas, y plantas "peak shaving" o en buques. En primer lugar, se realiza un análisis de los incidentes ocurridos "on shore", comenzando por nombrar el accidente más grave en la historia del GNL con la mayor cantidad de fatalidades (128 personas). Se analiza el hecho tal como fue ocurriendo y luego se analizan las conclusiones de la investigación, las cuales fueron muy importantes para las lecciones aprendidas aplicadas posteriormente.

Luego se analizan otros 4 incidentes ocurridos en diferentes plantas de licuefacción y "peak shaving" en la cual se produjeron fatalidades, detallando las conclusiones de las investigaciones. Algunos incidentes provocaron muertes directamente relacionadas con el producto GNL pero en la mayoría de los hechos, el Gas Natural Licuado no ha sido la razón principal o ni siquiera estuvo presente en el momento del siniestro.

Siguiendo con la lectura, se hace un apartado referido a los incidentes en los que se involucran buques en diferentes operaciones (carga-descarga, operación normal, navegando, etc.). No se mencionan todos, sino que se detallan los que son de referencia para la industria y los que sirven para tener una dimensión de los incidentes que se pueden encontrar y la gravedad de los mismos.

Utilizando la misma metodología, se analizan los eventos no previstos ocurridos sobre buques que a partir de causas similares, provocaron las mismas consecuencias. Los hechos ocurrieron en buques metaneros, aquellos que transportan la carga. Es de destacar que ninguno se produjo en buques denominados FSRU (unidades flotantes de almacenamiento y regasificación).

El último apartado trabajo hace referencia a los aspectos de Seguridad en las terminales de Regasificación de GNL que existen en Argentina, mediante la división de la terminal en: Buque, terminal y gasoducto; para poder distinguir más precisamente las acciones que se llevan a cabo para poder asegurar altos estándares de Seguridad.

Por último, se realiza una conclusión sobre la base de la evolución de la tecnología y las prácticas de la industria, aprovechando las lecciones aprendidas.

Note: Do not add page numbers. Do not refer to page numbers when referencing different portions of the paper

1. Breve reseña sobre el GNL

Ya en el siglo XVIII algunos físicos y químicos trabajaron experimentando con la licuefacción de gases. Entre ellos se destaca el físico-químico Michael Faraday, quien experimentó con metano, principal componente del gas natural.
En 1873, el ingeniero alemán Karl Von Linde construye la primera máquina práctica de refrigeración por compresión, siendo este principio el utilizado para construir la primera planta de licuefacción de gas natural en 1912 en West Virginia, Estados Unidos. Esta planta comenzó a operar en el año 1917.
Otro hito es la construcción en 1941 de la primera planta comercial de licuefacción de GNL en Cleveland, Ohio. En esta planta, 3 años después de comenzar a operar se produce el mayor accidente de la historia del GNL, en lo que se dio a conocer como la "Catástrofe de Cleveland".
En 1959, se realiza el primer transporte por mar de GNL de manera segura. El Methane Pioneer, un buque de la Segunda Guerra Mundial reconstruido y reacondicionado con 5 tanques aislados, parte con 7000 barriles de GNL desde Lake Charles, Louisiana, EEUU, hasta Canvey Island, Reino Unido.
En 1964 Argelia se convierte en el primer exportador del mundo y el Reino Unido en el primer importador, dando comienzo al transporte comercial. A su vez, se empiezan a construir muchas plantas de importación y exportación alrededor del mundo y principalmente en Estados Unidos, en donde se había parado la industria por la catástrofe de Cleveland.
En 1967 se establece la NFPA 59 A "norma de producción, almacenamiento y manejo del GNL".

2. Características y peligros del GNL

Antes de comenzar con el detalle de los accidentes, es conveniente conocer sobre las características del producto y los peligros que tienen asociados.
El GNL es un producto líquido que se encuentra a -160°C de temperatura a presión atmosférica. Es incoloro e inodoro. Su densidad con respecto al agua es de 0,45. Su relación volumétrica Liquido/Gas es de 600.
Con respecto a los peligros, estos son:
Por su temperatura criogénica se pueden producir con el contacto quemaduras en la piel o en los órganos respiratorios por su inhalación.
El GNL que se encuentra en contacto con la atmósfera se va evaporando produciendo gas natural y este tiene un rango de explosividad de entre 5 y 15%, pero además y quizás más importante, es que el metano desplaza al oxígeno lo que puede producir asfixia si el lugar no está ventilado.
El tercer gran peligro es la expansión de vapores y la probabilidad de ignición. En base a esto se pueden generar tres tipos de fuego:

- Fuego de charco (pool fire). Cuando en un derrame de GNL se produce la ignición de los vapores emanados (por el contacto con un punto caliente y siempre que se encuentre dentro del rango de explosividad).
- Fuego por nube de vapor (flash fire). Si bien el metano es más liviano que el aire, cuando empieza a evaporarse a -160°C y hasta los -110° el metano es más denso que el aire por

lo que se forma una nube temporal sobre el nivel del piso. Si se encuentra con un punto caliente la nube detona.

- Fuego tipo soplete (Jet fire). Este tipo de fuego es casi imposible ya que el GNL se encuentra a presión atmosférica. Solo puede ocurrir cuando se libera desde una fuente a presión.

Es importante recalcar que al estar a presión atmosférica este producto no genera una sobre presión, diferenciándolo de otros productos como el GNC (gas natural comprimido) y el GLP (gas licuado de petróleo).

3. Lista de incidentes

Esta lista de incidentes se conoce como "the Lloyd's List" y es un registro de todos los incidentes producidos. A partir de esta lista, se realizó este trabajo y se completó con incidentes ocurridos en los años 2013 y 2014. Algunos incidentes pueden faltar, sin embargo los de mayor gravedad y los que repercutieron en la industria están detallados. La lista se encuentra al final de trabajo en el anexo I.

Se contabilizan 35 incidentes de los cuales 14 se produjeron en plantas en tierra y los restantes 21 en buques. Se produjeron 5 incidentes fatales, todos en plantas en tierra, con un total de 194 muertes, pero no todas relacionadas directamente con el GNL.

2.1 Incidentes producidos en tierra

- *1944, Cleveland, Ohio, EEUU. "La catástrofe de Cleveland"*

Este incidente es el más grave en la historia del GNL.

Luego de 3 años de operación, se decide construir un nuevo tanque de almacenamiento de GNL. Poco tiempo después de ponerlo en operación, el tanque se parte y se derrama todo el GNL almacenado, vertiéndose por los desagües y llegando hasta los barrios linderos a la planta. El fuego resultante produjo la muerte de 128 personas. A continuación se detalla el incidente y las causas en base al reporte de investigación del U.S. Bureau of Mines. ("report on the investigation of the Fire at the Liquefaction, Storage and Regasification Plant of the East Ohio Gas Co., Cleveland, Ohio, October 20, 1944, "US Bureau of Mines, February, 1946)

El 20 de octubre de 1944, siendo aproximadamente las 14:30hs, el tanque cilíndrico cedió, perdiendo todo su contenido de GNL. El GNL llega hasta los desagües de los barrios vecinos. Se genera una nube de vapor producto de la evaporación que rápidamente se enciende provocando un incendio devastador que envuelve la planta, los tanques de almacenamiento y el vecindario. A los 20 minutos, cuando el fuego ya estaba por extinguirse, las patas de una esfera de almacenamiento se vencen y se produce otro derrame con el consecuente incendio. En total murieron 128 personas y 225 resultaron heridas. 200 hectáreas fueron alcanzadas por el incendio de las cuales 12 fueron totalmente devastadas.

La investigación arrojó los siguientes resultados:

En 1944 Estados Unidos estaba en plena guerra mundial y toda la industria de la acería estaba destinada a armamento y equipamiento bélico, por lo que el tanque que se partió fue construido con 3,5% de níquel. Esta aleación no llega a dar la ductilidad al acero para resistir temperaturas criogénicas (ahora se utilizan aleaciones de aceros - níquel 9%). Además, los tanques estaban ubicados cerca de una estación de trenes y de una planta de estampados donde se empleaban balancines. Esto provoco una aceleración en la propagación de la fractura del tanque. Una vez que se rompe el tanque interior, fue muy fácil romper el tanque de afuera ya que era de acero al carbono. El incidente se agravó porque los diques de contención no eran adecuados y el GNL llegó hasta lugares residenciales. La causa de la fracturas de las patas de la esfera fue que estas no estaban protegidas contra fuego, por lo que ante tanta radiación térmica terminaron cediendo, provocando el segundo derrame y posterior incendio.
Las explosiones que se produjeron fueron debidas a que la nube de vapor se encontraba en un espacio confinado (los desagües). No se registraron otras explosiones.
Este incidente dejo muchas lecciones aprendidas, que luego formaron parte de normativas obligatorias. Entre las que se encuentran:

- ✓ Aleaciones resistentes a temperaturas criogénicas. Actualmente las aleaciones de acero se componen de por lo menos 9% Ni y aluminio.
- ✓ Las patas de los tanques y esferas son revestidas con pinturas que protegen contra el fuego.
- ✓ Los diques de contención tienen una capacidad de por lo menos 110% de contenido de los recipientes.

- *1971, La Spezia, Italia. El primer incidente documentado sobre Roll Over.*

Luego de estar casi un mes amarrado en el puerto, el buque ESSO BREGA comenzó a descargar GNL a un tanque de almacenamiento en tierra. Luego de 18 horas se produce un repentino incremento de la presión interna del tanque lo que provoca que se activen los sistemas de seguridad (válvulas de seguridad y sistemas de venteo). El tanque resulta con daños menores en el techo. No se produjo la ignición de la nube generada por los venteos, ni hubo heridos.
La investigación arrojó que la causa fue provocada por un fenómeno llamado Roll over.
Este fenómeno se produce cuando se juntan dos cargas de GNL de diferente densidad. El GNL de mayor densidad queda debajo del de menor densidad. En cuanto a la capa de arriba, en esta se produce una evaporación natural que genera un incremento de la densidad (ya que se evaporan los componentes más livianos del GNL) pero se mantiene la misma temperatura. En cambio, en la capa de abajo, su densidad se mantiene constante, pero el contacto con las paredes y el piso del tanque produce un sobrecalentamiento del GNL, que no puede ser liberado por estar "encerrado" por la capa superior. Cuando la capa superior llega a la misma densidad que la capa inferior se produce una mezcla entre las dos capas y se libera el calor almacenado, provocando una mayor producción de vapores. Esto puede generar una sobre presión con daños en los materiales y hasta la detonación si encontrase un punto caliente.
Esto fue lo que se produjo en este incidente. El buque descargó una carga "pesada" sobre una carga con una densidad menor y más fría.

- *1973, Staten Island, New York, EEUU. Accidente de la construcción.*

La planta de peak shaving habia estado más de 3 años en operación cuando deciden sacarla de servicio para mantenimiento. El tanque en el cual se iba a trabajar ya estaba vacío, inertizado con nitrógeno y se había vuelto a llenar con aire fresco. 40 trabajadores estaban realizando tareas cuando de repente se enciende la espuma de poliuretano y el mylar (PET) que actuaban como aislantes dentro del tanque. Aunque los procedimientos establecían que se debía trabajar con equipos anti explosivos, se estaban utilizando equipos sin esta protección. Se asume que una chispa provocó el incendio. La alta temperatura provocó un incremento de la presión dentro del tanque, dañando el domo de concreto, colapsando el techo del tanque y matando a las 40 personas que se encontraban trabajando.
La investigación, sin embargo, concluyó que no fue un accidente de GNL sino que fue un accidente de la construcción.
("Report of Texas Eastern LNG Tank Fatal Fire and Roof Collapse, February 10, 1973" Fire Department of the City of New York, July, 1973)

- *1977, Argelia, una muerte por congelamiento*

Un operador fue rociado por GNL provocándole la muerte por congelamiento cuando se rompió el cuerpo de una válvula sobre un tanque de almacenamiento soterrado. Se calcula que hubo una fuga de 2000 m3 de GNL pero la nube que se generó no se encendió.
La investigación dio como resultado que la valvula no soporto la temperatura criogénica porque estaba construida íntegramente de aluminio. Hoy en día las válvulas utilizadas en la industria del GNL son de acero inoxidable.

- *1979, Cove Point, Maryland, EUU. Explosión en subestación eléctrica.*

Aproximadamente a las 15:00hs del día 6 de octubre de 1979 se produjo una explosión dentro de una subestación eléctrica en la terminal de importación de Cove Point, provocando la muerte de un operario y quemaduras importantes a otro.
De la investigación surge que una cantidad de GNL se filtró por el sello mal colocado de una bomba y ya vaporizado recorrió 60 metros por un conducto eléctrico subterráneo ingresando a la subestación. En el análisis de riesgo no se previó que podía ingresar gas a la subestación por lo que no había detectores de gases dentro del recinto. El arco generado por un interruptor fue lo que ocasionó la explosión.

- *2004, Skikda, Argelia. Destrucción de 3 trenes de licuefacción.*

En la planta de licuefacción y exportación de Skikda se produjo una fuerte explosión e incendio que produjo la muerte de 27 personas e hirió a otras 72.
La investigación arrojó que en uno de los sistemas de refrigeración de hidrocarburo que se utiliza dentro del proceso de licuefacción se produjo una pérdida de propano (gas utilizado para la refrigeración) El propano es más denso que el aire por lo que se formó una nube que se estacionó al nivel del piso y se introduce dentro de una caldera de vapor. Este aumento de combustible incrementa la presión dentro del domo de la caldera. Empezaron a actuar las válvulas de seguridad pero la presión excede su capacidad y se termina rompiendo el domo. Se produce una explosión que alcanza la fuente de la fuga lo que genera la ignición de la nube de vapor y una consecuente explosión por el espacio confinado en el que se encontraba. La explosión y el

incendio alcanzaron a destruir 3 trenes de licuefacción de los 6 que tenía la planta y además llegó a los edificios de Administración y Operaciones.
Las lecciones aprendidas de este suceso fueron:
Las distancias cortas entre los trenes y los edificios y el layout de la planta provocaron que el siniestro devastara gran parte de las instalaciones. Los otros 3 trenes no fueron afectados porque estaban alejados de este sector.
A partir de este siniestro no se utilizan más calderas a vapor, sino que han sido reemplazadas por calderas que usan gas como combustible, reduciendo costos y riesgos.
Por otro lado, la planta contaba con un deficiente sistema de shut down y falta de equipos de detección temprana.

2.2 Incidentes que involucran buques metaneros en diferentes operaciones

- *Jules Verne LNG Carrier – 25,500 m3 – 1964 y 1965*

En una carga de GNL, durante tareas rutinarias de ventilación a través del sistema de venteo cae un rayo cerca provocando la ignición de la nube de vapor. Si bien, la carga se había detenido al inicio de la tormenta, se estaban liberando a la atmósfera los vapores generados en el proceso. EL fuego se extingue rápidamente al inertizar con N2 la tubería.
En este mismo metanero, un tiempo después, durante otra carga, se produce un derrame de GNL que provocó la fractura de la cubierta y el puente. Aunque no se conoce con exactitud la causa, se asoció a una falla en el instrumento de medición de nivel y la falta de familiaridad por parte del oficial de carga con este instrumento y su deficiencia.

- *Methane Princess LNG Carrier – 27,400 m3 – 1965*

Se desconectaron los brazos de carga de GNL antes de que se vaciaran las líneas de líquido, lo que provocó que el GNL pase a través de una válvula cerrada que tenía una pequeña pérdida cayendo sobre la cubierta. Si bien la cubierta fue inundada con agua de mar, como es habitual hacerlo para evitar la fractura, igualmente se observó una pequeña falla.

- *Barge Massachusetts – 5,000 m3 – 1974*

El buque se encontraba en operaciones de carga de GNL. Luego de un corte de energía y el cierre automático de las válvulas de las líneas de líquido, una pequeña cantidad de GNL se filtró por una válvula globo de purga de nitrógeno de 1 pulgada. La investigación de la Guardia Costera concluyó que la presión generada por el corte de las líneas provoco la filtración de GNL por la válvula de purga sin producirse perdidas previas durante la carga. El derrame provocó la fractura de la cubierta en un área de 4 m2.

- *LNG Aquarius – 125,000 m3 - 1977 – Bontang*

Se desbordó GNL por el sistema de venteo del tanque que se estaba cargando. El incidente se produjo por fallas en el sistema de medición de nivel de líquido. Se había anulado la alarma de alto nivel para que no sonaran falsas alarmas. El derrame solo produjo una pequeña fractura en la cubierta del tanque.

- *Mostefa Ben Boulaid – 125,000 m3 – 1979 - Cove Point, Maryland, USA*

Mientras se estaba cargando, falló una válvula de retención liberando una pequeña cantidad de GNL, lo que provocó pequeñas fracturas en la cubierta del tanque. En esta válvula, la bisagra es retenida por un perno, que se incrusta en el cuerpo de la válvula. La vibración generada por la operación hizo que el perno se desajustara y la válvula fallara, liberando una lluvia de GNL sobre la cubierta.
A partir de este incidente, todas las válvulas de retención fueron modificadas colocándoles una traba de acero inoxidable a los pernos.

- *Pollenger – 87,600 m3 – 1979*

Mientras el buque descargaba GNL, se produce la pérdida de producto a través de una válvula, derramando el líquido en la cubierta y provocando daños en un área de 2 m2.

- *Isabella – 35,500 m3 – 1985*

Se produjo un derrame de GNL por sobre llenado del tanque al fallar el sistema de medición de nivel. Solo provocó fracturas en la cubierta.

4. Seguridad de procesos en las terminales de regasificación de Argentina

En este apartado se presentan todas las medidas que se han desarrollado en las terminales de Regasificación de Argentina, teniendo en cuenta todos los aspectos de Seguridad y tomando como input los incidentes ocurridos a lo largo de la historia de esta industria.
Las terminales se pueden dividir en tres partes a modo de poder simplificar y entender en qué medida se va asegurando un correcto control de los riesgos: el buque regasificador, el muelle principal y un gasoducto conectado al buque a través de un brazo de descarga de alta presión.

4.1 Buque regasificador

Para describir al buque desde el punto de vista de la Seguridad de Procesos, se puede hacer un análisis por capas de Protección pudiendo discernir 5 capas que determinan el grado de seguridad con que se ejecutan las actividades:

1- *Contención primaria.* Se denomina básicamente a los diseños de ingeniería que proporcionan seguridad. La calidad de los materiales empleados es vital para que no se produzca ningún incidente. La aleación que le da ductilidad al acero ante temperaturas criogénicas es de acero inoxidable con aluminio y por lo menos 9% de Níquel. Este tipo de aleaciones es muy costoso. Algunas tecnologías desarrolladas emplean hasta el 30% de níquel. Por su parte, los materiales que se utilizan como aislantes, en el caso del buque regasificador incluye perlita y nitrógeno, que además de inertizar la zona también sirve para detectar presencia de metano. Otro caso de esta primera capa de contención es el material con que están construidas las mangueras. 57 capas que evitan que ante una fuga de GNL, éste se escape al exterior, quedando atrapado entre las capas, provocando la inevitable evaporación.

2- *Contención secundaria.* Esta capa de protección depende del tipo de construcción de los tanques de almacenamiento de cada buque, como puede ser un mecanismo completo de diseño de contención secundaria para buques de diseño de membrana, que cuenta con un tanque exterior que contiene el derrame de GNL y los vapores generados. En el caso de buques con tanques esféricos o de estructura prismática esta barrera es una barrera guardafangos con un escurridero en el fondo donde el líquido acumulado es evaporado.
3- *Sistemas de seguridad.* Esta es la tercera capa de protección. Los buques cuentan con un sofisticado sistema de seguridad que incluye desde radar y posicionamiento que permite el manejo seguro de la embarcación y el monitoreo del tráfico y sistema global de socorro marítimo que transmite señales automáticas ante emergencias. Todos los buques cuentan con sistemas de detección temprana que incluyen detectores de gas, de fuego, de humo, de bajas temperaturas. Los tanques tienen censores de nivel y censores de presión. Todos estos instrumentos se encuentran dentro de un sistema de ESD, el cual está conectado entre ambos buques y con la terminal, de manera que cualquier emergencia que ocurriera en cualquiera de las 3 zonas (metanero, regasificador o terminal en tierra), el sistema repercute en toda la operación y lleva a condiciones de seguridad a todos los intervinientes. Los sistemas de ESD se encuentran interconectados a través de una conexión por fibra óptica y eléctricamente.
4- Distancias de Seguridad. Esta capa de protección implica el movimiento de los buques metaneros en el amarre y la zarpada, luego de la transferencia. Por un lado, los metaneros son asesorados por prácticos, personas que conocen bien la zona de tránsito y van acompañando al capitán. Por otro lado, son remolcados por remolcadores que ayudan al buque a posicionarse al lado del buque regasificador y lo amarran al muelle. Estos remolcadores tienen una potencia estipulada y la cantidad depende del buque que está entrando. Existen resoluciones de la Prefectura Naval Argentina que establecen las condiciones de tiempo y transito cuando un metanero debe entrar a los puertos de regasificación. Imponen reservas de navegación, horarios y condiciones meteorológicas.
5- *Marco regulatorio.* Esta capa enmarca todas las medidas y sistemas de seguridad mediante el establecimiento de normas y regulaciones internacionales que son de obligatorio cumplimiento. Las instituciones estatales, provinciales y municipales juegan un rol muy importante en la regulación de aspectos ambientales y de seguridad. Otorgan permisos y realizan inspecciones y auditorias establecidas en resoluciones y normativas específicas de GNL para tal fin.

En relación a esta última capa, se destaca que el mismo cuenta con certificados internacionales, cuyas validaciones periódicas son efectuadas por sociedades de clasificación avaladas por convenciones.

Como ejemplo de los certificados que obtiene el buque se pueden mencionar:

- Certificado Internacional de Aptitud para el Transporte de Gases Licuados a Granel ("International Certificate of Fitness for the Carriage of Liquefied Gases in Bulk").
- Certificado Internacional de Seguridad de Buques ("International Ship Security Certificate").
- Certificado de Auditoría de Gestión de Seguridad ("Safety Management Certificate").
- Certificado de Seguridad de Construcción de Buque ("Cargo Ship Safery Construction Certificate").
- Certificado de Equipo de Seguridad del Buque ("Cargo Ship Safety Equipment Certificate").
- Certificado de Navegabilidad ("Certificat de Navegabilite")
- Reporte de Inspecciones ("Survey Status Report")
- Certificado Internacional de Prevención de Contaminación ("International Oil Pollution Prevention Certificate")
- Detalle de Construcción y Equipos para Buques (no Buques Tanque de Petróleo) ("Record of Construction and Equipment for Ships Other than Oil Tankers")

Paralelamente, los buques cuentan con sistemas de gestión de seguridad y medio ambiente certificados de acuerdo a normas de la Organización Marítima Internacional (IMO, por sus siglas en inglés).

4.2 Muelle principal

Por su parte, el muelle está supeditado a normas regidas por la Prefectura Naval Argentina y se efectúan controles periódicos entre los que se encuentra por ejemplo el PLANACON, Plan Nacional de Contingencias, el cual debe ser convalidado una vez por año, y además, cumplen una función primordial en el tránsito de los metaneros que llegan para trasvasar producto. Cada buque que debe ingresar al puerto, debe obtener una "reserva de canal", un permiso que PNA concede para tener prioridad para navegar un tramo del canal, cumpliendo con las normas de seguridad impuestas. PNA ha establecido resoluciones y ordenanzas que regulan la distancia de seguridad que deben cumplir las demás embarcaciones cuando circula un metanero, la velocidad máxima del metanero, velocidades máximas de viento sobre la cual se prohíben las maniobras, planes de emergencias, sistemas de lucha contra incendio, simulacros, etc.

4.3. Gasoducto

Por último, el gasoducto se encuentra enmarcado dentro de las normas del ENARGAS, entre las destacables se citan: la NAG 100, que establece los estándares de diseño, operación y mantenimiento para las instalaciones de transmisión y distribución; y la NAG 153, que regula lo concerniente a la identificación y cuantificación de impactos ambientales, la formulación de medidas preventivas y correctivas de esos impactos y el establecimiento de pautas y procedimientos comunes a los distintos estudios e informes ambientales durante las etapas de planificación, diseño, construcción, operación, mantenimiento y abandono o retiro de sistemas de transmisión y de sistemas de distribución. El ENARGAS audita los gasoductos con una frecuencia semestral, durante la etapa de operación y mantenimiento.

Ilustración 1 - Vista del FSRU Exemplar, en Escobar

4.4. Otras consideraciones

Los buques de las dos terminales son similares y cuentan con la misma tecnología. Las dimensiones de ambos regasificadores son: 291m de eslora, 45m de manga y un calado de 9,3m. Tienen tanques criogénicos capaces de almacenar alrededor de 150 mil m3 de GNL, regasificando a una presión de hasta 90 bares, mediante un proceso que emplea:

- Bombas Criogénicas de Alta Presión. Las mismas llevan el GNL contenido en los tanques de carga a la presión previa a la vaporización.
- Vaporizadores de tubo y casco (Shell-and-Tube Vaporizers). Los mismos convierten el GNL en Gas Natural.
- Caldera. La misma provee el vapor y la energía necesaria para sostener las operaciones del buque y el proceso de regasificación.
- Tanques de GNL Reforzados y Torre de Bombeo. Diseñados para soportar los movimiento de la carga mientras el buque se encuentra en el mar.
- Manifold de descarga conectado al brazo de descarga y por donde el gas ya regasificado circula para llegar al gasoducto y ser entregado al cliente.

El esquema de la ilustración 6 muestra el proceso: Hay 4 tanques que tienen bombas sumergidas que succionan el GNL, los tanques, además, tienen una bomba de stripping para recirculación del GNL en el propio tanque. Las bombas de alta presión llevan el producto en estado líquido a la presión necesaria para inyección previamente a la vaporización. Los barcos tienen sobre cada banda 4 LNG y 1 VR y 1 gas HP. El cambio de estado se produce en vaporizadores S&T que operan a ciclo cerrado, no toman agua del río, aunque el barco puede trabajar también con ciclo abierto. Arriba de determinado umbral de vaporización, 5MMm3/día, se opera con un sistema de recuperación de calor, precalentando el agua del ciclo cerrado de vaporización a través de la integración con el condensador principal para enfriamiento del vapor de agua usado en los turbogeneradores.

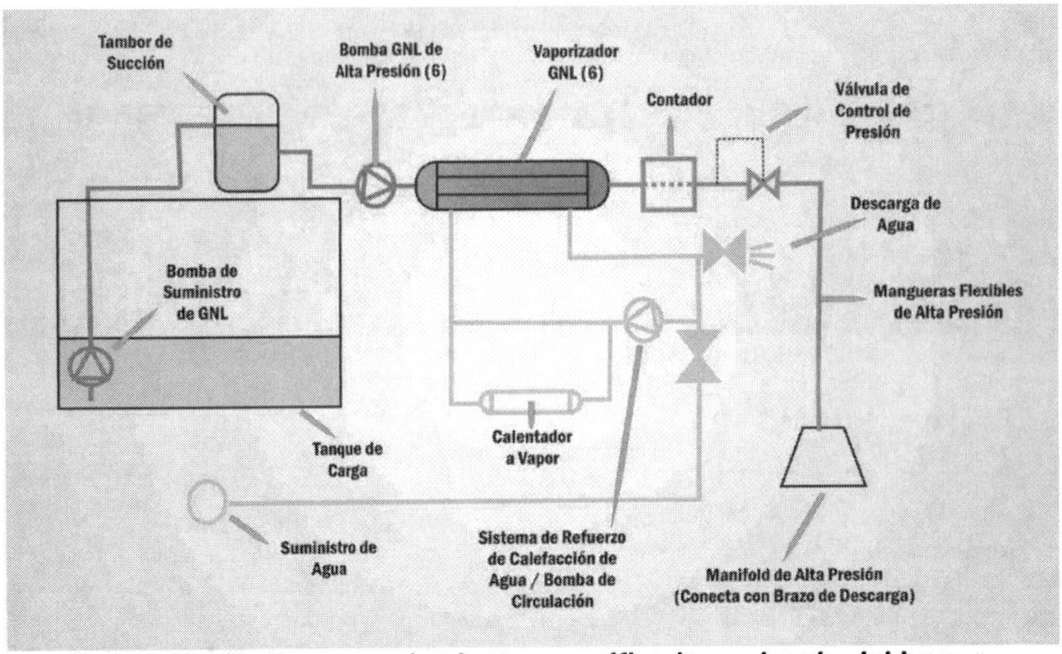

Ilustración 2 - Esquema de planta regasificadora a bordo del buque.

Tanto el buque como el muelle principal cuentan con similares medidas de prevención. Se emplean en la planta y a lo largo del gasoducto válvulas de parada de emergencias (ESDV), las cuales actúan automáticamente ya que se encuentran conectadas con un sistema de detección de fuego y gas, mediante un PLC. Y como complemento, se cuenta con un sistema que linkea al buque con los instrumentos de seguridad del muelle principal denominado Ship To Shore Link (SSL), el mismo también involucra al LNGC cuando se hace un STS.
El brazo de descarga a alta presión (HPLA, por sus siglas en inglés, High Pressure Loading Arm) tiene doble función: por un lado, ser el nexo entre el buque regasificador y el gasoducto para el flujo de gas natural; y a su vez, funciona como un sistema de seguridad que consiste en la neutralización de los movimientos del barco por mareas, corrientes y vientos. El brazo cuenta con unos movimientos limitados sobre los 3 ejes cartesianos. En cada dirección tiene un sistema de alarmas y sirenas que permiten a los operadores de la planta advertir una situación no deseada y poder estar preparados ante cualquier emergencia.

Ilustración 3 - Brazo de Descarga en Terminal Escobar

Otros sistemas implementados utilizados dentro del muelle son el sistema de lucha contra incendio y un sistema automático de ganchos de amarre que, ante una emergencia grave, se activa para liberar al buque.

Complementando estos sistemas y agregando valor al control de los riesgos, se deben citar otros métodos efectivos de control:

- Certificaciones. Tanto el buque como la planta en tierra cuentan con sistemas de gestión que son avalados por entidades certificadoras internacionales, en el caso de nuestras terminales, ellas tienen implementados el Sistema Integrado de Gestión que involucra a ISO 9001, ISO 14001 y OHSAS 18001; las cuales tienen una frecuencia de re certificación de 3 años y a su vez cuentan con auditorías de mantenimiento anuales. Los buques, como ya se mencionó están certificados por Bureau Veritas.
- Dentro de esos sistemas de gestión se encuentran Procedimientos y Planes de Contingencias. Tanto la planta en tierra como el buque regasificador cuentan con procedimientos para todas las actividades que desempeñan y cuentan con planes de contingencias, evaluando para cada escenario previsto los riesgos asociados. De manera frecuente, se realizan capacitaciones y simulacros al personal relacionado directamente con la operación.
- Se realizan análisis de riesgos periódicos con frecuencias bianuales o ante una modificación significativa en la planta.

Se llevan a cabo reuniones de comités de seguridad, lecciones aprendidas y de otras características entre los distintos agentes intervinientes con el fin de avanzar en la mejora continua.

5. Estudios de riesgo

Para la evaluación de los riesgos de estas terminales se han utilizado análisis cuantitativos y cualitativos. El QRA (Quantitative Risk Assessment, evaluación cuantitativa de riesgos), un

método estructurado y sistemático que consiste en evaluar numéricamente la probabilidad y la consecuencia de los riesgos.
El flujo de proceso de este método consiste fundamentalmente en la definición clara de un alcance y objetivos, ya que en base a ellos va a ser la profundidad del análisis. La segunda etapa es la recopilación de datos que vienen dados de acuerdo a lo que se estipuló en el primer punto. Estos datos se refieren básicamente a planos de ingeniería y procesos, puntos de ignición potenciales, datos de la población y datos meteorológicos.
El paso siguiente es la identificación de peligros (HAZID), también definidos por el alcance del estudio. Para ello, se obtienen datos de incidentes anteriores y estudios anteriores como HAZOP, what if, FMEA, etc.
Luego de una priorización de peligros, que permite descartar aquellos que tienen poca incidencia en el estudio (ya sea por la probabilidad o la gravedad de la consecuencia), se modelar los distintos escenarios posibles y se analizan las consecuencias. Para esta etapa se utilizan software que son capaces de modelar las consecuencias dividiendo en 3 fases este estudio:

1- Modelar las fuentes. Aquí se detalla las propiedades del fluido al momento de producirse la pérdida (temperatura, presión, fase, velocidad de propagación).
2- Modelar los efectos físicos. Se estudia cómo se comporta el fluido una vez que ha sido liberado. Por ejemplo, la dispersión de la nube, la dimensión en espacio y radiación térmica que produce el fuego o la dimensión de una explosión.
3- Modelar el impacto. Tanto a las personas como a los activos y al medio ambiente. En el caso de personas, se usa una función probabilística para determinar la relación entre la exposición ante un agente perjudicial y la probabilidad de fatalidad. Estos valores ya están tabulados y son inputs que se utilizan en el software que se utilizan para el estudio.

El punto siguiente es el análisis de frecuencia, usado para calcular:

1- la probabilidad de que se produzca una pérdida de un material peligroso.
2- Una vez producida la fuga, la probabilidad de que se sucedan los efectos físicos.
3- Dado un efecto físico resultante, la probabilidad de un hecho no deseado (dependiendo de la probabilidad de que se encuentre una persona, la dirección del viento, etc.).

El próximo análisis es de los riesgos. Aquí se obtienen:

- El índice de fatalidad
- El riesgo individual
- El riesgo social

El último punto es la evaluación del riesgo y su reducción. Se hace de acuerdo a ciertos criterios que pueden ser dados por agentes regulatorios o pueden tener su fuente en normas internas de la compañía, establecidas de acuerdo a la industria y la experiencia.
Estos criterios definen el nivel de riesgo inaceptable y aquel nivel de riesgo que es tan bajo que no es necesario implementar acciones para su reducción. Dentro de estos límites hay una zona donde se considera que el riesgo es tolerable, una vez que se han hecho todas las mediciones necesarias para la reducción del riesgo. Una vez que se ha postulado la reducción del riesgo, se debe calcular su efectividad, modificando los valores y re calculando los resultados. La decisión final de implementar o no la reducción del riesgo va a depender de la magnitud del riesgo, la magnitud de la reducción de ese riesgo y el costo que conlleva implementar tal acción (para la

evaluación del costo siempre de debe considerar desde el punto de vista de los beneficios en términos de seguridad).

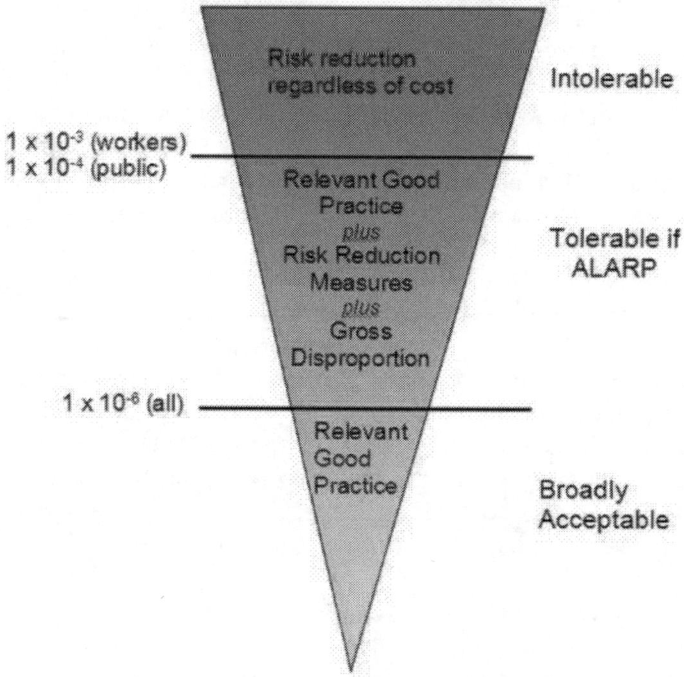

Ilustración 4 - Criterios de Riesgo

Para el caso de las dos terminales, como parte principal de esta metodología de cuantificación de riesgos, se realizaron estudios HAZID/HAZOP (Hazard Identification/Hazard and Operability) y QRA, en los cuales tuvieron participación muchos agentes relacionados con la industria y aquellos también que se encuentran cerca de las instalaciones (otras industrias) y que fuera coordinado por un consultor externo de reconocimiento internacional.

6. Conclusión

Aunque a lo largo de la historia del GNL han ocurrido incidentes, esta industria tiene unos altos índices de seguridad en comparación con otras industrias como la de refinación o petroquímica.

La industria del GNL ha cambiado radicalmente desde su incidente más grave (la catástrofe de Cleveland), desde el punto de vista de la evolución técnica y operativa, que fue desarrollándose con cada lección aprendida derivada de las investigaciones y con el desarrollo de nuevas herramientas tecnológicas (Sistemas de detección temprana, shut-down automáticos, etc.), y garantizando la continua evolución en el desarrollo de la seguridad mediante la incorporación de nuevos códigos, normas y reglamentos internacionales.

Las nuevas instalaciones son construidas siguiendo estrictas normas y regulaciones establecidas globalmente y en concordancia con muchos y diferentes estudios de riesgos y lecciones aprendidas, evitando que un accidente como el de Cleveland pueda volver a ocurrir. Algunos claros ejemplos son las aleaciones utilizadas en la construcción de tanques de almacenamiento, la capacidad de contención de los diques, las distancias de seguridad entre edificios en el layout de una planta, procedimientos de carga, descarga y maniobras con GNL, la protección contra fuego en las patas de las esferas, etc.

Durante más de 40 años se han realizado más de 30 mil viajes por mar transportando GNL. En todos estos viajes no se han producido fatalidades, incluso no ha habido heridos, tanto dentro de las tripulaciones como público en general. Tampoco hay registro de que se haya producido fuego por la emanación de vapores luego del derrame de GNL. Son todos incidentes considerados industriales, sin consecuencias significativas.

La tecnología aplicada en esta industria, como el brazo de descarga de alta presión o los tanques criogénicos dentro del FSRU, determinan la importancia que se le ha dado desde los comienzos de la actividad a la Seguridad, siempre incorporándola y tratándola como parte de la operación.

Existen controles de diversa índole: certificaciones obligatorias y voluntarias, comités de seguridad, simulacros y capacitaciones constantes que habilitan las instalaciones y aseguran una operación de acuerdo a todas las normas de seguridad aplicables en nuestro país.

[Escriba texto]

7. Fuentes consultadas

- "Safety History of international LNG operations" CH.IV International, Rev. 12, febrero 2012:
- http://www.chiv.com/pdfs/Safety%20History%20of%20International%20LNG%20Operations.pdf
- http://www.beg.utexas.edu/energyecon/lng/documents/CEE_Sistemas_de_Seguridad_y_Proteccion_de_GNL.pdf

- "LNG Spill Control and Fire Suppression", curso dictado en TEEX Texas Engineering Extension Service, College Station.

- http://co.williams.com/williams/news-media/plymouth-wa-update/ (Incidente en Plymouth, 2014).

- http://www.ptil.no/investigations/report-on-hydrocarbon-leak-at-hammerfest-lng-article10634-893.html (Incidente en Hammerfest, Noruega, 2014)

- The Lloyd's List. Lista de incidentes relacionados con el GNL, a partir de la cual se desarrolla este documento.

- Publicaciones de SIGTTO (the Society of International Gas Tanker and Terminal Operators.

- Los criterios de riesgos se tomaron en cuenta según la HSE UK.

- International Maritime Organization (IMO). http://www.imo.org/Pages/home.aspx

- Los datos estadísticos internacionales fueron tomados de la International Gas Union; IGU.org http://www.igu.org/gas-knowhow/publications/igu-publications/IGU_world_LNG_report_2013.pdf

- Fotos de los buques e información complementaria sobre los incidentes, causas y consecuencias

- encontradas mediante la búsqueda de páginas WEB a través del buscador GOOGLE.

A la lista se le suman 4 incidentes (los últimos) que fueron incorporados a la original que data de 2012. Los mismos fueron recopilados de varias publicaciones por el autor de este documento.

AÑO DEL INCIDENTE	BUQUE / INSTALACIÓN	UBICACIÓN	ESTADO DE LA OPERACIÓN	HERIDOS / FATALIDADES	DAÑOS A LA PROPIEDAD	DERRAME / PÉRDIDA DE GNL	COMENTARIOS
1944	East Ohio Gas LNG Tank	Cleveland	NA	128 muertes	NA	NA	Detallar
1965		Canvey Island, UK	transferencia	1 quemado seriamente		SI	
1965	Jules Vernet (buque)		Cargando	NO	SI	SI	Sobrecarga. Fracturas en la cubierta y puente del TK
1965	Methane Princess (Buque)		Desconectado despues de descarga	NO	SI	SI	Pérdida en válvula y fractura en puente
1971	Buque ESSO Brega, Terminal de importación de GNL La Spesia	Italia	Descargando GNL en TK de almacenamiento	NA	NA	SI	Primer accidente documentado de "roll-Over". Repentino incremento en la presión dentro del TK. El vapor de GNL escapa por válvula de seguridad del TK. Daños menores al techo del TK. No hubo incendio.
1973	Texas Eastern Transmission, TK de GNL	Staten Island	NA	40 muertos	NO	NO	Accidente industrial no relacionado a la presencia de GNL. Durante tareas de reparaciones, los vapores asociados con el proceso de limpieza habrían causado la ignición del recubrimiento de Mylar. El incendio causó una suba de T°, generando presión hasta dañar el techo de concreto de 6" derrumbandose sobre los operarios que estaban dentro del TK.
1973		Canvey Island, UK	NA	NO	SI	SI	Roturas de vidrio. Se derramó GNL sobre un charco de agua de lluvia, provocando RPT, sin daños.
1974	Massachusetts (Buque)		Cargando	NO	SI	SI	Pérdida en válvula y fractura en puente
1974	Methane Progress (Buque)		En puerto	NO	SI	NO	Encalló y tocó fondo en Arzew
1975	Philadelphia Gas Works		NA	NO	SI	NA	No causado por GNL. Se prendió fluido de un intercambiador de calor de iso-pentano, incendiando toda el área de regasificación.
1977	Arzew	Argelia	NA	1 muerte por congelamiento	NA	SI	Falla de la válvula al contacto con T° criogénicas. Aleación de Al erronea. Liberación de GNL sin ignición.
1977	LNG Aquarius (Buque)		Cargando	NO	NO	SI	TK sobrellenado
1979	Terminal de GNL Columbia Gas	Cove Point, Maryland	NA	1 muerte, 1 herido de gravedad	SI	SI	Explosión dentro de Subestación Eléctrica. El GNL se filtró por el sello de la bomba y recorrió 60m por un ducto electrico subterráneo y entro a la subestacion. Nunca se estimó presencia de gas en ese lugar, por lo que no había detectores de gas. Se produjo explosión.
1979	Mostefa BenBoulaid (Buque)	?	Descargando	NO	SI	SI	Pérdida en válvula y fractura en puente
1979	Pollenger (Buque)	?	Descargando	NO	SI	NO	Pérdida en válvula y fracturas en la cubierta del TK.
1979	El Paso Paul Kayser (Buque)		En El mar	NO	SI	NO	Varado. Daños severos al fondo, a los TK, agua a los motores y sistema de contención.
1980	LNG Libra (Buque)		En El mar	NO	SI	NO	El eje se movió contra el timón. Eje de cola fracturado.
1980	LNG Taurus (Buque)		En el puerto	NO	SI	NO	Varado. Inundación de los TK de balasto y daños en el fondo.
1984	Melrose (Buque)		En el mar	NO	SI	NO	Fuego en sala de maquinas. No hubo daños estructurales.
1985	Grandinia (Buque)		En el puerto	NO	sin reporte	NO	Fallas en el mecanismo de dirección. Sin reporte de daños.
1985	Isabella (Buque)		Descargando	NO	SI	SI	Falla en la vávula de carga. Sobrecarga. Fracturas en el puente.
1989	Tellier (Buque)		Cargando	NO	SI	SI	Amarras rotas. Roturas en casco y puente.
1990	Bachir Chihani		En el mar	NO	SI	NO	Fracturas estructurales causadas aparentemente por esfuerzo y fatiga en el casco interno.
1993	Instalación de licuefacción en Indonesia	Indonesia	NA	NO	NA	NA	Derrame de GNL en la línea de bajada durante un proyecto de modificación del ducto. El GNL entró por el sistema de drenaje de concreto bajo tierra, sufrió expansión rápida de vapor que provoco la rotura de las cañerías del drenaje.
2002	Norman Lady (Buque)	Al este del estrecho de Gibraltar	En el mar	NO	SI	NO	Colisión con un submarino nuclear de US Army. En condición de balasto. El buque sufrió un derrame de agua de mar en el área del TK seco del fondo doble.
2004	Terminal de GNL Skikda	Argelia	NA	24 muertos	SI	SI	La unidad 40 de la planta de GNL explotó, seguida de las unidades 20 y 30, en una aparente reacción en cadena. La explosión llegó hasta edificios linderos (incluyendo una central electrica, uno de los atraques en el puerto de Skikda y edificios de la comunidad).
2009	Terminal de GNL South Hook	UK	NA	NO	NO	SI	Se derramaron 10 l de GNL evaporandose inmediatamente, debido al disparo involuntario de ESD, lo que causo que se abrieran las ERC valves, en plena descarga de GNL.
2010	Terminal Montoir de Bretagne	Francia	Descargando	NO	SI	NO	El incidente ocurrió cuando el liquido pasó a la línea de descarga de gas durante las operaciones de descarga. El daño se extendióa la parte del manifold del buque y las líneas de alimentación.
2010	Instalación Withnell Bay	Australia	Cargando	NO	SI	SI	El buque sufrió quemaduras criogénicas al derramarse entre 2k y 4k litros de GNL
2011	Terminal de GNL Yung An	Taiwan	Descargando	NO	NA	NA	El capitan decide suspender la descarga y luego se retoman las actividades.
2011	Terminal de GNL Pyeongtaek	Corea del Sur	Descargando	NO	SI	SI	El barco desamarra luego de lo que fue descripto como una pequeña pérdida de GNL alrededor de una valvula de ERC, luego de finalizar una revisión programada de los brazos de descarga. Se reemplazó sello y valvula y se reanudo la descarga con los otros dos brazos.
2013	Hangin Italy (contenedores) y Al Gharrafa (metanero)	Singapur	Navegando	NO	SI	NO	El metanero colisiona el costado derecho de un buque de contenedores. No hay heridos ni se produce emision.
2014	Plymouth Plant	USA	NA				Investigación abierta. Se apunta a un recipiente presurizado de remoción de CO2
2014	Terminal Hammerfest LNG	Noruega	Operaciones cotidianas	NO	NO	SI	Se produjo derrame de hidrocarburo por rotura de una junta en una bomba. La planta estuvo parada 3 días. De la investigación surje que el incidente fue de alto potencial pudiendo haber matado a dos operarios que se encontraban en la línea de dispersión de explosión.
2014	Williams LNG	USA	Operaciones cotidianas	5 heridos leves, uno quemado	SI	SI	Explotó una línea de GNL y las esquirlas dañaron un tanque de GNL, provocando una pérdida. No hubo daños ni explosiones producto de la evaporación del GNL.

Tabla 1 - The Lloyd's List

Note: Do not add page numbers. Do not refer to page numbers when referencing different portions of the paper

Note: Do not add page numbers. Do not refer to page numbers when referencing different portions of the paper

11TH GLOBAL CONGRESS ON PROCESS SAFETY

Product Failure Rates vs Total Failure Rates at Specific Sites: Implications for Safety

Julia V. Bukowski, Ph.D.
Dept. of Electrical & Computer Engineering
Villanova University, Villanova, PA 19085
Julia.Bukowski@villanova.edu

William M. Goble, Ph.D., CFSE
exida
64 N. Main St., Sellersville, Pa 18960
wgoble@exida.com

Iwan van Beurden, MSc
exida
64 N. Main St., Sellersville, Pa 18960
vanbeurden@exida.com

Prepared for Presentation at
American Institute of Chemical Engineers
2015 Spring Meeting
11th Global Congress on Process Safety
Austin, Texas
April 27-29, 2015

Product Failure Rates vs Total Failure Rates at Specific Sites: Implications for Safety

Julia V. Bukowski, Ph.D.
Dept. of Electrical & Computer Engineering
Villanova University, Villanova, PA 19085
Julia.Bukowski@villanova.edu

William M. Goble, Ph.D., CFSE
exida

Iwan van Beurden, M.Sc.
exida

Keywords: total failure rate, product failure rate, site-generated failure rate, FMEDA, PFDavg

Abstract

To quantify the safety achieved through the use of automatic safety equipment, one must be able to determine the failure rates of the equipment which are then used to compute average probability of failure on demand (PFDavg). This is usually done by predicting failure rates via a failure modes, effects and diagnostics analysis (FMEDA) or by estimating the failure rates from failure data obtained from one or more specific end-user sites. When the predicted failure rates are compared to the estimated failure rates, there are occasions where they differ considerably. Furthermore, studies of field failure data (FFD) from different sites have shown the estimated total failure rate for a particular product model number has varied by up to four times between best and worst sites. An understanding of why this number varies from site to site has led to the conclusion that some random failures are caused by end-user practices at a particular site. This means that total failure rates computed from site-specific FFD consist of two parts: the inherent product failure rate and the specific site-generated failure rate. Inherent product failure rates which are predicted from failure rate and mode databases assume that the site end-user follows a maintenance process per IEC 61511 [1] and that actions to properly store, install, maintain, test and repair safety related equipment are performed accurately. Site(s)-generated total failure rates account not only for the inherent product failure rate but also, account for the effects of site-specific end-user practices that cause failures.

This paper

- Provides background information sufficient so that the paper is relatively self-contained
- Presents convincing evidence that the total failure rate estimated from FFD for a specific piece of mechanical equipment from a specific site in fact consists of an inherent product failure rate and a specific site-generated failure rate and that the site-generated failures may account for half or more of the total failures observed

Note: Do not add page numbers. Do not refer to page numbers when referencing different portions of the paper

- Shows that estimated total failure rates based on site-specific FFD may include failures from any life phase, i.e., initial failures, infant mortality failures, useful life failures, and end of life or aging failures and that all of these must be included in assessing the safety performance of a SIF
- Discusses the safety implications of omitting site-generated failure rates when assessing safety performance at a specific site and of using total failure rates based on multi-site FFD
- Proposes a method for extending prediction techniques to predict specific site-generated failure rates as well as inherent product failure rates

1. Introduction

Every electrical, electronic, mechanical or electromechanical product, including any equipment used to implement safety instrumented functions (SIF), has an inherent product failure rate due to the relative strengths and weaknesses of the product (as determined by the product's design and manufacture/assembly) and due to the environmental stresses in which the product is deployed. The authors contend that the inherent product failure rate is only one part of the total failure rate as experienced by the product's end user. In addition to the inherent product failure rate, there is an additional specific site-generated failure rate due to the actions (or inactions) of the end user. Since this additional failure rate is determined by human activities, both accidental and the result of deliberate decisions reflecting institutional safety culture, for the same product this additional failure rate may vary from one end-user site to another even within the same company.

For an end user to determine accurately the true safety performance of a SIF at a specific site, it is not sufficient to consider only the inherent product failure rates of each piece of equipment used to implement the SIF. The end user must also consider the site-generated failure rate for each piece of equipment which is determined by the actions or inactions taken at the site where the equipment is deployed.

As there is significant evidence that a product's total failure rate as experienced by the end user at a specific site is often two times or more the inherent product failure rate, it is clear that the site-generated failure rate is a significant contributor to the total failure rate and, consequently, to the safety achieved. The end user who does not include the site-generated failure rate in safety calculations risks seriously overstating the safety being achieved at a specific site.

2. Notation

exp[y]	notation equivalent to e^y
FFD	field failure data
FIT	one failure per 10^9 hours of operation
FMEDA	failure modes, effects and diagnostic analysis
FTO	fail to open
IEC	International Electrotechnical Commission
NBIC	National Board Inspection Code

PFD(t)	time varying probability that SIF is in a state of dangerous failure
PFDavg	average probability of dangerous failure on demand
PIF	probability of initial failure (may include infant mortality failure as explained in the paper)
PIF_{PROD}	inherent product PIF
RCA	root cause analysis
SIF	safety instrumented function
SIL	safety integrity level
SOPRV	spring operated pressure relief valve
T_P	time to proof test; time to periodic maintenance and inspection
λ_D	constant dangerous failure rate
λ_{DPROD}	inherent product constant dangerous failure rate
λ_{DSITE}	site-generated constant dangerous failure rate
λ_{DTOTAL}	total constant dangerous failure rate based on field failure data
$<<$	very much less than
$\approx$	approximately equal to

3. Background

3.1 What Is a SIF?

A SIF is a collection of equipment generally consisting of one or more input sensors, a logic solver, and one or more final elements such as switches or valves. SIFs are used to monitor processes such as power production, pharmaceutical manufacturing, or gasoline production and to shut down these processes safely in the event that such processes cease to operate within normal operating parameter ranges and enter operational regions known to be dangerous. When a process moves from a safe operational range into an unsafe operational range it is said that the process places a demand on the SIF or, more simply, that a demand has occurred.

A correctly operating SIF functions as follows. If the information from input sensors such as pressure transmitters or thermocouples indicates that the process is functioning normally, the SIF allows the process to continue. If, however, based on information from the input sensors, the logic solver within the SIF determines that a demand has occurred, then the logic solver instructs the final elements to safely shut down the process. For example, this shutdown may involve a switch opening to deprive the process of energy which would be of the "de-energize to trip" design.

The SIF itself can fail in two ways. In the event of a safe failure, the SIF unnecessarily causes a safe shutdown of a correctly operating process. If the SIF is in a state of safe failure, the consequences are primarily economic due to loss of production time; with a safe failure there is generally no damage to equipment or the environment nor injury to personnel. In the event of a dangerous failure the SIF is no longer able to intervene and shut down the process should a process demand arise. If the SIF is in a state of dangerous failure and no demand occurs, no harm is done. However, if the SIF is in a state of dangerous failure and a demand occurs, the possible consequences include equipment damage, environmental damage, injury, and/or loss of life.

Clearly, from a safety perspective, dangerous failures of the SIF are of prime concern; to maximize safety one needs to minimize the time that the SIF is unable to respond to demands. These dangerous failures can be further subdivided into two categories: dangerous failures detected by internal diagnostics and dangerous failures undetected by internal diagnostics. If a dangerous failure is detected by internal diagnostics and is promptly repaired, the time during which the SIF was in a state of dangerous failure is relatively short and does not significantly impact the safety performance of the SIF. If, however, the SIF is in a state of undetected dangerous failure, it is likely to remain in that state unless the dangerous failure is discovered and repaired at a time of periodic maintenance.

3.2 Quantifying the Safety Performance of a SIF via PFDavg

Safety standards, such as IEC 61511 [1], IEC 61508 [2] and ANSI/ISA SP84.00.01 [3], define a number of metrics which can be used to quantify the safety performance of a SIF. The principle metric used for equipment deployed in processes with low demand rates is PFDavg, the average probability that the SIF is in a state of dangerous failure when a demand occurs. PFDavg is defined as

$$PFDavg = (1/T_P)\ \int_0^{T_P} PFD(t)\, dt \qquad \text{[Eq. 1]}$$

where T_P is the time between periodic proof tests and PFD(t) is the probability that the SIF is in a state of dangerous failure as a function of time. Clearly PFDavg can vary depending on the reliability model that generates PFD(t). The standards allow PFDavg to be computed by any method generally accepted by the reliability and safety community as long as the computations can be justified. However, IEC standards require only that the failure models used account for failures which can realistically be modelled by a constant failure rate.

Approximately 70% of all process safety systems consist of a single channel, i.e., the components comprising the SIF are all connected in series. In this case, PFDavg is often approximated in practice as

$$PFDavg \approx PIF + (1 - PIF) \times \lambda_D \times T_P/2 \qquad \text{[Eq. 2]}$$

where λ_D is the total constant failure rate for dangerous failures computed by summing the constant failure rates for dangerous failures for each of the SIF components (sensors, logic solver, final elements), T_P is the proof test interval, and PIF is the probability of initial failure. When PIF equals zero, Eq. (2) reduces to the more familiar form

$$PFDavg \approx \lambda_D \times T_P/2. \qquad \text{[Eq. 2a]}$$

The accuracy of this approximation depends on several important assumptions, specifically, that $\lambda_D T_P << 1$ (which is usually the case) AND that the SIF reliability with respect to dangerous failures is exponential in form. This is true only if the SIF consists of a series combination of components AND if each component's dangerous failures are adequately modeled solely by a constant failure rate, i.e., that the failures are due to random stresses and are equally likely to occur during any operational time interval. Therefore, Eqs. (2) and (2a) do **not** apply to redundant architectures. Further, even for SIF's configured in series, Eqs. (2) and (2a) will **not** apply if any of the component failures are due to infant mortality failures, or failures due to aging or end of life. Also, Eq. (2) will not apply if any component failures are due to initial failures.

While it may be the case that most electrical/electronic components generally experience failures that are reasonably modeled by constant failure rates, this is not usually the case for mechanical components which often comprise the final element(s) of a SIF. Mechanical failures can generally include initial failures, infant mortality failures, and aging failures in addition to failures due to constant failure rates and there is significant evidence that they do [4, 5, 6, 7]. Thus, in this paper, mechanical equipment, specifically, spring operated pressure relief valves (SOPRV) form the basis of the data example and illustrate the various failure types.

3.3 Basics of Spring Operated Pressure Relief Valves

Many industrial processes use a SOPRV as a SIF to mitigate the hazards of a process overpressure event. The basic mechanics of a typical SOPRV are illustrated in the conceptual representation shown in Figure 1. While there are many possible design variations, Figure 1 along with its description provides sufficient background to understand the research presented in this paper.

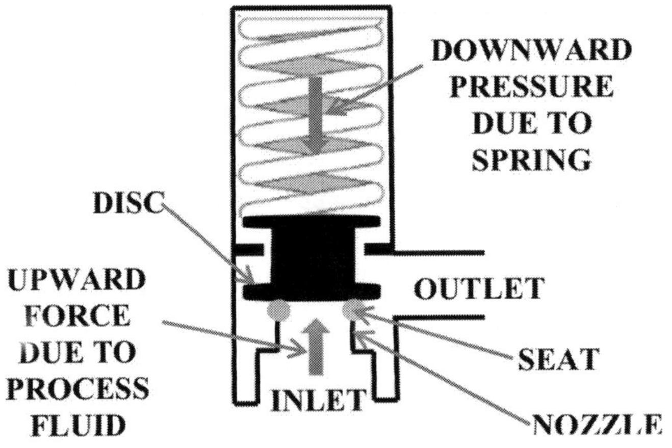

Fig. 1 Conceptual representation of a typical SOPRV

In a properly operating SOPRV a spring exerts a downward force/pressure on the disc pressing the disc against the seat. The seat base is generally the top surface of the wall of the nozzle and the seat itself is often a small raised structure of specific width on the seat base. The green circles in Figure 1 are not intended to describe the shape of the seat but merely to indicate its position. The orifice diameter (area) is the diameter (area) of the circular opening formed by the nozzle. The spring pressure on the disc results in the formation of a fluid tight seal preventing process fluid, which reaches the nozzle through the inlet, from leaving the nozzle through the orifice and outlet. The process fluid exerts an upward force/pressure on the disc. However, since the process *pressure* is nominally about 80-90% of the spring "*set pressure*" the disc remains closed.

During normal plant operation the SOPRV is in the closed position. If the process pressure increases beyond that of the spring set pressure, the disc will lift allowing process fluid to flow through the outlet thereby relieving excess process pressure. When the process pressure returns to the closing pressure of the SOPRV the disc once again seals the SOPRV and the process proceeds normally.

The SOPRV can fail in a number of ways. If the SOPRV either opens slightly or fails to form a fluid tight seal when the process pressure is within normal ranges, the valve is said to leak. This is usually a safe failure, provided that the unintended pressure relief and fluid release does not itself induce a safety hazard. On the other hand, if the SOPRV does not open under conditions of excessive process pressure, the valve is said to fail to open (FTO), or to be "stuck shut," and this is a dangerous failure. Because the SOPRV is normally closed while the SOPRV is in service, the FTO state is a state of undetected dangerous failure. This dangerous failure can only be observed in two ways: as the result of an accident when the SOPRV fails to respond to an overpressure event or in proof testing.

4. Evidence for the Existence of Site-Generated Failure Rates

Although it is easy to imagine site-specific actions or inactions that would increase the SIF total failure rate beyond that of the inherent product failure rate alone, it is nevertheless incumbent upon the authors to offer some actual evidence of instances where such practices clearly impact total failure rate and create a site-generated failure rate. In the following sections anecdotal evidence, some hypothetical scenarios, and actual field data evidence from proof test results are offered that clearly indicate that site-generated failure rates are a real phenomenon.

4.1 Anecdotal Evidence and Hypothetical Scenarios

In examining warranty data for a particular product, it was noted that when the data for that product were sorted by site there was a difference of greater than two times between the failure rate computed from failure data from the best site and the worst site [8]. Site visits provided an explanation. At the worst site some bad maintenance practices were routinely used. However, even when these practices were corrected, there still remained failure rate differences, albeit smaller ones, between sites. Even sites with excellent maintenance practices make mistakes that can result in dangerous failures and a site-generated failure rate.

This is not the only story of its kind and maintenance is not the only issue. For example, site-specific practices of delaying proof testing beyond the prescribed interval can lead to safety systems entering timeframes where aging failures rapidly accumulate, thereby increasing the number of failures of the equipment disproportionately to the constant failure rate assumption of equal likelihood of failure on all time intervals. This practice may not be under the control of maintenance but may depend on decision making in other areas of the company such as operational scheduling. Site-specific practices of failing to proof test equipment immediately after installation can result in initial failures if the installed equipment was in a state of dangerous failure or if installation was not performed correctly resulting in a dangerous failure. Sites that perform less than complete proof tests can generate initial failures that affect the safety performance of the equipment during the next in-service interval. End-users who do not calibrate equipment on schedule may induce failures. End-users may have official policies that support good practices but these may not always be followed. In short, end-user practices can affect the total failure rates of a SIF and create site-generated failure rates.

But it should also be noted that site-generated failure rates may relate not to particular practices or policies but simply to the fact that human beings make random mistakes. The technician that correctly calibrates equipment 999 times may perform the task incorrectly the next time. Sites that faithfully proof test equipment on schedule may occasionally miss a piece of equipment during a proof test cycle. Maintenance personnel may correctly place equipment in maintenance mode and correctly perform a maintenance task but fail to return the device to its operational mode. People make mistakes; it is impossible to prevent them all; some of these mistakes can result in dangerous failures.

4.2 Data Evidence That Site-Generated Failure Rates are Real Phenomena

Normally, a single site does not have sufficient failures on a particular equipment to permit useful estimation of total failure rates. However, a recent study [9] reported that a single site had a total of 13 dangerous failures on a total of 195 proof tests on very similar SOPRV. All the SOPRV in the study had stainless steel seats and discs, relatively low set pressures (less than 150 pounds per square inch gauge), and orifice diameters less than or equal to 1.0 inch. While not identical, they represent a remarkably similar group of equipment. The average proof test interval for this type of SOPRV at the data site was 3.7 years. Furthermore, all dangerous failures were subjected to root cause analysis (RCA) so that the underlying cause of each failure was known. This provided excellent data to compare the predicted inherent product failure rate to the site's estimated total failure rate.

Using existing failure rate and failure mode databases, FMEDA's were performed on the two most populous valve series in the study population. Together these two valve series accounted for 60% of all SOPRV in the study. For both valves the FMEDA's predicted an inherent product PIF (PIF_{PROD}) of 1.1% and an inherent product constant dangerous failure rate, λ_{DPROD}, of 437 FITS. (A FIT is one failure per 10^9 hours of operation.) Thus the FMEDA predicted

$$\lambda_{DPROD} = 437 \text{ FITS}$$

$$PIF_{PROD} = 0.011.$$

It should be noted that the safe failure rates of the two valve series differed by 5% and the residual rates (computed for failures that do not affect safety) differed by 9%. The similarity of the two inherent product failure rates across all types of failure rates (dangerous, safe, residual) confirms that the study population is sufficiently similar to be treated as a nearly identical group. Based on the average proof test interval of 3.7 years for this type of SOPRV at the data site, it is possible to project the expected number of dangerous failures at the proof test interval. The site does pre-installation testing on all SOPRV as post-installation testing is not possible for this equipment. Thus, it would be expected that there would be no initial failures and PIF would equal zero. However, as is explained in Appendix A, infant mortality failures which happen in-service can be conservatively modeled as PIF. Therefore, the FMEDA prediction accounts for infant mortality failures via PIF. The expected number of dangerous failures is

E[# of dangerous failures] =

$$=195x\{0.011+(1-0.011)\ [1 - \exp(-437x10^{-9}\text{failures/hr x 3.7 yrs x 8760 hr/yr})]\}$$

$$= 4.86 \approx 5 \text{ SOPRV dangerous failures due to product failure rates and PIF. [Eq. 3]}$$

Of these five failures, approximately two (1.1% x 195 SOPRV) would be expected to be infant mortality failures. But the actual number of dangerous failures was 13!

The most frequently used technique for estimating a total constant failure rate from FFD is to compute

$$\lambda_{DTOTAL} = \text{\# of failures / total \# of operating hours.} \qquad \text{[Eq. 4]}$$

Based on the data from the research study, this means that one would estimate λ_{DTOTAL} as

$$\lambda_{DTOTAL} = 13 \text{ failures/(195 SOPRV x 3.7 yrs/SOPRV x 8760 hrs/yr)} = 2.057 \times 10^{-6} \text{ failures/hr}$$

$$= 2057 \times 10^{-9} \text{ failures/hr} = 2057 \text{ FITS.} \qquad \text{[Eq. 4a]}$$

Note that the estimated total failure rate is more than four times that of the predicted inherent product failure rate but in assessing that difference one must remember that the estimated total failure rate does not include an estimate of initial failure whereas the FMEDA method also predicts a PIF for the product. Now, why is there such a difference?

Table 1 lists the dangerous failures and their causes as well as classifies the failures as inherent product failures or site-generated failures and identifies the possible life phases when the failures occurred. Note that there are six inherent product failures. However, the two manufacturing defect product failures were on the same physical valve. Although this specific SOPRV failed its first proof test after being in service because the first lift exceeded 50% of set pressure, subsequent lifts during the same proof test were normal and it was returned to service which was permitted by SRS maintenance program protocol. It failed its next proof test after additional in-service time by lifting for the first time at pressures exceeding 50% of set pressure and again returned to normal pressure lifts on subsequent lifts during the same proof test. Despite the fact that the SOPRV again met the SRS maintenance program protocol for being retagged and returned to service, the site decided to rebuild the valve and found that in the original assembly the wrong spring had been used. Thus one of the six SOPRV inherent product failures was a "duplicate" failure and five were individual inherent product failures. Note further that two of these product failures were due to sticking which is typically an infant mortality failure mechanism [10]. So the FMEDA's predicted λ_{DPROD} and PIF_{PROD} accounted appropriately for the number of inherent product failures and correctly divided those product failures between initial/infant mortality and useful life failures. This provides further validation of the FMEDA method and the associated databases.

Table 1. Categorization of Spring Operated Relief Valve Dangerous Failures

Failure #	Cause of Failure	Product or Site Failures	Possible Failure Life Phase
1	Sticking (Stainless Steel Adhesions)	Product	Infant mortality
2	Sticking (Stainless Steel Adhesions)	Product	Infant mortality
3	Galvanic corrosion cell	Product	Useful Life
4	Galvanic corrosion cell	Product	Useful Life
5	Manufacturer Defect	Product	Useful Life or Aging
6	Manufacturer Defect	Product	Useful Life or Aging
7	Corrosion; inappropriate design choice	Site	Useful Life or Aging
8	Corrosion; inappropriate design choice	Site	Useful Life or Aging
9	Corrosion; unprotected exposure to weather	Site	Useful Life or Aging
10	Corrosion; in service too long	Site	Useful Life or Aging
11	Corrosion; in service too long	Site	Useful Life or Aging
12	Corrosion; in service too long	Site	Useful Life or Aging
13	Sticking; in service too long	Site	Useful Life or Aging

The total failure rate estimate from Eq. (3) differs from the predicted product failure rate due to the site-generated failure rate which, based on seven site-generated failures, can be now be estimated as

$$\lambda_{DSITE} = 7 \text{ failures}/(195 \text{ SOPRV} \times 3.7 \text{ yrs/SOPRV} \times 8760 \text{ hrs/yr}) = 1.108 \times 10^{-6} \text{ failures/hr}$$

$$= 1108 \times 10^{-9} \text{ failures/hr} = 1108 \text{ FITS.} \qquad \text{[Eq. 5]}$$

Note that λ_{DSITE} is a little more than half the size of λ_{DTOTAL}, which is expected because the number of site-generated dangerous failures account for a little more than half of all the dangerous failures. These site-generated failures are important because they represent failures over which the site *may* have some control and *may* be able to eliminate by changes in site practices.

In this example, none of the site generated failures were due to initial failures or to infant mortality failures. However, it is entirely possible that these types of failures could be generated by certain kinds of end-user activities and therefore may need to be accounted for in particular cases.

4.3 Why All Failures Must Be Accounted for in Estimating Failure Rates and Computing PFDavg

Safety standards such as [1, 2, 3] require that the assessment of safety performance via failure rates include failures due to failure mechanisms which can be described by a constant failure rate. The standards do not require inclusion of failures due to mechanisms that are not described by constant failure rates. However, while the standards do not require the inclusion of all failures in the computation of PFDavg, it is also true that they do not prohibit inclusion of all failures either. Dangerous failures that exist when a demand occurs lead to significant consequences. If the consequences of a dangerous failure are the same regardless of how the corresponding dangerous failure rate is *modeled* then there is little justification for eliminating certain failure types from the analysis of failure rates and the computation of PFDavg. After all, if one is a resident whose home is lost to an uncontrolled oil refinery fire, that individual does not care what type of dangerous failure caused the fire. If a family member is disabled or dies as the result of an industrial accident unmitigated by safety equipment that was in a state of dangerous failure, again, the family will not care what type of dangerous failure failed to prevent the accident. Further, most FFD does not contain sufficient information to correctly distinguish between different underlying causes of failure. Thus, when failures are eliminated from analysis and computation, safety *appears* to be better but in reality is not. Consequently, it is the opinion of the authors that it is important to include all failures discovered in FFD or reasonably anticipated/expected/predicted to occur (even if not observed in FFD) in the computation of PFDavg.

4.4 Safety Implications of Omitting Site-Generated Failure Rates When Assessing Safety Performance

If no FFD were available for the SOPRV subpopulation above, then the computation of safety performance would have to be based on predictions about λ_D. If an analyst chose to limit the analysis only to the predicted constant failure rate, λ_{DPROD}, (and therefore omitted the predicted PIF_{PROD} as permitted by standards [1, 2, 3]), the PFDavg at the time of proof test would be computed from Eq. (2a) as

$$\text{PFDavg} = 0.5 \times 437 \times 10^{-9} \text{ failures/hr} \times 3.7 \text{ yrs} \times 8760 \text{ hr/yr} = 0.007 \qquad \text{[Eq. 6]}$$

whereas PFDavg based on λ_{DTOTAL} would be

$$\text{PFDavg} = 0.5 \times 2057 \times 10^{-9} \text{ failures/hr} \times 3.7 \text{ yrs} \times 8760 \text{ hr/yr} = 0.033 \qquad \text{[Eq. 7]}$$

and finally PFDavg based on PIF_{PROD}, λ_{DPROD}, and λ_{DSITE} would be

$$\text{PFDavg} = 0.011+(1-0.011) \times 0.5 \times (437+1108) \times 10^{-9} \text{ failures/hr} \times 3.7 \text{ yrs} \times 8760 \text{ hr/yr} = 0.036. \qquad \text{[Eq. 8]}$$

Table 2 shows the correspondence between PFDavg and safety integrity levels (SIL) for low demand operation which is characteristic of processes using SOPRV. Using the value of PFDavg from Eq. (6) which is based on inherent product failure rate alone provides a rating of SIL 2 for the SOPRV. Using the value of PFDavg computed from either Eqs. (7) or (8), which both include the site-generated failure rate (though in different ways), provides a rating of SIL 1 for the SOPRV. Excluding site-generated failure rates can cause an order of magnitude difference in assessed SIL level.

Table 2 Correspondence Between PFDavg and SIL

SIL per IEC61508[1]	PFDavg
1	$\geq 10^{-2}$ to $< 10^{-1}$
2	$\geq 10^{-3}$ to $< 10^{-2}$
3	$\geq 10^{-4}$ to $< 10^{-3}$
4	$\geq 10^{-5}$ to $< 10^{-4}$

As was previously mentioned, it is rare for a single site to have sufficient FFD on a particular piece of equipment to produce accurate estimates of dangerous failure rates. Furthermore, if multiple-site FFD is pooled in order to acquire sufficient data for an estimate, the resulting λ_{DTOTAL} will containing the product failure rate along with an *average* site-generated failure rate. This will benefit the assessed safety performance of sites that engage in practices that are likely to increase site-generated failure rates while penalizing sites that engage in practices that are likely to decrease site-generated failure rates.

5. Incorporating All Failure Mechanisms into the Computation of PFDavg

Clearly the above example demonstrates that an accurate evaluation of safety performance requires inclusion of all failures and that the different failure mechanisms be accounted for appropriately at both the product and site levels. Based on the RCA for the 13 failures in Table 1, it is clear that both inherent product and site-generated failures were not limited to failure mechanisms consistent with constant failure rates. Failures were identified as being consistent with infant mortality failure, useful life failure, and failure due to aging. At a site that does not perform pre-installation proof testing of SOPRV (or immediate post-installation proof testing of equipment that can be tested after installation) the possibility of initial failure also exists.

If an analyst is to appropriately include in the computation of PFDavg initial failures (which require one parameter), infant mortality failures (which require a minimum of two parameters), useful life failures (which require one parameter) and failures due to aging (which require a minimum of two parameters), the task seems overwhelming. Further, if there is insufficient information in failure data to classify the failures, how can these different failure types be accounted for and appropriately incorporated into the computation of PFDavg? Fortunately, PFDavg can be *conservatively* approximated by Eq. (2) using only two parameters while still including all failure mechanisms. This is accomplished by including in PIF the fraction of equipment population expected to exhibit infant mortality failure and by including in the predicted constant failure rate the failures expected due to aging. These predictions can be made separately for inherent product failures (as is already done with the FMEDA technique) and for site-generated failures (as is proposed below as an extension to the FMEDA technique). The reasons why all failure mechanisms can be conservatively modelled with only two parameters is outlined in the Appendix.

6. Extending the FMEDA Technique to Predict Site-Generated Failure Rates & PIF

Most sites do not have sufficient FFD of their own for a given model of equipment to estimate a λ_{DSITE} for that specific site. However, the FMEDA technique [11] which is a proven method for product failure rate prediction, and for some equipment product PIF prediction also, offers a practical solution to the problem of insufficient data. The FEMDA technique bases prediction of a constant failure rate on the assumptions that the SIF components are configured in series with each component's failure rate modeled as a constant. A validated database [12, 13] which includes component failure rates for each component failure mode serves as a basis for predicting the failure rates by mode of equipment comprised of various components in the database. The component failure rates are adjusted for specifics of equipment usage and environment.

There is no reason why the FMEDA component databases cannot be extended to include probabilities of initial and infant mortality failures for additional equipment and to include adjustments to the component constant failure rates that would account for different site practices as these impact useful life and aging failures. In computing PFDavg based on FMEDA predictions, some site practices, such as performing incomplete proof testing, are already included.

The current FMEDA component databases were constructed by initializing the failure rate values of a large number of frequently used components for each of the various component failure modes. Equipment failure rates were then predicted using the component database. As FFD became available for the equipment with failure rate predictions, the predicted equipment failure rates were compared to the FFD equipment rates, component failure rates were adjusted if necessary, and new predictions were made. The databases were also supplemented with additional component failure modes as these were discovered. This feedback process was repeated over many different equipment products until the component failure rate and failure mode database stabilized.

It is proposed to proceed in a similar fashion to extend the FMEDA component database to include the effects of site practices. Five levels of site practices are being defined from the lowest level where little or no effort is made to avert failures chargeable to site actions to the highest level where site practices include documentation and analysis of all field failures with understanding of root cause and a written, auditable plan for improvement. Each level of site practice will result in a site-generated constant failure rate for each component's failure mode. Additionally for each level of site practice, each component's failure mode may have a non-zero probability of initial and infant mortality failure. Using this database, new predictions will be made for various equipment and these predictions will be compared to existing FFD on a site by site basis because site practices can vary from location to location. Adjustments will be made to the extended database until it stabilizes.

This procedure will allow sites not only to predict the failure rates of equipment at their particular locations but also to quantify how much improvement in safety can be realized by changes in specific site practices.

7. Conclusions

This paper clearly shows that the safety performance of equipment is a function not only of the inherent product failure rates and PIF but also of site-generated failure rates determined by end-user practices as well. This means that for an end user to assess realistically the safety being achieved by a particular piece of safety equipment at a specific site, λ_{DPROD}, as provided by the manufacturer or assessed by other means meeting IEC minimum requirements, is insufficient. Information about the site-generated failure rate must also be included. The methods presented in this paper provide a framework for accomplishing this task.

8. References

[1] IEC 61511, *Application of Safety Instrumented Systems for the Process Industries*, Geneva, Switzerland, 2nd edition, 2004.

[2] IEC 61508, *Functional Safety of electrical / electronic / programmable electronic safety-related systems*, Geneva, Switzerland, 2010.

[3] ANSI/ISA SP84.00.01 – 2004 (IEC 61511-1 Mod.), *Application of Safety Instrumented Systems for the Process Industries*, Raleigh, NC, ISA, 2004.

[4] J. V. Bukowski, R. E. Gross and W. M. Goble, "Probability of Initial Failure for Spring Operated Relief Valves," *Proceedings of the ASME 2011 Pressure Vessels and Piping Division Conference*, Baltimore, MD, July 2011.

[5] J. V. Bukowski, R. E. Gross and W. M. Goble, "The Adhesion Failure Mode in Stainless Steel Trim Spring Operated Pressure Relief Valves," *ASME Journal of Pressure Vessel Technology*, Vol 135 No 6, December 2013.

[6] L. Stewart, J. V. Bukowski and W. M. Goble, "Improving Reliability & Safety Performance of Solenoid Valves by Stroke Testing," *Proceedings AIChE 9th Annual Global Conf on Process Safety – Process Plant Safety Symposium*, San Antonio, TX, April-May 2013.

[7] J. V. Bukowski, R. E. Gross and W. M. Goble, "Investigation of Adhesion Formation in New Stainless Steel Trim Spring Operated Pressure Relief Valves," *ASME Journal of Pressure Vessel Technology*, Paper 061602 1-9, Vol 136 No 6, December 2014.

[8] W. M. Goble, "Maintenance capability: Does it matter in the long term?" Hydrocarbon Processing, June 1, 2013.

[9] J. V. Bukowski, W. M. Goble, R. E. Gross and S. P. Harris, "The Effects of Maintenance Actions on the PFDavg of Spring Operated Pressure Relief Valves," *Proceedings of the ASME 2014 Pressure Vessels And Piping Division Conference*, Anaheim, CA, July 2014.

[10] Z. Zhao and B. Bhushan, "Effect of Lubrication Thickness and Viscosity and Rest Time On Long-Term Stiction in Magnetic Thin-Film Rigid Disks," *IEEE Transactions On Magnetics*, Vol 34. No 4, July 1998.

[11] W. M. Goble and A. C. Brombacher, "Using a Failure Modes, Effects and Diagnostic Analysis (FMEDA) to Measure Diagnostic Coverage in Programmable Electronic Systems," *Reliability Engineering and System Safety*, Vol. 66, No. 2, Nov 1999, pp. 145-148.

[12] W. M. Goble and J. V. Bukowski, "Development of a Mechanical Component Failure Database," *Proceedings of the 2007 Annual Reliability and Maintainability Symposium*, Orlando, FL, January 2007, pp. 451-455.

[13] J. V. Bukowski and, W. M. Goble, "Validation of a Mechanical Component Constant Failure Rate Database," 2009 *Proceedings Annual Reliability and Maintainability Symposium*, Fort Worth, TX, January 2009, 338-343.

9. Acknowledgement

The authors wish to acknowledge with gratitude the following individuals who have been involved in and/or have supported the Savannah River Site SOPRV data collection and RCA analyses: Robert E. Gross, PE, Site Infrastructure Engineering; James Fulmer and Chester Enlow, SRS Valve Shop Mechanics; Duane Edington and Albert (Chip) Jenison, SRS Valve Shop Management; Bob Davis and Jim Broome, Site Services SOPRV Coordinators; SRS Pressure Protection Committee Members.

10. Appendix

In order to show that all four failure mechanisms can be *conservatively* modeled using only two parameters, first consider computing PFD(t) assuming that the times to failure of the failed units **are known**. Figure A1a shows PFD(t) data used for computing PFDavg of a hypothetical population of 1000 identical SIF. In Figure A1a, the state (successful or failed) of ten units of the total population is displayed as a function of time over the time interval [0, T_P] where T_P equals 5 years. Note, for each unit's timeline, the dashed line indicates that the SIF is successful. If the line ends in S at t equals 5 years this indicates the system was successful on the entire time interval. A dashed line that changes to a series of F's indicates that a dangerous failure occurred at the time of the change and that the SIF remained in the dangerous failure state until T_P. Figure A1a shows that SIF units 9 and 10 have always been successful on the time interval [0, T_P] and for the purposes of this illustration it is also assumed that 990 other SIF units not shown in Figure A1a were also successful over the entire time interval. Figure A1a also shows eight other SIF units that were discovered failed in proof test.

Based on the information in Figure A1a it is possible to construct an empirical PFD(t) as follows: for any point in time between 0 and 5 years, PFD(t) is the percent of units failed dangerously recalling, of course, that there are a total of 1000 units. At t equal 0, PFD(0) equals 0 because there are no failed units. At t equals 1 year, PFD(1 year) = 0.001 because there is one failed unit (unit 7) out of the total population of 1000 SIF units. Continuing in this manner one can construct the empirical PFD(t) associated with Figure A1a which is displayed in Figure A1b. Now to find PFDavg from Figure A1b, PFD(t) is integrated over the interval [0, 5] and divided by 5 years. The area under the step function is easily computed noting that the discontinuities occur at times 0.77, 1.23, 2.21, 2.65, 2.98, 3.20, 4.10, and 4.44 years, respectively. This results in an empirical value of PFDavg of 0.0037.

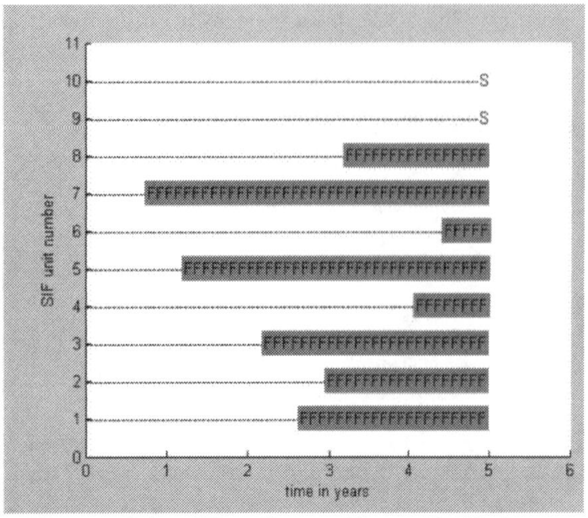

Figure A1a. Timelines of hypothetical SIF states for 10 SIF units consistent with a constant failure rate model.

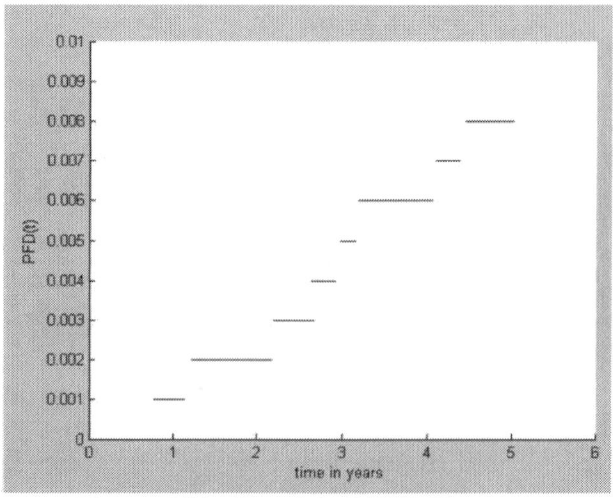

Figure A1b. Empirical PFD(*t*) corresponding to the hypothetical timelines in Figure A1a.

In general, the times to failure will **not** be known. Therefore, some assumptions must be made about the failure rate describing the data so that mathematical models can be used to compute PFDavg. Assuming that the failures are governed by a constant failure rate, and knowing that there were a total of 8 failures in 1000 units with about 5 years of operating time on each unit, λ_D would commonly be estimated as

λ_D = total # failures / total operating time [Eq. A1a]

$= 8 / (1000 \text{ units} * 5 \text{ yrs} * 8760 \text{ hrs/yr}) = 1.83 * 10^{-7}$ failures/hr. [Eq.A1b]

Then, from Eq. (2a), PFDavg $\approx 1.83 * 10^{-7} * (5 \text{ yrs}*8760 \text{ hrs/yr})/2 = 0.004$. The empirical PFDavg from Eq. (A1b) and the PFDavg calculated from Eq. (2a) match well because the failure data exhibited in Figure A1a were generated under the assumption of a constant failure rate. Note that PFDavg is directly related to the amount of time the SIF population is in a state of dangerous failure so that PFDavg is related to the amount of red shown in Figure A1a. Clearly it is true that the more red shown in a figure of timelines, the higher the value of PFDavg and, correspondingly, the less red, the smaller the value of PFDavg.

Because, in general, the times to failure are **not** known and based on proof test data of 8 dangerous failures in 1000 proof tests, the failure times *could* just as well have occurred at the times shown in Figure A2. In this figure it is clear that the times-to-failure pattern is more consistent with an increasing failure rate mechanism, i.e., failures due to aging. In this case, the empirical PFDavg would be substantially smaller than the value of 0.0037 computed above, consistent with the substantially smaller amount of time the SIF population is in a state of dangerous failure and the, consequently, smaller amount of red in Figure A2. However, without knowing the times to failure, if a constant failure rate were assumed to describe the data, then the mathematical model for λ_D would be identical to that of Eq. (A1b) above and again the calculated PFDavg would have been 0.004. This would lead to a conservative value of PFDavg

Note: Do not add page numbers. Do not refer to page numbers when referencing different portions of the paper

for the failures due to aging. Thus it can be concluded that, in general, combining failures that are due to mechanisms described by a constant failure rate with failures due to aging and modeling them together by an equivalent constant failure rate, λ_D, leads to an acceptably conservative approximation for PFDavg when Equation (2a) is then used.

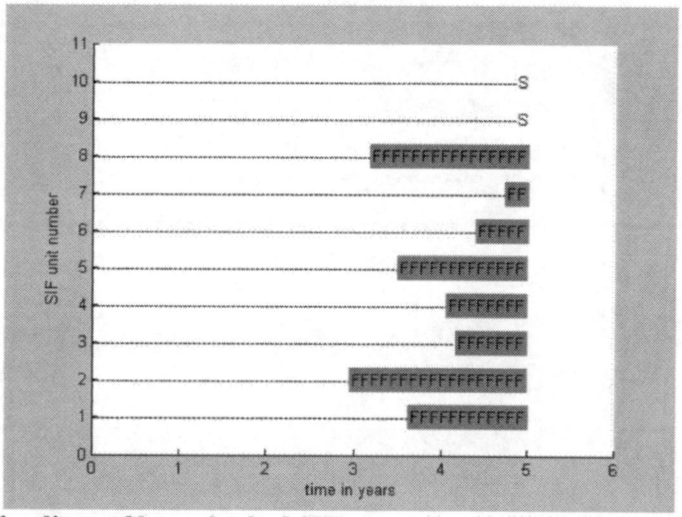

Figure A2. Timelines of hypothetical SIF states for 10 SIF units consistent with an increasing failure rate model due to aging.

Now consider Figure A3 which shows 8 failed units where the failures were all due to initial failures. Clearly in this case, PFDavg is equal to 0.008. Lastly, consider Figure A4 which shows the 8 failed units to have failed at times consistent with a combination of initial failures and infant mortality failures. In this case it is clear that the empirical PFDavg would be less than the 0.008 from Figure A3 but greater than the 0.004 that would be computed under the assumption of a constant failure rate model. Further, it is clear that the empirical value would be closer to 0.008 because the SIF population spends only slightly less time in a state of dangerous failure in Figure A4 when compared to Figure A3, i.e., there is only slightly less red in Figure A4 than in Figure A3. Therefore, if all initial failures and infant mortality failures are combined and modeled as initial failures this leads to a conservative approximation for PFDavg for those two types of failures.

Since all failure mechanisms can be conservatively modelled using only two parameters, PFDavg can be conservatively approximated using only two parameters. This makes the problem of including all failures in the computation of PFDavg both tractable and practical. It requires only that all failures be divided into one of two categories: initial or infant mortality failures vs constant failure rate or end-of-life/aging failures.

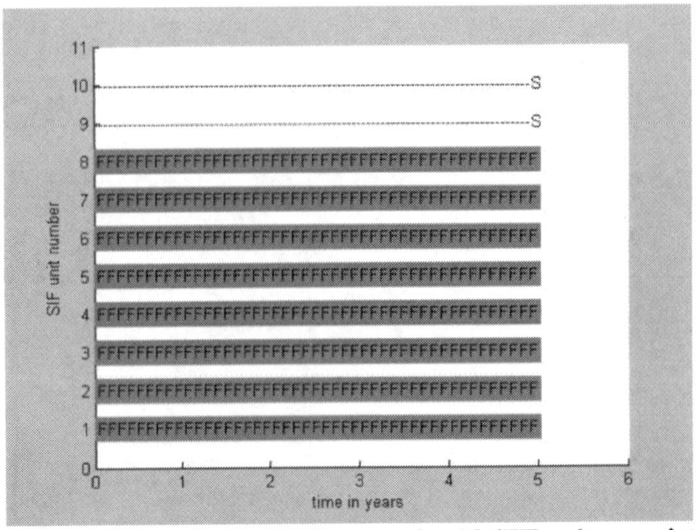

Figure A3. Timelines of hypothetical SIF states for 10 SIF units consistent with initial failures.

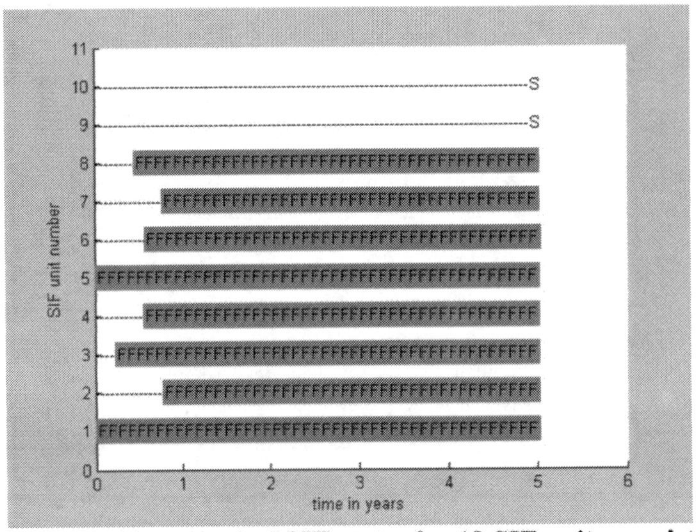

Figure A4. Timelines of hypothetical SIF states for 10 SIF units consistent with initial failures and infant mortality failures.

397151 Exploring Dependencies of Ignition Locations on Probabilistic Explosion Analysis

Abstract:

CFD based Probabilistic Explosion Analysis has become a common method for estimating structural design accidental load (DAL) specifications for offshore capital projects. This approach has been shown to yield more reasonable results than traditional analysis i.e. solely consequence based which generally result in highly conservative values. As expected, results from probabilistic based approach may have dependencies based on chosen methodologies. Previous research and studies have generally focused on exploring methodology changes which are considered to result in significant differences i.e. transient vs. non-transient dispersion modeling, different levels of 3-D model detail, etc. The purpose of this paper is to explore the dependency that can come from changes to ignition locations (often viewed as small adjustment) within a probabilistic explosion analysis. Commonly a limited number of ignition points are chosen for each simulated cloud location/size, per NORSOK Standard Z-013 [2] *"Ignition can in principle take place anywhere in the gas cloud. Explosion simulations shall include at least three different ignitions: central, edge and other."* This paper will be broadly divided in two parts where the first part will explore the use of synthesized ignition locations based on data from CFD simulations conducted using a fewer number of ignition points. This will be compared to actual simulation data from a larger set of ignition locations for establishing convergence. Having established criteria from the first part, the second part of the paper will explore the use of different ignition probability distribution functions (equal, linear, Gaussian, etc.) and compare the exceedance results on specified targets. It is the intent that by understanding how the locations of ignition sources affect the sensitivities of the forthcoming QRA work performed to internally or externally demonstrate ALARP for most facility design.

397853 Process Safety Near-Miss Management System in Takreer Ruwais Refinery: An Effective Tool for Applying Prevention Strategies

Abstract:

The primary focus of any process safety management system is always to reduce the risk of a major incident by controlling major accident hazards. Process safety events (which include near misses and minor & major incidents) are key indicators of a company's process safety performance that demonstrate how effectively a system is working to deliver its objective of abating major accident hazards. While analysing various process safety events, we observe that there is a close relationship between near-miss events and major incidents as both have often common causes. A near-miss could be defined as a hazardous condition where the event sequence could lead to an accident if it had not been interrupted. An effective Process Safety Near-miss Management System aims to quickly recognize signals from the operational fields in order to apply prevention strategies.

TAKREER Ruwais Refinery has been reviewing process safety near miss events for many years but has developed within the last two years a detailed and systematic Process Safety Near Miss Program which is modelled after API and CCPS guidelines. This paper introduces some practical strategies adopted in Ruwais Refinery for implementing a model near miss reporting system which has contributed to excellence in health and safety. It is hoped this paper will initiate further consideration and conversation about ways to enhance involvement in Near-miss reporting and thereby reduce the occurrence of unintentional injuries, fatalities, and property-damage incidents.

How Maintaining Current Heat & Material Balances Commits to Process Safety

Dustin Smith
Smith & Burgess, LLC
7600 W. Tidwell Rd., Ste. 600, 77040
Dustin.Smith@smithburgess.com

Dick Baum
Smith & Burgess, LLC
7600 W. Tidwell Rd., Ste. 600, 77040
Dick.Baum@smithburgess.com

Nicholas Cristea
Smith & Burgess, LLC
7600 W. Tidwell Rd., Ste. 600, 77040
Nicholas.Cristea@smithburgess.com

Juan Lopez
Smith & Burgess, LLC
7600 W. Tidwell Rd., Ste. 600, 77040
Juan.Lopez@smithburgess.com

Prepared for Presentation at
American Institute of Chemical Engineers
2015 Spring Meeting
11th Global Congress on Process Safety
Austin, Texas
April 27-30, 2015

UNPUBLISHED

How Maintaining Current Heat & Material Balances Commits to Process Safety

Juan Lopez
Smith & Burgess, LLC
7600 W. Tidwell Rd., Ste. 600, 77040
Juan.Lopez@smithburgess.com

Nicholas Cristea
Smith & Burgess, LLC

Keywords: LOPA, PHA, Relief Systems Design, Heat and Material Balance.

Abstract

A Heat & Material Balance (H&MB) is a valuable component of Process Safety Management (PSM) because it helps to follow the four pillars of risk-based process safety. It is important to ensure H&MBs are updated on a regular basis to accurately represent how the process is running. An H&MB is a valuable source of information for many aspects of PSM such as; Relief Systems Analysis, Process Hazard Analysis (PHA), and Layers of Protection Analysis (LOPA). If the information is not accurate and current, the analysis may lead to inaccuracies and inconsistencies that could result in: increased expense (rework), increased exposure (citations), and greater potential for incidents. This paper will look at some of the possible consequences of not having up-to-date H&MB in regards to a Relief System Analysis, PHA, or a LOPA.

1. Introduction

OSHA's Process Safety Management standard cites Heat and Material Balances (H&MBs) for any process built after May 26, 1992 as a requirement for PSM Compliance for the oil and gas industries.[1] Any modification to an existing facility that requires installing a new line, unit, or train to supplement the process could be considered as a new process and therefore requires H&MBs.[2]

Even though a plant may have been built before 1992, the plant owner should still actively attempt to keep updated documentation at least as a means to decrease the probability of incidents, which in turn may result in additional expenses or citations. Chances are even if the process was built before 1992, a small change or the installation of additional equipment may both be interpreted by the authorities as a change in the process requiring updated H&MBs. In many cases, the owner of the plant may not even be aware of these small changes that may occur over time, hence, highlighting the importance of updated and organized documentation.

It is a common occurrence in today's oil and gas industry that an operating facility does not have updated or readily available documentation regarding its process or equipment. Over time, changes in the process may occur, additional pieces of equipment may be added, or modifications may be made to existing pieces of equipment. All of these changes have the potential to adversely affect previous analyses. Not only is it a requirement by OSHA to regularly update the applicable PSM Compliance documentation, but inaccuracies in the documentation may also lead to inaccurate results in other aspects of PSM such as relief systems analysis, LOPA, and PHA. Because they normally provide compositions, temperature, pressure, and flow rates related to the process, H&MBs are a major part of PSM Compliant documentation that can help in obtaining accurate results in the aforementioned areas of PSM. Keeping updated H&MBs speaks volumes about the commitment of a plant in keeping process safety as a priority.

2. Discussion

2.1 PHA and LOPA

A process hazard analysis (PHA), also referred to as process hazard evaluation, is a major part of the PSM program. A PHA is performed by a group of qualified engineers that leads to identifying the potential of hazards associated to the process or handling of a material. It is used as a tool to make risk based decisions on their process. A Layer of Protection Analysis (LOPA) is often used as an addition to a PHA in order to facilitate the decisions to be made for a plant. Both of these types of analyses are risk based and attempt to make important modifications or installation of additional layers of protection based on the probability and consequence of any given scenario.

OSHA's standard 1910.119 App. C highlights the importance of having knowledge about the process and how innovation or change may affect the methodology:

> *"The selection of a PHA methodology or technique will be influenced by many factors including the amount of existing knowledge about the process. Is it a process that has been operated for a long period of time with little or no innovation and extensive experience has been generated with its use? Or, is it a new process or one which has been changed frequently by the inclusion of innovative features [?]..."* [3]

Often times an H&MB is not available at the time of performing a Process Hazard Analysis or Layer of Protection Analysis. However, due to the nature of these analyses, which attempt to measure probability and consequence as a means to assess risk, the lack of an H&MB during the analysis could contribute to inaccuracies when measuring the consequence of an event. An H&MB includes information such as temperature, pressure, and composition of the fluid within a system; these variables can make a significant difference when measuring the consequence of a release.

For example, a large enough leak in a pipeline containing a liquid hydrocarbon at high pressure and temperature may result in an immediate loss of pressure and lead to a

Boiling Liquid Expanding Vapor Explosion (BLEVE). If one does not have the pressure, temperature, and composition of the fluid, the consequence of such an event could be miscalculated. The PHA analyst may reach the conclusion that a leak in the pipeline may only result in a vapor cloud that dissipates quickly. Having an H&MB during the PHA or LOPA could be the difference between assessing the consequence of the leak as a BLEVE, vapor cloud release, or the release of toxic fluid.

In a LOPA, the consequences of events are categorized depending on their severity. Examples of consequence categorization that can be used in a LOPA are shown in Table 1 (shown below). In the table, the event is categorized depending on the characteristics and the quantity of the fluid being released.[4] As can be seen from the example, the quantity and characteristics of a certain release can affect the categorization of an event, and can directly influence whether a decision is made to add additional safety measures to reduce the probability of an incident.

Table 1. Example Consequence Categorization[4]

Release Characteristic	Size of a release (beyond a dike)					
	1 to 10 pound release	**10 to 100 pound release**	**100 to 1,000 pound release**	**1,000 to 10,000 pound release**	**10,000 to 100,000 pound release**	**>100,000 pound release**
Extremely toxic above BP*	Category 3	Category 4	Category 5	Category 5	Category 5	Category 5
Extremely toxic below BP or highly toxic above BP	Category 2	Category 3	Category 4	Category 5	Category 5	Category 5
Highly toxic below BP or flammable above BP	Category 2	Category 2	Category 3	Category 4	Category 5	Category 5
Flammable below BP	Category 1	Category 2	Category 2	Category 3	Category 4	Category 5
Combustible liquid	Category 1	Category 1	Category 1	Category 2	Category 2	Category 3

*BP = atmospheric boiling point

Consequence Characteristic	Magnitude of Loss					
	Spared or non-essential equipment	**Plant Outage <1 month**	**Plant Outage 1-3 months**	**Plant Outage >3 months**	**Vessel rupture 3,000 to 10,000 gal 100-300 psi**	**Vessel rupture >10,000 gal >300 psi**
Mechanical damage to large main product plant	Category 2	Category 3	Category 4	Category 4	Category 4	Category 5
Mechanical damage to small by-product plant	Category 2	Category 2	Category 3	Category 4	Category 4	Category 5

2.2 Relief Systems Design

Relief systems design is also recognized as part of PSM compliant documentation. During relief systems design, the entire plant, one piece of equipment at a time, is analyzed for the potential scenarios that could lead to overpressuring a piece of equipment over the safe limit. All potential applicable scenarios are considered, and calculations are performed in order to determine the rate, temperature, and thermodynamic properties of the fluid that needs to be relieved, so as to keep the vessel from being pressurized over the acceptable limit. A relief device is then sized and recommendations can be made if any changes should be performed. Any change in the conditions of the process, including temperature, pressure, composition, or process rate, may lead to a relief device of different size.

As part of this research paper, a blocked vapor outlet scenario was simulated for a depropanizer column. The purpose of the simulation was to find the effect of changing feed compositions in relief valve sizing. The column was simulated with different propane compositions; the other hydrocarbons in the feed stream were ethane, isobutane, n-butane, isopentane, n-pentane, n-hexane, and n-heptane in the same ratio. The composition of the aforementioned components was lowered equally with increasing compositions of propane. The diagram of the Aspen HYSYS Simulation © is shown in Figure 1.

The following specifications were used in order to simulate the column:

- Feed charge of 200,000 lb/hr
- Relieving pressure of 300 psig
- Constant reflux rate of 80,000 lb/hr
- Reboiler duty of 20 MMBtu/hr

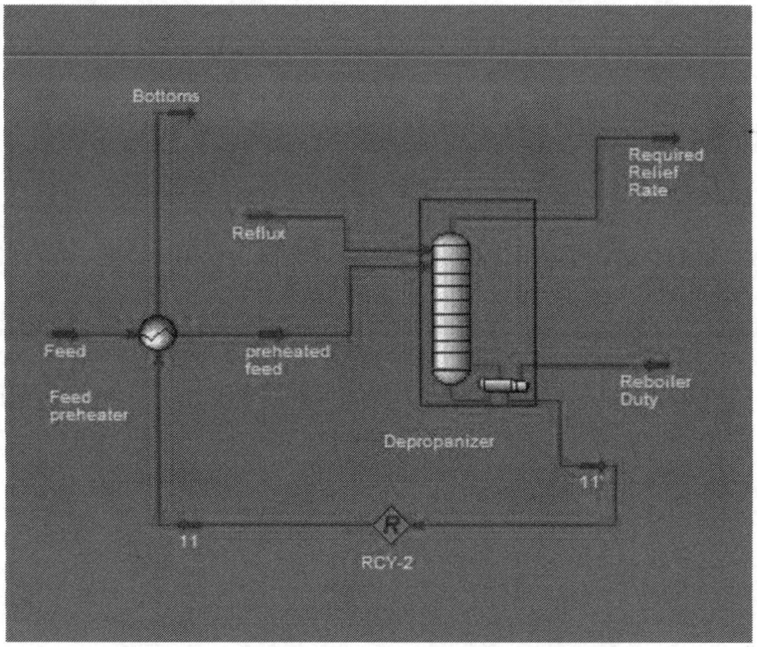

Figure 1. Depropanizer column simulation

The resulting effective discharge area required to achieve the required flowrate was found for each composition of propane using the derived critical flow sizing equations from API Standard 520 Part I, shown as Equation 1 below.[5]

$$w = A \times C \times K_d \times K_B \times P \times \sqrt{\frac{M}{TZ}} \qquad \text{Eq. (1)}$$

Where

$w =$ mass flow through the orifice
$A =$ Effective orifice area
$C =$ function of the ratio of specific heats
$K_d =$ Effective coefficient of discharge
$K_B =$ Backpressure correction factor
$P =$ Relieving pressure
$M =$ Molecular Weight
$T =$ Relieving Temperature
$Z =$ Compressibility Factor

The function of the ratio of specific heats, *C*, is defined by Equation 2.

$$C = \sqrt{k\left(\frac{2}{k+1}\right)^{\frac{(k+1)}{(k-1)}}} \qquad \text{Eq. (2)}$$

Where
$k =$ ideal ratio of the specific heats

Figure 2 shows the effective discharge area as a function of propane compositions. As shown in the figure, the effective discharge area has a direct correlation with increasing compositions of propane. While the composition of a feed stream is only a portion of all the information contained in an H&MB, it can make a difference in the results of the design.

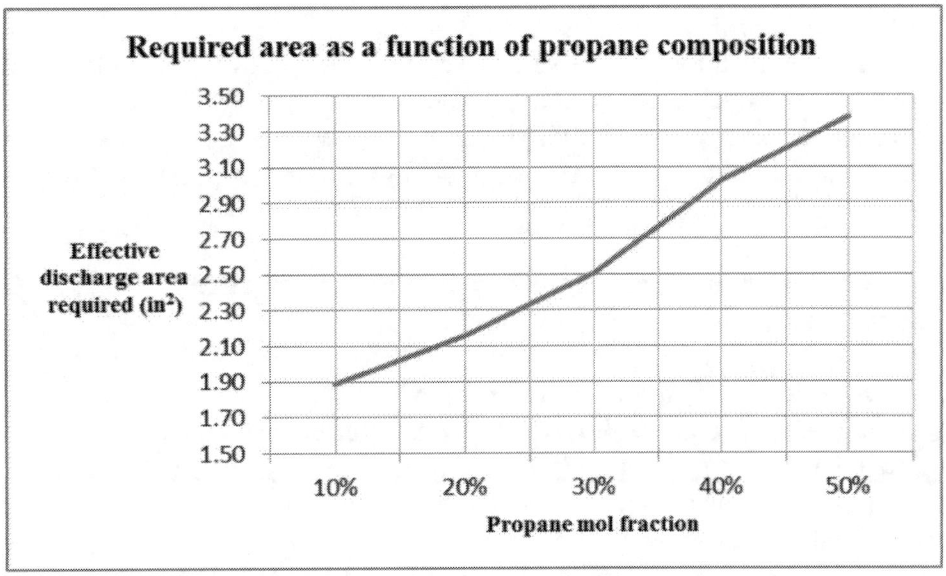

Figure 2. Depropanizer relief valve effective discharge area vs propane composition.

It is common in the industry that the process safety analyst who performs the relief systems design is provided with an outdated H&MB or no H&MB at all as supporting documentation. If this situation occurs, the analyst is forced to make conservative assumptions for the process that can often result in oversizing a relief device; this can lead to relief device chattering and additional cost.

If the relief systems analysis is performed with an inaccurate H&MB, the results of the work performed may be unreasonable to the plant owner. When the analysis is performed based on outdated information, the result of the analysis may not be what the client initially expects due to the conservative measures that need to be assumed by the engineer. A significant number of action items may be generated due to the conservative measures, and the plant may have to choose to perform the entire analysis again after they have updated the corresponding documentation; this decision may depend on the cost associated with mitigating the extensive list of action items versus the combined cost of updating H&MBs, performing the analysis for a second time, and dealing with any leftover action items. Therefore, in order to avoid rework and additional expenses, keeping updated H&MBs should be a priority.

3. Conclusion

The main reason to maintain current H&MBs is to make the plant safer. It is common in the process safety industry that the H&MBs, provided as supporting documentation for performing PSM related analyses, are outdated or not provided at all. Flow rates, compositions, pressures, and temperatures shown in H&MBs can affect the sizing during relief systems design, decision making during a PHA, and consequence categorization during a LOPA, all of which can contribute to process safety.

Current H&MBs are an important aspect of PSM documentation that relates directly to the commitment to process safety. Any process built after 1992 must have updated H&MBs in order to comply with the PSM standard. In addition, modifications made to a plant could be interpreted by OSHA as a change to the process requiring updated H&MBs.

In addition to being an integral part of process safety, maintaining current Heat and Material Balances can help in:

- Providing accurate sizing for relief devices
- Avoiding rework related to PHA, LOPA, and relief systems design
- Reducing action items from relief systems design
- Preventing citations related to outdated documentation
- Reducing the potential for incidents
- Accurately measuring the consequence of incidents

4. References

[1] "Process Safety Management of Highly Hazardous Chemicals -- Compliance Guidelines and Enforcement Procedures", 29 CFR 1910.119, 09/28/1992

[2] "Clarification of the Process Safety Management standard with regard to material and energy balances", 29 CFR 1910.119, 09/25/1995

[3] "Compliance Guidelines and Recommendations for Process Safety Management (Nonmandatory)", 29 CFR 1910.119 App C

[4] "Layer of Protection Analysis – Simplified Process Risk Assessment", Center for Chemical Process Safety of the American Institute of Chemical Engineers

[5] Sizing, Selection, and Installation of Pressure-relieving Devices Part I – Sizing and Selection, API Standard 520, 9th Edition, July 2014

397930 Safety Culture and Leadership- Independent of Language

Abstract:

The existence of safety in the workplace is mandatory for a sustainable business. Strong safety cultures are characterized by workforce engagement and fully committed safety leadership. With that being said, increased diversity in work environments and international recruitment can oftentimes complicate the process of implementing a common safety culture.

More specifically, those who come from different types of work environments than their current environment may have trouble interpreting the procedures and guidelines that are enforced at their new place of employment. Differences in languages and dialects may also lead to misinterpretations of intended safety guidelines. Lastly, there may be some instances where the workplace follows a non-flexible and hierarchical structure that is not very open to change.

In order to best develop safety practices in the workforce, the senior leaders need to first develop an environment based on the understanding of the different rationales and expectations of a multi-cultural workforce. Once that has been established, communicating with the workforce on safety expectations of the organization and optimizing interpretation of this message is critical.

This paper will continue the discussion from a previous publication; how safety leadership can consider linguistic differences when implementing a safety plan for a diverse workforce. The discussions will emphasize the learnings of the Bradley Curve, showing how important it is to follow safe practices to reach the proximity of an almost 100% incident free area, and will provide real examples from international work assignments and diverse work teams, highlighting this need and offering potential approaches.

Best Practice in Application of CAA CAP 437 Standards for Offshore Helicopter Landing Areas

Chris Robinson PhD, CEng, MIMechE.
Principal Consultant. MMI Engineering, Apollo House, Eboracum Way, York, YO31 7RE

Tim Jones, MEng, CEng, MIMechE.
Principal Consultant. MMI Engineering, The Brew House, Wilderspool Park, Greenall's Ave, Warrington, WA4 6HL

Andrew Grand, PhD.
Senior Engineer. MMI Engineering, Corum Office Park, Crown Way, Warmley, Bristol, BS30 8FJ.

Morgan Reed, D. Eng.
Process Safety Engineer. MMI Engineering, Houston, Texas, USA.

A review has been carried out on methods for calculating turbulence with regard to the UK Civil Aviation Authority CAP437 standards for offshore helicopter landing areas. Information is provided on the implementation of CAP437 limits for turbulence conditions and appropriate assumptions which may be made. A discussion of turbulence generation and transport is included to demonstrate that these are not linear processes; and a best practice guideline for generating results to demonstrate CAP437 compliance is provided.

Keywords: CAP437, CFD, turbulence, helideck, helideck availability

Introduction

The UK Civil Aviation Authority (CAA) publishes a set of guidelines for offshore helidecks – *"CAP 437 Standards for Offshore Helicopter Landing Area"*. The first publication was released in 1981 and is now in its 7th edition; as well as being used as the standard for offshore installations in UK waters. The standard is also used by US based companies to provide guidance on their worldwide projects. It provides guidance on helideck layout and equipment, placing limits on environmental variables over the helideck for turbulence, temperature and hydrocarbon gas concentrations. CAP 437 stipulates that modeling should be used to demonstrate these limiting criteria are met, generally carried out using computational fluid dynamics (CFD).

The specific requirement for turbulence over the helideck is that the standard deviation of the vertical airflow velocity should be less than 1.75 m/s. Some practitioners struggle to interpret this requirement properly and give poor quality or incorrect advice to duty holders regarding the ability to meet the criterion in different wind speeds and hence the availability of the helideck for operations. There are several points which are commonly misunderstood. Firstly, this applies to turbulent fluctuations in the wind, not the mean velocity. Secondly, the ambient wind itself will be turbulent with the turbulence intensity dependent on the atmospheric stability conditions and wind speed; this must be taken into account along with any additional turbulence generated over the platform topsides and helideck. Finally, the amount of turbulence over the helideck is not linearly dependent on the wind speed. In some instances, practitioners have used CFD to determine the degree of turbulence at one or two wind speeds and simply scaled these over a wide range of wind velocities to determine the helideck availability. This would be appropriate if the turbulence over the helideck were linearly dependent on the mean wind velocity; however, some of the turbulence generating effects over a helideck are non-linear with mean wind and this assumption cannot be made.

The International Standard "*ISO 15138 Petroleum and natural gas industries - Offshore production installations - Heating, ventilation and air-conditioning*" also discusses requirements for turbulence modeling over helidecks. It requires that a Differential Stress Model, DSM (sometimes also referred to as a Reynolds Stress Model, RSM) is used for the CFD model calculation of turbulence over the helideck. This is an additional requirement to CAP437 but in keeping with the spirit of the CAP437 standard.

The following sections provide a review of turbulence, how turbulence is calculated in computational fluid dynamics (CFD) models and the best practices for applying these in CAP437 offshore helideck studies. Since this is a very specific and important aspect of the CAP437 standards, information is provided on the potential consequences when incorrect methods are employed.

Turbulence

The term "turbulence" is generally understood to mean flow that doesn't follow straight lines, or recirculates or swirls in some manner. Lewis Fry Richardson (1881 – 1953) a Fellow of the Royal Society and early pioneer of flow modeling particularly for weather forecasting described this in an aphorism: "Big whorls have little whorls that feed on their vorticity and little whorls have lesser whorls and so on to viscosity." This simple statement describes some important features of turbulence – that there exist turbulent "whorls" or "eddies" of different scales; and that flow energy is transferred for the largest eddies to the smallest eddies and is ultimately dissipated as heat by the action of viscosity. Large turbulent eddies typically have a length scale the same order of magnitude as the objects that create them. So in atmospheric physics turbulent eddies in weather patterns may be kilometers across – though for engineering applications the largest eddies are typically associated with the wake from buildings or the order tens of meters. The smallest eddies are defined by the Kolmogorov length scale at which the energy from turbulent motion is in balance with the energy dissipated by viscosity (Davidson, 2004). In the atmosphere this is typically in the range 0.1 to 10 mm.

In engineering applications, Reynolds number provides the definition for whether a flow is turbulent or laminar. This is a non-dimensional term which expresses the ratio of momentum to viscous forces:

$$Re = \frac{\rho U L}{\mu} \tag{1}$$

ρ – fluid density; U – mean velocity; L – characteristic length scale; μ - fluid viscosity

For boundary layer flows, transition from laminar to turbulent flow is generally understood to start at a critical Reynolds number of the order 1000, although this depends on local conditions such as roughness and vibration. For free shear flow the transition to turbulence begins to occur much sooner – for jets the order of magnitude for the critical Reynolds number is typically 10 (Tritton, 1988).

For analysis it is often useful to decompose a velocity to its mean and turbulent components:

$$\hat{U} = U + u' \tag{2}$$

$\hat{U}$ – instantaneous velocity; U – mean velocity; u' – fluctuating (turbulent) component of velocity.

In this way, the contributions to the flow of the mean and turbulent fluctuations can be separated which is often useful for calculation, computation and measurement techniques. An example is shown in Figure 1 for a velocity measured over time at a single point. The turbulent fluctuation appears to provide a random variation to the mean velocity. However it is not truly random as the turbulent velocity fluctuations are associated with the energy of individual eddies which are carried by the flow. Note that in the figure, there is also a periodic, non-turbulent variation in the mean flow which is a separate phenomenon to the turbulent fluctuation.

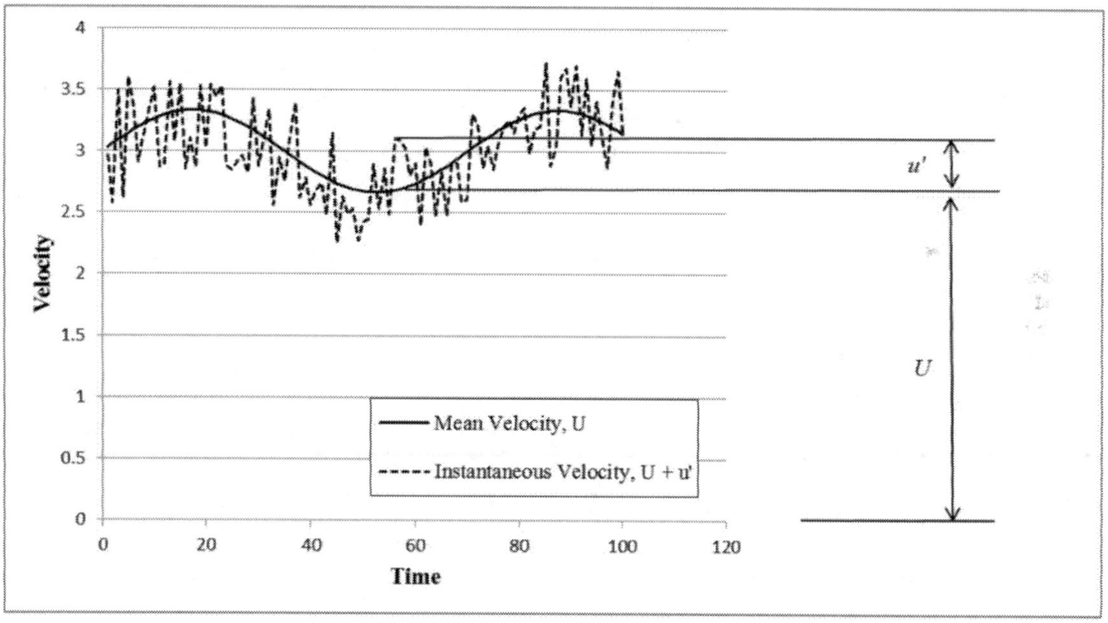

Figure 1. Example of a flow velocity measured at a single point showing the separate mean and turbulent fluctuation contributions to velocity

This breakdown of the flow into component part is important when carrying out CFD model calculations to determine whether the limits imposed by CAP437 have been achieved. In CAP437 this is stated: "*As a general rule, a limit on the standard deviation of the vertical airflow velocity of 1.75 m/s should not be exceeded*". The standard deviation of the vertical airflow is the root-mean-square value of v' (where the vertical components of velocity are normally represented by $\hat{V} = V + v'$).

Modeling Turbulence

Computational Fluid Dynamics

Modeling turbulence is normally carried out within the framework of a computational fluid dynamics (CFD) analysis. Both CAP437 and ISO 15138 describe the use of CFD modeling for analyses of helideck turbulence. The basic principal of CFD software is to calculate flows using the conservation of momentum equations – these are effectively an expression of Newton's second law of motion which when expressed for fluids includes viscous action. The equations are more commonly referred to as the Navier-Stokes equations, after Claude-Louis Navier (1785-1836) and Sir George Gabriel Stokes (1819–1903) who, working independently, were the first to define the equations.

The Navier-Stokes equations are written:

$$\rho \frac{\partial \widehat{U}_i}{\partial t} + \rho \widehat{U}_j \frac{\partial \widehat{U}_i}{\partial x_j} = -\frac{\partial P}{\partial x_i} + \mu \frac{\partial^2 \widehat{U}_i}{\partial {x_j}^2} + \rho g_i \quad (3)$$

t – time; x – distance; P – pressure; ρ – density; μ - viscosity; g – gravity

i, j = 1, 2, 3 represent the three coordinate directions along the primary orthogonal x-,y- and z-axes where the Einsteinian convention is followed that repeated indices indicate summation over the three coordinate directions. Hence equation 3 is in fact three equations for the x,y and z velocities, $\widehat{U}, \widehat{V}, \widehat{W}$..

These provide a complete description of the flow through the entire range of turbulent energy length scales. Using the Navier-Stokes equations in this form in a CFD model is called Direct Numerical Simulation (DNS). For engineering purposes this is not particularly practicable as the model would need to resolve all turbulent eddies. In the atmosphere these are around 0.1 - 10 mm in size and have characteristic times of around 0.1 seconds. For a typical helideck on an offshore platform this would need a computational mesh size of more than 1,000 billion cells and 10,000 time steps, clearly beyond the capabilities of today's computers.

A more useful approach is to express the Navier Stokes equations in terms of the mean and fluctuating velocity components. Essentially the process is to substitute equation (2) for the three coordinate directions into equation set (3). The result is known as the Reynolds Average Navier Stokes equations (RANS).

$$\rho \frac{\partial U_i}{\partial t} + \rho \frac{\partial U_j U_i}{\partial x_j} = -\frac{\partial P}{\partial x_i} + \frac{\partial}{\partial x_j}\left[\mu\left(\frac{\partial U_i}{\partial x_j} + \frac{\partial U_j}{\partial x_i}\right) - \rho\overline{u_i' u_j'}\right] + \rho g_i \quad (4)$$

In any averaging process information is lost and to complete the model this information must be accounted for by some means. In the RANS equations, this is manifested by the appearance of the term: $-\rho\overline{u_i' u_j'}$. This has the same units as viscous stress and acts on the mean flow velocity in the same way as viscosity μ - it is usually called the turbulent stress or Reynolds stress.

The aim of a turbulence model is to provide values for the turbulent stresses – note that when taking the indices into account for the three coordinate directions, $-\rho\overline{u_i' u_j'}$ describes six separate "stresses": $-\rho\overline{u'u'}$, $-\rho\overline{v'v'}$, $-\rho\overline{w'w'}$, $-\rho\overline{u'v'}$, $-\rho\overline{v'w'}$, $-\rho\overline{w'u'}$. For CAP437 analysis the term $-\rho\overline{v'v'}$ is the most useful as noted earlier v' is the standard deviation of vertical velocity which should be limited to 1.75 m/s.

Closure Models

There are a large number of turbulence models which have been developed for specific applications and general purposes since the 1970's. These have different levels of "closure" – i.e. use different levels of detail to calculate the turbulence stresses. This paper does not set out to review a wide range of models only to comment on the most commonly used ones for helideck airflow modeling.

Eddy Viscosity Modeling

Eddy Viscosity Models (EVM) use the Boussinesq assumption that as the turbulent stresses act in the same way as viscous stresses, they can also be calculated from the mean velocity gradients in much the same way as viscous stresses act (Versteeg & Malalasekera, 1995):

$$-\rho\overline{u_i' u_j'} = \mu_t\left(\frac{\partial U_i}{\partial x_j} + \frac{\partial U_j}{\partial x_i}\right) - \frac{2}{3}\rho k \delta_{ij} \quad (5)$$

μ_t – turbulent viscosity; g – gravity; k – turbulent kinetic energy

The turbulent viscosity is defined from turbulent kinetic energy and the dissipation rate of turbulent kinetic energy, ε:

$$\mu_t = \rho C_\mu \frac{k^2}{\varepsilon} \tag{6}$$

C_μ - a constant

Partial differential equations (PDE) can be formulated for the transport of k and ε through the flow and these are solved alongside the RANS momentum equations (4). This is the basis for all EVMs which include the "standard" k-ε model, RNG k-ε model, k-ω model, SST model and other similar derivations.

The turbulent kinetic energy is itself defined from the normal turbulent stresses as:

$$2k = \overline{u'u'} + \overline{v'v'} + \overline{w'w'} \tag{7}$$

Once the PDE for k has been calculated it is possible to make an estimation of the magnitude of v' over the helideck. Typically, around half of the turbulent kinetic energy will be associated with the principal flow direction; hence $k = \overline{u'u'}$. If the remaining turbulent kinetic energy can be assumed to be evenly split between the two cross-stream directions, then $0.5k = \overline{v'v'} = \overline{w'w'}$ and the required value of v' can be extracted from the CFD model calculations.

However, these are not robust assumptions and there are a number of arguments that can be made against them – e.g. the stream-wise direction is not always in the horizontal plane and may have a vertical component – in which case the amount of turbulent kinetic energy that should be assigned to $\overline{v'v'}$ should be greater. A more conservative approach is to set $(2/3)\,k = \overline{u'u'} = \overline{v'v'} = \overline{w'w'}$, but again, this is an assumption that cannot readily be tested.

In the absence of any other method, this does provide a fair means to estimate v' and in the past it has been used reasonably widely in CAP437 helideck assessments. However, sufficient computing power is common place for more detailed approaches with higher levels of turbulence closure – this is the basis of the requirement in ISO 15138 that a Differential Stress Model of turbulence should be used.

Differential Stress Modeling

Differential Stress Models (DSM), alternatively referred to as Reynolds Stress Models (RSM) use the same RANS framework as EVMs to calculate the mean velocity in a CFD model. The conservation of momentum equation shown in equation (4) is used, but instead of calculating the turbulent stress $-\rho\overline{u_i'u_j'}$ by proxy with models for k and ε, a separate PDE for the transport (convection, diffusion, generation and destruction) of each of the six turbulent stresses is solved (Versteeg & Malalasekera, 1995).

This requires more computing power than the EVM approach, as there are more PDEs to solve within the CFD model and the CFD model will be more numerically "stiff" with dependences in each of the PDEs on the values calculated by the other PDEs. A consequence of this is that there may be poor convergence of the CFD model, particularly the convergence of the numeric for the turbulence equations. If this is the case then the DSM result for v' will still contain inaccuracies.

However, the principal benefit is that no additional assumption for the distribution of turbulent kinetic energy amongst the turbulent stresses is required; values for v' can be extracted directly from the CFD model solution.

Large Eddy Simulation

Large Eddy Simulation (DES) has the potential to become the new preferred method where turbulent fluctuations need to be resolved in cases such as CAP437 helideck studies. LES is based not on the RANS equations, but the Navier-Stokes equations (3) which calculate instantaneous velocities, rather than mean velocities. Recall from above that this is referred to as Direct Numerical Simulation if all the turbulent eddies are resolved down to the smallest length and time scales. If only the largest eddies are resolved, the technique is called Large Eddy Simulation (LES). This is viable as the largest turbulent eddies carry the greatest proportion of the turbulent energy. A filter or "grid" is used in LES to define which eddies are included explicitly in the CFD flow calculation and a "sub-grid-scale" model is used to account for the lesser action of the smaller eddies on the velocity field.

LES can be carried out reasonably efficiently in free-stream regions away from walls. However at walls, turbulent eddy size decreases linearly with distance and a finer computational grid is required to calculate their effects and this increases the computational expense of the calculation. A number of approaches are possible to improve efficiency which are based on the concept of using LES to calculate the free-stream regions away from walls and RANS models to calculate the wall-affected flows. Two such approaches are Detached Eddy Simulation (DES) and Wall-Modeled LES (WMLES) and these are now incorporated in some commercial software (Menter, 2011).

Turbulence Relationship with Mean Velocity

The relationship between mean velocity and turbulence is important in a CAP437 study as it determines both the minimum standard of turbulence model that should be used and the number of different wind speeds that need to be calculated for each wind direction. The number of wind speeds is important as it is likely that the availability for the helideck will be a key output from the study: to calculate this, the turbulence over the helideck must be determined for a large number of wind speeds. It is impracticable to calculate all the required wind speeds explicitly, yet reasonable to infer these from a limited number of calculations. If the relationship between mean velocity and turbulence is linear, then only two wind speeds are required to determine the relationship between mean velocity and turbulence. However, if the relationship is non-linear then at least three wind speeds must be calculated for each wind direction.

Homogeneous shear flow occurs in simple boundary layers without cross flow or separation – this can be found in straight pipe flow of flow over a plane surface. The only velocity gradient in such a flow is $\partial U/\partial y$ and there are no non-linear dependencies between mean velocity and turbulence. A curved flow– which will occur at an offshore installation as the wind is "bent" around the structures – does not contain only homogeneous shear. There is likely to be an asymmetric velocity profile across the flow which will lead to non-linear dependence of turbulence on the mean velocity (Ellis & Joubert, 1974). Similarly experiments (Cheah, et al, 1993) show that in swirling shear flows – e.g. where any vortices are developed around or over the helideck – the mean swirl velocity varies non-linearly with distance from the center of the vortex. Others (Craft et al 1996) have shown that linear EVMs as described above cannot calculate this non-linear variation.

To summarize these points, the non-linear dependencies between mean velocity and turbulence values mean that multiple wind speeds (at least three) should be calculated for each wind direction, if results are to be interpolated for other wind speeds in the availability study. Also that an EVM (such as k-ε, k-ω, SST models) will not be able to calculate the non-linear dependencies in the mean velocity over the helideck, and a model that can calculate non-linear turbulence effects such as a DSM is required. It is useful to know that the k-ε model (Launder & Spalding, 1974) was originally only intended for simple shear flows (e.g. flow without separation or cross-wind over a flat plate) and to expect it to calculate accurately the swirling and curved flows over a helideck is unreasonable. Any model using the same stress-strain relationship as k-ε (e.g. k-ω, SST models) will have the same limit on performance (Robinson, 2001)

Example of Helideck Model for Turbulence

A typical CFD model for a helideck analysis is shown in Figure 2 - in this particular example the helideck is located on a Living Quarters platform. Incorporated in the model are all the weather deck features which will cause flow disturbances and hence generate turbulence. This includes the crane, crane rest, telecoms mast, stairs and AHU units. The items shown in pale blue are represented in the model as porous distributed resistances as the geometry features are too small to be included in the model explicitly, but the sum effect of the small components is still significant to the flow field. For reference the mesh lines on the significant surfaces are shown; note that although these appear as tetrahedral elements; prism cells are used over the sea surface and helideck surface.

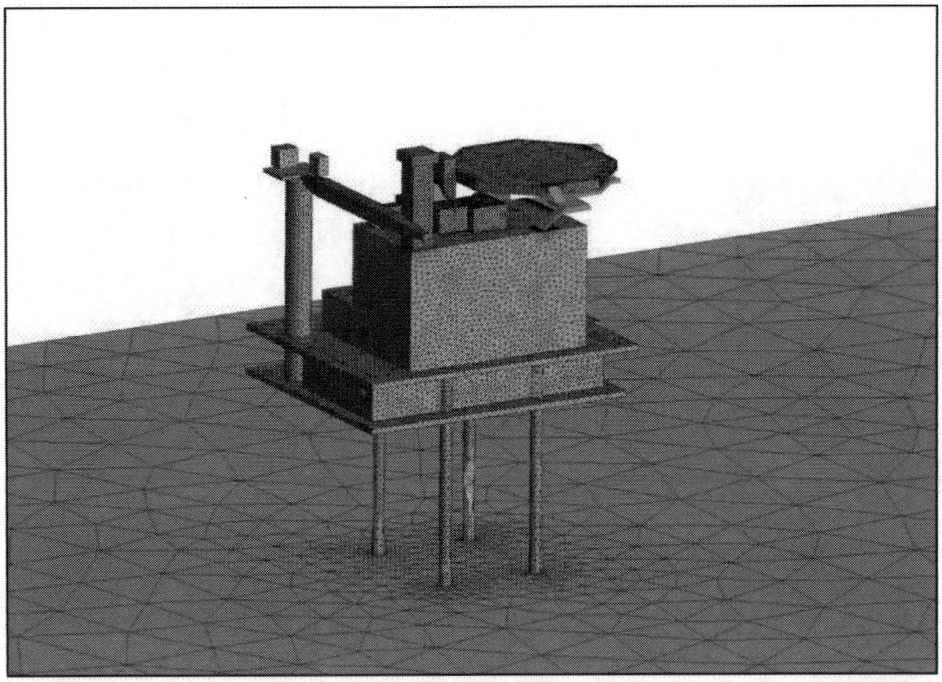

Figure 2. Example of the geometry and mesh used in a CFD model for a CAP437 helideck study on an offshore Living Quarters platform.

Results for 15 m/s wind flow in neutrally stable conditions are provided for the most onerous direction; this occurs when the wind passes over the platform before reaching the helideck and there is the largest opportunity for flow disturbance. Two calculations have been carried out using different turbulence model approaches: (i) firstly, with the SST model (Menter, 1994) which blends a k-ε model in the free stream with a k-ω model near walls and calculates turbulence with the same limitations as the EVM technique as described in this paper; (ii) secondly, with a DSM, in this case the "SSG" model (Speziale et al, 1991).

Figure 3 shows results from the CFD model calculations – the wind is passing from left to right as shown on the Figure and streamlines are used to demonstrate the degree of curvature in the wind passing over the center line of the helideck. To demonstrate the degree of turbulence calculated, an iso-surface is shown in both results for $v' = 1.75$ m/s – i.e. the limiting value in a CAP437 study. It is apparent that the EVM approach (to the left, here using the SST model) produces a much larger region of turbulence in excess of the CAP437 limit than the DSM approach (to the right, here using the SSG model).

Figure 4 shows values of v' for the same cases on a plane at 5 m above the helideck. It is clear that when using the EVM approach large areas over the helideck would be deemed non-compliant with CAP437 standard with the v' value in excess of the 1.75 m/s limit. However, the DSM approach shows that the helideck would compliant to CAP437 standard. Note that there are no differences in the cases other than the turbulence models. Figure 5 shows the plots of v' on a vertical line through which is located at the position of the maximum value of v'.

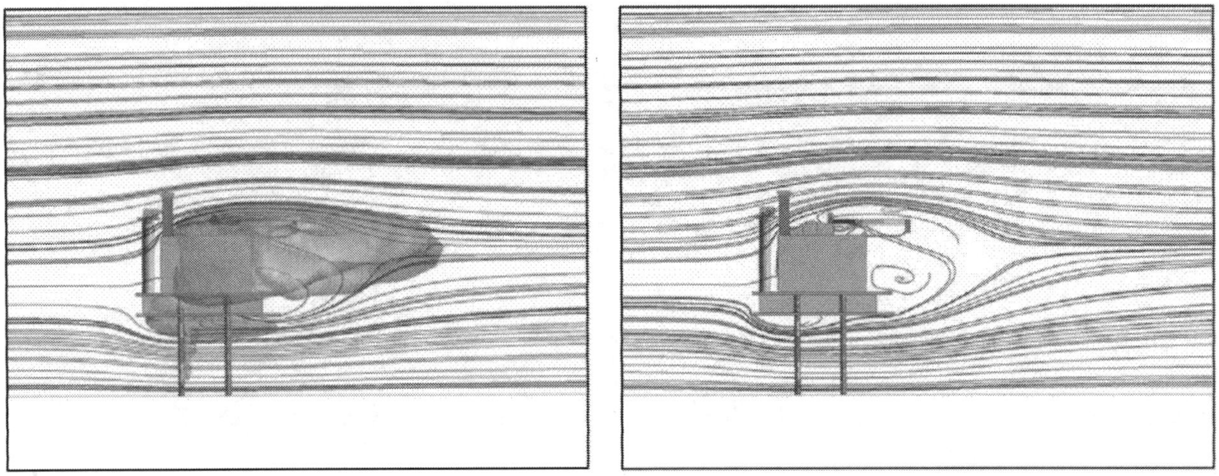

Figure 3. CFD calculations showing streamlines and an iso-surface for the value v' = 1.75 m/s.
Left – Image from SST (EVM) model; Right – Image from SSG (DSM) model.

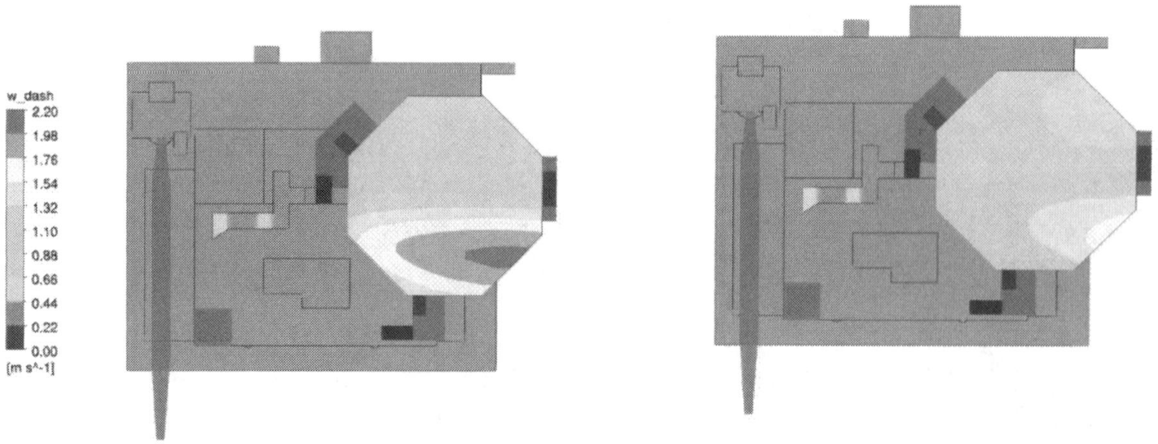

Figure 4. CFD calculations showing contours of *v'* on a plane 5 m above the helideck
Left – SST (EVM) model; Right – SSG (DSM) model.

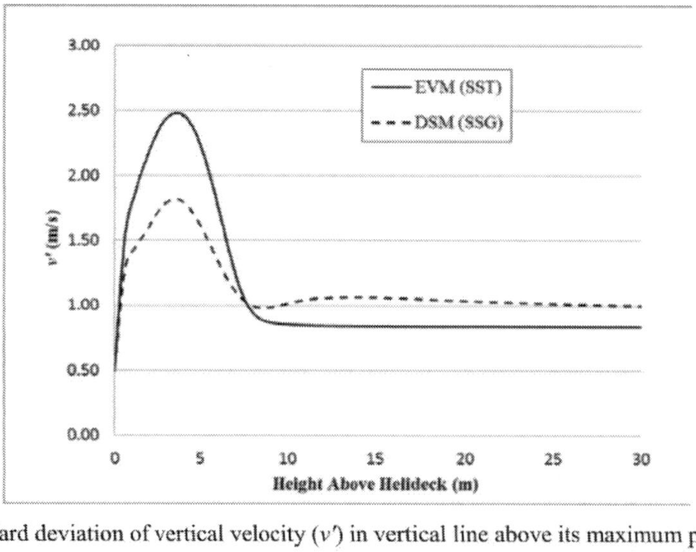

Figure 5. Plots of standard deviation of vertical velocity (*v'*) in vertical line above its maximum position on the helideck

An important point to note with these results is that independent of any measured data to validate the calculations, it is not possible to state definitively that one is more accurate than the other. However, the detail provided in this paper for calculating curved and swirling flow effects describes how DSM methods have better accuracy than EVM; there is also a large amount of published literature for curved and swirling flows which supports this.

Best Practice Guidelines

The following best practice guidelines are intended not as instructions for how to carry out CFD modeling for offshore structures in general – this is available elsewhere. In particular ISO 15138 provides good advice on CFD calculations for offshore structures and NAFEMS provides information on checking quality in CFD calculations. The purpose of the guidelines is to provide information on how a CFD model can be made suitable for CAP437 helideck assessment and how the results are used.

1. Aim of calculation

The aim of the calculation should be considered during commencement of the work. CAP437 describes limits on hydrocarbon and thermal plumes over the helideck in addition to the turbulence limits. If all these are to be included in the analysis, it may have implications for the model definition, number of cases and winds (speed and/direction) considered.

For the assessment of turbulence common requests from the duty holder will be the "availability" of the helideck and the maximum wind speed which is permissible before turbulence limits are exceeded. It is normally necessary and sufficient to calculate winds from eight directions - usually selecting N, NE, E, SE, S, SW, W, NW. If however, the wind rose for the installation is skewed such that very little of very significant amounts of wind come from a particular quarter, then fewer or more wind directions may be considered, or greater detail concentrated on a particular range of directions.

2. Turbulence Model and Software Selection

ISO 15138 states that a DSM turbulence model should be used for proper resolution of the turbulence stress and determination of the standard deviation on vertical air velocity relative to the 1.75 m/s limit in CAP437. This is supported by the description of turbulence models in this paper, and important to reduce the reliance on assumption on the distribution of turbulence energy between the turbulent stresses.

It is important to note however, that not all commercial CFD software have a DSM capability. For example, FLACS (GexCon) and KFX (ComputIT) are both commonly used in offshore analysis; these are specialized CFD codes which are optimized for efficient explosion and fire analysis respectively. Neither of these codes currently employ a DSM capability and if the CAP437 helideck analysis is to be carried out with either of them, the EVM approach must be used as described above.

3. Background Turbulence

The wind approaching the helideck over the sea will contain "naturally occurring" turbulence due to physical processes in the atmosphere. The sea surface acts as a wall with low wind velocities in the boundary layer immediately above the sea level and higher velocities at higher elevations. This velocity gradient induces turbulence. Additionally, sea surface temperatures greater than the ambient atmosphere promotes convection and rising air above the sea to various degrees depending on strength of the temperature difference, relatice humidity, cloud cover, etc. These rising air currents modify the wind profile and turbulence it contains as it interacts with the helideck.

As an example over open sea with some waves, in a neutrally stable atmosphere with a 10 m/s (approximately 20 knots) wind measured at a reference height of 40 m, the standard deviation of the vertical component of wind, $v' = 1.2$ m/s.

This demonstrates how important it is to include the background turbulence in the model as under this particular condition it is only permissible for an additional $v' = 0.55$ m/s to be induced by the installation and helideck construction.

4. Wind Speeds

At least three non-zero wind speeds should be determined for the CFD calculation in each wind direction. As demonstrated previously, curvature and swirl effects in the wind have a non-linear relationship with mean velocity. If only two wind speeds are calculated and the turbulence at other wind speeds is extrapolated from these, it will not be possible to quantify the non-linear effects. With three wind speeds calculated, curves can be fitted for the relationship between mean wind velocity and turbulence values to allow the vertical velocity fluctuation to be determined at intermediate wind speeds which have not been calculated in the CFD model.

A well-designed helideck, which has a good air gap and is sufficiently remote from other structures, will sit in the undisturbed "free shear" part of the wind profile. In this instance the turbulence will vary linearly with mean velocity. However, if this is assumed from the onset of the analysis, then it reduces the likelihood of finding deficiencies in the helideck design through modeling.

5. Result Heights above Helideck

The highest turbulence values, including v', are typically found immediately above the helideck surface due to wall shear effects. However these are not significant to the CAP437 analysis. The purpose of limiting v' is to prevent adverse and unanticipated loads (positive or negative lift forces) being placed on the helicopters rotors and hence increasing the pilots work load during the crucial stages of landing and take-off (CAA, 2004 & 2008). Turbulence fluctuations need only be assessed at and above the position of the helicopters' rotors, typically around 3 meters above the helideck.

6. Availability

Helideck availability can be determined from the number of cases with different wind speeds and directions which fall within and outside the $v' = 1.75$ m/s limit stated in CAP437. If the frequency of these winds is taken from the wind rose, the overall availability of the helideck can be obtained. The acceptable frequency will usually be determined by the installation duty holder and may be dependent on a number of operational factors.

Conclusion

Computational fluid dynamics has become the principal means to determine the turbulence environment over offshore helidecks – this has been driven by the requirements of the Civil Aviation Authority CAP437 standard and advances in computing power which allows the CFD calculations to be carried out routinely. However, the level of detail in CFD models used to find the specific turbulence quantities required in the CAA standard is not always applied correctly. This paper has reviewed the CAA standard, turbulence physical effects and turbulence models to provide best practice guidance on applying the standard.

The consequence of not applying appropriate detail in the CFD models will have some dependence on the helideck design itself. For a well-designed helideck, with good air gap and well-remote from other weather deck structures, the wind flow over it should be relatively undisturbed, exhibit simple shear and may indeed be reasonably calculated with basic turbulence models (EVMs). However, for a poorly designed helideck this will not be the case and a properly detailed turbulence modeling approach is required. As it is generally the case that the aim of the analysis is to determine whether a helideck is "well-designed" or not, then it will generally be necessary to adopt the more detailed turbulence modeling approach.

Similarly, as the availability of the helideck is usually the output of a CAP437 study, using an inappropriate CFD modeling approach may reduce the accuracy of this calculation. The example calculation provided in this paper demonstrates that even with a well-designed helideck, if an EVM model for turbulence is used rather than a DSM method, it will likely over-predict the level of turbulence over the helideck by a large degree. The consequence for the duty holder in this event is that the availability may be calculated much lower than is actually the case.

References

Civil Aviation Authority, 2013. Standards for Offshore Helicopter Landing Area. Safety Regulation Group Report CAP437, Seventh edition incorporating Amendment 01/2013

Civil Aviation Authority, 2004. Helicopter Turbulence Criteria for Operations to Offshore Platforms. Safety Regulation Group Paper 2008/02

Civil Aviation Authority, 2009. Offshore Helideck Research. Safety Regulation Group Paper 2008/02

Civil Aviation Authority, 2009. Helideck Design Considerations – Environmental Effects. Safety Regulation Group Paper 2008/03

Cheah S.C. Cheng L. Cooper D. & Launder B. E. 1993. On the structure of turbulent flow in spirally fluted tubes. Proc 5th Int. Symp Refined Flow Modeling and Turbulence Measurements.

Davidson P. A. 2004. Turbulence – An introduction for Scientists and Engineers. Oxford University Press. First published 2004.

Ellis L. B. & Joubert P.N. 1974. Turbulent shear flow in a curved duct. J. Fluid Mech (56) pp 337-351

International Standards Organisation, 2007 Petroleum and natural gas industries - Offshore production installations - Heating, ventilation and air-conditioning. ISO 15138. Second Edition 2007

Launder B.E & Spalding D.B. 1974 The numerical computation of turbulent flows. Computer Methods in Applied Mechanics and Engineering (3) pp 269-289

Menter, F.R., 1994. Two-equation eddy-viscosity turbulence models for engineering applications, AIAA Journal, Vol. 32, No. 8, pp. 1598-1605

Menter, F. 2012. Best Practice: Scale-Resolving Simulations in ANSYS CFD, ANSYS Best Practice Guide v. 1.0

Robinson C. M.E. 2001 Advanced CFD Modeling of Road-Vehicle Aerodynamics, PhD These, UMIST

Speziale C.G., Sarkar, S. and Gatski T. B. 1991. Modeling the pressure-strain correlation of turbulence: an invariant dynamical systems approach. J Fluid Mech 227 pp245-272.

Tritton D. J. 1988. Physical Fluid Dynamics. Oxford Science Publications, Second Edition 1988

Versteeg H.K. & Malalasekera W. 1995An Introduction to Computational Fluid Dynamics Longman Scientific and Technical First published 1995

Introducing a Dynamic Calculation Method for Pressure Relief Load and Orifice Sizing

Steve Streblow
Contek Solutions LLC
6221 Chapel Hill Boulevard, Suite 300
Plano, TX, 75093
sstreblow@conteklc.com

Jay Riha
Contek Solutions LLC
6221 Chapel Hill Boulevard, Suite 300
Plano, TX, 75093
jriha@conteklc.com

Prepared for Presentation at
American Institute of Chemical Engineers
2015 Spring Meeting
11th Global Congress on Process Safety
Austin, Texas
April 27-29, 2015

UNPUBLISHED

Introducing a Dynamic Calculation Method for Pressure Relief Load and Orifice Sizing

Steve Streblow
Contek Solutions LLC
6221 Chapel Hill Boulevard, Suite 300
Plano, TX, 75093
sstreblow@contekllc.com

Jay Riha
Contek Solutions LLC
6221 Chapel Hill Boulevard, Suite 300
Plano, TX, 75093
jriha@contekllc.com

Keywords: dynamic pressure relief

Abstract

This paper presents a low cost, dynamic calculation method for pressure relief load and orifice sizing, as well as several case studies detailing the additional information that can be found with using this more detailed method and the more realistic information that can be determined versus the method typically used by many. Using this more detailed method will allow for greater insight into the overpressure event, resulting in safer emergency response and pressure relief system design.

1. Introduction

For many simple systems, such as single component mixtures operating close to the relief valve set pressure, the well-established simplifying assumptions and equations in API 521, such as constant level, heat input, heat of vaporization, etc. are adequate. However, using these well-established methods for steady-state relief does not give one much insight into the time aspects, such as how long it takes to get to relieving conditions, how long it remains there, time the boil the vessel dry or all flow regimes.

For more complex systems, such as multiple components, multiple phases, operating conditions much different than relieving conditions, using the above mentioned methods may result in less safe designs due to failure to understand the dynamic relief event.

This paper presents a more detailed, dynamic method for overpressure work. Several real life examples for a pool fire scenario will be discussed, showing some of the significantly different results obtained with the more detailed method and the profound insights gained. Use of this method will bring many of the benefits of dynamic simulation to the majority of engineers, without the associated cost, resulting in safer relief system design.

2. Background

When complex systems relieve pressure due to excessive heat input such occurs during an external pool fire, engineers have made several simplifying assumptions to allow for rapid determination of relief load, orifice sizing and subsequent activities such as pipe sizing in the over-pressure protection field. For example, with a partially liquid-full vessel, assumptions of constant composition, heat input, heat of vaporization, level, molecular weight, phase, properties, and temperature are common. How good are these assumptions? Oftentimes, these are not good and result in systems which are undersized based on calculations which provide little insight into what is physically going on during the transient event.

For example, from API 521 [1], pool fire heat input is calculated from:

$$\mathbf{Q = C2 \times F \times Aws^{0.82}} \quad \text{[Eq. 1]}$$

Where

C2	= 34,500(drainage/firefighting are inadequate)
F	= 1 (environmental factor for uninsulated vessel)
Aws	= Wetted Surface Area

For a single component boiling fluid, energy output is calculated from Q= mass flowrate x latent heat of vaporization at relief conditions.

In real life, such as an initial oil-water-gas separator, it is much more complicated, with multiple phases, such as heavy liquid-light liquid-vapor, multiple components, liquid expansion upon heating, and increasing liquid level and pressure. This problem then becomes like a batch reactive distillation problem, where heavy components crack, light components are relieved first, and the relieving molecular weight and latent heat of vaporization are changing.

This is important because an undersized over-pressure protection system can result in catastrophic issues such as loss of containment resulting in fire, explosion or toxic effects. Also, lack of insight of the transient event does not allow one to know how long it will take before loss of containment is expected to occur. Gaining knowledge of the dynamics allows for effective emergency response to better mitigate impact of the event.

This problem is well-suited to dynamic simulation. Unfortunately, many engineers do not have access to dynamic process simulation software, which is generally more expensive than the more commonly available steady-state process simulation software.

This method is suitable for complex systems such as multi-phase, multi-component mixtures in pool fire scenarios such as initial oil-water-gas separation in the upstream area, but can also be extended to other types of problems, such as different types of equipment (i.e. process and storage tanks), and different types of scenarios, such as maximum heat input from a heater.

Contek has seen many companies with pressure vessels that were not properly engineered. Many companies simply order a vessel, with standard openings, and ask the vessel manufacturer to also supply a relief valve, but they do not provide all, most, or even any of the information needed to make sure that the relief system is adequate. We encourage you to review this paper and if you notice or suspect that the vessels at your company are not properly engineered, contact a qualified company to study them and provide you with the best path forward.

3. Method and Benefits

This method is one in which a base spreadsheet is used to determine the constant volume parameters, initial conditions and boundary conditions. An appropriate time step is then chosen, and through transformation of variables, the problem is converted to a constant flow rate process which then allows for solution of thermodynamic properties in a steady-state process simulator. The process simulation results are then placed back in the spreadsheet, and the calculation is complete for that time step, and the starting conditions for the next time step are calculated. After the pressure relief event occurs, the orifice size is automatically calculated at the end of each time step. This process is repeated until enough results are available to have confidence that the maximum orifice size has been calculated and is the one which should be specified.

By using this method, one can determine the time it takes until the pressure relief event starts, the fluid being relieved during the event, the liquid level in the vessel, and the time versus temperature profile, which is critical to the evaluation of vessel integrity during the relief event. Sometimes the liquid in the vessel may initially swell and compress the vapor. Therefore, the initial relief may be vapor, but if the liquid then completely fills the vessel, the relieving fluid will be liquid. As boiling becomes more vigorous, it may next relieve a two-phase mixture, with a final period of vapor release. There may also be a supercritical relief phase. This sequence will not be understood and properly designed for unless a dynamic analysis is conducted.
Many times when pressure relief work by others is reviewed, the documentation is lacking or the underlying thoughts are not stated. This can be problematic when the facility is expanded, since the basis needs to be recreated. This method is thoroughly documented, starting at the beginning with the objective, then proceeding through the references, assumptions, basis, constants, formulas, scenarios, calculations, conclusions and finishing with the recommendations. It includes an extensive section on scenario selection which has about 100 scenarios identified and requires a written response to each one.

4. Case Studies

Thirteen case studies are presented. They cover different geographical areas, different operating companies, several different fluids, pressure ranges, and vessel orientations.

4.1 Case Study 1 – Three Phase Separator

The first case study is for a pool fire impacting a horizontal three phase oil, water, and gas separator, which operates at 80% liquid full at 120 degrees F and 500 psig. Since the operating pressure is much lower than the relieving pressure of 1718.2 psig, the temperature of the fluid will need to be increased substantially before the relief starts. Due to the large temperature increase to 388 degrees F, the fluid expands, compressing the vapor. Due to the large pressure increase, the vapor condenses to a liquid. Therefore, at the start of the relief event, the fluid relieving will be liquid. Note that taking into account the change in liquid level in the vessel means that the heat input from the fire increases from the initial conditions. By using the dynamic method, it is determined that it takes twenty seven minutes before the relief event starts. At seventy two minutes, the maximum relieving vapor flowrate is reached, at a temperature of 606 degrees F. This data can be used to determine if the vessel will survive this event, since it experiences an internal temperature well above the design temperature for more than one hour. Using a simpler approach, as is commonly done by many other companies in the oil and gas industry will underestimate the relief load, not identify the fact there will be liquid and two-phase relief in addition to vapor relief, and does not evaluate how long it will take before the relief starts or how long the vessel will remain in the over-temperature condition. See Tables 1 and 2 below for further details.

Table 1. Case Study 1 Parameters

Vessel	Production Unit
Drainage	Inadequate
Fire Fighting	Inadequate
Insulation	Bare
Case	External Pool Fire
Orientation	Horizontal
Seam to Seam [Feet]	12
Diameter [Feet]	3
Normal Liquid Level [Feet]	2.4
Liquid Volume	50% water, 50% hydrocarbon
Operating Temperature [Degrees F]	120
Design Temperature [Degrees F]	130
Operating Pressure [PSIG]	500
Design Pressure [PSIG]	1420
Relief Valve Set Pressure [PSIG]	1420
Relieving Pressure [PSIG]	1718.2

Table 2. Case Study 1 Details

Time [min]	Temperature [Degrees F]	Pressure [PSIG]	Vapor Flow [Ft^3/ Hr]	Oil Flow [Ft^3/ Hr]	Water Flow [Ft^3/ Hr]	Relief Flow [Ft^3/ Hr]	Relief Phase	Vessel Liquid Level, Ft
0.0	120	500	12.4	38.4	38.5			2.40
26.8	388	1420	0.0	45.7	43.7			3.00
33.7	448	1420	0.0	47.8	45.7	35.8	Liquid	3.00
40.7	505	1420	0.0	45.9	48.1	40.1	Liquid	3.00
47.6	554	1420	11.3	42.0	50.2	122.0	2 Phase	3.00
54.5	587	1420	42.6	36.8	49.5	342.0	Vapor	2.77
62.5	601	1420	86.9	31.3	43.9	546.8	Vapor	2.35
71.7	606	1420	122.8	24.6	34.8	607.1	Vapor	1.89
82.4	607	1420	146.6	18.2	2.2	437.9	Vapor	

4.2 Case Study 2– Two Phase Separator

The second case study is for a pool fire impacting a vertical two phase oil-gas separator, which operates at 80% liquid full at 120 degrees F and 15 psig. Since the operating pressure is much lower than the relieving pressure of 151.3 psig, the temperature of the fluid will need to be increased substantially before the relief starts. Due to the large temperature increase to 448 degrees F, the fluid expands, compressing the vapor. Due to the large pressure increase, the vapor condenses to a liquid. At the start of the relief event, the fluid relieving will be liquid. By using the dynamic method, it is determined that it takes 3 minutes before the relief event starts. At four minutes, the relief phase changes from liquid to two-phase vapor-liquid. At five minutes, the relief phase changes from two-phase vapor-liquid to vapor. At this point the heat input starts decreasing. After this time, the relief will be vapor. At six minutes, the maximum relieving vapor flowrate is reached, at a temperature of 793 degrees F. This data can be used to determine if the vessel will survive this event, which experiences an internal temperature well above the design temperature for several minutes. See Tables 3 and 4 below for further details.

Table 3. Case Study 2 Parameters

Vessel	Fuel Gas Scrubber
Drainage	Inadequate
Fire Fighting	Inadequate
Insulation	Base
Case	External Pool Fire
Orientation	Vertical
Seam to Seam [Feet]	2
Diameter [Feet]	0.67
Normal Liquid Level [Feet]	1.6
Liquid Volume	100% hydrocarbon
Operating Temperature [Degrees F]	120
Design Temperature [Degrees F]	150
Operating Pressure [PSIG]	15
Design Pressure [PSIG]	125
Relief Valve Set Pressure [PSIG]	125
Relieving Pressure [PSIG]	151.3

Table 4. Case Study 2 Details

Time [min]	Temperature [Degrees F]	Pressure [PSIG]	Vapor Flow [Ft^3/ Hr]	Oil Flow [Ft^3 /Hr]	Water Flow [Ft^3/ Hr]	Relief Flow [Ft^3/Hr]	Relief Phase	Vessel Liquid Level, Ft
0.0	112	15	0.14	0.56				1.60
2.9	448	151	0.00	0.70				2.00
3.9	551	151	0.18	0.76		15.47	2 Phase	2.00
4.8	654	151	0.44	0.76		32.23	2 Phase	2.00
5.3	707	151	0.38	0.72		51.57	2 Phase	2.00
5.7	755	151	0.80	0.66		97.73	Vapor	1.89
6.2	793	151	2.04	0.36		208.77	Vapor	1.02
7.0	874	151	2.59	0.00		148.46	Vapor	0.00

4.3 Case Study 3 - Two Phase Separator

The third case study is for a pool fire impacting a vertical two phase oil-gas separator, which operates at 80% liquid full at 120 degrees F and 0 psig. Since the operating pressure is much lower than the relieving pressure of 121 psig, the temperature of the fluid will need to be increased substantially before the relief starts. Due to the large temperature increase to 353 degrees F, the fluid expands, compressing the vapor. Due to the large pressure increase, the vapor condenses to a liquid. At the start of the relief event, the fluid relieving will be liquid. By using the dynamic method, it is determined that it takes three minutes before the relief event starts. At six minutes, the relief phase changes from liquid to two-phase vapor-liquid. At seven minutes, the heat input starts to decrease. At eight minutes, the relief phase changes from two-phase vapor-liquid to vapor. After this time, the relief will be vapor. At nine minutes, the maximum relieving vapor flowrate is reached, at a temperature of 788 degrees F. This data can be used to determine if the vessel will survive this event, which experiences an internal temperature well above the design temperature for several minutes. See Tables 5 and 6 below for further details.

Table 5. Case Study 3 Parameters

Vessel	Blowcase
Drainage	Inadequate
Fire Fighting	Inadequate
Insulation	Bare
Case	External Pool Fire
Orientation	Horizontal
Seam to Seam [Feet]	1.92
Diameter [Feet]	1
Normal Liquid Level [Feet]	0.8
Liquid Volume	100% Hydrocarbon
Operating Temperature [Degrees F]	120
Design Temperature [Degrees F]	650
Operating Pressure [PSIG]	0
Design Pressure [PSIG]	150
Relief Valve Set Pressure [PSIG]	100
Relieving Pressure [PSIG]	121

Table 6. Case Study 3 Details

Time [min]	Temperature [Degrees F]	Pressure [PSIG]	Vapor Flow [Ft^3/ Hr]	Oil Flow [Ft^3/ Hr]	Water Flow [Ft^3/ Hr]	Relief Flow [Ft^3/Hr]	Relief Phase	Vessel Liquid Level, Ft
0.0	120	0	0.23	1.45				0.80
3.2	353	121	0.00	1.68				1.00
4.5	461	121	0.00	1.83		6.79	Liquid	1.00
5.8	569	121	0.62	1.83		35.38	2 Phase	1.00
7.1	673	121	2.05	1.76		97.55	2 Phase	1.00
7.8	724	121	2.00	1.62		178.47	Vapor	0.92
8.5	764	121	4.93	1.13		352.54	Vapor	0.64
8.9	778	121	3.80	0.70		430.07	Vapor	0.44
9.4	788	121	5.29	0.08		443.33	Vapor	

4.4 Case Study 4- Three Phase Separator

The fourth case study is for a pool fire impacting a horizontal three phase oil, water, gas separator, which operates at 80% liquid full at 82 degrees F and 102 psig. Since the operating pressure is much lower than the relieving pressure of 1742.4 psig, the temperature of the fluid will need to be increased substantially before the relief starts. As is quite common for these vessels in upstream and midstream oil and gas, the design temperature of 120 degrees F was only slightly above maximum ambient temperature. The operator wanted the relief valve set high, so that it would minimize probability of it ever flowing. Due to the large temperature increase to 415 degrees F, most of the liquid water vaporizes. Due to the large pressure increase, most of the hydrocarbon gas condenses. Overall, the liquid level decreases substantially. At the start of the relief event, the fluid relieving will be vapor. By using the dynamic method, it is determined that it takes thirty five minutes before the relief event starts. For the next several minutes, the liquid level decreases. Then it increases, before reaching a maximum and decreasing again. At sixty five minutes, the maximum relieving vapor flowrate is reached, at a temperature of 585 degrees F. This data can be used to determine if the vessel will survive this event, which experiences an internal temperature well above the design temperature for thirty minutes. See Tables 7 and 8 below for further details.

Table 7. Case Study 4 Parameters

Vessel	Production Unit
Drainage	Inadequate
Fire Fighting	Inadequate
Insulation	Bare
Case	External Pool Fire
Orientation	Horizontal
Seam to Seam [Feet]	10
Diameter [Feet]	2.5
Normal Liquid Level [Feet]	2
Liquid Volume	50% water, 50% oil
Operating Temperature [Degrees F]	82
Design Temperature [Degrees F]	130
Operating Pressure [PSIG]	102
Design Pressure [PSIG]	1440
Relief Valve Set Pressure [PSIG]	1440
Relieving Pressure [PSIG]	1742.4

Table 8. Case Study 4 Details

Time [min]	Temperature [Degrees F]	Pressure [PSIG]	Vapor Flow [Ft^3/ Hr]	Oil Flow [Ft^3/ Hr]	Water Flow [Ft^3/ Hr]	Relief Flow [Ft^3/Hr]	Relief Phase	Vessel Liquid Level, Ft
0.0	82	102	31.60	95.21	95.61			2.00
34.6	415	1742	112.39	109.19	1.06			1.24
46.6	501	1742	116.78	127.07	1.07	26.32	Vapor	1.40
55.6	553	1742	152.22	97.35	1.07	44.01	Vapor	1.14
61.6	576	1742	126.98	96.71	89.55	212.45	Vapor	1.95
65.3	585	1742	154.35	72.25	82.07	324.99	Vapor	1.64
69.5	593	1742	188.70	55.03	72.95	314.91	Vapor	1.40
72.5	597	1742	176.34	46.20	66.14	310.02	Vapor	

4.5 Case Study 5- Three Phase Separator

The fifth case study is for a pool fire impacting a vertical three phase oil, water, gas separator, which operates at 80% liquid full at 200 degrees F and 60 psig. Since the operating pressure is much lower than the relieving pressure of 211.8 psig, the temperature of the fluid will need to be increased substantially before the relief starts. Initially the liquid level increases, and then proceeds to decrease. By using the dynamic method, it is determined that it takes fourteen minutes before the relief event starts. At thirty nine minutes, the maximum relieving vapor flowrate is reached, at a temperature of 366 degrees F. This data can be used to determine if the vessel will survive this event, which experiences an internal temperature somewhat above the design temperature for thirty minutes. See Tables 9 and 10 below for further details.

Table 9. Case Study 5 Parameters

Vessel	Vapor Recovery Tower
Drainage	Inadequate
Fire Fighting	Inadequate
Insulation	None
Case	External Pool Fire
Orientation	Vertical
Seam to Seam [Feet]	25
Diameter [Feet]	4
Normal Liquid Level [Feet]	20
Liquid Volume	10% water, 90% oil
Operating Temperature [Degrees F]	200
Design Temperature [Degrees F]	300
Operating Pressure [PSIG]	60
Design Pressure [PSIG]	175
Relief Valve Set Pressure [PSIG]	175
Relieving Pressure [PSIG]	211.8

Table 10. Case Study 5 Details

Time [min]	Temperature [Degrees F]	Pressure [PSIG]	Vapor Flow [Ft^3/ Hr]	Oil Flow [Ft^3 /Hr]	Water Flow [Ft^3/ Hr]	Relief Flow [Ft^3/ Hr]	Relief Phase	Vessel Liquid Level, Ft
0.0	199	59	214.8	773.5	85.9			20.00
13.6	323	212	159.1	827.2	87.7			21.29
23.5	359	212	4208.3	758.7	63.2	5629.2	Vapor	19.13
31.6	372	212	4258.6	701.4	34.6	6812.3	Vapor	17.13
39.4	366	212	4632.9	653.9	0.9	7524.4	Vapor	15.24
47.0	443	212	690.7	670.9	0.9	534.5	Vapor	15.63

4.6 Case Study 6- Three Phase Separator

The sixth case study is for a pool fire impacting a vertical three phase oil, water, gas separator, which operates at 80% liquid full at 200 degrees F and 60 psig. Since the operating pressure is much lower than the relieving pressure of 211.8 psig, the temperature of the fluid will need to be increased substantially before the relief starts. Initially the liquid level increases, and then proceeds to decrease. By using the dynamic method, it is determined that it takes eight minutes before the relief event starts. At twenty three minutes, the maximum relieving vapor flowrate is reached, at a temperature of 370 degrees F. This data can be used to determine if the vessel will survive this event, which experiences an internal temperature somewhat above the design temperature for fifteen minutes. See Tables 11 and 12 below for further details.

Table 11. Case Study 6 Parameters

Vessel	Vapor Recovery Tower
Drainage	Inadequate
Fire Fighting	Inadequate
Insulation	None
Case	External Pool Fire
Orientation	Vertical
Seam to Seam [Feet]	25
Diameter [Feet]	2.5
Normal Liquid Level [Feet]	20
Liquid Volume	10% water, 90% oil
Operating Temperature [Degrees F]	200
Design Temperature [Degrees F]	300
Operating Pressure [PSIG]	60
Design Pressure [PSIG]	175
Relief Valve Set Pressure [PSIG]	175
Relieving Pressure [PSIG]	211.8

Table 12. Case Study 6 Details

Time [min]	Temperature [Degrees F]	Pressure [PSIG]	Vapor Flow [Ft^3/ Hr]	Oil Flow [Ft^3/ Hr]	Water Flow [Ft^3/ Hr]	Relief Flow [Ft^3/ Hr]	Relief Phase	Vessel Liquid Level, Ft
0.0	200	60	83.93	302.15	33.57			20.00
7.8	323	212	62.332	323.14	34.26			21.29
13.9	360	212	1785.52	294.35	23.73	3911.72	Vapor	18.95
18.5	372	212	1671.19	272.31	12.47	4633.58	Vapor	16.97
22.9	370	212	1695.57	255.02	0.38	4921.05	Vapor	15.21
26.9	441	212	259.607	261.17	0.37	357.98	Vapor	15.58

4.7 Case Study 7- Three Phase Separator

The seventh case study is for a pool fire impacting a horizontal three phase oil, water, gas separator, which operates at 80% liquid full at 200 degrees F and 69.09 psig. Since the operating pressure is much lower than the relieving pressure of 151.3 psig, the temperature of the fluid will need to be increased substantially before the relief starts. Initially the liquid level increases, and then proceeds to decrease. By using the dynamic method, it is determined that it takes twenty four minutes before the relief event starts. At ninety minutes, the maximum relieving vapor flowrate is reached, at a temperature of 358 degrees F. This data can be used to determine if the vessel will survive this event, which experiences an internal temperature somewhat above the design temperature for sixty five minutes. See Tables 13 and 14 below for further details.

Table 13. Case Study 7 Parameters

Vessel	Heater Treater
Drainage	Inadequate
Fire Fighting	Inadequate
Insulation	None
Case	External Pool Fire
Orientation	Horizontal
Seam to Seam [Feet]	20
Diameter [Feet]	6
Normal Liquid Level [Feet]	4.8
Liquid Volume	50% water, 50% oil
Operating Temperature [Degrees F]	200
Design Temperature [Degrees F]	400
Operating Pressure [PSIG]	69.09
Design Pressure [PSIG]	125
Relief Valve Set Pressure [PSIG]	125
Relieving Pressure [PSIG]	151.3

Table 14. Case Study 7 Details

Time [min]	Temperature [Degrees F]	Pressure [PSIG]	Vapor Flow [Ft^3/ Hr]	Oil Flow [Ft^3/ Hr]	Water Flow [Ft^3/ Hr]	Actual Relief [Ft^3/ Hr]	Relief Phase	Vessel Liquid Level, Ft
0.0	200	69	360.8	1108.4	1108.4			4.80
24.3	281	151	280.2	1154.3	1144.6			5.00
62.1	345	151	20940.9	956.0	1078.6	4771.9	Vapor	4.41
89.8	358	151	24066.5	826.1	947.6	5442.0	Vapor	3.89
105.5	366	151	12388.6	828.4	886.7	2696.4	Vapor	3.78

4.8 Case Study 8- Three Phase Separator

The eighth case study is for a pool fire impacting a horizontal three phase oil, water, gas separator, which operates at 80% liquid full at 65 degrees F and 30 psig. Since the operating pressure is much lower than the relieving pressure of 90.8 psig, the temperature of the fluid will need to be increased substantially before the relief starts. Initially the liquid level increases, and then proceeds to decrease. By using the dynamic method, it is determined that it takes forty nine minutes before the relief event starts. At one hundred and nineteen minutes, the maximum relieving vapor flowrate is reached, at a temperature of 332 degrees F, which is below the design temperature. See Tables 15 and 16 below for further details.

Table 15. Case Study 8 Parameters

Vessel	Heater Treater

Drainage	Inadequate
Fire Fighting	Inadequate
Insulation	Bare
Case	External Pool Fire
Orientation	Vertical
Seam to Seam [Feet]	20
Diameter [Feet]	6
Normal Liquid Level [Feet]	16
Liquid Volume	67% water, 33% oil
Operating Temperature [Degrees F]	65
Design Temperature [Degrees F]	400
Operating Pressure [PSIG]	30
Design Pressure [PSIG]	75
Relief Valve Set Pressure [PSIG]	75
Relieving Pressure [PSIG]	90.8

Table 16. Case Study 8 Details

Time [min]	Temperature [Degrees F]	Pressure [PSIG]	Vapor Flow [Ft^3/ Hr]	Oil Flow [Ft^3 /Hr]	Water Flow [Ft^3/ Hr]	Actual Relief [Ft^3/ Hr]	Relief Phase	Vessel Liquid Level, Ft
0.0	65	30	434.9	573.8	1166.1			16.00
48.8	223	91	333.2	622.0	1219.6			16.93
77.2	311	91	2582.1	640.8	1266.7	1142.0	Vapor	17.54
91.0	330	91	12078.1	637.8	1232.9	11957.4	Vapor	17.20
119.1	332	91	40299.1	633.1	1068.6	19903.1	Vapor	15.65

4.9 Case Study 9- Two Phase Separator

The ninth case study is for a pool fire impacting a vertical two phase water gas separator, which operates at 19% liquid full at 70 degrees F and 40. Since the operating pressure is much lower than the relieving pressure of 1742.4 psig, the temperature of the fluid will need to be increased substantially before the relief starts. Initially the liquid level increases, and then proceeds to decrease. By using the dynamic method, it is determined that it takes ninety seven minutes before the relief event starts. At one hundred nineteen minutes, the maximum relieving vapor flowrate is reached, at a temperature of 611 degrees F, which is well above the design temperature. See Tables 17 and 18 below for further details.

Table 17. Case Study 9 Parameters

Vessel	Inlet Separator
Drainage	Inadequate
Fire Fighting	Inadequate
Insulation	None

Case	External Pool Fire
Orientation	Vertical
Seam to Seam [Feet]	10
Diameter [Feet]	2.5
Normal Liquid Level [Feet]	1.92
Liquid Volume	100% water
Operating Temperature [Degrees F]	70
Design Temperature [Degrees F]	125
Operating Pressure [PSIG]	40
Design Pressure [PSIG]	1440
Relief Valve Set Pressure [PSIG]	1440
Relieving Pressure [PSIG]	1742.4

Table 18. Case Study 9 Details

Time [min]	Temperature [Degrees F]	Pressure [PSIG]	Vapor Flow [Ft^3/ Hr]	Oil Flow [Ft^3 /Hr]	Water Flow [Ft^3/ Hr]	Relief Flow [Ft^3/Hr]	Relief Phase	Vessel Liquid Level, Ft
0.0	70	40	187.40	0.00	57.70			1.92
96.9	580	1440	174.83	0.00	70.24			1.92
104.1	606	1742	174.50	0.00	70.58			2.33
118.8	611	1742	353.79	0.00	53.53	154.32	Vapor	2.34
137.3	614	1742	382.50	0.00	34.83	130.61	Vapor	1.78

4.10 Case Study 10- Two Phase Separator

The tenth case study is for pool fire impacting a vertical two phase water gas separator, which operates at 25% liquid full at 70 degrees F and 5 psig. Since the operating pressure is much lower than the relieving pressure of 1210 psig, the temperature of the fluid will need to be increased substantially before the relief starts. Initially the liquid level increases, and then proceeds to decrease. By using the dynamic method, it is determined that it takes fifty nine minutes before the relief event starts. At seventy six minutes, the maximum relieving vapor flowrate is reached, at a temperature of 567 degrees F, which is well above the design temperature. See Tables 19 and 20 below for further details.

Table 19. Case Study 10 Parameters

Vessel	Inlet Separator
Drainage	Inadequate
Fire Fighting	Inadequate
Insulation	None
Case	External Pool Fire

Orientation	Vertical
Seam to Seam [Feet]	11
Diameter [Feet]	2
Normal Liquid Level [Feet]	2.75
Liquid Volume	100% Water
Operating Temperature [Degrees F]	70
Design Temperature [Degrees F]	125
Operating Pressure [PSIG]	5
Design Pressure [PSIG]	1000
Relief Valve Set Pressure [PSIG]	1000
Relieving Pressure [PSIG]	1210

Table 20. Case Study 10 Details

Time [min]	Temperature [Degrees F]	Pressure [PSIG]	Vapor Flow [Ft^3/Hr]	Oil Flow [Ft^3/Hr]	Water Flow [Ft^3/Hr]	Relief Flow [Ft^3/Hr]	Relief Phase	Vessel Liquid Level, Ft
0.0	70	5	119.46	0.00	45.82			2.75
58.7	541	1000	108.79	0.00	56.54			2.75
62.6	565	1210	107.97	0.00	57.36			3.39
75.5	567	1210	353.65	0.00	43.04	251.20	Vapor	3.44
91.9	568	1210	373.75	0.00	28.17	202.98	Vapor	2.58

4.11 Case Study 11- Two Phase Separator

The eleventh case study is for a pool fire impacting a vertical two phase water gas separator, which operates at 18% liquid full at 80 degrees F and 850 psig. Since the operating pressure is much lower than the relieving pressure of 1742.4 psig, the temperature of the fluid will need to be increased substantially before the relief starts. Initially the liquid level increases, and then proceeds to decrease. By using the dynamic method, it is determined that it takes thirteen minutes before the relief event starts. At thirty three minutes, the maximum relieving vapor flowrate is reached, at a temperature of 330 degrees F, which is well above the design temperature.
See Tables 21 and 22 below for further details.

Table 21. Case Study 11 Parameters

Vessel	Inlet Separator
Drainage	Inadequate
Fire Fighting	Inadequate
Insulation	None
Case	External Pool Fire
Orientation	Vertical
Seam to Seam [Feet]	10
Diameter [Feet]	2
Normal Liquid Level [Feet]	1.83
Liquid Volume	100% Water
Operating Temperature [Degrees F]	80
Design Temperature [Degrees F]	125
Operating Pressure [PSIG]	850
Design Pressure [PSIG]	1440
Relief Valve Set Pressure [PSIG]	1440
Relieving Pressure [PSIG]	1742.4

Table 22. Case Study 11 Details

Time [min]	Temperature [Degrees F]	Pressure [PSIG]	Vapor Flow [Ft^3/ Hr]	Oil Flow [Ft^3/ Hr]	Water Flow [Ft^3/ Hr]	Relief Flow [Ft^3/ Hr]	Relief Phase	Vessel Liquid Level, Ft
0.0	80	850	119.04	30.04	3.60			1.83
12.7	256	1440	129.78	19.39	3.51			1.83
18.5	308	1742	149.41	3.28	0.00			1.25
33.1	330	1742	163.25	3.05	0.00	13.05	Vapor	0.18
48.6	352	1742	163.32	2.76	0.00	12.12	Vapor	0.17
65.5	374	1742	163.63	2.40	0.00	11.12	Vapor	0.15
84.4	395	1742	164.11	1.95	0.00	9.92	Vapor	0.13
106.9	415	1742	164.40	1.40	0.00	8.21	Vapor	0.11

4.12 Case Study 12- Glycol Absorber

The twelfth case study is for a pool fire impacting a vertical glycol absorber, which operates at 16% liquid full at 80 degrees F and 850 psig. Since the operating pressure is much lower than the relieving pressure of 1210 psig, the temperature of the fluid will need to be increased substantially before the relief starts. The liquid level increases throughout the event.

By using the dynamic method, it is determined that it takes eight minutes before the relief event starts. At eighty one minutes, the maximum relieving vapor flowrate is reached, at a temperature of 739 degrees F, which is well above the design temperature. See Tables 23 and 24 below for further details.

Table 23. Case Study 12 Parameters

Vessel	Glycol Absorber
Drainage	Inadequate
Fire Fighting	Inadequate
Insulation	None
Case	External Pool Fire
Orientation	Vertical
Seam to Seam [Feet]	19
Diameter [Feet]	2.5
Normal Liquid Level [Feet]	3
Liquid Volume	Mostly TEG
Operating Temperature [Degrees F]	80
Design Temperature [Degrees F]	100
Operating Pressure [PSIG]	850
Design Pressure [PSIG]	1000
Relief Valve Set Pressure [PSIG]	1000
Relieving Pressure [PSIG]	1210

Table 24. Case Study 12 Details

Time [min]	Temperature [Degrees F]	Pressure [PSIG]	Vapor Flow [Ft^3/ Hr]	Oil Flow [Ft^3 /Hr]	Water Flow [Ft^3/ Hr]	Relief Flow [Ft^3/ Hr]	Relief Phase	Vessel Liquid Level, Ft
0.0	92	845	353.46	0.00	80.44			3.00
7.6	160	1000	350.96	0.00	82.95			3.00
18.2	249	1210	346.70	0.00	87.20		Vapor	3.09
40.7	421	1210	483.08	0.00	96.79	90.79	Vapor	3.25
61.5	584	1210	467.68	0.00	106.68	95.17	Vapor	3.61
80.6	739	1210	464.95	0.00	110.99	104.23	Vapor	3.98
99.1	812	1210	415.30	0.00	96.56	59.10	Vapor	4.14

4.13 Case Study 13- NGL (Natural Gas Liquids) Vessel

The thirteenth case study is for a pool fire impacting a horizontal NGL vessel, which operates at 32% liquid full at 80 degrees F and 150 psig. Since the operating pressure is much lower than the relieving pressure of 266.2 psig, the temperature of the fluid will need to be increased substantially before the relief starts. The liquid level increases initially and then decreases. By using the dynamic method, it is determined that it takes fourteen minutes before the relief event starts. At sixty six minutes, the maximum relieving vapor flowrate is reached, at a temperature of 268 degrees F, which is well above the design temperature. See Tables 25 and 26 below for further details.

Table 25. Case Study 13 Parameters

Vessel	NGL vessel
Drainage	Inadequate
Fire Fighting	Inadequate
Insulation	None
Case	External Pool Fire
Orientation	Horizontal
Seam to Seam [Feet]	64
Diameter [Feet]	9.07
Normal Liquid Level [Feet]	2.92
Liquid Volume	Hydrocarbon
Operating Temperature [Degrees F]	60
Design Temperature [Degrees F]	150
Operating Pressure [PSIG]	150
Design Pressure [PSIG]	250
Relief Valve Set Pressure [PSIG]	220
Relieving Pressure [PSIG]	266.2

Table 26. Case Study 13 Details

Time [min]	Temperature [Degrees F]	Pressure [PSIG]	Vapor Flow [Ft^3/ Hr]	Oil Flow [Ft^3 /Hr]	Water Flow [Ft^3/ Hr]	Relief Flow [Ft^3/ Hr]	Relief Phase	Vessel Liquid Level, Ft
0	60	150	14,015	-	5322.03			2.92
14	120	220	13,856	-	5480.35			2.92
21	152	266	13,790	-	5546.18			3.00
35	193	266	25,227	-	4959.90	10,904.33	Vapor	3.04
50	232	266	27,983	-	4022.17	12,058.26	Vapor	2.72
66	268	266	31,376	-	2631.63	12,640.33	Vapor	2.20
86	303	266	35,499	-	655.92	11,926.00	Vapor	1.44
96	315	266	24,256	-	0.00	7,107.82	Vapor	0.36

5. Conclusions

What conclusions can be drawn from comparing the thirteen cases?

When the operating pressure is much less than relieving pressure, the vapor may actually be compressed back into the liquid phase. When the operating pressure is much less than relieving pressure, the relieving temperature may be much higher than the operating temperature and design temperature. By using this method, one can get a temperature versus time curve, which can be used to determine if the vessel is likely to fail due to exceeding the rated temperature.

When the operating level is high in a vessel, the liquid may expand enough at the relieving conditions that there will be liquid relief, followed by two-phase flow, and then finally vapor. Without a dynamic model, one may not realize that there could be relief cases other than vapor. The cases other than vapor may mean that the outlet should go to a closed system to prevent personnel exposure to hot liquids and potential sources for pool fires and may result in larger orifice and pipe sizes.

With the liquid expanding due to the increase in temperature, the heat input actually increases, due to a larger wetted area. Without using this method, one may undersize the relief valve. Note that in NFPA 30, there is no credit given for normal liquid level. Steady-state external fire relief evaluated using NFPA 30 is based on full straight-side height.

By using this method, one can determine how long it takes before the relief occurs, thus allowing for proper emergency response by determining time available for evacuation or attempting to extinguish the fire. By using this method, one can determine how long it takes before the vessel is no longer wetted by liquid and therefore at increased potential for failure.

By using the dynamic method, one can determine when during the event the maximum relief load occurs. Sometimes it is right at the beginning, sometimes at the end, and sometimes at an intermediate point.

6. References

[1] API Standard 521, Pressure-relieving and De-pressuring Systems, Sixth Edition, January 2014

Quantifying Human Error Probabilities through a Bayesian Network using Expert Opinion and an Auditing Tool: An Application to the Tokai-Mura Accident

J. P. Duarte
Department of Telecommunications and Control Engineering
Polytechnic School, University of São Paulo
Av. Prof. Luciano Gualberto, Travessa 3, 158, 05508-900 São Paulo, SP, Brazil
julianapduarte@usp.br

A. C. O. Ribeiro[1], A. L. Sousa[2], P.F. Frutuoso e Melo[3]
[1]Bayer Crop Science Brazil S.A.
Estrada Boa Esperança 650, 26110-100 Belford Roxo, RJ, Brazil
antonio.ribeiro@bayer.com

[2]National Commission of Nuclear Energy (CNEN)
Rua General Severiano 90, 22290-901 Rio de Janeiro, RJ, Brazil
alsousa@cnen.gov.br

[3]Graduate Program of Nuclear Engineering, COPPE, Federal University of Rio de Janeiro
Av. Horácio Macedo 2030, Bloco G, sala 206, 21941-914 Rio de Janeiro, RJ, Brazil
frutuoso@nuclear.ufrj.br

Prepared for Presentation at
American Institute of Chemical Engineers
2015 Spring Meeting
11th Global Congress on Process Safety
Austin, Texas
April 27-29, 2015

UNPUBLISHED

Quantifying Human Error Probabilities through a Bayesian Network using Expert Opinion and an Auditing Tool: An Application to the Tokai-Mura Accident

J. P. Duarte
Department of Telecommunications and Control
Polytechnic School, University of São Paulo
Av. Prof. Luciano Gualberto, Travessa 3, 158, 05508-900 São Paulo, SP, Brazil
julianapduarte@usp.br

A. C. O. Ribeiro[1], A. L. Sousa[2], P.F. Frutuoso e Melo[3]
[1]Bayer Crop Science Brasil S.A.

[2]National Commission of Nuclear Energy (CNEN)

[3]Graduate Program of Nuclear Engineering, COPPE, Federal University of Rio de Janeiro

Keywords: Tokai-Mura accident, Human Reliability Analysis, THERP, CREAM, Expert opinion elicitation

Abstract

This paper presents a human reliability analysis (HRA) model for allowing the incorporation of elements that represent the facility conditions in determining human error probabilities (HEP) used in probabilistic safety analyses of process plants. It is based on an approach presented in an earlier paper by the same authors and the difference is that the model has been implemented in a Bayesian network and run by means of the Netica™ tool. The results in terms of HEPs are the same as those of the earlier paper and a sensitivity analysis was performed by means of the network developed.

1. Introduction

This paper is based on the model presented in [1]. The idea here is to take the same model and develop it by means of a Bayesian network. The reasons for doing so is to take advantage of the tools Bayesian networks [2], like easiness for modeling dependent events, which is the case when one models human failure analyses for risk analysis purposes of process plants.

In a previous work [1] a model for assessing human error probabilities for process plants was developed by considering, as a starting point, human failure probabilities taken from THERP [3] and CREAM [4].

Those human error probabilities do not take into account elements that represent the facility conditions in determining human error probabilities (HEP) used in probabilistic safety analyses of process plants.

An approach to show the predominance of human factors as accident cause was presented, as well as existing methodologies for HEP determination and their deficiencies in incorporating socio-technical elements that influence them. Such elements are: control center design, remote operations, human-machine interface, training, communications, environmental factors, workloads and staffing levels, safety culture, procedures, maintenance, management of change, and incident investigation [5].

A mathematical model, based on Bayesian networks, is now proposed to incorporate these elements in an easier way.

This paper is organized as follows: section 2 briefly discusses Bayesian networks. Section 3 presents the application of a Bayesian network for the Tokai-Mura failure event, as had been in the previous paper [1] and section 4 presents the conclusion reached.

2. Bayesian Networks

Given the complexity involved in Bayesian inference when it comes to systems with more than two variables, Bayesian networks are recommended [2]. Bayesian networks (BN) are directed acyclic graphs that, in a probabilistic way, represent dependencies between variables. Network nodes represent random variables (discrete or continuous) and directed arcs illustrate the dependency relationships among variables [6]. The relationship between cause and effect is expressed by conditional probabilities.

BNs are useful for aggregating expert opinions. The complexity of a Bayesian network depends on the level of information that can be obtained and the importance that the analyst gives to such information.

Each node has an associated conditional probability table (CPT) that quantifies the effects that parents exert on a node, i.e., the probability of the node being in a specific state, given its parent states. For each variable A that has as parents $X_1, ..., X_i$, there is a table of conditional probabilities $P(A | X_1, ..., X_i)$.

3. Tokai-Mura Human Failure Analysis by a Bayesian Network

It should be recalled from [1] that the purpose of our model is to modify basic human failure probabilities by considering plant characteristics.

The purpose it to take a ground human failure probability (HEP) and modify it by considering a proposal originally set in [7]. Notice that the difference between our model and that of [7] is the way we have modified the ground human error probabilities: we propose this modification by means of three grades (1, 2, and 3) and an auditing factor (r_i).

Grade # 1 measures the relative importance of the factor influences and is achieved through an array of relationship factors by which it is possible to analyze the distribution and intensity of factor relationships.

Grade # 2 measures the weight of each factor or element influence through the elicitation of expert opinion [8]. Twelve experts attributed a degree 1-5, where 1 is the lowest contribution degree and 5 is the largest one to the importance of each factor. The grade of each element was calculated by the average of the grades assigned by the experts.

As seen, the fourth factor (training) was considered the most relevant (grade equal to 4.7). It should be noted that these weights were attributed by the experts prior to the model application to the Tokai-Mura event (as will be seen, factors 2 and 10 have no influence on the analysis). The participants were senior engineers working in the nuclear, chemical, and petrochemical industries and Brazilian regulatory agencies.

Grade 3 represents the incidence weights of factors or elements as root causes or contributors, being established from an analysis of abnormal events (incidents, accidents and near misses, [9], [10]) in the plant, which shows the number of times that each element contributed as a root cause. The incidence of the factors or elements shown in Fig. 1 (boxes at the left) appears as a root cause of the abnormal events analyzed and this normalized figure was used as a weight for Grade 3. If an event history is unavailable, either data from a similar plant is used or Grade 3 is set equal to 1.

The r_i factor measures the degree of implementation of each factor assessed by the plant auditing process. To measure the degree of implementation of the elements a questionnaire was used to assess compliance of each factor. Each verification item is scored from 1 – 5, where 1 means noncompliance and 5 means full compliance. No scoring for elements 2 and 10 were considered because they did not apply to the Tokai-Mura plant. Modifications and grouping of elements in the original OGP questionnaire have been made to simplify the scoring and model application.

In order to correct the original HEP values of existing HRA techniques we proposed a quantitative model that is taken from [7] and adapted to our purposes. The correction to the ground human error probabilities are performed by means of the discussed Grades 1, 2, and 3 and also the auditing factor r_i. Notice that the ground probabilities are taken from THERP [3] and CREAM [4], as already done in our earlier paper [1].

Figure 1 displays the Bayesian network developed for analyzing the Tokai-Mura event. The boxes on the first column (left) display the OGP elements that are central to estimate Grade # 1. As discussed in [1], elements 2 and 10 do not apply to the Tokai-Mura plant. The second column displays the failure events that appear in the human failure event tree of [2] and finally the box in the third column displays the Tokai-Mura human failure probability. It should be noticed that the results of both analyses match. This means that Bayesian networks may be a useful tool for performing human reliability analyses in a systematic way and also perform a sensitivity analysis of the results in order to shed light on the most important contributors to the final result.

The data displayed in the first column of Fig. 1 is related to the results of the auditing procedure that indicates the degree of implementation of each element (that is, r_i). Next, for each failure event of Fig 1 we connect to it all elements that impact it and in the box for each failure event we can write down an equation for calculating the failure probability for each event. Again, failure probabilities are displayed in percentages. The first percentage is the failure probability and its complement is the success probability. For example, for event A (Failure of correct directing and/or consideration of mass, volume and geometry in the safe preparation of batch (P2 inadequate planning) the failure probability is 2.51×10^{-2}. The equations in each box of a failure event are exactly the ones developed in the earlier paper for each case. Table 1 describes all failure events for the analysis. Events B, C, E, and G are not displayed here because their contribution is negligible [1].

Table 1. Description of failure events

Failure Event	Event Description
A	Failure of correct directing and/or consideration of mass, volume and geometry in the safe preparation of batch (P2 inadequate planning)
D	Failure to conduct a thorough briefing in batch preparation (I2 - decision error with P2 - planning error)
F	Failure of the supervisor and the production head to detect and correct the operator group decision on the tank mode use and number of batches
H	Supervisor fails to stop production

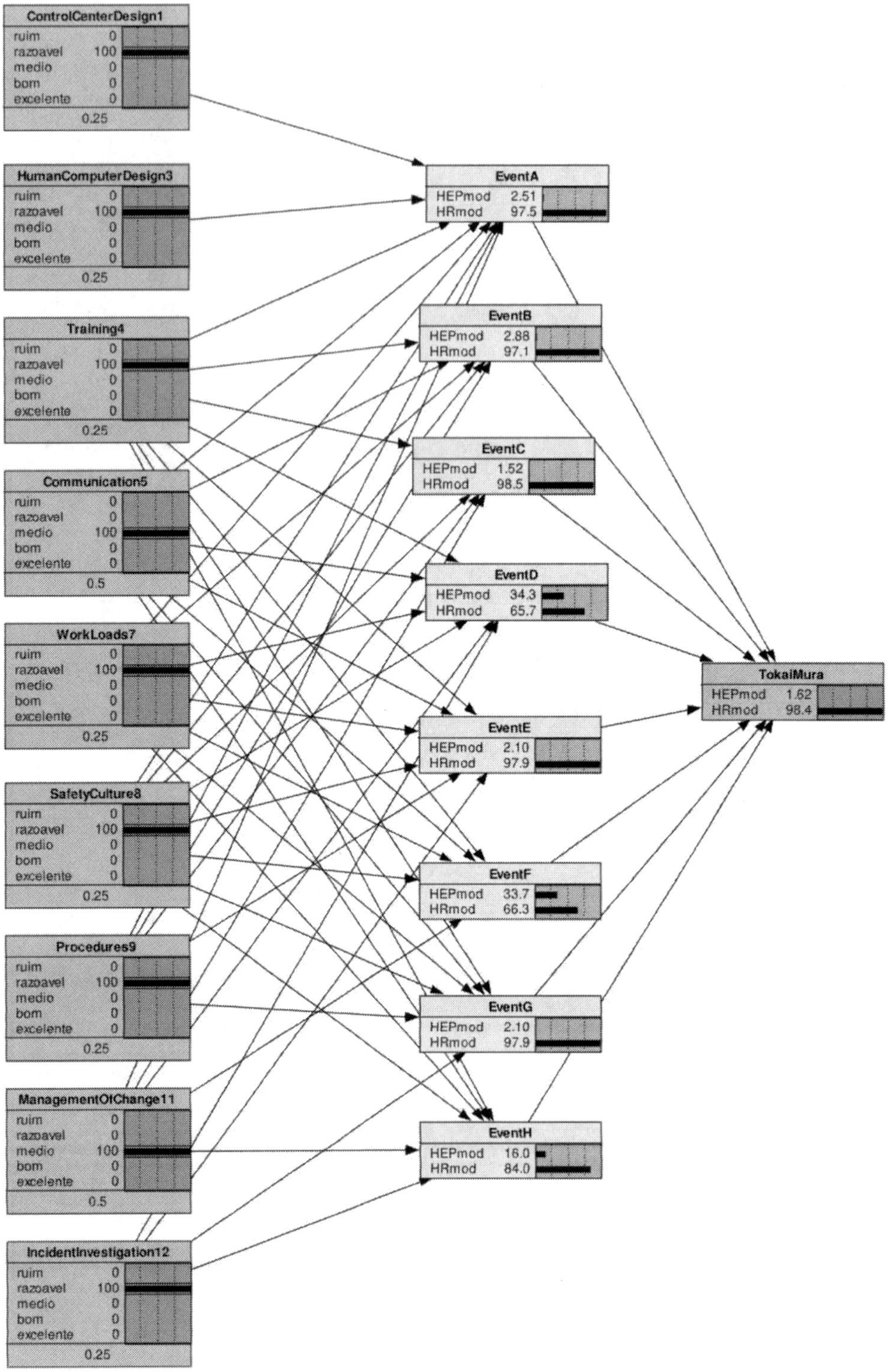

Figure 1. Bayesian Network for modeling the Tokai-Mura event

Table 2 displays an initial sensitivity analysis related to the r_i factor. Notice that the first column displays the OGP elements, the second one displays the human failure probability range and the last one displays the relative percentage variation. The range of column 2 was obtained as follows: we varied the degree of implementation from 'Bad' (which means a degree of implementation equal to zero) to 'Excellent' (which means a degree of implementation equal to 100%). The first figure in the second column is the failure probability for 'Bad' implementation degree and the second one is the failure probability for the 'Excellent' degree of implementation.

Table 2. Sensitivity Analysis for the r_i Factor

Element	Range for HEP (x 10^{-2})	Relative variation (%)
1	1.65 – 1.58	4.3
3	1.62 – 1.62	0.0
4	3.45 – 0.19	201.2
5	2.06 – 1.29	47.5
6	1.62 – 1.62	0.0
7	1.87 – 1.08	48.8
8	1.75 – 1.29	28.4
9	2.03 – 0.84	73.5
11	2.85 – 0.94	117.9
12	1.74 – 1.32	25.9

It can be seen from Table 1 that elements 4, 11, 9, 7, and 5 are the most important for the Tokai-Mura.

Table 3 shows the sensitivity analysis for the human failure event *A*. A similar sensitivity analysis was performed for human failure events D, F, and H. Notice that human failure events B, C, E, and G were not considered here since their contribution to the global human failure event is negligible, as can be seen from Fig 1.

Table 3. Sensitivity Analysis for the Human Failure Event *A*

Element	Range for HEP (x 10^{-2})	Relative variation (%)
1	3.38 – 2.04	53.4
3	2.54 – 2.42	4.8
4	3.48 – 0.95	100.8
5	2.80 – 2.25	21.9
7	2.68 – 2.07	24.3
8	2.59 – 2.28	12.4
9	3.04 – 1.41	64.9
11	3.57 – 1.77	71.7

It may be seen from Table 3 that Element # 4 is the most important since it doubles the event human failure probability. Element # 9 is the second most important. A similar behavior may be found in Table 4, 5, and 6. Notice that these results may help making decisions as to where to implement modification when resources are limited.

Table 4. Sensitivity Analysis for the Human Failure Event *D*

Element	Range for HEP (x 10^{-2})	Relative variation (%)
4	49.9 – 11.2	112.8
5	38.9 – 30.3	25.1
7	37.0 – 27.5	27.7
8	35.6 – 30.7	14.3
11	51.4 – 22.9	83.1
12	36.3 – 29.1	21.0

Table 5. Sensitivity Analysis for the Human Failure Event *F*

Element	Range for HEP (x 10^{-2})	Relative variation (%)
4	50.3 – 10.1	119.3
5	38.5 – 29.4	27.0
7	36.4 – 26.5	29.4
8	35.0 – 29.9	15.1
11	51.9 – 21.8	89.3

Table 6. Sensitivity Analysis for the Human Failure Event *H*

Element	Range for HEP (x 10^{-2})	Relative variation (%)
4	30.3 – 2.34	174.8
5	19.8 – 12.9	43.1
7	18.1 – 10.9	45.0
8	17.0 – 13.2	23.8
11	31.9 – 7.98	149.5
12	17.6 – 12.0	35.0

4. Conclusions

It has become clear at safety conferences and congresses in the nuclear and chemical and petrochemical fields that existing laws and regulations, especially some requirements of international regulatory bodies such as the CSB (Chemical Safety Board, USA), are more and more explicit in regarding the implementation of human reliability analysis (HRA) as a way of risk reduction. However, most organizations still do not have efficient mechanisms to understand and implement policies for human factors analyses. This work offers a contribution to include in a comprehensive manner the elements that influence human error. Also, improvements on plant management that can be taken into account by the 12 factors considered can be easily evaluated by the Bayesian network, thus allowing for estimating human reliability improvements.

A benefit of the Bayesian network model is the fact that a sensitivity analysis can be easily performed to analyze the impact of each of the 12 factors mentioned and thus allowing for a more realistic plant behavior modeling in face of abnormal events.

A contribution of the proposed model is to allow seeing how elements relate and how they influence HEP quantification, which allows directing efforts in the short and long term to reduce HEPs or even review the effectiveness of the efforts being made to reduce them.

5. References

[1] A. C. Ribeiro, A. L. Sousa, J. P. Duarte, and P. F. Frutuoso e Melo, Human Reliability Analysis of the Tokai-Mura Accident through a THERP-CREAM and Expert Opinion Auditing Approach, submitted to the Nuclear Engineering and Design journal, 2014.

[2] F. V. Jensen, An Introduction to Bayesian Networks, UCL Press, University College, London, 1996.

[3] A. D. Swain and H. E. Guttman, Handbook of Human Reliability Analysis with Emphasis on Nuclear Power Plant Applications, NUREG/CR-1278, US Nuclear Regulatory Commission, Washington, DC, 1983.

[4] E. Hollnagel, Cognitive Reliability and Error Analysis Method (CREAM), Elsevier, Oxford, 1998.

[5] CCPS, Human Factors Methods for improving Performance in the Process Industries, Center for Chemical Process Safety, American Institute of Chemical Engineers, Wiley, New York, 2007.

[6] J . Pearl, Probabilistic Reasoning in Intelligent Systems. Morgan Kaufmann Publishers, San Francisco, CA, 1988.

[7] S. G. Kariuki, Integrating Human Factors into Chemical Process Quantitative Risk Analysis, PhD dissertation, Berlin Technical University, Berlin, 2007.

[8] B. M. Ayyub, Elicitation of Expert Opinions for Uncertainty and Risks. CRC Press, Boca Ratton, FL, 2001

[9] CCPS, Guidelines for Chemical Process Quantitative Risk Analysis, Center for Chemical Process Safety, American Institute of Chemical Engineers, Wiley, New York, NY, 2000.

[10] IAEA, IAEA Safety Glossary, Terminology Used in Nuclear Safety and Radiation Protection, International Atomic Energy Agency, Vienna, Austria, 2007.

396889 Numerical Modeling of Pool Fire inside an Offshore Methanol Storage Tank

Abstract:

Methanol is widely used in the oil and gas industry as a reactant for the production of methyl tertiary butyl ether (MTBE) and as down-hole fluid additive to reduce hydrate formation and prevent freezing of solutions during injection into gas wells. Methanol is classified as a flammable liquid with a flammable range from 6% to 36% bounding the lower and upper flammable limits, respectively.

There are many experimental and numerical studies carried out for the methanol liquid pool fires in an open air environment, or a wind tunnel, with multiple pool fire sizes. However, few numerical studies have been performed for methanol pool fires inside or emanating from a storage container (typical storage units containing man-way entrances). The inadequate oxygen inside the container will limit the duration of the fire, and extinction and re-ignition of sectional small fires propose a challenge for numerical simulation of such kind of fire. In this paper, the Fire Dynamics Simulator (FDS) tool is used in the analysis of methanol pool fires. FDS is a computational fluid dynamics (CFD) fire model using large eddy simulation (LES) turbulence model. The basis for FDS has been developed from the mathematical background common to many CFD models with an emphasis on the slow moving flow and heat transfer caused by fires.

Methanol tanks typically have a pressure vacuum safety valve (PVSV) installed to relieve static pressure inside the tank when the pressure difference between the vessel pressure and the ambient pressure is small. If the methanol pumps continue to run and the tank PVSV or man-way is opened, air (e.g. oxygen) may ingress and begins mixing with the denser methanol vapors to a flammable range. Assuming the vapors find an ignition source, a pool fire is possible with the flammable vapors burning above the methanol liquid surface. This paper considers two scenarios for air ingress that lead to pool fires, a PVSV opening during normal venting operations and a man-way opening during maintenance to the inside of the tank. The relationship between the fire duration and tank fill levels are obtained and discussed.

Credibility of Designing Refrigerated Atmospheric Storage Tank Pressure Relief Systems for a Tornado

Syed Usman Ahmed
Agrium Wholesale – Engineering & Risk Management
11751 River Road, Fort Saskatchewan, Alberta, Canada, T8L-4J1
Syed.Ahmed@agrium.com

Keywords: Plant Operations, Pressure Relief Devices, Pressure Relief Systems, Process Safety and Storage Tanks

Abstract

Atmospheric pressure changes occur due to varying weather patterns. As per API 2000, a change in barometric pressure is a possible cause of overpressure or vacuum in refrigerated storage tanks due to the potential for a substantial amount of vaporization or condensation. To serve as a valid layer of protection, relief systems are required to be sized to handle the worst case scenario with the largest required relief area. An extreme case of a drastic change (sudden decrease followed by a sudden increase) in barometric pressure is the passing of a severe electrical storm front (i.e. tornado). This causes a sudden atmospheric pressure drop and results in a loss of power to the site (all electrically driven equipment would cease operation) as the system tries to re-establish vapor-liquid equilibrium within the storage tank.

This paper will analyze Agrium's anhydrous ammonia storage tank, 100-T at Leal Distribution Terminal in North Dakota and delve into the reasons a tornado would not cause a realistic overpressure scenario for the storage tank's pressure relief devices.

1. Introduction

Severe weather changes can result in the following:

- High winds
- Precipitation
- Rapid temperature change
- Sudden change in atmospheric pressure due to storm front
- Loss of site-wide electrical power due to storm

If the material being stored is refrigerated liquid at or near its boiling point, low atmospheric pressure (decrease in barometric pressure) during a severe storm will cause the stored material to re-establish vapor-liquid equilibrium. Liquids that are not at or near their boiling point are not affected by weather changes.

API 2000 Sections 4.2.5.11 and 5.2.1.2 state: *A rise or drop in barometric pressure is a possible cause of vacuum or overpressure in a tank. This should be considered for refrigerated storage tanks. A change in atmospheric pressure can result in a substantial amount of vaporization or condensation* [1]. The worst credible case is the basis for relief system design. As such, the design basis scenario is the passing of a severe electrical storm front (i.e. tornado) causing a sudden atmospheric pressure drop and resulting in a loss of power to the site (all electrically driven equipment would cease operation); this will result in an overpressure scenario caused by:

- Flash vapor generated from liquid inventory
- Expansion of vapor in the enclosed vapor space (head of tank)
- Normal boil-off

2. Designing Storage Tank Pressure Relief Systems for a Tornado

2.1 Background Information

Determining accurate barometric pressure measurements in tornado cores is difficult due to the harsh physiology of the tornado itself. API 2000 Section 5.2.1.2 suggests using 2000 Pascals/hr as the absolute rate of pressure change (d_{patm}/dt) if local atmospheric pressure change data is not available [1]; however, this typically is not a concern as weather networks have historical weather data that is monitored and collected by local airports through the use of pressure probes. For a pressure probe to remain stationary during a tornado passage, special considerations are applied to the physical geometry of the device in order to increase the accuracy of the measurements [2]. From Figures 1 and 2 below, it is evident that the pressure plunge that takes place as a tornado passes is very transitory (i.e. a few seconds).

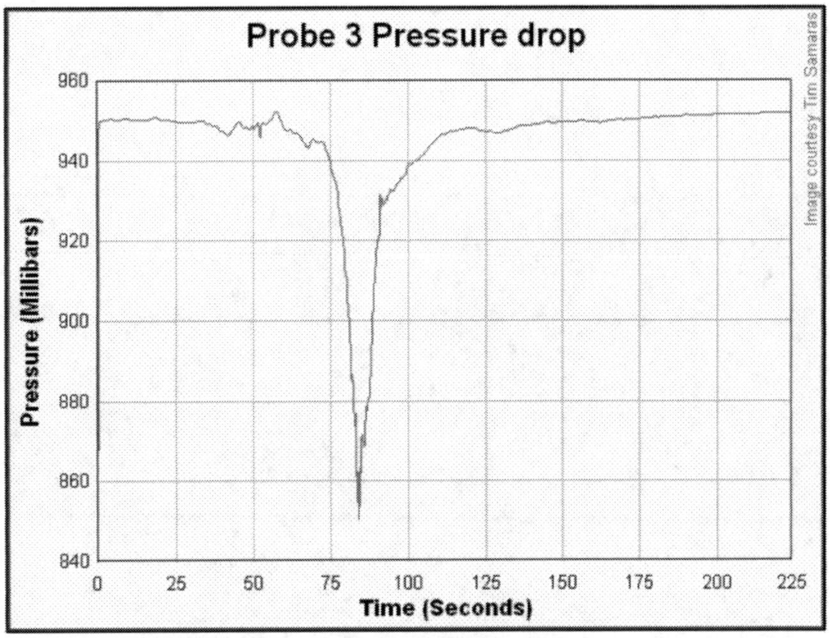

Figure 1: Tornado on June 24, 2003 (Manchester, SD) [3].

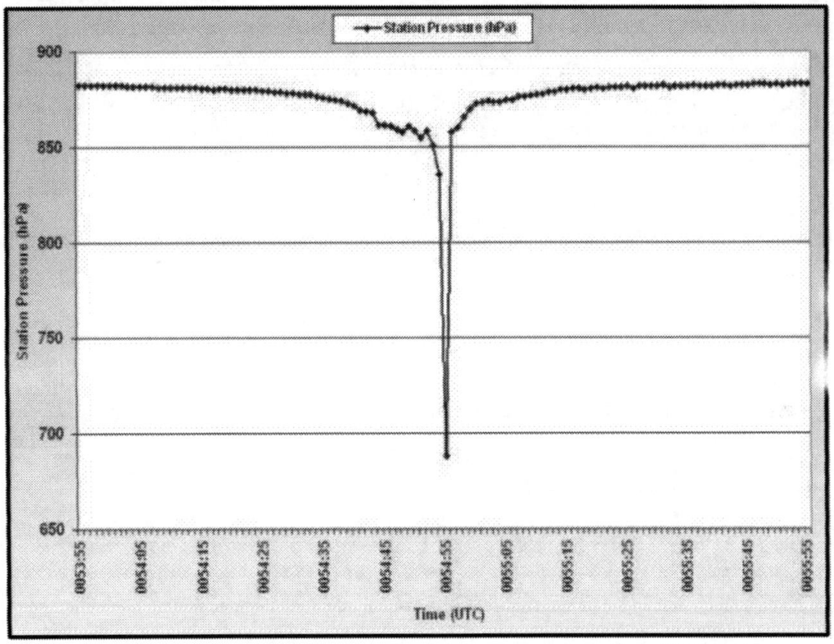

Figure 2: Tornado on April 21, 2007 (Tulia, TX) [3].

2.2 Design Basis

The storage tank must be designed so that it will remain in a safe condition under severe meteorological events, including those that could result in the most severe tornado that could to occur at the site. Referring to Figure 3, the U.S. Nuclear Regulatory Commission divides the Contiguous United States into three tornado intensity regions at the 10^{-7} probability level and each of these regions is characterized by a pressure drop, which is related to the weather conditions during the tornado as provided in Table 1. It is evident from Table 1 that the pressure drop is not large, but occurs very rapidly (2 – 3 seconds).

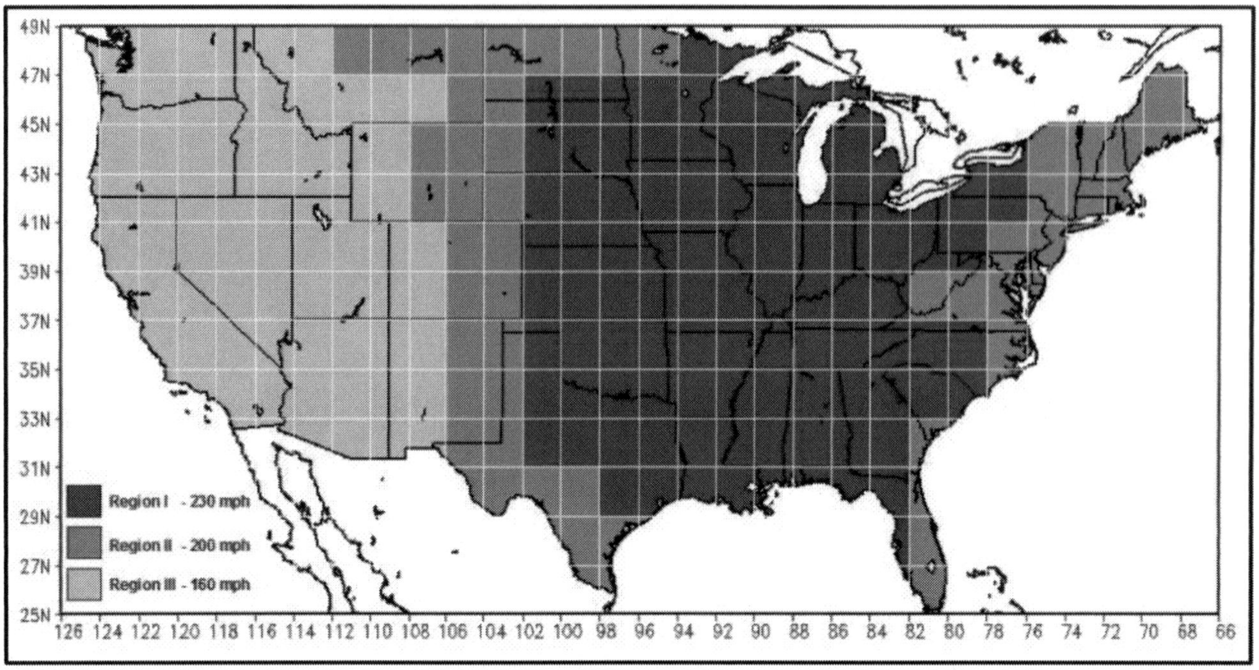

Figure 3: Tornado Intensity Regions for the Contiguous United States for Exceedance Probabilities of 10^{-7} per year [4]

Table 1: Design-Basis Tornado Characteristics [4]

Region	Maximum wind speed m/s (mph)	Translational speed m/s (mph)	Maximum rotational speed m/s (mph)	Radius of maximum rotational speed m (ft)	Pressure drop mb (psi)	Rate of pressure drop mb/s (psi/s)
I	103 (230)	21 (46)	82 (184)	45.7 (150)	83 (1.2)	37 (0.5)
II	89 (200)	18 (40)	72 (160)	45.7 (150)	63 (0.9)	25 (0.4)
III	72 (160)	14 (32)	57 (128)	45.7 (150)	40 (0.6)	13 (0.2)

2.3 Required Relief Rate

As per API Standard 2000, the total required relief rate is the sum of that caused by vapor expansion and that caused by flashing. The description of the calculation method holds true for only certain cases (i.e. does not take into consideration the required relief due to normal boil-off). Further, the actual equations in API 2000 shall not be used as it is not valid for a refrigerated liquid at or near its boiling point. The total required relief rate is described in Equation 1 as:

$\dot{W} = \dot{W}_{AL} + \dot{W}_{AG} + \dot{W}_{AB}$ (1)

where,

$\dot{W}_{AL}$ = Flash vapor generation due to atmospheric pressure drop, lb/hr

$\dot{W}_{AG}$ = Vapor expansion due to atmospheric pressure drop, lb/hr

$\dot{W}_{AB}$ = Normal boil-off (must be taken into account due to loss of power), lb/hr

In order to calculate the required relieving capacity the following parameters shall be used:

- The tank is filled to 100% of its design liquid capacity
- No inlet flow[1]
- The absolute rate of pressure change of atmospheric pressure (d_{patm}/dt) for the site as determined from Table 1

Refer to Section 6.1 for sample calculation.

2.3.1 Flash Vapor Generation

Apply the conditions below to 100% liquid inventory in the storage tank to determine the flash vapor generation over a one hour period (using an adiabatic flash):

- The initial conditions in the tank are as follows:
  - Site design atmospheric pressure
  - Storage tank design pressure
  - Liquid inventory at bubble point

- The final conditions are as follows:
  - Site design atmospheric pressure less the absolute rate of pressure change of atmospheric pressure
  - Storage tank design pressure
  - Bubble point

2.3.2 Vapor Expansion

The drop in atmospheric pressure results in a change in density in the vapor space. Hence, the change in density results in vapor expansion which requires relief and is described in Equation 2 as:

[1] There is little difference in the overall relief requirements if inlet flow continues compared to assuming no inlet flow. It has been determined that the overall relief requirements for the storage tank contents are primarily dictated by the liquid inventory in the tank and not the inlet flow stream. Between 0.25% and 0.5% of the total vapor generated is from the liquid inflow (including flash), compared to that which flashes from the liquid in the tank. Therefore, 100% liquid inventory with no inlet flow shall be used as the relief system design basis.

$$\dot{W}_{AG} = V\frac{(\rho_1-\rho_2)}{1\ \text{hr}} \quad (2)$$

where,

$\dot{W}_{AG}$ = Mass flow rate requiring venting due to vapor expansion, lb/hr

V = Volume of vapor space above maximum liquid level, ft^3

ρ_1 = Density of vapor at P_1 and its dew point temperature, lb/ft^3

ρ_2 = Density of vapor at P_2 and its dew point temperature, lb/ft^3

P_1 = Design pressure of storage tank

P_2 = Design pressure of storage tank less the absolute rate of pressure change (d_{patm}/dt)

2.3.3 Normal Boil-Off (Evaporation due to Ambient Heat Input)

The normal boil-off rate is equal to the vapor generation rate as a result of the normal heat input to the tank. The vapor generation rate is found using the normal ambient heat input and the heat of vaporization of the stored liquid at relief conditions (i.e. pressure and temperature)[2] .

$$\dot{W}_{AB} = \frac{Q}{\Delta H_{vap}} \quad (3)$$

where,

Q = Heat input (as per Heat & Material Balance), Btu/hr

ΔH_{vap} = Heat of vaporization of the stored liquid at the reliving pressure and temperature, Btu/lb

2.4 Pressure Relief Devices

There are various types of pressure relief devices that can be used for pressure and vacuum relief: open vents, direct-acting vent valves, and pilot-operated vent valves.

Open vents are always open and provide overpressure or vacuum protection for atmospheric storage tanks by allowing the tank to in-breath and out-breath at any pressure differential. Open vents are often equipped with a weather hood or turndown (see Figure 4) to prevent rain or snow from entering the tank.

---

[2] The heat of vaporization is obtained from simulation software or thermodynamic and physical properties charts/tables for pure liquids.

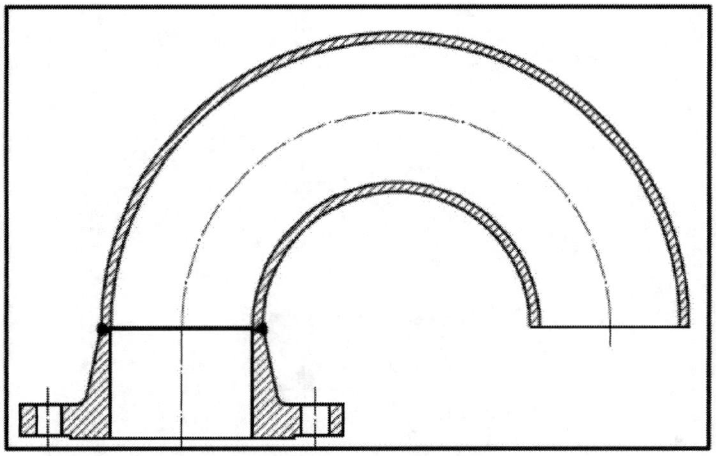

Figure 4: Open vent [1].

Direct-acting vent valves are available to provide pressure relief, vacuum relief, or a combination of pressure and vacuum relief. Combination vent valves may be of a side-by-side configuration. Side-by-side vent valves or pressure-relief vent valves are available with flanged outlets for pressure discharge when it is necessary to vent the pressure-relief vapors away.

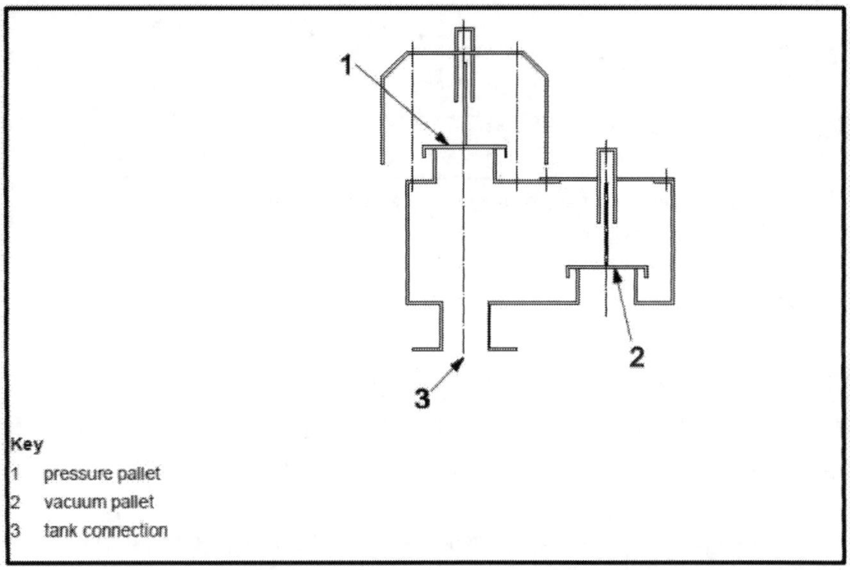

Figure 5: Side-by-side pressure/vacuum vent [1].

Pilot-operated vent valves are available to provide pressure relief, vacuum relief or a combination of pressure and vacuum relief. Unlike side-by-side direct-acting vent valves, pilot-operated vent valves relieve pressure or vacuum through the same opening to atmosphere.

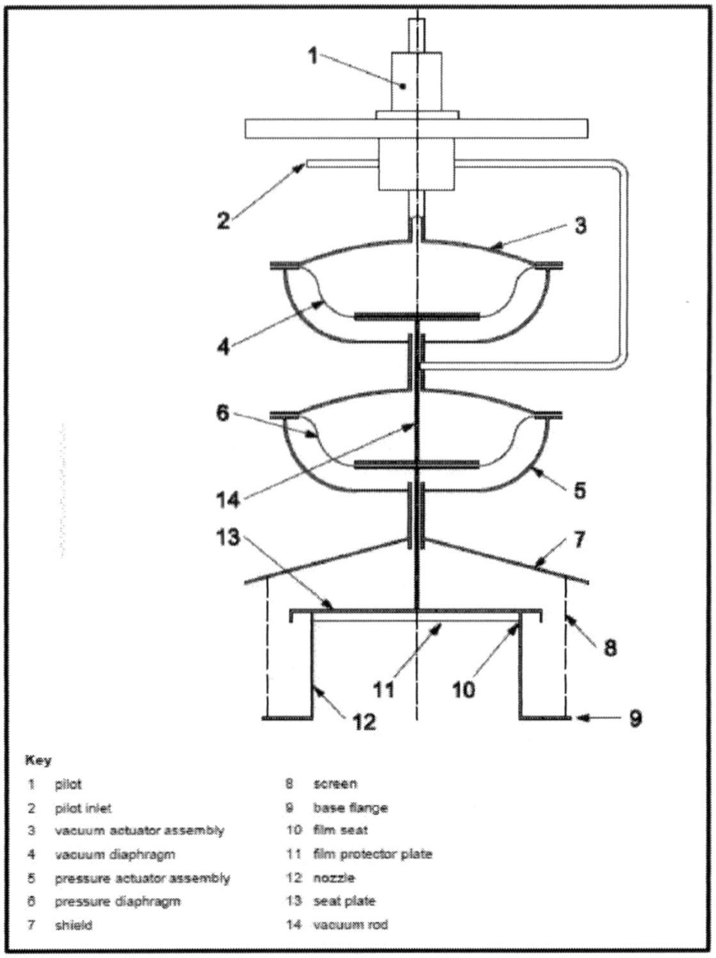

Figure 6: Pilot-operated pressure/vacuum vent [1].

2.4.1 Orifice Area – Anderson Greenwood Low Pressure POPRVs

To size a pressure relief valve, the required relief rate for the controlling contingency must be known, along with the fluid properties at relief conditions. The required orifice area is calculated from vendor supplied formulae to determine the size of the relief device.

Referring to Formula 11 on page 33 of the Anderson Greenwood Low Pressure Pilot Operated Pressure Relief Valves (POPRV) Catalog [5], the required orifice area can be calculated based on the mass flow rate at actual conditions or volumetric flow rate at standard conditions as shown in Equation 4:

$$A_{(in^2)} = \frac{W\sqrt{TZ}}{735 K_d P_1 F \sqrt{M}} \tag{4}$$

where,

W = Required relief rate, lb/hr

T = Absolute relieving temperature[3], °R
P_1 = Pressure at valve inlet during flow, psia
M = Molecular weight of the flowing gas, MW
Z = Compressibility factor of the flowing gas
K_d = Coefficient of discharge
F = Subsonic flow factor

Refer to Section 6.2 for sample calculation.

3. Results & Conclusion

Assuming that vapor-liquid equilibrium could be instantly established, 3.95 x 10^6 lb/hr of NH_3 vapor would have to be relieved if a tornado came in direct contact with 100-T, Ammonia Storage Tank, causing a rapid decrease in atmospheric pressure. The corresponding required orifice area is 10,396 in^2. Therefore, the system would require 92 POPRVs as each POPRV has an orifice area of 113 in^2.

4. Conclusion

Based on the analysis conducted at Leal Distribution Terminal, it is evident that a tornado which results in a sudden decrease in barometric would not cause a realistic overpressure scenario for the storage tank's pressure relief devices to be sized due to the following:

- The unlikely probability (10^{-7} chances per year) of encountering such a tornado that exceeds the intensity of Region II (200 mph winds)
- The near impossibility of the system (ammonia within the storage tank) to re-establish vapor-liquid equilibrium in a few seconds (i.e. time span of pressure drop) thus resulting in an overpressure scenario
- The more likely damage mechanism of very high wind velocities, heavy rain, and wind-driven debris renders overpressure protection a moot point.

Therefore, when determining the required relief rate for the Weather Changes / Atmospheric Pressure Drop contingency the absolute rate of pressure change of atmospheric pressure (dp_{atm}/dt) used should not be that for a tornado in the region, rather the greater dp_{atm}/dt for the site over at least 50 years, or that provided in API Standard 2000 Section 5.2.1.2 (2000 pascals/hr), whichever is greater.

---

[3] Storage tank relief pressure for overpressure and normal maximum temperature at site for vacuum

5. References

[1] *ANSI/API STANDARD 2000: Venting Atmospheric and Low-pressure Storage Tanks*. 6th ed. N.p.: American Petroleum Institute, 2009. Print.

[2] Samaras, Timothy M., and Julian J. Lee. *Pressure Measurements Within a Large Tornado.* Proc. of Eighth Symposium on Integrated Observing and Assimilation Systems for Atmosphere, Oceans, and Land Surface, Washington State Convention and Trade Center, Seattle. Applied Research Associates, Inc., n.d. Web.

[3] Edwards, Roger. "The Online Tornado FAQ." *Storm Prediction Center*. NOAA / National Weather Service, 07 Mar. 2014. Web. 24 July 2014. <http://www.spc.noaa.gov/faq/tornado/index.html#pressure1>.

[4] Regulatory Guide 1.76, "Design-Basis Tornado and Tornado Missiles for Nuclear Power Plants," *U.S. Nuclear Regulatory Commission*, Washington, DC, March 2007.

[5] Pentair. *Anderson Greenwood Low Pressure POPRV Catalog*. N.p.: Pentair Valves & Controls, 2013. Print.

6. Appendix – Sample Calculation

6.1 Required Relief Rate

Data and Constants:

		Reference
Tank MAWP	1 psig	Site General Arrangement
Tank Design Temperature	-39°F	Site General Arrangement
P_{atm}, Atmospheric Pressure	13.8 psia	Atmospheric Pressure at site
P_{set}, Set pressure of PSV	1 psig	P_{set} = Current set pressure of PSVs
P_1, Relief Pressure	14.9 psia	$P_{relief} = (1.10)\ P_{set} + P_{atm}$
T_{relief}, Relief Temperature	-26.27°F	Simulation Software
P_2, Absolute Operating Pressure of Tank (includes 10% overpressure) - Absolute Rate of Pressure Change	14.5 psia	$P_{relief} - d_{patm}/dt$
ρ_1, Density of vapor at P_1 and dew point Temperature	0.0554 lb/ft^3	Simulation Software
ρ_2, Density of vapor at P_2 and dew point Temperature	0.0540 lb/ft^3	Simulation Software
Q, Ambient Heat Input	680000 Btu/hr	Site General Arrangement
ΔH_{vap}, Heat of Vaporization	607.9675 Btu/lb	Simulation Software
r, Radius of head	128 ft	Site General Arrangement
h, Height of head	28.72 ft	Site General Arrangement
d_{patm}/dt, Absolute Rate of Pressure Change	0.4 psi/sec	[4]
$\dot{W}$, Total relief rate due to vapor expansion, flashing & normal boil-off	Calculate	
$\dot{W}_{AB}$, Relief rate due to normal boil off	Calculate	
$\dot{W}_{AL}$, Relief rate due to flashing	Calculate	Simulation Software
$\dot{W}_{AG}$, Relief rate due to vapor expansion	Calculate	
V_{head}, Volume of Head/Dome section	Calculate	

Calculation:

- The volume of the head/dome section is:

$$V_{head} = \frac{\pi}{6}(3r^2 + h^2)h = \frac{\pi}{6}(3(128\text{ft})^2 + (28.72\text{ft})^2)28.72\text{ft}$$

$$V_{head} = 751539\ \text{ft}^3$$

- The drop in atmospheric pressure results in a change in density in the vapor space. Hence, the change in density caused by an atmospheric pressure drop results in vapor expansion which requires relief and is described in Equation 3 as:

$$\dot{W}_{AG} = V\frac{(\rho_1 - \rho_2)}{1\ \text{sec}} = 751539\ \text{ft}^3\frac{\left(0.0554\frac{\text{lb}}{\text{ft}^3} - 0.0540\frac{\text{lb}}{\text{ft}^3}\right)}{1\ \text{sec}}$$

$$\dot{W}_{AG} = 1052\frac{\text{lb}}{\text{sec}}\ \text{of NH3 Vapor at} - 26.27°\text{F and } 14.9\ \text{psia}$$

- A flash simulation was performed to determine the amount of vapor generated due to the flashing of liquid in the storage tank caused by an atmospheric pressure drop. The inlet stream was evaluated at the relieving pressure of the tank (14.9 psia) and the corresponding saturation temperature (-26.27 °F). The flash unit was set to a pressure drop of 0.4 psia and was operating adiabatically (Duty = 0 Btu/hr).

$$\dot{W}_{AL} = 804\ \frac{\text{ft}^3}{\text{sec}}\text{at} - 27.70°\text{F and } 14.5\ \text{psia}$$

The equivalent mass flow rate at relieving conditions was obtained from simulation software:

$$\dot{W}_{AL} = 804\ \frac{\text{ft}^3}{\text{sec}}\bigg|0.0554\frac{\text{lb}}{\text{ft}^2}$$

$$\dot{W}_{AL} = 45\ \frac{lb}{sec}\ \text{of NH3 Vapor at} - 26.27°\text{F and } 14.9\ \text{psia}$$

- The normal boil-off rate is calculated dividing the ambient heat input by the heat of vaporization.

$$\dot{W}_{AB} = \frac{Q}{\Delta H_{vap}} = \frac{680000\frac{\text{Btu}}{\text{hr}}}{607.9675\frac{\text{Btu}}{\text{lb}}} = 1119\frac{\text{lb}}{\text{hr}}$$

$$\dot{W}_{AB} = 0.31\ \frac{\text{lb}}{\text{sec}}\ \text{of NH3 Vapor at} - 26.27°\text{F and } 14.9\ \text{psia}$$

- Total flow rate is the sum of that caused by vapor expansion, flashing and normal boil-off:

$$\dot{W} = \dot{W}_{AG} + \dot{W}_{AL} + \dot{W}_{AB} = 1052\frac{lb}{sec} + 45\frac{lb}{sec} + 0.31\frac{lb}{sec}$$

$$\mathbf{\dot{W} = 3.95 \times 10^{6}\ \frac{lb}{hr}\ of\ NH3\ Vapor\ at - 26.27°F\ and\ 14.9\ psia}$$

6.2 Required Orifice Area

Data and Constants:

		Reference
$A_{(in2)}$, Orifice Area	Calculate	[5]
F, Subsonic Flow Factor	Calculate	[5]
$k = C_p/C_v$	Calculate	Simulation software
K_d, Subsonic Valve Coefficient	Calculate	[5]
M, NH3 Molecular Weight	17.0306	Simulation software
Z, Compressibility Factor	0.9853	Simulation software

Calculations:

$$A_{(in^2)} = \frac{W\sqrt{TZ}}{735K_dP_1F\sqrt{M}} \rightarrow T = T_{relief}$$

where

$$K_d = 0.650\left(\frac{P_2}{P_1}\right)^{-0.349} \rightarrow P_1 = P_{relief} \text{ and } P_2 = P_{atm}$$

$$K_d = 0.650\left(\frac{13.8}{14.9}\right)^{-0.349} = 0.6676$$

$$k = \frac{C_P}{C_V} = \frac{0.4868}{0.3643}$$

$$k = 1.3363$$

$$F = \sqrt{\frac{k}{k-1}\left[\left(\frac{P_2}{P_1}\right)^{\frac{2}{k}} - \left(\frac{P_2}{P_1}\right)^{\frac{k+1}{k}}\right]}$$

$$F = \sqrt{\frac{1.3363}{1.3363-1}\left[\left(\frac{13.8\text{ psia}}{14.9\text{ psia}}\right)^{\frac{2}{1.3363}} - \left(\frac{13.8\text{ psia}}{14.9\text{ psia}}\right)^{\frac{1.3363+1}{1.3363}}\right]} = 0.2602$$

$$A_{(in^2)} = \frac{3.95 \times 10^6 \frac{\text{lb}}{\text{hr}}\sqrt{433.40°\text{R} \cdot 0.9853}}{735 \cdot 0.6676 \cdot 14.9\text{ psia} \cdot 0.2602\sqrt{17.0306}}$$

$$\mathbf{A_{(in^2)} = 10396\ in^2}$$

Results:

Assuming that vapor-liquid equilibrium could be instantly established, 3.95 x 10^6 lb/hr of NH3 Vapor would have to be relieved if a tornado came in direct contact with 100-T, Ammonia Storage Tank, causing a rapid decrease in atmospheric pressure. The corresponding required orifice area is 10,396 in^2. Therefore, the system would require 92 POPRVs as each POPRV has an orifice area of 113 in^2.

397003 Cultura De Liderazgo y Seguridad – Independientemente Del Idioma

Abstract:

La seguridad en el ámbito de trabajo es vital para el bienestar y sostenibilidad de cualquier negocio. No se trata de un requerimiento, o meramente un lineamiento corporativo; sino es algo que debe ser abrazado por el personal y su liderazgo en cualquier sector productivo. Una vez dicho esto, prevalece una gran diversidad en regiones y ambientes de trabajo, lo cual puede complicar el proceso de crear e implementar una cultura de seguridad común. Dado que los empleados son originarios de diferentes geografías y culturas traídas de sus empleos anteriores, la definición del término "seguridad en el lugar de trabajo" se hace subjetiva. En adición a la variedad de culturas; pueden confluir lenguajes y dialectos en grupos de trabajo por lo que la barrera lingüística frecuentemente juega un importante rol.

Este papel discutirá como un liderazgo de cultura segura puede considerar las diferencias lingüísticas para promover una cultura de seguridad común; al momento de implementar un plan de seguridad para una fuerza laboral diversa. Las discusiones enfatizaran aprendizajes sobre la curva Bradley y demostrarán la importancia de seguir prácticas de trabajo seguras para alcanzar aproximarse a un área 100% libre de accidentes. Ejemplos reales de regiones de trabajo internacionales y equipos de trabajo diversos se presentaran para subrayar esta necesidad y compartir lecciones aprendidas.

398031 Ensure Compliance to NFPA Standards through Corporate Level Combustible Dust Audits

Abstract:

The Occupational Safety and Health Administration (OSHA) CPL 03-00-008 Combustible Dust National Emphasis Program (reissued) has been in effect since March 11, 2008. The traditional strategy of regulating only major industry sectors has long passed. Currently industries handling hazardous combustible solid particulates/fibers, including small manufacturing businesses, are subject to Instruction compliance. Industries covered by this Instruction must also comply with applicable OSHA and related National Fire Protection Association (NFPA) Standards. Managing compliance is a necessity but can sometimes be perceived as an overwhelming task. After conducting a number of corporate audits for compliance to the combustible Dust NEP, some common technical and administrative themes will be shared that could help an EH&S audit group and initiate a revision of their current protocols especially for legacy facilities and new design/construction.

Using NFPA 654 as the reference standard, audit protocol items related to facility dust hazard assessments will be presented. These include but are not limited to: dust control, ignition control, damage control, and administrative issues such as training for new and veteran affected employees and contractors, Management of Change, Process Hazard Analysis (PHA), Inspections and Preventive Maintenance, and Safety Data Sheets. Reports from individual facility findings and recommendations should follow a standard format, and allow for local management to obtain a clear understanding of the issues related to their specific site and compare findings with plants of similar technology. In legacy facilities, which were built or acquired prior to current combustible dust regulations, there are additional unique findings that are specific to industry type, age of the facility, and vendor installed equipment.

Corporate audits can be planned to identify common strengths and opportunities across the organization to allow senior management to understand the prioritized recommendations in a systematic action plan based on safety needs. A "report card" has been developed to provide an overview of a corporation's combustible dust current program, which compares actual compliance to Instruction guidance and NFPA standards. Common field trends will be presented as well as guidance for establishing and properly maintaining engineering and administrative systems for a safer combustible dust program.

11TH GLOBAL CONGRESS ON PROCESS SAFETY

Combining Traditional UOP HAZOP Analysis With Dynamic Simulation A New Process Safety / Risk Assessment Tool

Scott M. Wozniak
UOP, A Honeywell Company
25 E. Algonquin Road, Des Plaines, IL. 60017-5017
Scott.Wozniak@Honeywell.com

Bill Wiede
UOP, A Honeywell Company
25 E. Algonquin Road, Des Plaines, IL. 60017-5017
Bill.Wiede@Honeywell.com

Prepared for Presentation at
American Institute of Chemical Engineers
2015 Spring Meeting
11th Global Congress on Process Safety
Austin, Texas
April 27-29, 2015

UNPUBLISHED

Combining Traditional UOP HAZOP Analysis With Dynamic Simulation A New Process Safety / Risk Assessment Tool

Scott M. Wozniak
UOP, A Honeywell Company
25 E. Algonquin Road, Des Plaines, IL. 60017-5017
Scott.Wozniak@Honeywel.com

Bill Wiede
UOP, A Honeywell Company
25 E. Algonquin Road, Des Plaines, IL. 60017-5017
Bill.Wiede@Honeywell.com

Keywords: HAZOP, Dynamic Simulation, Process Safety, Risk Assessment and Flowsheets

Abstract

There is a clear link between safety and reliability in process design and operation. The traditional Hazard and Operability (HAZOP) analysis identifies potential operational failures and weak points in a process design; evaluates risks introduced by them and develops effective risk reduction strategies before a customer takes delivery of a process design.

This paper shows how our current HAZOP methodology is supported by dynamic simulation and modeling techniques. By leveraging this process technology, more robust HAZOP studies result by integrating our current UOP Process Technology library of **UNISIM®** flowsheet templates in all phases of the HAZOP work process.

We will show how dynamic simulation can determine consequences of deviations from normal design exactly and how this tool can potentially identify new unknown hazards and safety issues.

1. Introduction

Process Safety can help provide better assurance for communities and for the refinery, petrochemical and chemical plants through a HAZOP study, a method of identifying both deviations to process design that could cause potential hazards, and safeguards required to minimize potential hazards.

The HAZOP study is traditionally a qualitative study. To give it more quantitative character dynamic simulation is sometimes used to complement the HAZOP study. Combining HAZOP with dynamic simulation provides the means for investigating (and demonstrating) the consequences of deviations from normal operating conditions. This paper shows a few examples using dynamic simulation in a HAZOP.

2. HAZOP

Companies and the public are increasingly asking for better assurance that sites and communities will not be adversely impacted by fires, explosions and releases from oil refineries and chemical plants. The banking and the insurance industries are also increasingly asking for this assurance. Process Safety can help provide that assurance through a HAZOP study, a method of identifying both deviations to process designs that could cause potential hazards, and the safeguards required to minimize potential hazards. The HAZOP study is recognized worldwide as a primary methodology for hazard identification conducted for the oil, petrochemical and chemical industries for satisfying two basic requirements. The first is checking a new design for safety and operability problems whilst it is at the process instrumentation diagram stage. The second is the identification of latent safety and operability problems in existing plant which have yet to be revealed during operations.

A HAZOP study provides benefits to the owner and contractor as it:

- Identifies improvements for the safe operation of the process unit at an earlier stage in the project making it easier and usually significantly less expensive to make those changes.
- Reduces the chances of unplanned shutdowns.
- Significantly reduces time and costs for future HAZOP studies due to changes made to the process unit during construction or later revalidations (a government requirement in some parts of the world) with the use of the electronic version of the HAZOP study report.
- Helps answer questions during the training of operators and maintenance personnel about deviations or unusual scenarios that may occur in the operation of the process unit.
- Provides information for developing process-unit-specific operating and maintenance procedures.
- Identifies links to process equipment "outside battery limits"
- Provides guidance for developing mechanical integrity programs, including information required by the ANSI/ISA (USA) or IEC 1508 (Europe) instrumentation standards.
- Contributes toward demonstrating to the communities that potential hazards have been identified.

3. HAZOP Study

3.1 HAZOP Methodologies

UOP HAZOP studies are well documented and include all of the supporting information used during the HAZOP sessions in a way that is useful to the owner and the contractor.
HAZOP studies of UOP process units are conducted with a team of experts including a Chairman/HAZOP Team Leader, Process Design Engineer, Process Engineer, Technical Service Engineer, Control Engineer, and the Project Engineer. Specialists to be consulted on specific points may include: Electrical Engineer, Piping Engineer, Research Scientist, Civil Engineer, Chemist, and the Flowsheeting Specialist, as necessary (Figure 1).

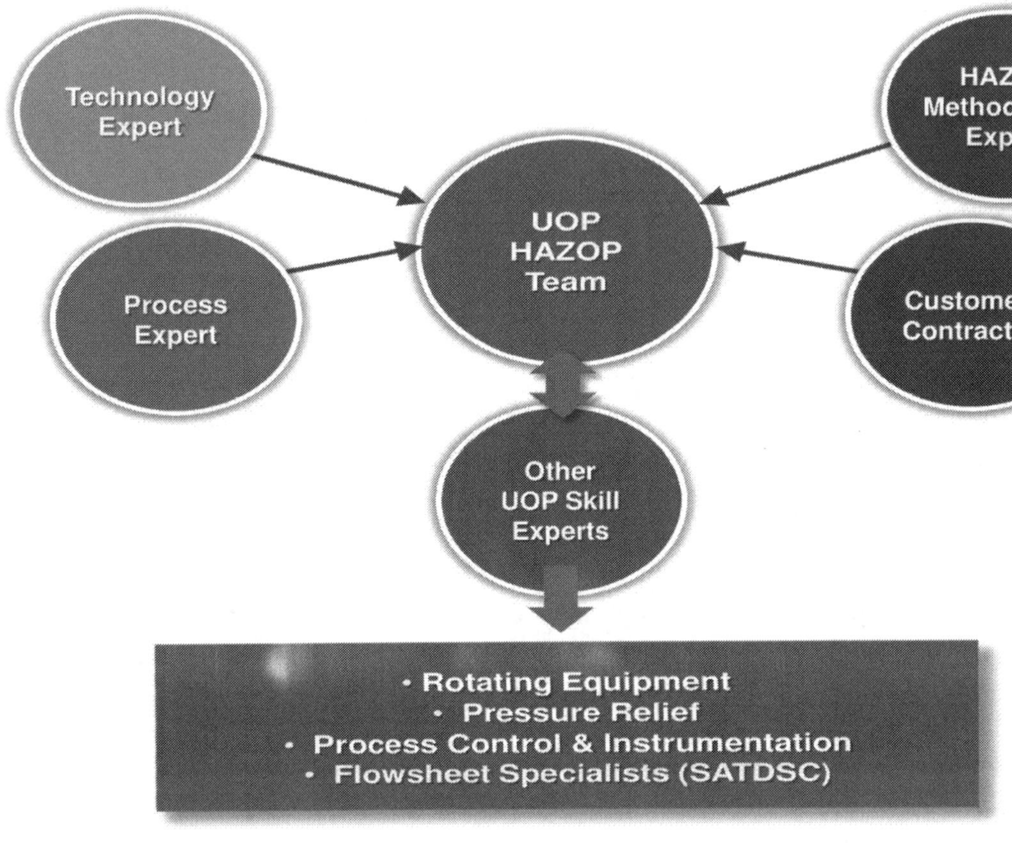

Figure 1. HAZOP Team

HAZOP is carried out as a structured study. The method relies on using guidewords (such as, no, more, less) combined with process parameters (e.g., temperature, flow, pressure) that aim to reveal deviations (such as less flow, more temperature and so on) of the process intention or normal operation. HAZOPs are best to be done early in the project. Figure 2 shows the cost of changes vs. project life cycle and one can see the ability to influence changes decreases over time while the cost of changes increases over time as the project progresses. The potential for adding value to the products of a project are obviously highest during the conceptual phase of the project and lowest during the finishing phase. Between these two extremes, the curve tends to follow a reverse "S" curve as shown in the figure. Conversely, the cost of making changes is lowest in the planning phases, but rises more and more steeply as the project progresses through the two production phases. In construction, for example, it has been suggested that the cost to make a change, or fix a safety issue, increases by ten times through each succeeding major phase.

If the Cost-to-Change curve is considered in conjunction with that of Adding-Value, the implications to management decision making become readily apparent. The intersection of the two curves probably represents the point at which a change in scope changes from a constructive opportunity into a destructive intervention.

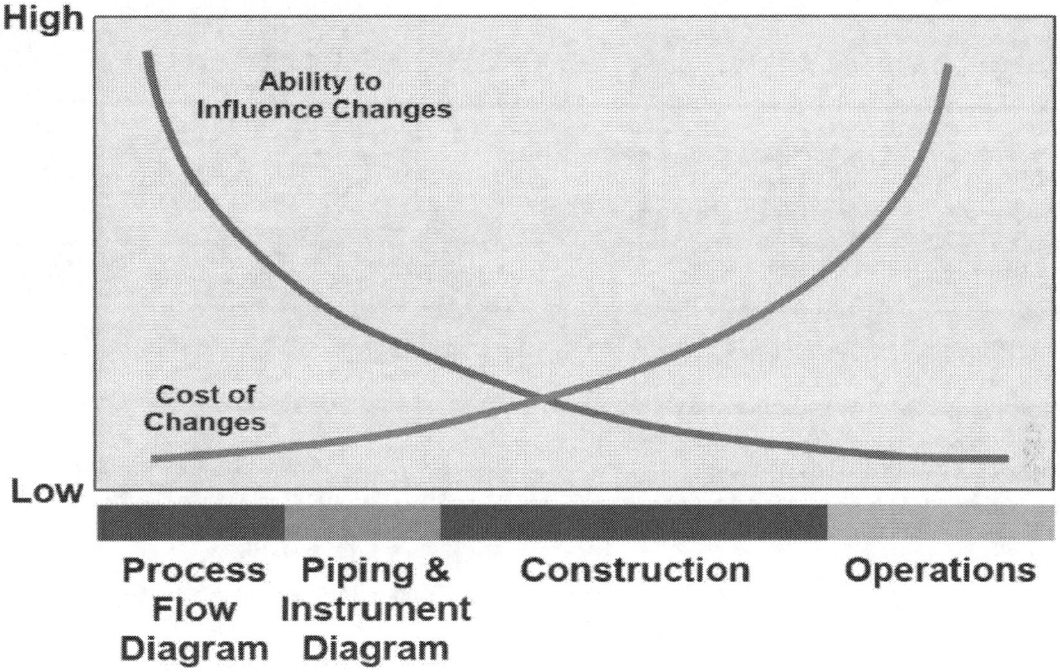

Figure 2. Cost of Changes vs. Project Life Cycle

3.2 HAZOP Consequences

The consequences are then risk rated based on their severity and likelihood. Figure 3 shows a Risk Ranking Matrix that is used in all of our HAZOP reviews. The Risk Ranking Matrix has four levels of risk which are the following:

1. Risk Ranking D: Risk not acceptable
2. Risk Ranking C: Risk acceptable with additional administrative or engineering controls which lower the risk ranking to "A" or "B"
3. Risk Ranking B: Risk acceptable with additional administrative or engineering controls
4. Risk Ranking A: Risk acceptable even if no further action taken.

If during the HAZOP a Risk Ranking of "D" or "C" is determined by the HAZOP team the HAZOP leader may engage the Flowsheeting Specialist to see if a dynamic simulation could determine a better understanding of the issue and the severity. The dynamic simulation may determine that the severity may not be as severe as the HAZOP team determined by qualitatively assessing the severity quantitatively and potentially move the Risk Ranking to an "A" or "B".

Severity:

1. Catastrophic (Death or sever Occupational illness; Major environmental impact; Production loss of more than six months; Facility/equipment damage of more than $10MM)
2. Severe (Injury or moderate occupational illness; Serious environmental impact or permit violation; Production loss of between one month and six months; Facility/equipment damage between $1MM and $10MM)
3. Moderate (Minor injury or occupational illness; Minor environmental impact; Production loss of between one week and one month; Facility/equipment damage between $100M and $1MM)
4. Low (No injury or occupational illness; No environmental impact; Production loss of less than one week; Facility/equipment damage of less than $100M)

Likelihood:

1. Frequent (Potential to occur frequently, many times per year)
2. Occasional (Potential to occur occasionally, once per year)
3. Moderate (Potential to occur under unusual circumstances, once or twice within the facility life time)
4. Unlikely (Could possibly occur or known to have occurred within the petroleum/petrochemical industry, but not likely over the facility lifetime)

Ranking:

Likelihood \ Severity	1	2	3	4
1	D	D	C	A
2	D	C	B	A
3	C	B	A	A
4	B	A	A	A

Risk Ranking	Definition
D	Risk not acceptable
C	Risk acceptable with additional administrative or engineering controls which lower the risk ranking to "A" or "B"
B	Risk acceptable with additional administrative or engineering controls
A	Risk acceptable even if no further action taken
NR	Not Ranked (insufficient information)

Figure 3. Risk Ranking HAZOP

The application of the HAZOP technique to a detailed process unit design is a complex and time consuming task. UOP has created a model where the HAZOP worksheets are pre-populated by an experienced HAZOP leader who during consultations with our skill experts will allow us to make the HAZOP reviews more efficient and less time consuming during the HAZOP meetings.

3.3 Using Dynamic Simulation in UOP HAZOP Reviews

In the past 30 years, much research effort has been dedicated to the development of the dynamic analysis. The dynamic analysis approach for a HAZOP of a process unit seems to be the most straightforward procedure which gives quantitative results.

Usually, the HAZOP analysis does not consider the duration and magnitude of the deviations generated during the operation. However, what exactly does the deviation 'less flow' mean: 80% or 10% of the usual operation value? Does the deviation occur as an immediate (step) decrease of the flow lasting five minutes or more, or is it only an impulse? Is this decrease continuous at some rate? Answers to these questions can be obtained using dynamic simulation which is one of the newer features. Consider the example where a valve is left open on a line going to a storage tank. Dynamic simulation will assist on the duration of how long it will take to overfill that tank even if there are multiple inlets to the storage tank. In dynamic simulation the extent of the deviations can be easily incorporated and possible consequences investigated.

The dynamic analysis serves for analyzing the time and direction of shifting from one steady state to another one due to a failure deviation. It is also very useful for the investigation of dynamic behavior of the system with respect to the time and length of the failure. The results of safety analysis can be used in HAZOP studies. HAZOP, with dynamic simulation, has the potential to become a very practical and robust tool. We will illustrate this by presenting examples of dynamic simulation for conventional column on loss of reflux, for UOP's Phenol™ Process operating pressure evaluation.

However, it is important to point out that the implementation of dynamic simulation for any chosen system is dependent on an adequate steady state model that is dependent on physical properties, kinetics, mass and heat transfer behaviors. During a HAZOP study, possible deviations are generated by rigorous questioning, prompted by a series of standard "guidewords" applied to the intended design. After the guideword is matched with a parameter, a deviation is generated. The next step in the HAZOP study is to look for potential consequences. At this moment, the application of dynamic analysis is useful in order to find quantitative consequences.

The HAZOP study is traditionally a qualitative study. By definition, a dynamic simulation is the imitation of the operation of a real-world process or system over time, which means that in principle it should be a realistic way of representing an actual process. Combining HAZOP with dynamic simulation does provide the means for investigating and demonstrating the consequences of deviations from normal operating conditions. Dynamic simulation now provides the UOP HAZOP team the ability to quickly investigate and test the effectiveness of various suggested strategies dealing with high risk situations (intolerable levels of risk). Figure 4 shows the magnitude of risk from intolerable risk to ALARP "as low as reasonably practicable" to tolerable risk. To move from an intolerable risk to ALARP decreases the risk with increasing cost. In process safety we cannot accept intolerable risk so one must move from the intolerable level of risk to a risk level of ALARP or tolerable risk. In the past this would involve increasing cost, but we will demonstrate that by using dynamic simulation we may be able to move from an intolerable risk to ALARP by showing a less severe consequence with no significant additional cost.

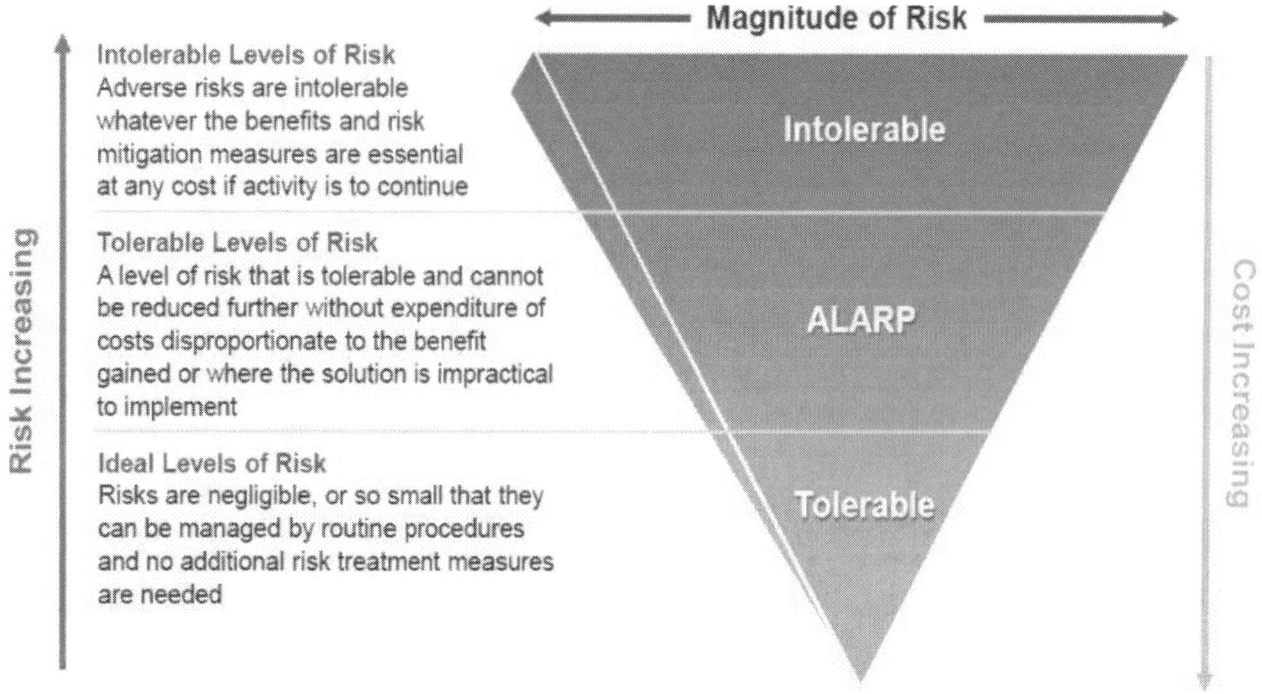

Figure 4. ALARP

By having a dynamic simulation of the process as a support tool, an extensive, easier and more complete study can be accomplished. That provides a systematic screening of process deviations associated with possible high risk events, determining the threshold values that may lead to such events and enabling the examination of a particular design for the adequate safe range of operation. Dynamic simulation should be seen as a tool that complements the traditional HAZOP procedure, it does not replace it. There are still many processes that cannot be modelled accurately enough due to a lack of enough quantitative information, particularly in emergency situations.

Process safety is paramount in the secure and viable operation of any refinery, petrochemical and chemical process plant. Many techniques and methodologies have been defined over the years to undertake steps that will ensure a safe operating environment at a chemical facility. However, until now, process simulation has found very little or no use in safety-related studies such as HAZOP. In this paper, we introduce a systematic framework based on a quantitative HAZOP (that is, HAZOP supported by dynamic simulation related to process malfunctions.) The quantitative HAZOP differs from standard HAZOP in documenting results, classification of frequency and consequences of process deviations, and application of a risk potential matrix (Risk Ranking Matrix).

Dynamic simulations can be used to support the HAZOP in the following ways:

- An important part of the HAZOP procedure is to think of the consequences of a certain deviation

- For complex and nonlinear systems, it is not straightforward to assess the consequences. Hence, the use of dynamic simulation with the deterministic models can be helpful in assessing the effect of faults on the operations and dynamics of the process.
- The purpose of the methodology is to determine the effect of operational disturbances on the safety of the plant and devise ways to reduce the risk of the consequences.

HAZOP can provide qualitative answers regarding the magnitude of the deviations that will lead to severe consequences. It cannot provide the time it takes to reach a "no return" stage of an accident after the deviation has occurred, or the action that can be taken in order to prevent the accident. Dynamic simulation can provide this information that a traditional HAZOP could not provide. Dynamic simulation in HAZOP can provide quantitative assessment of the consequences of abnormal operating conditions.

4. Dynamic Simulation and HAZOP Analysis

4.1 Background

Steady-state process templates (flowsheets) are used extensively in the engineering design workflow for all UOP process technologies. UOP engineers utilize the Honeywell **UNISIM®** Design Suite of process modeling tools (See References [1-3]) in the preparation of a process design package to our customers called Schedule A. This suite of tools allows for very robust development of steady state and dynamic process models.

In the process phase of the Schedule A workflow we have developed a large database of validated steady-state process templates for most product offerings in the refining and downstream petrochemical areas (see Figures 5 and 6). Modifications to these templates are made based on customer specific requirements. All the equipment and piping associated with a project are sized based on flow rates calculated from steady-state heat and mass balances which interface all of our engineering sizing tools. Simulation results are downloaded to our standalone hydraulic tools to determine line sizes and generate tabulation reports. If needed, various parts of the steady-state template can be checked using the equivalent dynamic simulation template.

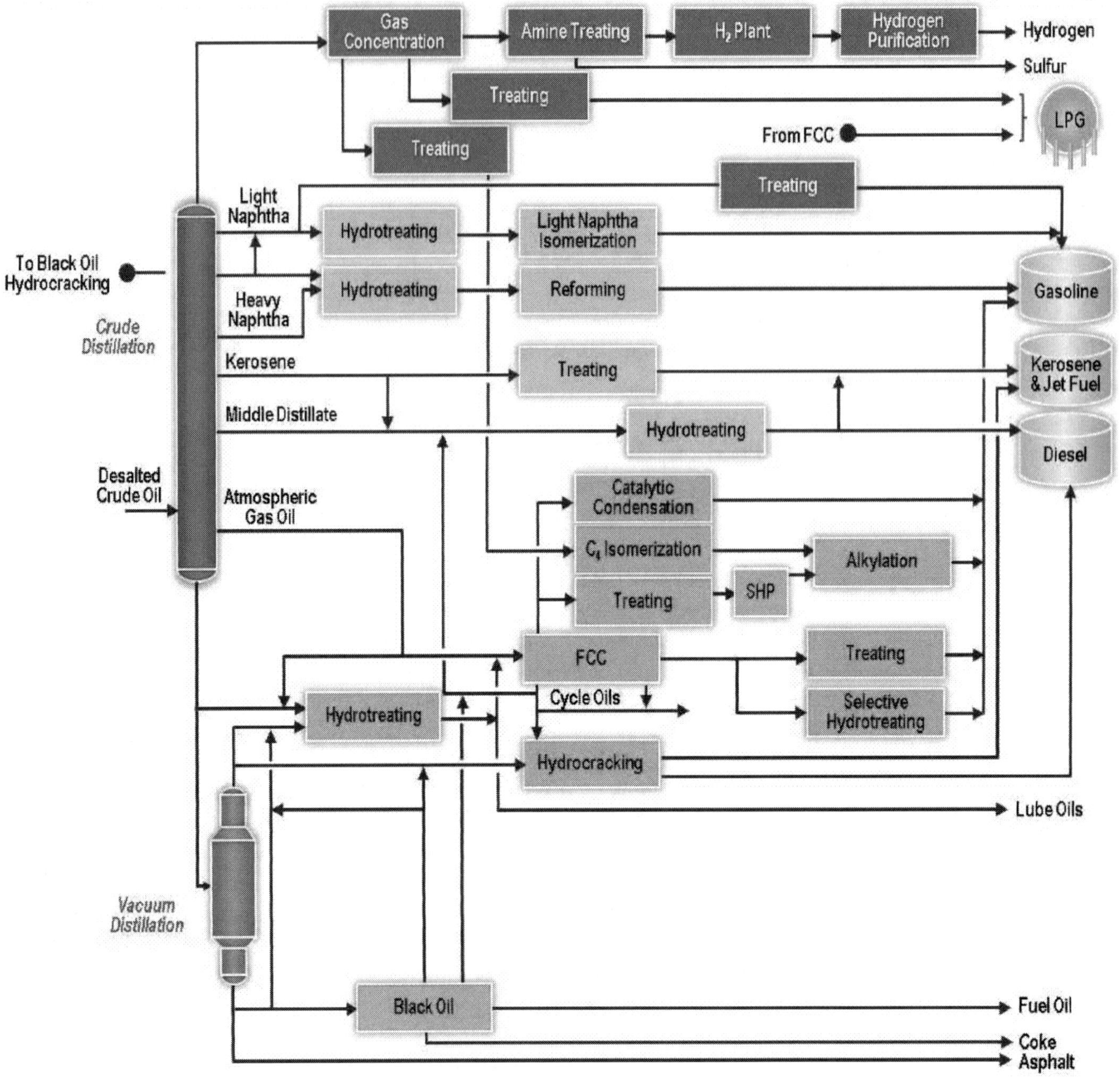

Figure 5. UOP Refining Processes

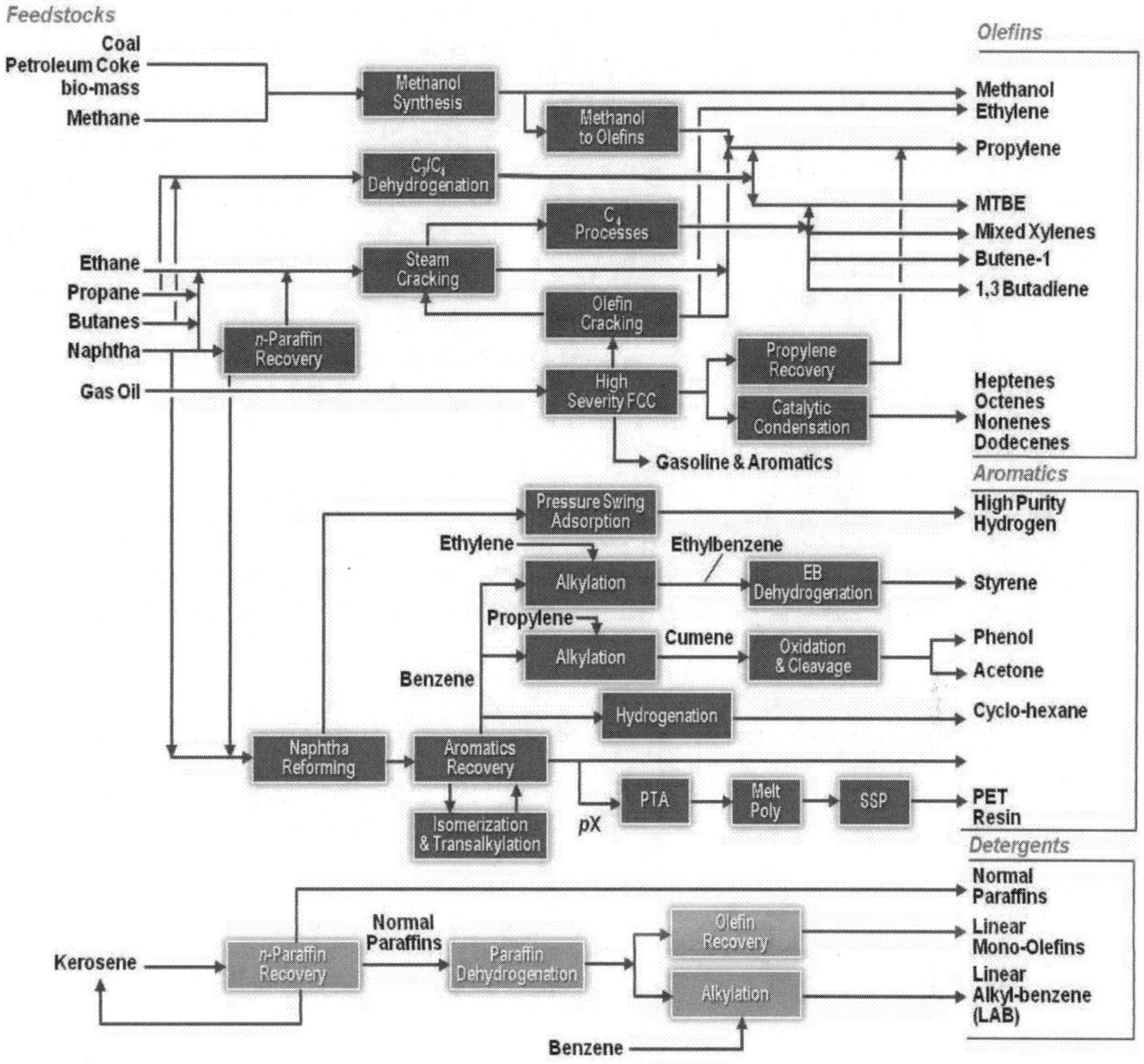

Figure 6. UOP Petrochemical Processes

4.2 Dynamic Simulation Templates

Transitioning the converged steady template to a dynamic simulation template requires a different approach towards specifying unit operations. In our implementation of dynamic templates a sequence of important actions are required to insure a successful high fidelity dynamic analysis:

1) Converge a realistic steady-state simulation of the problem to be studied.
2) Add valves and pipes to ensure pressure-driven flow between unit operations.
3) Set terminal feed and product pressures.

4) Determine approximate or include exact dimensions of all process equipment.
5) Specify sizing data for all pipe runs (lengths, diameters, etc.). Size the CVs.
6) Include nozzle elevations for all valves, piping and process equipment.
7) Add all necessary controllers (FICs, TICs, etc) to all control loops.
8) Run the dynamic simulation: tune the controllers, check pressure drops, and line out the model to match the original steady-state performance.
9) Add events to the Event Scheduler to model upsets and other conditions.

The above steps need to be completed only once and serve as the starting point for dynamically evaluating all normal and upset operating scenarios.

Each of our dynamic models is preconfigured to include upset conditions, strip charts and tabulate historical data. In **UNISIM®**, the event scheduler is used to model the preconfigured dynamic upset scenarios. Additional upsets are also manually activated as a dynamic run is proceeding during time integration.

Over the years we have established a strategy and timetable to implement dynamic simulation methods into our current UOP process work flow. We use dynamic process templates to supplement the work as needed during the Schedule A projects. Various sections of these dynamic templates are used to identify specific design issues for each project (e.g. a project HAZOP review, emergency shutdown events, customer inquiries, etc.) that normally would not be recognized through the use of steady state simulators.

Sections of the dynamic templates are utilized to explore transient behavior around individual equipment, e.g. reactors, columns, etc. There are many different levels of fidelity possible when creating dynamic flow sheets. In UOP engineering we use dynamic templates that implement the basic regulatory control schemes shown on all our process flow diagrams for all UOP process units.

From a full dynamic template of a UOP process unit, the process engineer is able to break down the process flows and create smaller process snapshots for analysis. To consider and evaluate time dependent behavior in our Schedule A work process, a separate group of flowsheeting skill specialists are utilized on Schedule A projects to assist in preparing and designing these dynamic flowsheets.

We have realized substantial benefits in leveraging our expertise in steady-state design modeling to include dynamic design modeling as well. We have recently included our flowsheeting specialists to support our internal HAZOP team reviews. This support includes the preparation and evaluation of dynamic models as needed from the working HAZOP worksheets. Supporting data (e.g. strip charts, tabular historical data of key events / new hazards / safety issues) from these dynamic runs are included in the reports. Additional dynamic runs are made to address any customer concerns after the reports are issued.

With the use of our dynamic templates from the steady-state design templates we can now quantitatively test the effectiveness of proposed strategies for emergency situations. Furthermore, dynamic analysis has helped us identify new unknown hazards and safety issues.

We can now accurately determine the consequences of any deviations from normal design. We have also included dynamic analyses in our HAZOP reviews of new technologies proposed by the R&D Process Design and Development groups.

4.3 Dynamic HAZOP Analysis

Over the years we have performed many dynamic analyses to support our internal HAZOP reviews. Most of these are very detailed and require specific customer information such as "as built" equipment and line size data as well as detailed valve performance data. Other flowsheets are more simply configured and are used occasionally in our HAZOP analysis to confirm and/or support the qualitative conclusions normally encountered during the reviews. The following examples illustrate some process scenarios typically encountered during HAZOP reviews.

4.3.1 Conventional Column – Loss of Reflux

The majority of process units utilize two and three phase conventional separation columns. Figure 7 shows a conventional two phase column with a full reflux condensing circuit with vapor overhead product and a vertical thermosiphon reboiler.

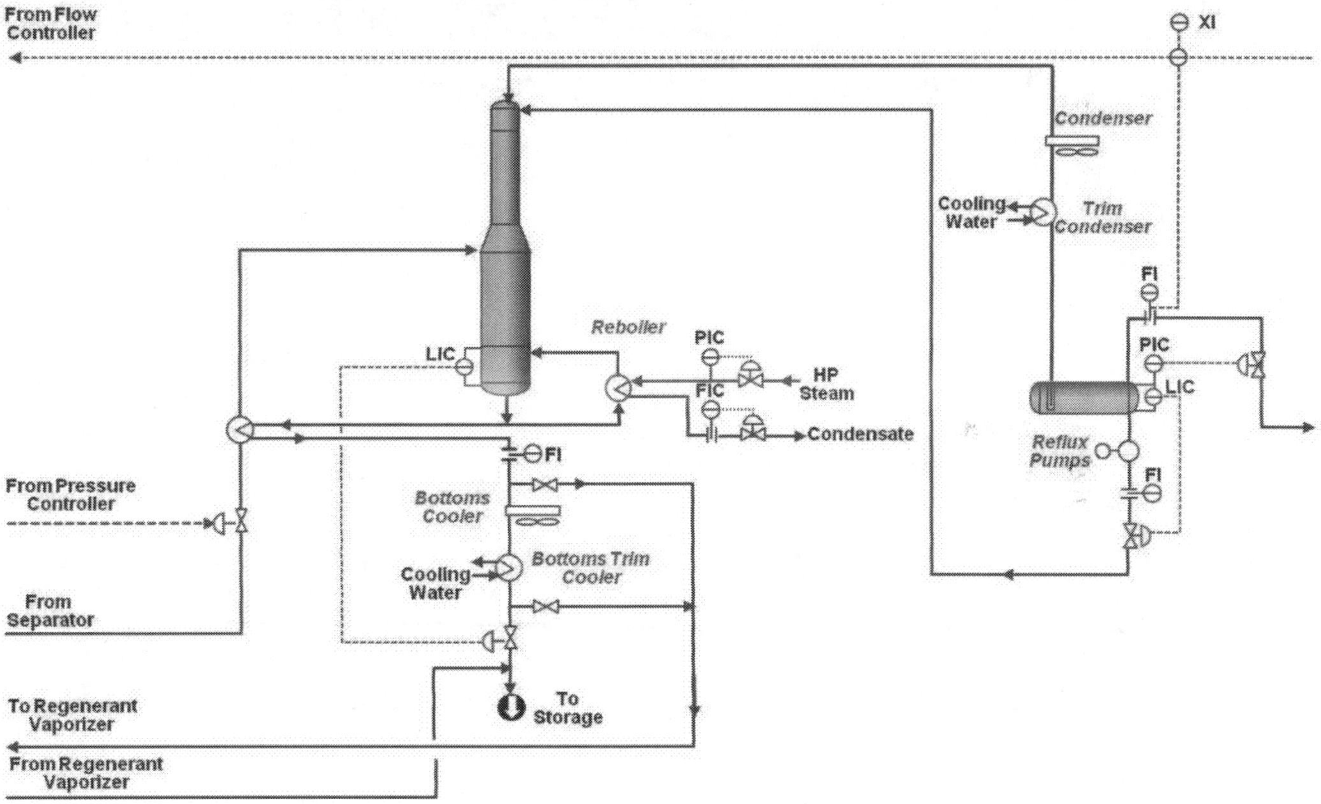

Figure 7. Conventional Column

In a typical HAZOP analysis, there could be potential hazards associated with the installation of engineer controlled emergency isolation valves. In those situations, an evaluation will be conducted to determine whether the installation of the engineer controlled emergency isolation valves provides the safer option for personnel, the environment or surrounding equipment than if these valves were not installed. As shown in Figure 8, these isolation valves can be local or remote manually operated and are typically not interlocked to other instrumented devices.

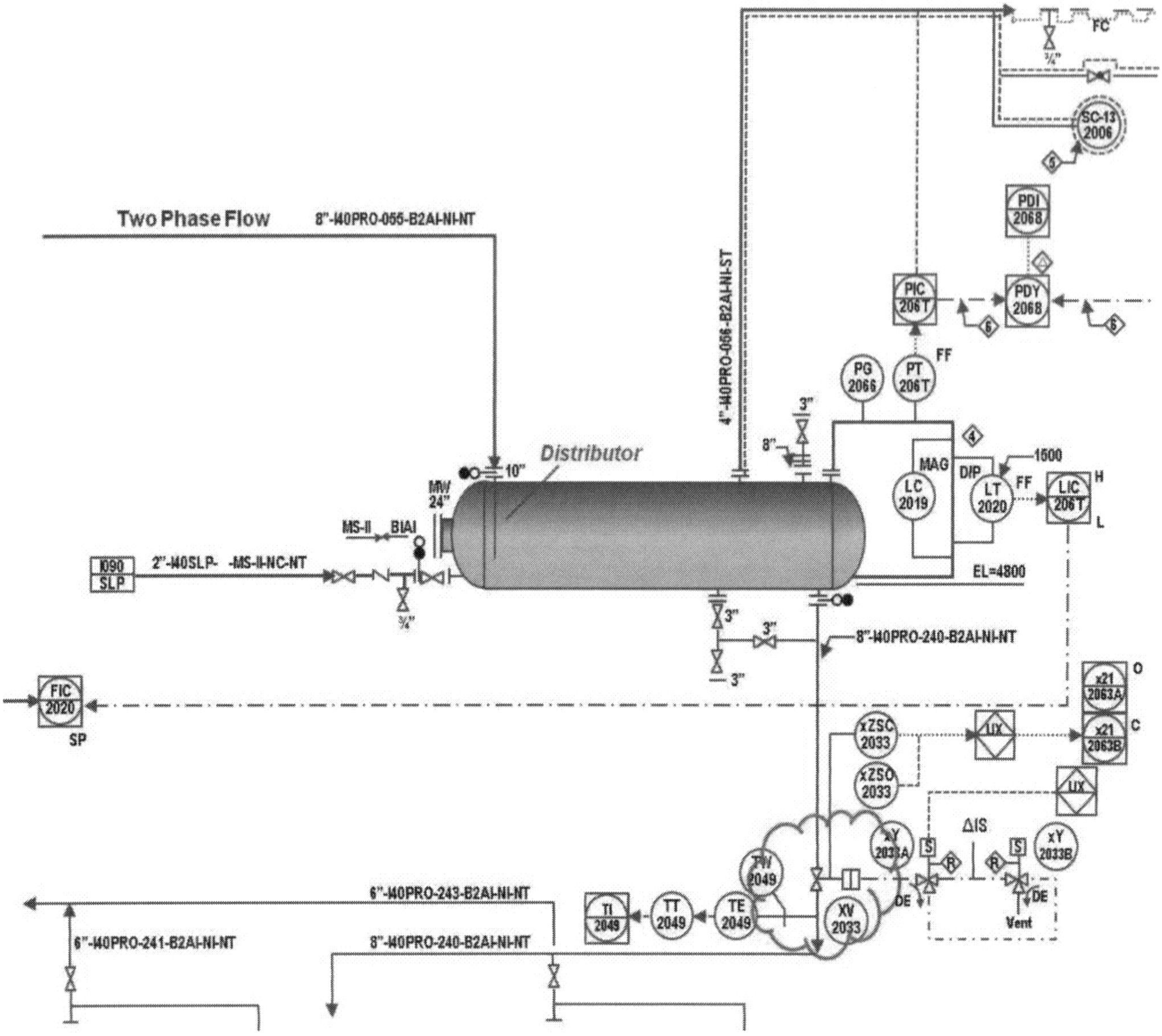

Figure 8. Expanded Piping and Instrument Diagram View – Column Receiver

From our database of **UniSim®** templates, a dynamic model in of the conventional column shown in Figure 9 follows our standard process flow diagram view of equipment, piping and basic piping and instrument diagram control designations. This model has event schedules preconfigured with a number of upset scenarios. If needed, adhoc scenarios can easily be added to the model during the HAZOP team analysis.

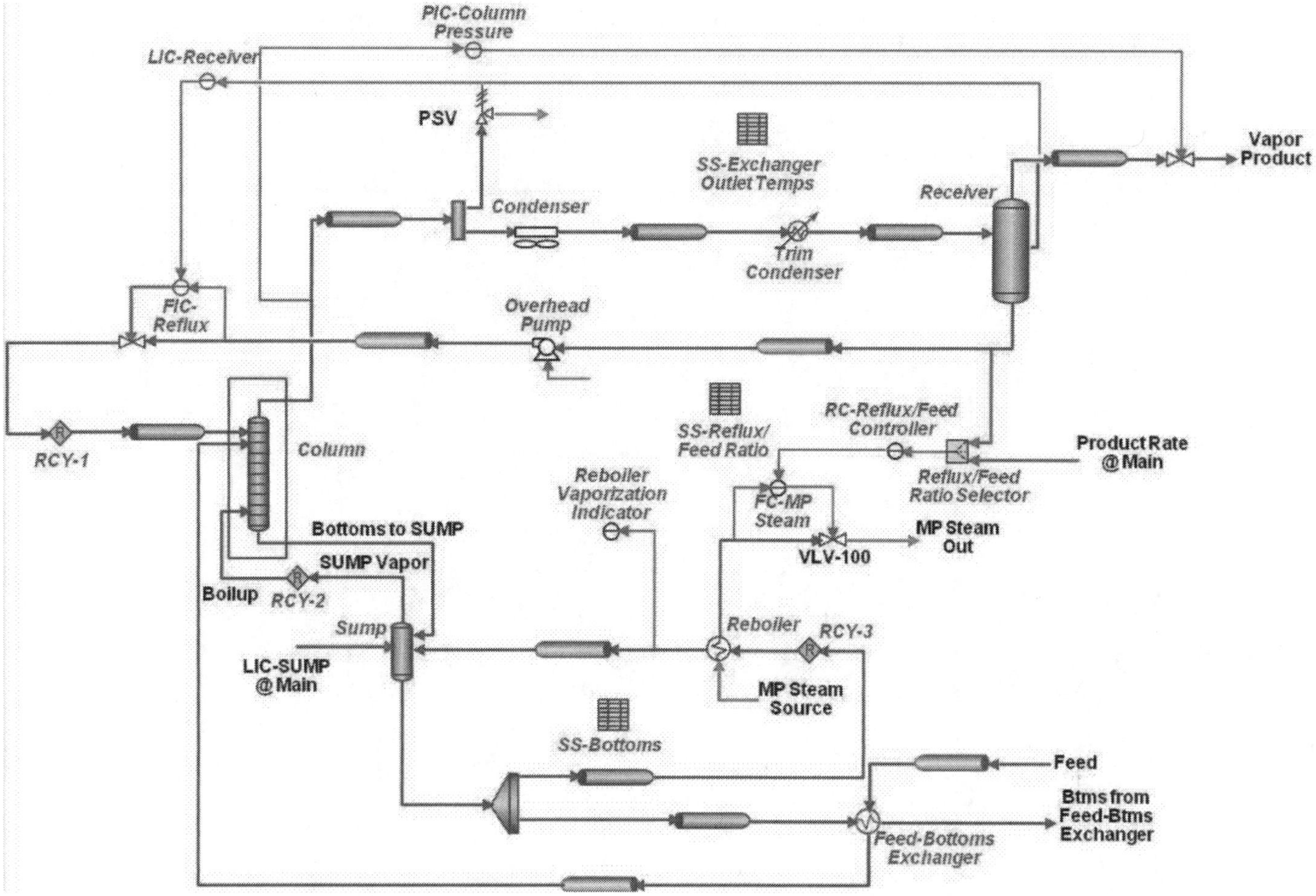

Figure 9. UNISIM® Dynamic Conventional Column Model

Figures 10 through 12 illustrate via strip charts, a time sequence of events when a manually operated isolation valve is inadvertently closed during normal operation. First the column is lined out to steady state conditions; then the isolation valve is triggered to fully close. As Figure 10 indicates, the overhead vapor from the column immediately spikes upward then moves downward rapidly as it is being displaced by the increasing liquid level in the receiver. After 3.5 minutes the liquid level completely fills the receiver and overflows through the vapor line at a large rate (see Figure 11). Also, with no reflux flow returning to the column, the sump liquid level in the column starts to decrease. As indicated in Figure 12, the liquid level in the sump ultimately drops to zero after another 7.5 minutes. During this upset event the column operating pressure increased close to the pressure safety valve set pressure but did not open.

Obviously during normal operation, the appropriate trip activation or operation actions in response to safety alarms would have been in place to mitigate any severe operating scenarios such as the one performed in this simple example. As this example shows, the continued use of dynamic simulation in our HAZOP reviews has allowed us to quantitatively evaluate the effect of time on process safety issues particularly as they relate to equipment failures, startup and shutdown scenarios. We can also readily incorporate vendor data and validate their performance. For instance we use actual valve flow coefficients C_V and stroke time data for all control and isolation valves whenever dynamic simulation work is being implemented in designs.

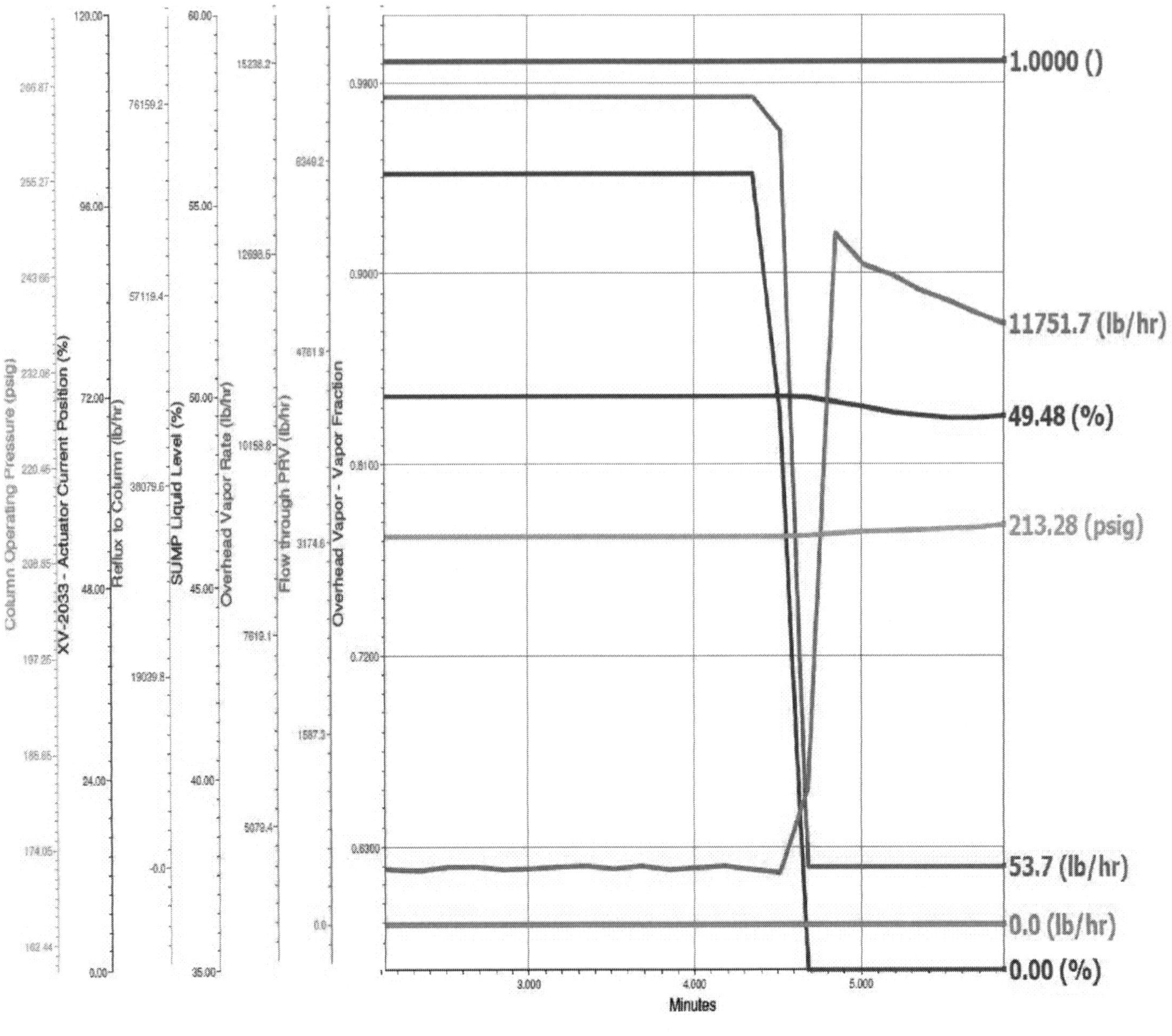

Figure 10. Dynamic Upset – Loss of Reflux: Isolation Valve Closes

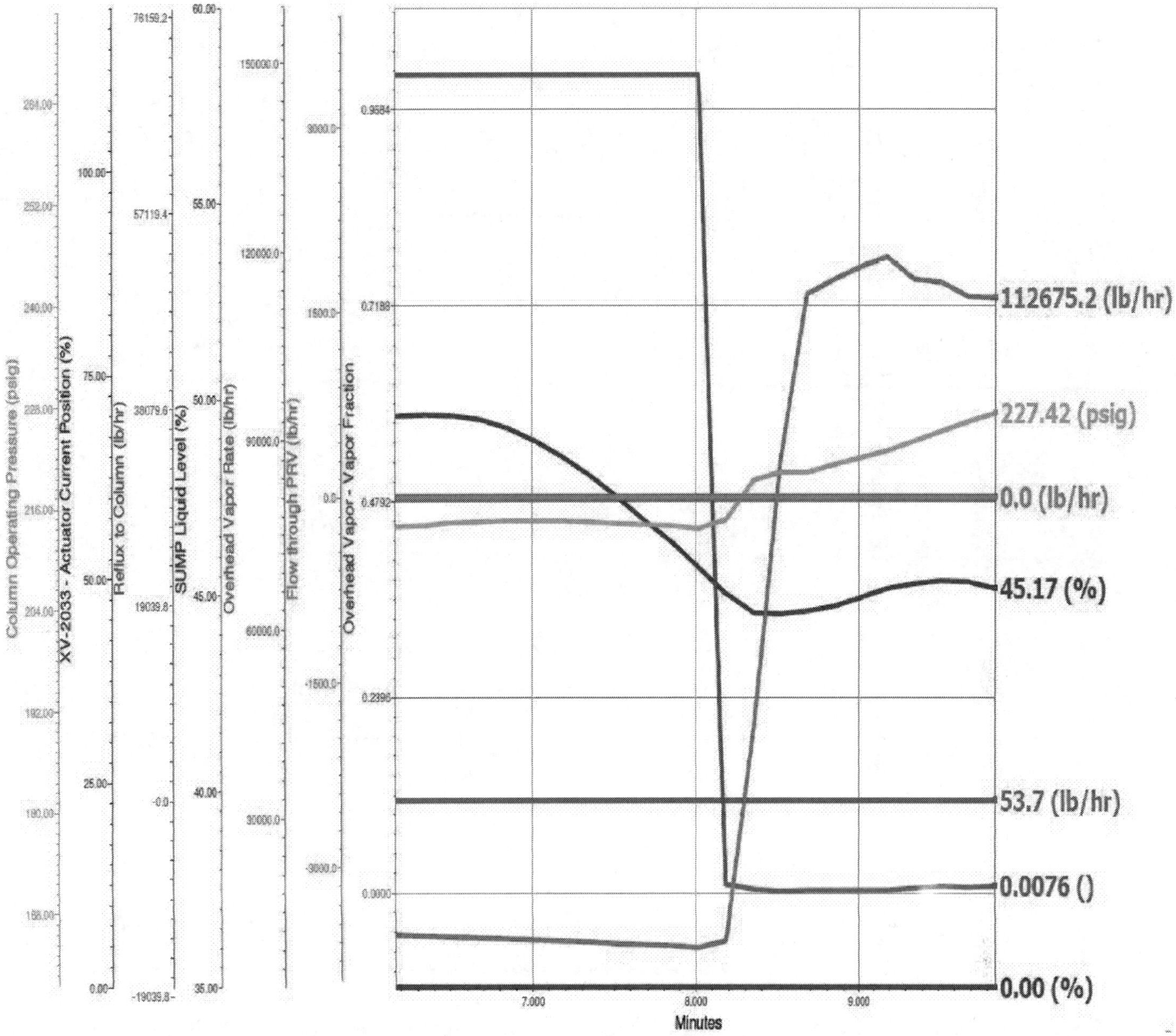

Figure 11. Dynamic Upset – Loss of Reflux→Receiver Liquid Full > 3.5 minutes

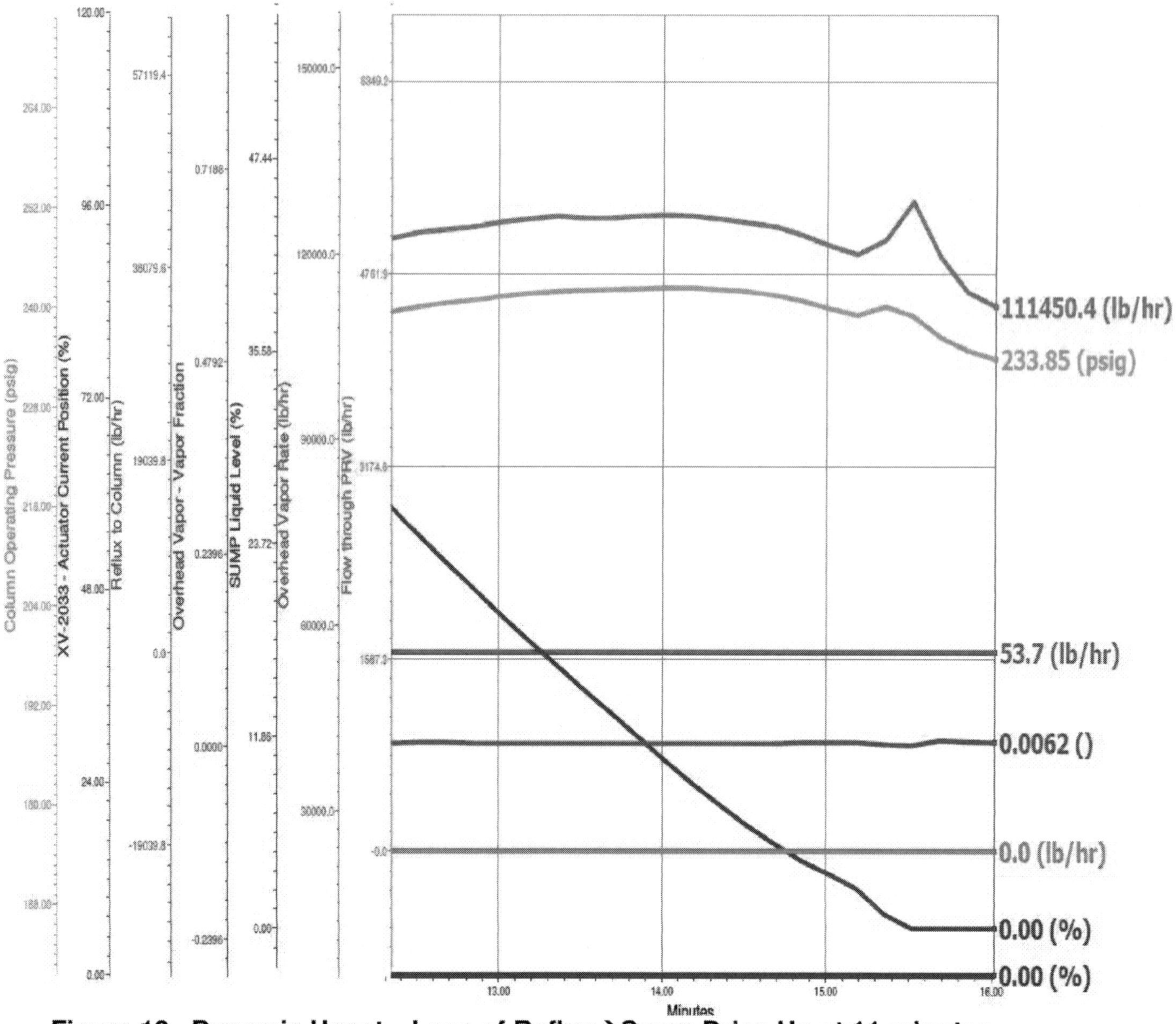

Figure 12. Dynamic Upset – Loss of Reflux→Sump Dries Up at 11 minutes

4.3.2 UOP Phenol Process - Operating Pressure Evaluation

Dynamic studies are also performed for customers based on their own internal HAZOP reviews. One of our customers requested that we evaluate the possibility of running oxidizer tanks at higher pressures so as to increase product flow.

The process flow diagrams for the oxidizer section were taken from our library of design templates and modified to include all customer specific information. Figure 13 shows a dynamic model representation of a section of the oxidizer feed train. Piping and instrument diagrams were also used to accurately define the placement of all control and isolation valves as well as control logic systems in the dynamic simulation. Some elevations and line sizing information were also extracted from these drawings. The remaining data was obtained from previous hydraulic reports. Equipment sizes were also incorporated into the model.

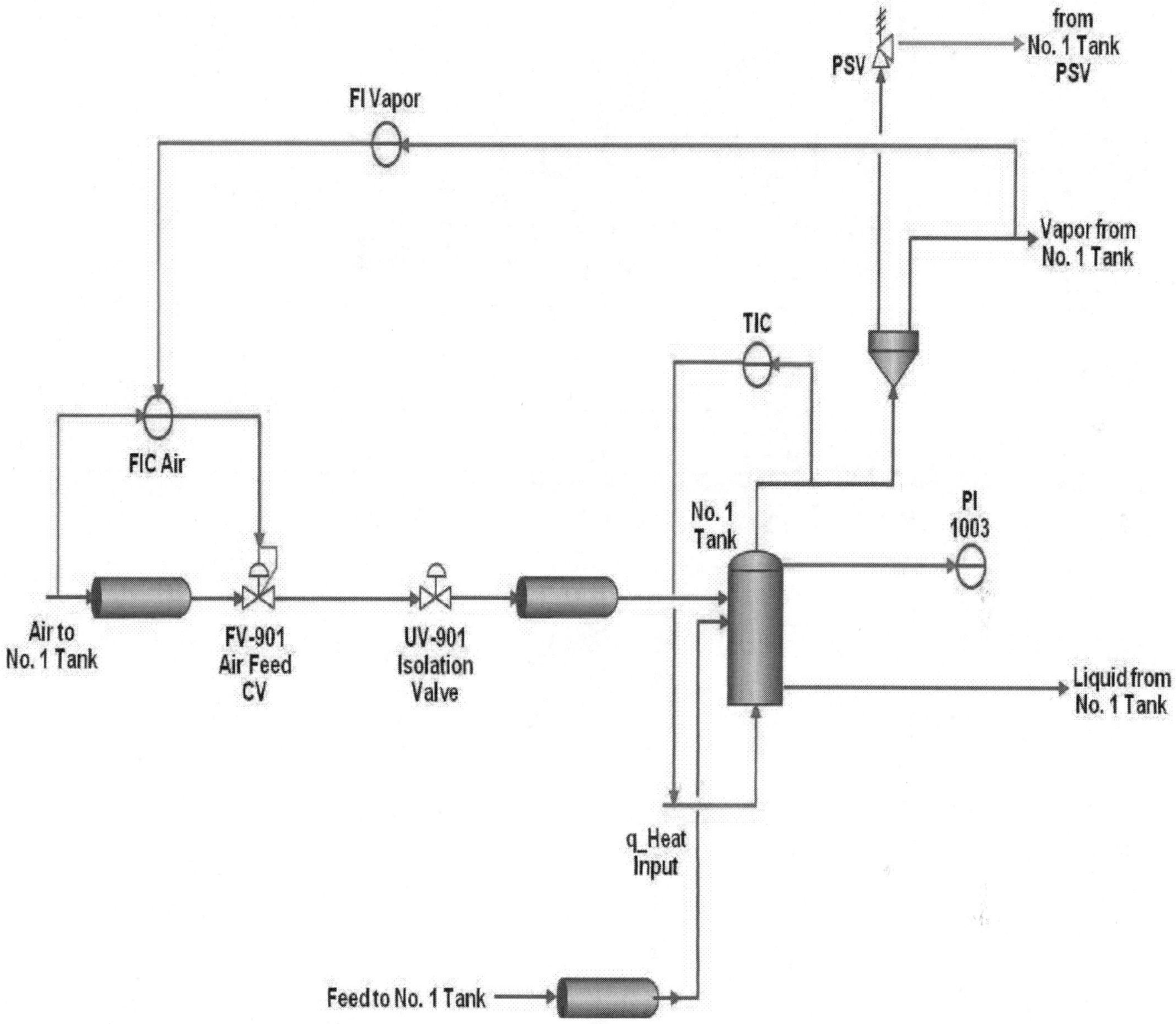

Figure 13. UNISIM® Dynamic Oxidizer Model

Figure 14 shows a typical air feed valve / isolation valve configuration. These valves utilized the "as built" detailed valve performance data provided by the customer.

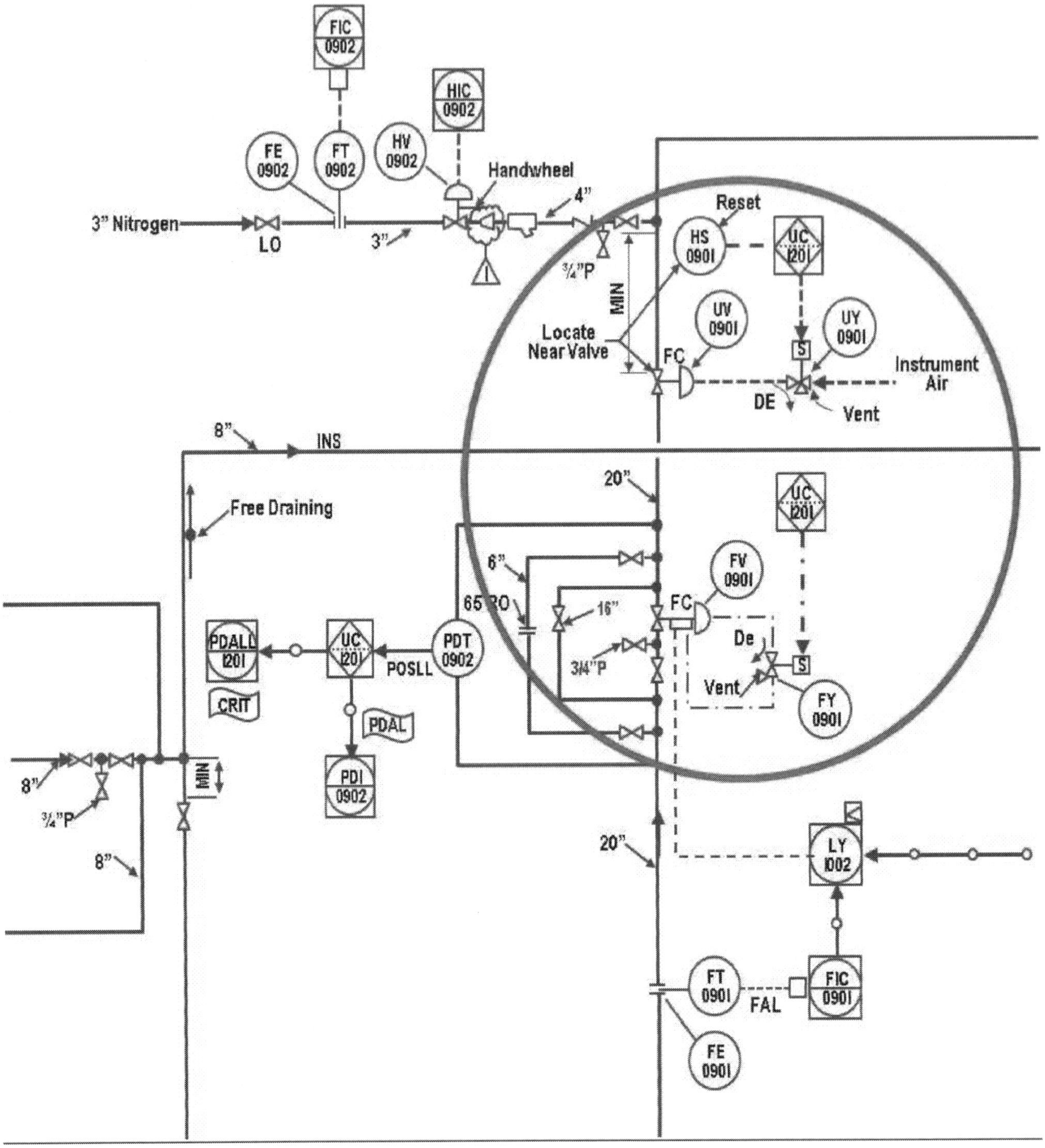

Figure 14. Piping and Instrument Diagram View – Air Feed / Isolation Valves

The results of the dynamic analysis showed that the customer can indeed run their unit at higher operating pressures. A higher pressure trip point was recommended for the oxidizer shutdown system which would still prevent overpressure of the oxidizer tanks from reaching their set relief pressures. Based on the "as built" stroke time valve data obtained from the customer, the results shown in Figure 15 indicate that the valves can perform quickly in avoiding pressure relief of the oxidizer tanks at the highest possible operating pressure.

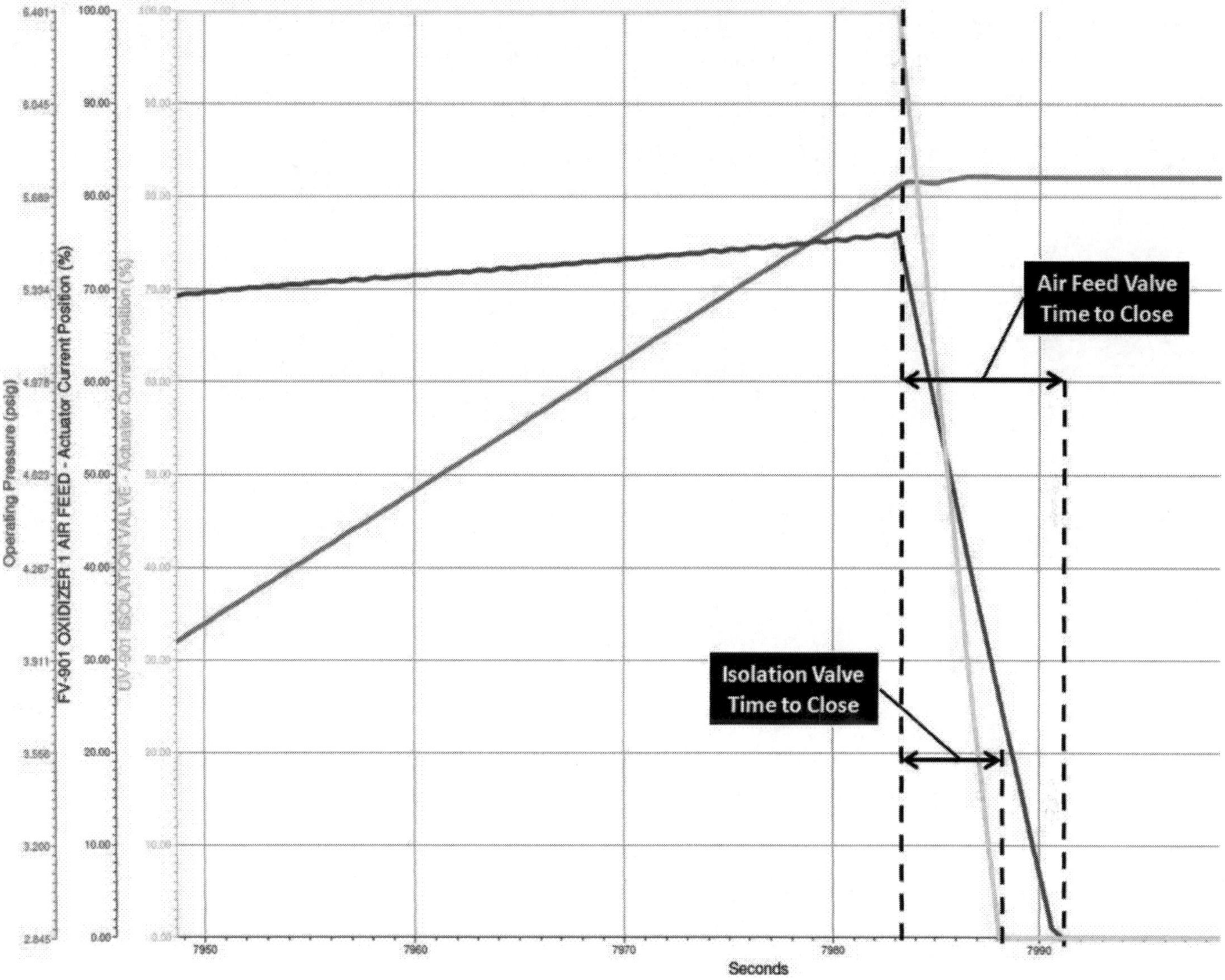

Figure 15. Dynamic Upset – Valve Stroke Time

5. Conclusion

The combination of traditional HAZOP analysis with dynamic simulation facilitates accurate determination of the consequences of deviations from normal design. Potential hazards and safety issues can be better identified. By dividing the problem model into smaller, independent flowsheet sections, the level of complexity associated with the dynamic simulations can be reduced.

HAZOP teams can now test the effectiveness of mitigation strategies for many emergency situations. The use of dynamic simulation as a process safety / risk assessment tool provides significant added value to HAZOP reviews and reports.

6. References

[1] Process Design Simulation: Fundamentals – Dynamic Modeling Using **UNISIM®** Design Suite Implementation, PDS-4528 R380 Student Guide, Rev 01.0, 07/2010.

[2] **UNISIM®** Design Dynamic Modeling-Reference Guide, May 2012 R410 Release.

[3] **UNISIM®** Operations Guide, May 2012 R410 Release.

How close is my operation to a Major Accident?

Abouzar Yousefi
Repsol
Mendez Alvaro, 44, 28045 Madrid, Spain
abouzar.yousefi@repsol.com

Manuel Rodriguez Hernández
Universidad Politécnica de Madrid
José Gutiérrez Abascal, 2 28006 Madrid, Spain
manuel.rodriguezh@upm.es

Jaime Martin Juez
Repsol
Mendez Alvaro, 44, 28045 Madrid, Spain
jmartinj@repsol.com

Prepared for Presentation at
American Institute of Chemical Engineers
2015 Spring Meeting
11th Global Congress on Process Safety
Austin, Texas
April 27-29, 2015

UNPUBLISHED

How close is my operation to a Major Accident?

Abouzar Yousefi
Repsol
Mendez Alvaro, 44, 28045 Madrid, Spain
abouzar.yousefi@repsol.com

Manuel Rodriguez Hernández
Universidad Politécnica de Madrid

Jaime Martin Juez
Repsol

Keywords: Major Accident, Process Safety, Safety Indicator, Accident Model,

Abstract

Despite many efforts made to prevent major accidents in upstream industry, these accidents continue to occur. It seems that our thinking about how accidents happen and all the management systems based on that approach, while reasonably successful, does not enable us to achieve the goal of zero accidents.

Different accident models have been introduced to explain how accidents happen. Accident models have evolved over time from being linear and deterministic to be non-linear and non-deterministic. These models not only describe how potentially hazardous situations develop and what processes form the basis for this development, but also serve the goal of helping to define measures to prevent and control them.

To understand how effective are these preventive and control measures, safety performance indicators can be used. These indicators are of a reactive, or a proactive nature, depending on the characteristics of the indicators. Reactive indicators or lagging indicators are used more commonly in upstream operations. These indicators are post-event types and generally show the safety level of the past time.

Safety level is not static and depends on many risk factors that change in presence and intensity over location and time. On the other hand major accidents have a very low frequency. Therefore reactive indicators based on these events cannot neither indicate how safe the operation is in real time nor warn how close the operation is to a major accident. For this reason proactive indicators or leading indicators are necessary to provide enough information at the right time to decision makers before a major accident happens.

Considering various accident models, it is recommended to use an adequate model for upstream operation and develop methods for the establishment of indicators, which can unveil early warnings of major accidents by measurements of underlying causes and contributing factors to major accidents.

1. Introduction

Major accident is defined in different sources. As an example International Association of Oil & Gas Producers (IOGP) has defined a major incident as an unplanned event with escalation potential for multiple fatalities and/or serious damage, possibly beyond the asset itself. Typically there are hazardous releases, but also include major structural failure or loss of stability that could put the whole asset at risk [1]. Also, it has been defined by governing regulations in UK [2], Norway [3], Australia [4] and others. While each definition differs slightly, a major accident is generally described as a fire, explosion, or release of a hazardous chemical substance that involves the death or serious injury to multiple individuals, severe environmental damage and/or loss of major financial assets [5].

Major accidents continue to occur in the process industry and often have serious consequences. Over the last generation, successive major accidents, from the explosion at Flixborough (1974), the deadly toxic gas release in Bhopal (1984), the explosion at Piper Alpha (1988) to more recent examples including the explosions at BP Texas City (2005), Buncefield (2005) and the blow-out and subsequent environmental disaster in the Gulf of Mexico (2010) have raised concerns with the public, stakeholders and regulators [6]. The Gulf of Mexico accident (2010) has highlighted a number of serious problems in scientific thinking about safety. One of these is that our current thinking about how accidents happen and all the management systems based on that approach, while reasonably successful, does not appear to enable us to achieve the goal of zero accidents [7].

Undoubtedly the main reason we want to understand how accidents happen is because we wish to prevent future ones [8].The best way to do this is to have a solid theoretical basis that can be shown both to describe what actually happens and to serve the goal of helping define effective preventive measures; the categorization of causal and contributing factors and the understanding of the mechanisms should be explicitly related to opportunities to prevent, remedy and improve. Theories about how accidents happen have evolved over time and different accident models have been introduced from being linear and deterministic to be non-linear and non-deterministic. In section 2 a summary of accidents models is provided.

On one hand, to know how close an operation is to a major accident, it is necessary to recognize how developed the path to a major accident is in that operation considering accident models. On the other hand, safety is often defined as the absence of unwanted events [9]. In other words, when safety is missing unwanted events occur. Such definition of safety is, however, indirect rather than direct since safety is defined by what happens when it is absent or missing. Properly speaking, it is therefore definition of lack of safety (or unsafety) rather than of safety. One consequence of this is that safety management relies on measurements that refer to the absence of safety rather than to the presence of safety [10]. Counting accidents or calculating frequency rates of accidents are examples of measurement based on the absence of safety. In section 3 a summary of safety measurements is provided.

Although these measurements are useful, they are not able to indicate how close an operation is to an accident in general or to a major accident in particular. Therefore considering various accident models, in section 4 it is recommended in detail how to use an adequate model for upstream operation and develop methods for the establishment of indicators, which can unveil early warnings of major accidents by measurements of underlying causes and contributing factors to major accidents.

2. Accident models

The historical development of accident models and various approaches for accident analysis have been discussed by engineers, scientists, cognitive psychologists, and sociologists. One of the earliest accident causation models is the Domino theory proposed by Heinrich in the 1940s [11], which describes an accident as a chain of discrete events which occur in a particular temporal order. This theory belongs to the class of sequential accident models or event-based accident models, which underlie most accident models such as Failure Modes and Effects Analysis (FMEA), Fault Tree Analysis (FTA), Event Tree Analysis, and Cause-Consequence Analysis [12]. These models work well for losses caused by failures of physical components or human errors in relatively simple systems. Typically, in these models, causal factors in an accident which are not linked to component failures are classified as human error as a kind of catchall or "garbage can" [13]. These models are limited in their capability to explain accident causation in the more complex systems that were developed in the last half of the 20th century.

In the 1980s, a new class of epidemiological accident models tried to explain accident causation in complex systems. Epidemiological models regard events leading to accidents as analogous to the spreading of a disease, i.e., as the outcome of a combination of factors, some manifest and some latent, that happen to exist together in space and time. Reason's Swiss cheese model [14, 15] is a major contribution to this class of models, and has greatly influenced the understanding of accidents by highlighting the relationship between latent and immediate causes of accidents.

Sequential and epidemiological accident models are inadequate to capture the dynamics and nonlinear interactions between system components in complex sociotechnical systems. New accident models, based on systems theory, classified as systemic accident models, endeavor to describe the characteristic performance on the level of the system as a whole, rather than on the level of specific cause-effect "mechanisms" or even epidemiological factors [16]. Systemic safety models have their roots in systems theory and cybernetics. Systems theory includes the principles, models, and laws necessary to understand complex interrelationships and interdependencies between components (technical, human, organisational and management) of a complex system. Safety models based on systems theory view accidents as emergent phenomena that arise from interactions among system components, where the interactions may be nonlinear and involve multiple feedback loops [17].

A major difference between systemic accident models and sequential/epidemiological accident models is that systemic accident models describe an accident process as a complex and interconnected network of events while the latter describes it as a simple cause-effect chain of events. Two notable systemic modelling approaches, Rasmussen's hierarchical sociotechnical framework (AcciMap) [18] and Leveson's Systems-Theoretic Accident Model and Processes (STAMP) model [19], endeavor to model the dynamics of complex sociotechnical systems.

Modern technology and automation has significantly changed the nature of human work from mainly manual tasks to predominantly knowledge intensive activities and cognitive tasks. This has created new problems for human operator performance (such as cognitive load) and new kinds of failure modes in the overall human-machine systems. Cognitive systems engineering has emerged as a framework to model the behavior of human-machine systems in the context of the environment in which work takes place [20]. Human factors should be evaluated in three hierarchical levels [21]: The first level should concern the cognitive behavior of human beings during the control of processes that occur through the man-system interface. At this level human errors are evaluated through human reliability techniques of first and second generation. In the second level, the focus is on the cognitive behavior of human beings when they work in groups, as in nuclear power plants. At this level the focus is on the anthropological aspects that rule the interaction among human beings. In the third level, the interest is in the influence of organizational culture on human beings as well as on the tasks being performed. At this level the economical and political aspects that shape the company organizational culture are added to the factors of the second level. Based on the abovementioned principles, various tools, techniques, models and methods have been introduced: THERP - Technique for Human Error Rate Prediction [22], HCR - Human Cognitive Reliability [23], ASEP - Accident Sequence Evaluation Program [24], HEART - Human Error Assessment and Reduction Technique [25], CREAM - Cognitive Reliability and Error Analysis Method [26], HFACS- Human Factors Analysis and Classification System [27], FRAM- Functional Resonance Accident Model [16], ATHEANA - A Technique for Human Error Analysis [28], etc.

3. Safety related indicators

As mentioned already safety is a concept often understood on the absence of undesired events and not easy to define directly. It is even more difficult to define what constitutes an adequate level of safety. In other words it is difficult to assess safety directly, because its appearance is indicated by absence of accidents. Then the only direct measure of achieved performance is in terms of absence of harm or loss, and reducing the cost of losses provides direct evidence of performance improvements.

The term Indicator in the safety field is rather new, but safety measurements were performed in the 1980s and before with terms like Index, Rate and Measurements [29]. In addition, there are terms such as Performance Measure, Performance Indicator and Metric, but these terms are not always used in a consistent way [30]. An indicator is a measurable representation of an aspect of reality [29]. According to Hale [31] safety indicators are used to: (1) monitor the level of safety in a system; (2) decide where and how to take action; and (3) motivate those in position to take the necessary actions to take them.

Until recently, the primary safety indicators have been things that go wrong like the number of injuries, the cost of accidents and claims, etc. Lost time injury frequency (LTIF) and total recordable incident rates (TRIR) are more common used safety indicators. Over the last decades, substantial improvements in the areas of LTIF and TRIR are reported [32]. It was for a long time assumed that personal injuries reflected a facility's major hazard risk, but since, personal (occupational) safety is about avoiding cuttings, trips and falls type hazards among employees, it does not represent management of major hazards that have the potential to cause major accidents [33]. Therefore, good personal safety performance does not ensure the satisfactory condition of an asset and its operation [34]. The recent undesired incidents in oil and gas facilities are clear signs of this, for example the explosion at a Shell Chemical Company plant in Deer Park, Texas in 1997 [35], the BP Texas City refinery disaster in 2005 [36] and Gulf of Mexico blowout in 2010 [5]. Relying on injury rates as an indicator for safety level significantly hindered BP's perception of process risk. As a result, BP's corporate safety management system for its US refineries did not effectively measure and monitor process safety performance [36]. There are different opinions to answer why some people do not properly distinguish between process safety and personal safety indicators. Hale [37] argues that the tendency to conflate process safety and personal safety, which is in relation to the major and minor accidents, goes back to a misinterpretation of the Heinrich triangle.

3.1 Personal Safety Indicator versus Process Safety Indicator

Hopkins makes a clear distinction between personal and process safety [38]: It is really a distinction between different types of hazards. Process safety hazards are those arising from the processing activity in which a plant may be engaged. Typical process safety incidents involve the escape of toxic substances and the release of flammable material which may or may not result in fires or explosions. Many process safety incidents either damage the plant or have the potential to damage the plant. Moreover, they have the potential to generate multiple fatalities. The term process safety originates in the US; in some other parts of the world process safety is referred to as asset integrity or technical integrity. Personal safety hazards, on the other hand, affect individuals but may have little to do with the processing activity of the plant. Typically they give rise to incidents such as falls, trips, crushings, electrocutions and vehicle accidents. It turns out that most injuries and fatalities are a result of personal safety hazards rather than process hazards and, as a result, injury and fatality statistics tend to reflect how well an organisation is managing personal safety hazards rather than process safety hazards. Any organisation that seeks to assess how well it is managing process safety hazards cannot therefore rely on injury and fatality data; it must develop indicators that relate specifically to process hazards. That is one of the primary conclusions of recent major accident reports [39].

In recent years, many organizations have recognized the importance of measuring process safety performance in order to improve process safety management. In 2003, the Organisation for Economic Cooperation and Development (OECD) developed and published the initial version of the Guidance on Developing Safety Performance Indicators related to Chemical Accident Prevention Preparedness and Response which has been revised in 2008 [40]. In 2006, the UK Health and Safety Executive (HSE) published a guide for development of leading and lagging process safety metrics: Developing Process Safety Indicators, A Step-By-Step Guide for Chemical and Major Hazard Industries [41]. In 2007, the Center for Chemical Process Safety (CCPS) released a document on Process Safety Leading and Lagging Metrics [42] which has been updated later. In 2010, American Petroleum Institute (API) published a recommended practice called Process Safety Performance Indicators for the Refining and Petrochemical Industries [43] which is under revision. In 2011, International Association of Oil & Gas

Producers (IOGP) published the Report number 456 called Process Safety - Recommended Practice on Key Performance Indicators [44]. In 2011, the European Chemical Industry Council (Cefic) published a document called Guidance on Process Safety Performance Indicators with a focus on measuring process safety incidents as a lagging indicator for process safety performance [45].

3.2 Lagging Indicator versus Leading Indicator

Reactive indicators ('lagging indicators') are after-the-event type of indicators, while proactive indicators ('leading indicators') assume measurements of underlying causes and contributing factors to accidental events, such as (inadequate) training and supervision, and thus providing early warnings [46].

OECD Guidance on Developing Safety Performance Indicators [40] divides safety performance indicators into two types: "outcome indicators" and "activities indicators."

- Outcome indicators are designed to help assess whether safety-related actions (policies, procedures and practices) are achieving their desired results and whether such actions are leading to less likelihood of an accident occurring and/or less adverse impact on human health, the environment and/or property from an accident. They are reactive, intended to measure the impact of actions that were taken to manage safety and are similar to what are called "lagging indicators" in other documents. Outcome indicators often measure change in safety performance over time, or failure of performance. Thus, outcome indicators tell you whether you have achieved a desired result (or when a desired safety result has failed). But, unlike activities indicators, they do not tell you why the result was achieved or why it was not.
- Activities indicators are designed to help identify whether enterprises/organisations are taking actions believed necessary to lower risks (e.g., the types of policies, procedures and practices described in the Guiding Principles). Activities indicators are proactive measures, and are similar to what are called "leading indicators" in other documents. They often measure safety performance against a tolerance level that shows deviations from safety expectations at a specific point in time. When used in this way, activities indicators highlight the need for action when a tolerance level is exceeded. Thus, activities indicators provide enterprises with a means of checking, on a regular and systematic basis, whether they are implementing their priority actions in the way they were intended. Activities indicators can help explain why a result (e.g., measured by an outcome indicator) has been achieved or not.

Traditionally, the focus of safety indicators for major hazards has been on incident indicators or lagging indicators, which means recording the number of accidents or near-misses. Major accidents are rare and occur so infrequently specially at a site level. Many companies have not seen one at all and some have seen only just. The frequency of major accidents therefore is not a good indicator for the state of safety of a particular plant [47]. Also it has been claimed that these indicators may not be useful as early warnings [36], and the focus on developing indicators that provide feedback before an accident occurs (so-called leading indicators) has increased [31]. Leading indicators are generally intended to help in identifying early signals of deteriorating performance and thereby to provide means for warning of impending problems before an actual incident or accident occurs. Early warnings from these indicators should initiate a further investigation of the causes of the symptoms to ensure that corrective actions will be timely initiated and rightly directed [30].

After distinguishing two dimensions of safety indicators: personal safety versus process safety indicators and lead versus lag indicators it is necessary to highlight another issue about safety indicators. Maybe it is not relevant whether the indicator is current or after the fact. The issue is whether, in the relevant time period, there are sufficient instances of the events being counted to be able to talk meaningfully about a rate. If there are, then charting this indicator over time will indeed provide evidence of whether the safety management system is improving. On the other hand, if years can go by without the occurrence of a single countable event, it is not possible to compute a meaningful annual rate, nor is it possible to conclude from one occurrence that safety is deteriorating. This is what Hopkin called a zoom effect at work [38]. Fatality rate can be sensibly computed for an industry, and possibly even for a very large company. Regulators can therefore treat the industry fatality rate as an indicator of how well safety is being managed in that industry and seek to drive it down by encouraging various industry-wide safety initiatives. The same may be true for a very large company. However, once we zoom in on a particular company site we may find that it goes for many years without a fatality. Under these circumstances it will not make sense to treat the fatality rate at the site as an indicator of safety. Management at such a site will need to identify more frequently occurring events, such as injuries, in order to be able to judge how well they are managing safety over time. Similarly, catastrophic accidents occur so infrequently at any one major hazard site that they cannot be used as an indicator of how well process safety is being managed at that site, whereas loss of containment events may be frequent enough to serve the purpose. In short, where harmful events are occurring frequently enough to be able to talk about a rate, this rate can be used to measure safety; where harmful events are rare, we must look to more frequently occurring precursor events in order to be able to measure safety. Putting this another way, there are circumstances in which so-called lag indicators are perfectly good indicators of how well safety is being managed, and circumstances in which they are not. It depends, in part, on the zoom's level. As mentioned by Hopkins [38], if one is interested in how well safety is being managed, the distinction between lead and lag indicators becomes largely irrelevant.

4. Discussion and Conclusion

Different approaches have been followed to evaluate operational safety at process plants and consequently different process safety indicators have been introduced. Although process safety indicators provide important information on the effectiveness of safety systems and try to provide early warnings of impending catastrophic failures, most of the existing indicators are considered as "lagging" indicators and they are expected to show an impact only when a downward trend has already started. Therefore these indicators are not able to declare how close an operation is to a major accident and development of new indicators can help warning when an operation is out of safe zone.

For early warnings, we need to look further back in the causes and the underlying conditions and factors that lead to a major accident in process industry. The acknowledgement of industrial installations as complex systems points to systems thinking theory as the most suitable tool to deal with major accident in process industry. Hence it is recommended to select a system theory based accident model like STAMP which best fits the process and accordingly map the process into the model in order to define a set of indicators which could demonstrate the progress of a major accident through its modeled path. Required information for indicators has to be obtained from the operation and processed accordingly in order to indicate the progress toward major accident. This can answer questions such as: How close is an operation to a major accident? And meanwhile it will support management decision making about required actions to restore safety.

No doubt, realization will be accompanied by numerous problems and unforeseen difficulties. Essential will be sensible choosing and observing a set of indicators based on the selected model as no single indicator can be a sure sign of imminent major accident.

5. References

Reference in order of first citation

[1] "Report No. 415, Asset integrity - the key to managing major incident risks", International association of Oil and Gas Producers, Dec 2008

[2] "Offshore Installations, the offshore installations (Safety Case) regulations", UK HSE, 2005

[3] "Guidelines Regarding the Management Regulations", Norway PSA, December 16, 2014

[4] "Offshore Petroleum and Greenhouse Gas Storage (Safety) Regulations", 2009 (Select Legislative Instrument 2009 No. 382 as amended), Australia, prepared on 1 January 2012

[5] "Investigation report overview, explosion and fire at the Macondo well", US Chemical and Safety and Hazard Investigation Board (CSB), June 5, 2014

[6] ''Corporate governance for process safety, Guidance for senior leaders in high hazard industries'', OECD Environment, Health and Safety Chemical Accidents Programme, June 2012

[7] B. Ale et al, Towards BBN based risk modelling of process plants, Safety Science, 69 (2014) 48–56

[8] Wagenaar, W.A. & Hudson, P.T.W., The analysis of accidents with a view to prevention. Report for Shell International SIPM, The Hague. Department of Experimental Psychology, Leiden University, The Netherlands (1987)

[9] Hollnagel, Risk + barriers = safety?, Safety Science 46 (2008) 221–229

[10] Hollnagel, Is safety a subject for science, Safety Science 67 (2014) 21–24

[11] Heinrich, H. W., Peterson, D., and Roos, N. (1980). Industrial accident prevention. McGraw-Hill, New York.

[12] Leveson, N.G., 1995. Safeware: System Safety and Computers. Addison Wesley, Reading, MA.

[13] Hollnagel, E. (2001). Anticipating Failures: What should Predictions be About? In The Human Factor in System Reliability – Is Human Performance Predictable? RTO Meeting Proceedings 32, RTO-MP-32, January, Research and Technology Organization, North Atlantic Treaty Organization, Cedex, France: RTO/NATO.

[14] Reason, J. (1990). Human Error. Cambridge: Cambridge University Press.

[15] Reason, J. (1997). Managing the Risks of Organizational Accidents. England: Ashgate.

[16] Hollnagel, E., 2004. Barriers and Accident Prevention. Ashgate Publishing Limited, Aldershot, UK.

[17] Perrow, C. Normal Accidents: Living with High-Risk Technologies. 1984, New York: Basic Books

[18] Rasmussen, J. Risk Management in a Dynamic Society: A Modelling Problem, Safety Science, 27 (1997), 2, 183-213.

[19] Leveson, N. A New Accident Model for Engineering Safer Systems, Safety Science, 42 (2004), 237-270.

[20] Hollnagel, E. & Woods, D. D. Cognitive Systems Engineering: New wine in new bottles. International Journal of Man-Machine Studies, 18 (1983), 583-600.

[21] Alvarenga et al, A critical review of methods and models for evaluating organizational factors in HRA - Human Reliability Analysis - Progress in Nuclear Energy 75 (2014) 25-41

[22] Swain, A.D., Guttman, H.E., 1983. Handbook of Human Reliability Analysis with Emphasis on Nuclear Power Plant Applications. NUREG/CR-1278. U.S. Nuclear Regulatory Commission, Washington, D.C.

[23] Hannaman, G.W., Spurgin, A.J., Lukic, Y., 1984. Human Cognitive Reliability Model for PRA Analysis, NUS-4531. Draft EPRI Document. Electric Power Research Institute, Palo Alto, CA.

[24] Swain, A.D., 1987. Accident Sequence Evaluation Program Human Reliability Analysis Procedure. NUREG/CR-4772/SAND86-1996. Sandia National Laboratories, U.S. Nuclear Regulatory Commission, Washington, DC.

[25] Williams, J., 1988. A data-based method for assessing and reducing human error to improve operational performance. In: IEEE Conference on Human Factors in Power Plants. Monterey California.

[26] Hollnagel, E., Cognitive Reliability and Error Analysis Method (CREAM). Elsevier Science (1998) New York.

[27] Wiegmann, D.A., Shappell, S.A., 2003. A Human Error Approach to Aviation Accident Analysis: The Human Factors Analysis and Classification System. Ashgate Publishing Company, Burlington, VT.

[28] NRC, 2007. ATHEANA's User Guide Final Report. NUREG-1880. US Nuclear Regulatory Commission, Washington, D.C.

[29] Øien et al., Building Safety indicators: Part 1 – Theoretical foundation, Safety Science 49 (2011) 148–161

[30] Lehtinen, E. (1995). "A Concept of Safety Indicator System for Nuclear Power Plants", Technical Research Centre of Finland, Espoo, Research Notes 1646, 26 p. + app. 106 p.

[31] Hale Andrew, Why safety performance indicators - Safety Science 47 (2009) 479–480

[32] ''Safety performance indicators – 2010 data'' IOGP, Report No. 455, 2011

[33] Eskogdalen et al, Developing safety indicators for preventing offshore oil and gas deepwater drilling blowouts - Safety Science 49 (2011) 1187–1199

[34] Hassan and Khan, Risk-based asset integrity indicators, Journal of Loss Prevention in the Process Industries 25 (2012) 544-554

[35] EPA/OSHA Joint Chemical Accident Investigation Report, Shell Chemical Company plant in Deer Park, Texas, 1998

[36] Baker et al, The report of the BP U.S. refineries independent safety review panel, 2007

[37] Hale, A., 2001, Conditions of occurrence of major and minor accidents. Institution of Occupational Safety and Heath Journal 5(1), 7–21.

[38] Hopkin Andrew, Thinking about process safety indicators, Safety Science 47 (2009) 460–465

[39] CSB, 2007. Investigation Report, Refinery Explosion and Fire. US Chemical Safety and Hazard Investigation Board.

[40] ''Guidance on developing safety performance indicators related to chemical accident prevention, preparedness and response'' OECD - 2nd edition 2008.

[41] HSE, 2006. Developing process safety indicators: a step-by-step guide for chemical and major hazard industries, UK Health and Safety Executive.

[42] ''Process Safety – Leading and lagging Metrics'', CCPS 2007

[43] ''Process Safety Performance Indicators for the Refining and Petrochemical Industries'', API-2010

[44] ''Process Safety - Recommended Practice on Key Performance Indicators'', Report number 456, IOGP – 2011

[45] ''Guidance on Process Safety Performance Indicators'', European Chemical Industry Council (Cefic) - 2011

[46] Øien et al., Building Safety indicators: Part 2 – Application, practices and results, Safety Science 49 (2011) 162-171

[47] Ale, Ben, More thinking about process safety indicators, Safety Science 47 (2009) 470–471

393997 An Approximate Model for Discharge Density Associated with Critical Flow

Abstract:

The purpose of this presentation (paper) is to present an approximate model derived from the entropy balance associated with isentropic, choked flow utilizing the van der Waal equation of state and an approximate form of the van der Waal equation for high pressures that will provide the stagnation density. This will provide the form of the model where a non-dimensional parameter related to the deviation from isothermal behavior will be estimated via three non-dimensional numbers related to pressure, temperature and ratio of specific heats of the system. The model will be validated against three gases which are ideal and thermodynamically perfect, ideal but not thermodynamically perfect, and a real gas. Pressures will range up to 4 MPa. The relationship of this model against other models in the literature will be explored and its relationship to the Omega method will be discussed.

396653 A Multilevel Safety EARLY Warning Method Based on Bayesian Networks with Applications for Petrochemical Process

Abstract:

Process systems in petrochemical industries have becoming increasingly large and automatic. There exist strong interdependencies between various facilities and components. Once any unit in the system fails, it often triggers cascade failures as a chain reaction, resulting in significant production losses, and even leads to catastrophic accidents. They are the result of a number of failures at different technical and organizational levels.

Proactive methodologies for the development of early warning methods can unveil early deviations in the causal chain. Two main issues exist that are how to describe the interdependency in the complex process system, and how to ratiocinate and demonstrate the reasons at each level of the current abnormal event and also foresee the possible consequence. In this way, a multilevel safety early warning method is proposed for petrochemical process to overcome above problems. The approach combines the advantages of multilevel flow modeling (MFM), HAZOP analysis and dynamic Bayesian network (DBN) as well, which become a special and effective mechanism of safety early warning.

Firstly MFM is used to represent a system in terms of goals, objectives, functions and components each of which can be described at different levels of part-whole decomposition. Secondly HAZOP study is carried out based on the MFM models, by which all the possible deviations and their corresponding potential fault causes and consequences are analyzed carefully. Thirdly dynamic Bayesian Network (DBN) is used to build the fault causal relationships which represent the fault interdependencies in the complex process system. Finally by the inference mechanism of DBN, the most possible initial reason(s) happened in the fault interdependency network with multi reasons and consequences can be found out accurately and also the future possible consequences can be predict for proactive maintenance or emergency decision making.

The analysis shows that the proposed model fits its purpose enabling causal reasoning that explains causes and consequences derived from deviations. Examples from case studies with respect to the Fluidized Catalytic Cracking unit (FCCU) in petrochemical process system are included to illustrate the effectiveness and accuracy of the proposed approach. Based on the proposed multilevel safety early warning method, an expert system is developed and applied to practice in the petroleum industry. Its application is also illustrated to show the effect of accident prevention in the petrochemical enterprises. The salient features of this expert system and its performance on practical petrochemical systems is presented.

Summary of Macondo Inquiries and Recommendations

Dr Robin Pitblado
DNV GL
1400 Ravello Dr, Katy TX 77441
robin.pitblado@dnvgl.com

Prepared for Presentation at
American Institute of Chemical Engineers
2015 Spring Meeting
11th Global Congress on Process Safety
Austin, Texas
April 27-29, 2015

UNPUBLISHED

Summary of Macondo Inquiries and Recommendations

Dr Robin Pitblado
DNV GL
1400 Ravello Dr, Katy TX 77441
robin.pitblado@dnvgl.com

Keywords: Inquiries, Lessons Learned, Macondo, Process Safety

Abstract

The Macondo event, which commenced on April 20 2010 was the largest oil spill event in history and caused 11 fatalities, 17 injuries and the total loss of the Deepwater Horizon drilling rig. Not surprisingly, it is also amongst the most studied of events, with close to 20 major inquiries having been undertaken in the USA and elsewhere. The purpose of this paper is to summarize the main inquiries findings and recommendations that are publicly available. While this involves some selection, the aim of the paper is to be neutral and to simply report the facts as stated. The aim is that readers will be able to identify what body recommended what and to match this to changes to regulations and good practices post the accident.

1. Introduction

The Macondo accident on April 20 2010 was the largest oil spill event in history and caused 11 fatalities, 17 injuries and the total loss of the Deepwater Horizon drilling rig. It is important that the industry learns key lessons from this accident. While many lessons have been incorporated into new regulations, updated standards, and revised company practices, many engineers are unaware of the full range of findings from the over 209 inquiries that have been undertaken. Also many good recommendations have not been adopted and these deserve fresh consideration.

A list of the most important inquiries since Macondo is provided in Table 1, and there are further reports expected. The CSB has noted it plans two further reports addressing human factors and organizational issues, and we expect more detailed environmental findings as the environmental impacts become more fully understood.

Table 1. Major Macondo Investigation reports summarized in this study

No.	Author	Title	Release Date
1	DWH Commission	National Commission on the BP Deepwater Horizon Oil Spill and Offshore Drilling – Main Report and multiple topic papers	11-Jan 2011
2	DWH Commission	Chief Counsel's Report	Feb 2011
3	BP	DWH Accident Investigation Report	8-Sep-2010
4	Transocean	Macondo Well Incident: Transocean Investigation Reports, Vol. I and II	June 2011
5	USCG	Report of Investigation into the Circumstances Surrounding the Explosion, Fire, Sinking and Loss of Eleven Crew Members Aboard the Mobile Offshore Drilling Unit DWH Vol.1 (MISLE 3721503)	22-Apr-11

6	BOEMRE	JIT - BOEMRE Final Report regarding Macondo Well Blowout	14-Sep-11
7	Republic of the Marshall Islands	Deepwater Horizon Marine Casualty Investigation Report	17-Aug-11
8	Adm Thad Allen	Incident Commanders Report	Oct 2010
9	DNV GL	Forensic Examination of Deepwater Horizon Blowout Preventer, Vol. I and II (Appendices).	March 2011
10	Chemical Safety Board	Investigation Report (Vols 1-2)	5-Jun-14
11	Center for Catastrophic Risk Management (UC Berkeley)	Final Report on the Investigation of the Macondo Well Blowout	March 2011
12	National Academy of Engineering	Macondo Well-Deepwater Horizon Blowout: Lessons for Improving Offshore Drilling Safety	2012
13	National Research Council	Ecosystem Services Approach to DWH	2013
14	US District Court Eastern Louisiana	Court Judgment	4-Sep-2014
15	Norway Petroleum Safety Authority	Interim report – DWH Accident	June 2011
16	Norway Petroleum Safety Authority	PSA's Concluding report on its follow-up for the DWH accident	5-Feb-2014
17	OGP	Getting it Right	Jan 2013
18	OLF	DWH Lessons Learned and Follow-up	June 2012
19	SINTEF	Report on DWH (A19148)	May 2011
20	UK HSL	DWH Fire & Explosion Issues	Apr 2014
21	US Transportation Research Board	Evaluating the Effectiveness of Offshore Safety and Environmental Management Systems (Special Report 309)	2012

DNV GL (2015) has published a fuller review of these inquiries, and this paper provides a shorter summary of key findings and recommendations.

2. National Commission – BP Deepwater Horizon Oil Spill

The National Commission was established by President Obama soon after the event with a mandate to report within 6 months. The Commission was chaired by Bob Graham and William K. Reilly, along with 5 other members and supported by a chief counsel and technical team. The main report was issued in January 2011 and was structured into 3 main sections: I) Path to the Tragedy, II) Explosion and Aftermath, and III) Lessons Learned for Industry, Government, and Energy Policy.

In Part I, the history of development of drilling in the Gulf of Mexico (GoM) and particularly in Deepwater was summarized, along with the technical challenges.

In Part II, the causes and consequences of the disaster were reviewed. Part III addressed lessons learned for industry, government and for energy policy and contained the Commission recommendations. The report emphasized safety culture as a key factor for enhancing safety and discussed culture issues affecting BP, its contractors, and the GoM industry generally. The Commission also identified a conflict of interest in API, which serves as coordinator for industry expertise and standards, but also acts as the industry chief lobbyist. API standards were characterized as "lowest common denominator" rather than "best practice" standards, and since these standards underpin the Department of Interior regulations, the whole regulatory system was compromised. Commission recommendations were grouped as follows:

A. Improving the Safety of Offshore Operations (A1-A5)
B. Safeguarding the Environment (B1-B3)
C. Strengthening Oil Spill Response, Planning, and Capacity (C1-C6)

D. Advancing Well-Containment Capabilities (D1-D3)
E. Overcoming the Impacts of the Deepwater Horizon Spill and Restoring the Gulf (E1-E7)
F. Ensuring Financial Responsibility (F1-F4)
G. Promoting Congressional Engagement to Ensure Responsible Offshore Drilling (G1-G2)

3. National Commission – Chief Counsel's report

The Chief Counsel's report complements the Commission's main report and contains significantly more technical details. This report focuses on the direct technical and managerial causes of the accident and does not address response activities or environmental, social or economic impacts. Main chapters deal with deepwater drilling topics and the Macondo well, deepwater well design, cementing, the temporary abandonment, the negative pressure test, kick detection and response, and the blowout preventer and its maintenance. Many findings are made on technical and management issues, but the report does not make recommendations – these are presented in the Commission main report (1). But the findings do highlight important issues and lessons learned.

It is not possible to list all the findings in this lengthy report, but some main findings are presented. The use of a long production casing made the cementing job more difficult and the use of rupture disks to prevent annular pressure build-up compromised later containment efforts as these were believed to have failed inwards. BP selected the long-term benefits of using the long string without adequate assessment of the short-term risks. The report is clear that the cement job failed but there were several possible reasons and it was not possible to select amongst these. Management findings focused on BP's systems that did not force the Macondo team to identify and evaluate all cementing risks; and changes to the well plan that were not adequately assessed. The BP team did not assess contractor data in sufficient detail and made too simple analysis of well data as an indicator of cementing success.

A final chapter dealt with overarching failures of management. The Chief Counsel's team observed at least the following management failures: (1) ineffective leadership at critical times; (2) ineffective communication and silo'ing of information; (3) failure to provide timely procedures; (4) poor training and supervision of employees; (5) ineffective management and oversight of contractors; (6) inadequate use of technology; and (7) failure to appropriately analyze and appreciate risk. Ultimately, the companies placed undue reliance on timely intervention and human judgment, in light of their failure to provide individuals with the information, tools, and training necessary to be effective.

The Chief Counsel's report concluded: "the blowout occurred in large part because the companies diffused knowledge, responsibility for, and ownership of safety among themselves and among groups of people. The people onshore and on the rig had a false sense of security. They did not recognize the need for individual leadership in addressing the multiple anomalies and uncertainties that they observed. Instead, they relied on many ambiguous "dotted line" relationships within and between the companies and personnel involved."

4. BP Report: Deepwater Horizon Accident Investigation Report

The BP team commenced their investigation very soon after the blowout event commenced. They focused on the direct technical causes of the incident and did not address management related issues or alternative risk management strategies that might have been employed nor the sequence of emergency actions immediately after the explosion. They identified a sequence of eight key barriers that failed and these permitted the accident to occur. The BP team made recommendations in eight areas that would apply to BP and to its contractors:

1) Procedures and technical practices - cementing, well control, well design, negative-pressure testing
2) Capability and competency - strengthen the Technical Authority role and enhance competency
3) Strengthen audit and verification program
4) Process safety performance management - Establish leading and lagging indicators for drilling
5) Develop cementing services assurance
6) Confirm that essential well control and monitoring practices are clearly defined and applied
7) Rig process safety – require HAZOP studies on key parts of drilling systems
8) Establish minimum levels of reliability for BOP systems, develop ROV intervention strategies

5. Transocean: Macondo Well Incident

The Transocean investigation had three main objectives – to establish 1) how the reservoir fluids reached the rig floor, 2) why the BOP failed to stop the flow, and 3) how did the fluids ignite. The Transocean report did not make specific recommendations. Their main conclusion was: "The Macondo incident was the result of a succession of interrelated well design, construction, and temporary abandonment decisions that compromised the integrity of the well and compounded the risk of its failure."

6. USCG JIT - Vol.1 Report into the Deepwater Horizon (Ref: 3721503)

This investigation was conducted under a Memorandum of Agreement with BOEMRE. The USCG made a large number of detailed recommendations related to its areas of responsibilities (these are listed in Table 2).

Table 2. Summary of USCG Recommendations – related to fire & explosion

Ref	Recommendation summary
1A	Labelling of rated electrical equipment in hazardous areas.
1B	Guidance on design of detection and alarm systems.
1C	Guidance on fire and explosion strategies for units with DP.
1D	Specification of minimum explosion design loads.
1E	Explosion risk analysis for each existing facility.
1F/G	Control of location of ventilation inlets. Shut down of ventilation in the event of gas detection.
2A	Fire pumps to be self-contained.
2B	H-60 rated separation between drill floor and adjacent accommodation areas.
2C	Guidelines on engineering evaluations of fire partitions.
2D	Fixed deluge or monitors provided for drill floor.
2E	Fire risk analysis to supplement the prescriptive requirements of the MODU code.
3A	Standards for maximum radiant heat allowable at muster stations and lifeboat stations.

Related to the recommendations related to fire and explosion the USCG (2014) issued guidance for offshore floating or fixed installations recommending modeling of fire and explosion scenarios consistent with API RP 2FB or equivalent.

7. BOEMRE – Report into Causes of Macondo Well Blowout

The BOEMRE investigation was carried out as part of the Joint Investigation Team (JIT) with USCG. It focused on BOEMRE responsibilities, especially the drilling activity and the cause of the blowout. A final section provided 41 recommendations and was divided into six sections

(with recommendations counts in each): wells (6), kick detection and response (8), ignition source (4), BOP (6), regulatory agency (11), and OSC Companies (6).
Some of the more significant recommendations included:

- require the negative pressure testing of wells where appropriate
- require at least two barriers (one mechanical and one cement barrier) for a well that is undergoing temporary abandonment procedures, and to clarify that a float collar/valve is not to be considered to be a "mechanical barrier."
- revise the incident reporting rule at 30 CFR§ 250.188 to capture well kick incidents
- several specific recommendations to enhance BOP operation and intervention
- expand 30 CFR § 250.446 to include documentation and record keeping requirements for major (3-5 year) inspections as required by BOEMRE's adoption of API RP 53

Note, there was no recommendation regarding safety and environmental management systems as BOEMRE had already decided by this time to require adoption of SEMS based on API RP 75.

8. Marshall Islands Registry – DWH Marine Casualty Investigation Report

REPUBLIC OF THE MARSHALL ISLANDS
Office of the Maritime Administrator

The Flag State for the DWH was the Republic of the Marshall Islands. Under international conventions, the Flag State must report the causal factors of all serious and very serious marine casualties.
The report concludes that the causal factors were: deviations from standards of well control engineering; deviation from the well abandonment plans submitted to and approved by the Minerals Management Service; and failure to react to multiple indications that a well control event was in progress.
The report contains 17 recommendations, but the Marshall Islands highlights four: enhanced communications between the Flag State and the Coastal State; IMO should review the MODU code related to aspects of this incident; there should be better definition of responsibilities of the Master and OIM especially during emergencies; and better procedures for actuating the Emergency Disconnect System.

9. Adm Thad Allen: Incident Commander's Report

Once the scale of the incident was established, the event was declared to be a Spill of National Significance and Thad Allen, then Commandant of the USCG, was declared the National Incident Commander. These were the first applications of these terms and the event tested existing laws and regulations very hard. The Oil Pollution Act (OPA 90) and the National Contingency Plan (NCP – 40 CFR Part 300) provided the guiding regulations but there was overlap with a Homeland Presidential Directive for domestic incidents. The NCP had been deployed successfully for thousands of spill events over the past 20 years.
It started to break down during the Deepwater Horizon spill as some aspects were not accepted by the public or by state officials – notably the role of the *responsible party* (here BP) was not understood and some states/local authorities did not accept federal authority.
The National Incident Commander (NIC) concluded that the Oil Pollution Act of 1990 and the NCP provided a sound basis for response and any changes should not change its fundamental governance structure. But many changes are required. There must be better integration of all necessary government agencies before there is an emergency, actions are needed to de-conflict and reconcile the role of the NIC, the NIC must have adequate authorities to execute the role

properly and work with the National Response Team (15 relevant federal agencies), and the private sector must be encouraged to develop modern oil spill response capabilities, matching the challenges of deepwater activities.

10. DNV Forensic Examination of Deepwater Horizon BOP

The Justice Department commissioned DNV (now DNV GL) to carry out a forensic examination of the Transocean BOP recovered from the sea floor once the well had been killed. The BOP was transported to NASA's Michoud facilities in Louisiana where tests were conducted. The report was designed to support legal proceedings, if required, and thus contains a very detailed set of information with 149 figures and 30 tables.

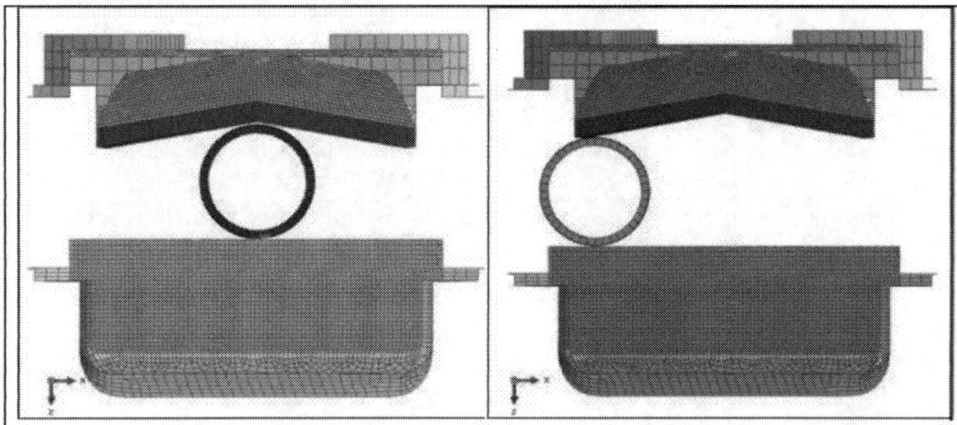

Drill pipe properly centered in BSR | Drill pipe buckled beyond BSR cutting zone

Figure 1. Location of drill pipe inside BSR and its effect on BSR cutting ability

The DNV report made a number of detailed recommendations to industry regarding BOP design and operation issues. These included:

- Study of elastic buckling if drill pipe – as this is what prevented the BSR from functioning properly
- Study of the shear blade surfaces of shear rams – as they could not shear the off-center drill pipe
- Study of well control procedures or practices – as the timing of the closure of the Upper Annular and the Upper VBR's contributed to the drill pipe buckling and moving off center
- Better testing of back-up control systems – as some system components did not perform as intended.
- Better processes to prevent common mode failure of back-up control systems
- Provide better means to verify the operation or state of various components of BOPs in an emergency
- Study of the effectiveness of Remotely Operated Vehicle interventions
- Stipulating requirements for back-up control system Performance – as now for primary control functions

11. CSB Investigations (Vols 1-2)

The US Chemical Safety Board, which is well known for its onshore investigations, undertook an investigation into the Macondo offshore event at the request of Congress. They issued two reports in June 2014. Vol 1 addresses the relationship between the parties involved and describes the many technical challenges associated with the drilling at Macondo. Vol 2 is much more detailed and this focuses on the CSB technical findings on the Deepwater Horizon BOP and its ongoing management as a safety critical element (SCE).

The CSB participated in some of the forensic examinations of the BOP and carried out its own additional tests. The CSB noted several opportunities to improve US regulations. These should include risk based concepts with a focus on major accident events; that safeguards for such accidents be based on a sound philosophy; that there should be an explicit target for risk reduction (e.g. ALARP); and that Safety Critical Elements (SCE's) be identified and managed through life to attain defined performance standards.

The report addressed a number of specific faults for the BOP, but the CSB decided not to make detailed BOP design recommendations, rather to focus on a higher level on issues relating to all Safety Critical Elements. These include that BSEE should develop regulations requiring SCE's to be identified and performance standards defined for each, and that risks be reduced to an ALARP target level. SCE's should be monitored through life; that there should be an independent verification scheme for these; and deficiencies identified be documented and corrected. The CSB recommended that API support this process by developing guidance documents for identifying SCE and reducing risks to ALARP. More specifically also, the API should revise Blowout Preventer Equipment System for Drilling Wells (API Standard-53, 4th edition) to establish additional component level tests to supplement the current system level tests.

12. Center for Catastrophic Risk Management (UC Berkeley)

A group of experienced risk management and organizational specialists resides at the UC Berkeley Center for Catastrophic Risk Management. The final report was issued in March 2011, and this was preceded by 3 progress reports. The study group addressed both technical and organizational causes.

In the three progress reports the Center concluded: 1) this disaster was preventable had existing progressive guidelines and practices been followed; 2) the failures (to contain, control, mitigate, plan, and clean-up) appear to be deeply rooted in a multi-decade history of organizational malfunction and shortsightedness; and 3) thus, as a result of a cascade of deeply flawed failure and signal analysis, decision-making, communication, and organizational - managerial processes, safety was compromised to the point that the blowout occurred with catastrophic effects. At the time of the Macondo blowout, BP's corporate culture remained one that was embedded in risk-taking and cost-cutting.

The Center recommendations included:

- Develop technology systems between government and industry to develop high hazard wells sustainably.
- Develop institutions capable of effective risk assessment and management of high hazard wells.
- Implement such risk based systems in the context of high reliability organizations and capable of responding to major blowout events.
- Utilize validation projects to show such characteristics.

13. National Academy of Engineering

The National Academy of Engineering convened a review panel of 15 experts in response to a request from the Department of Interior. Their report included main sections on well design and construction, the BOP system, MODUs, industry management of offshore drilling, and regulatory oversight; each of these with multiple findings, observations and recommendations. Appendix C in the report groups together all 38 findings, 23 observations, and 68 recommendations. While these are all important, space does not permit listing all of these here, however the Summary chapter highlights the most important and those 13 recommendations are listed here:

1) Develop guidelines for drilling and completion that address credible risks and safety margins
2) Tests should be defined with acceptance criteria for primary cement and mechanical barriers
3) BOP systems need to provide reliable cutting, sealing and separation, with maintenance and training
4) Instrumentation / expert systems should warn of loss of well control with automatic functions
5) Loss of well control should be mitigated by access to well capping and containment capabilities
6) US regulations should be updated to be a hybrid of prescription with proactive goal-setting
7) BSEE should identify and approve safety critical points in well construction and abandonment
8) A single US agency should have responsibility for an integrated approach to system safety
9) The Lessee should have responsibility for well integrity and the drilling contractor for equipment
10) Industry should carry out more R&D on all aspects of drilling safety
11) There needs to be formal education and training of offshore staff to properly implement system safety
12) Develop industry-wide systems for incident reporting and share lessons learned openly
13) Enhance safety culture through training, addressing human factors, system safety, and indicators

14. Norway PSA – Interim Report DWH

The Petroleum Safety Authority is the safety regulator for offshore developments in Norway. It established an internal review team after the Macondo event, but considered also Montara (2009), Snorre A (2004) and Gullfaks C (2010) events and that work generated an interim report in 2011.

The PSA does not believe that the Macondo accident challenged the critical philosophies underpinning the Norwegian regulatory system such as the division of responsibility for regulatory compliance and the demand for systematic and risk-based compliance with functional requirements. But it did challenge whether the PSA is continuously improving the way it seeks to influence safety. An overall observation is that industry needs to significantly improve how it manages major accident risk and that safety culture must be enhanced.

The need for effective barrier management covering all barrier elements is emphasized. There should be an integrated approach to well barriers, including the principle of two independent and tested well barriers as well as barrier monitoring. The DWH accident confirms the importance of defining the ambition level of performance standards for barrier elements and ongoing monitoring of these. In Norway, the effects of cold climate and aging facilities are important in this regard. The BOP may need to have an associated SIL level to ensure its reliability.

The DWH accident has highlighted the need for an effective capping and containment capability. Furthermore there is a need for better training of key staff in dealing with emergencies. Better systems are needed for disconnection from the well, and for enhanced ignition controls especially related to air intakes of power generators. Maintenance of the BOP was an issue and the PSA reemphasized the need to maintain safety critical equipment.

The PSA highlighted that offshore safety needs a systems perspective – complex systems fail in complex ways and simplistic solutions are not likely to be effective. Initiatives are needed to enhance safety culture and the development of a culture of accountability. Companies need to demonstrate their commitment to safety by participating actively in standards development.

The Interim report was issued in 2011 and the PSA decided that a follow-up report was justified in 2014 addressing new information that appeared after that date. The central findings did not change, but extra emphasis was given in several areas, and some recent changes are documented.

The PSA studied well control events on the Norwegian Continental Shelf and identified four main challenges:

- the need for a stronger commitment to technical measures to improve safety
- a bigger commitment to planning, barrier management and better-adapted risk analyses
- paying greater attention to major accident risk – more investigation of incidents
- creating operational parameters for good collaboration in the operator/supplier hierarchy.

In the area of risk assessment, the PSA noted that the DWH accident demonstrated a need for better risk management and processes which lead to more robust solutions and operations. Barrier management has been a PSA focus area for several years, but there is an ongoing need to maintain industry commitment to barrier management and to link this to overall risk management. Finally, with respect to safety culture and organization, the DWH accident highlighted the need for companies to be open to the unknown and paying attention to signals of vulnerability. The Norwegian industry has initiated several projects relating to safety culture, organization and management. The NORSOK D-010 standard for well control has been updated, including key performance indicators and for change management.

PSA agreed with BSEE's four main expectations for the BOP:

- that shear rams have the capacity to cut the equipment being run into the well through the BOP
- that the BOP is maintained as a critical safety system
- that real-time condition monitoring is established by instrumenting the BOP
- that requirements are developed for the competence of BOP maintenance personnel.

15. OLF – DWH Lessons Learned and Follow-up

Norsk olje&gass

The Norwegian Oil Industry Association (OLF), now renamed Norwegian Oil and Gas Association, in collaboration with NOFO (the Norwegian Clean Seas Association for Operating Companies) carried a review of the DWH incident, compared regulations in the US and Norway, and made recommendations to enhance Norwegian operations and engineering standards. The group made 45 recommendations for improvements, summarized below.

Prevention: The focus for improvement is to enhance the Norwegian NORSOK drilling standards (D-001: drilling facilities and D-010: well integrity in drilling and well operations). Specific recommendations address critical cementing jobs, lockdown requirements for tubing and casing hangers, negative pressure testing, fluid displacement requirements, well control exercises, BOP backup control systems, and enhanced BOP testing requirements. Enhancement

to well management systems, process safety, and teamwork and communications were also proposed. Action on drill crew expertise is being addressed by OGP (17).
Intervention and Response: Capping and containment is being addressed through the OGP SWRP Project (17). The Macondo unified command system proved to be effective and is now considered a best practice approach and it is recommended for Norway. There are lessons to be learned about protecting workers from chemical exposure during a response through protective equipment and training. OLF commissioned the Norwegian Institute for Water Research to assess the environmental impact to fisheries and of dispersant usage. Fortunately, environmental damage from DWH was less than expected and the use of in-situ burning and underwater use of dispersants was considered beneficial. There remain gaps in knowledge regarding Corexit dispersant and its potential to delay natural degradation of oil. While fisheries have recovered, there were very large costs to the fishing industry due to an extensive closure of the resource. OLF provided a useful summary of recommendations related to well control and response issues (Figure 2).

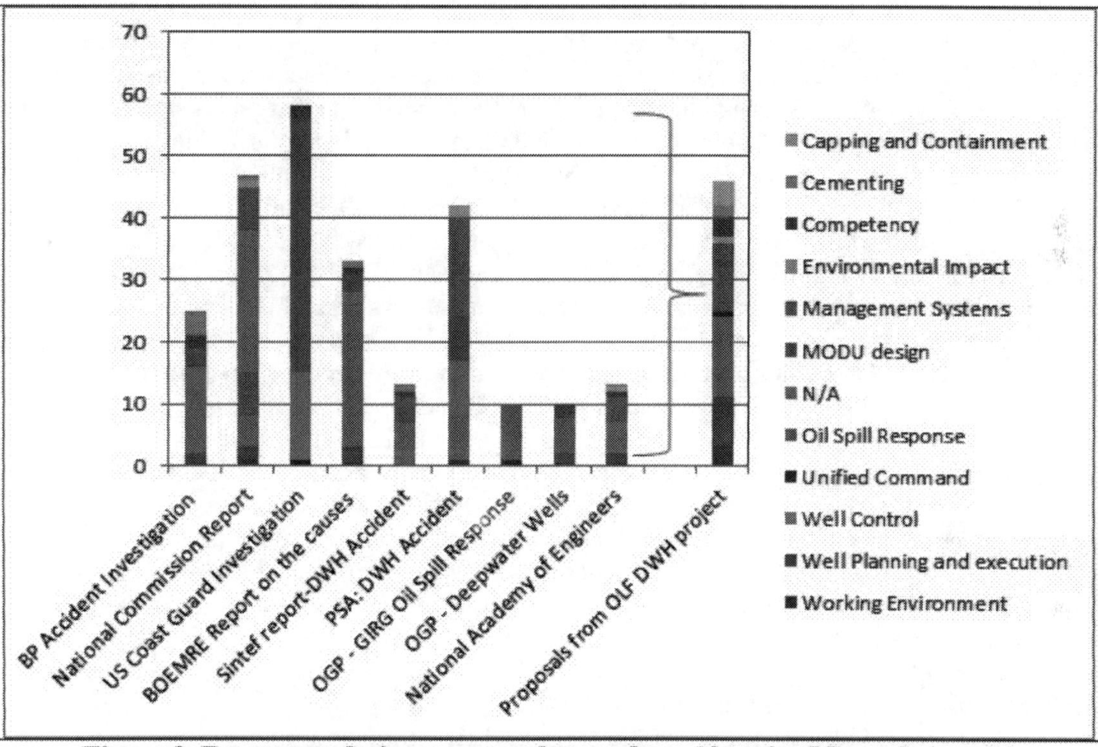

Figure 2. Recommendation count and types from 10 major Macondo inquiries

16. UK Health & Safety Laboratory - DWH Fire & Explosion Issues

The HSE requested the HSL to study fire & explosion issues related to the Macondo event. HSL relied on several key investigations including the BOEMRE, USCG, BP, Transocean, and RMI reports (all summarized above). The HSL focus was on technical issues related to fire and explosion. Based on their analysis, the HSL made several recommendations:

- Control of ignition sources across the rig in the event of blowout should be considered in ALARP demonstrations

- Substantial reductions in risk can be achieved by improving the reliability of detection systems and simplifying or eliminating reliance on human responses
- Preventing the ingestion of gas into large spaces deep within the rig should be a priority
- If there is a residual risk of gas accumulation in internal spaces, the consequences of explosion should be considered. The risks of explosions venting through vulnerable areas of the rig should be minimized; this may involve strengthening some internal partitions and/or providing explosion relief to the outside.

The HSL report summarized fire and explosion related recommendations in other Macondo investigations and found that these generally match UK offshore recommendations. Some later chapters work through examples of mitigations of ignition likelihood, flammable gas dispersion, and explosion damage.

17. Conclusions

As can be seen there have been many investigations into the Macondo accident and these have addressed several different aspects of the accident. Some investigations took a purely technical cause perspective whereas others emphasized more compliance with regulations or addressed organizational or human causes. Early investigations provided very detailed tutorials on deepwater drilling challenges. An indication of the range of recommendations is shown in Table 3 and a more specific focus on technical recommendations appears in the OLF Figure 8.
Many of the key recommendations have been adopted in one form or another. The USA has seen major changes to the MMS regulator, now BSEE and BOEM, with an updated style of operation emphasizing goal-based safety and increased inspector staff to allow greater presence offshore. BSEE has issued new requirements for drilling safety, for BOP recertification, negative pressure tests, and professional engineer sign-offs on casing and cement, and a requirement to estimate worst case blowout events. API and IADC have worked on an interface requirement between lessee and contractor, and several new API standards have been developed. SEMS I and SEMS II have been implemented with the Center for Offshore Safety defining the protocols and approving third party audit service providers. BSEE has provided guidance on safety culture and is working with another federal agency to implement a confidential reporting system. BSEE has established the Ocean Energy Safety Institute to research several longer term problems (e.g. risk, reliability data, BAST). USCG has issued guidance on additional fire and explosion assessments it would like to see. There have been consortia established in the USA and by OGP at 4 international locations to provide emergency response support. There have been standards updates for drilling and well control by API and by NORSOK, and they have made their safety standards freely available. Detailed assessments of fire and explosion lessons have been made and these are starting to find their way into designs.

Table 3. Range of Findings and Recommendations

No.	Title	Prevention	BOP Design & Operation	Contain & Respond	Manageme nt & culture	US Regulatory	International Regulatory
1	DWH Commission Main	X	X	X	X	X	
2	DWH Chief Counsel's report	X	X		X	X	
3	BP Investigation	X	X	X			
4	Transocean Investigation	X	X	X			
5	USCG	X	X	X		X	X
6	BOEMRE	X	X	X		X	
7	Republic of the Marshall	X	X	X		X	X

	Islands						
8	Adm Thad Allen Report			X		X	
9	DNV GL Forensic Investigation		X				
10	Chemical Safety Board	X	X		X	X	
11	Center for Catastrophic Risk Management (UC Berkeley)	X	X	X	X	X	
12	National Academy of Engineering	X	X	X	X	X	
13	National Research Council			X			
14	US District Court Eastern Louisiana	X	X	X			
15	Norway Petroleum Safety Authority – Interim	X	X	X	X		X
16	Norway Petroleum Safety Authority – Final	X	X	X	X		X
17	OGP	X	X	X	X		
18	OLF	X	X	X	X		
19	SINTEF	X	X	X	X		
20	UK HSL			X			
21	US Transportation Research Board				X	X	

Environmental findings of long term damage are too soon to be available, but initial indications are that the damage is less than might have been feared.

There have been changes overseas. The EU has decided to adopt a safety case approach (similar to that in the UK) for all offshore developments in the EU, and Australia expanded coverage of its regulator (NOPSEMA) to address drilling and environmental impacts to achieve close to a single offshore regulator. The UK and Norway both believe their goal-setting approach was not challenged by the Macondo accident, but that many detailed aspects of drilling safety did need enhancement and HSE, PSA and OGP are working on these issues.

There are some important recommendations for the US regulatory system which have not yet been adopted. These include suggestions to adopt a safety case approach with a greater focus on risk assessment with a risk target, and to nominate safety critical items with defined performance standards. It is not yet known if these will be adopted, or perhaps addressed in some other equivalent manner.

References

Macondo Inquiry references – see Table 1 in above text.

DNV GL (2015) The Macondo Event: Summary of Inquiry Findings and Recommendations, DNV GL, Oil & Gas Advisory Services, Houston.

USCG (2014) Outer Continental Shelf Units-Fire and Explosion Analyses, Notice Of Recommended Interim Voluntary Guidelines, Federal Register, May 2, 2014.

Operational Risk Management Aided by Continuous Monitoring of Barriers

11th Global Congress on Process Safety
Apr 27th, 2015

Ed Silva
Sergio Saad

Summary

"In the process industry, post-Macondo (...), it is essential that operators move away from just demonstrating that they understand the level of risk that personnel and the environment are exposed to. We all need to move towards demonstrating that we have adequate and effective barriers and controls in place to manage that risk effectively."
(Bower-White, 2013)

Business challenge	Solution	Results and benefits
Stop working with bow-tie diagrams statically by starting to continuously monitor the barriers' states, in order to	Use of event-processing technique to automatically update the states of barriers of bow-tie diagrams, based on data from	Successful prototype implementation and execution; Possible application to real situations; Aid in the decision-making in process safety.

Problem statement

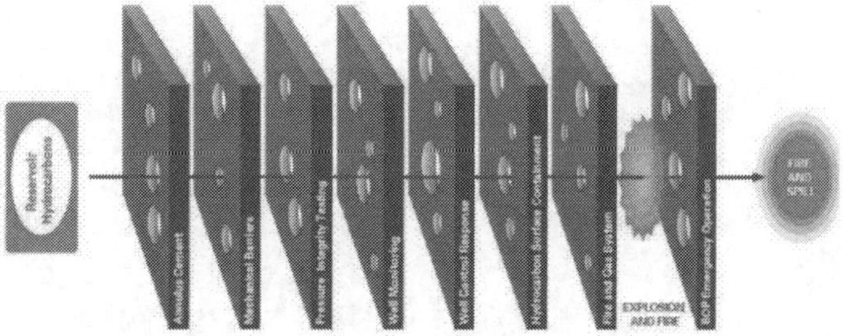

- Continuous monitoring of the states of barriers as a way of as a way to manage operational risk more effectively – no concrete solutions
- Difficulties
  - Information about barriers' states may have origin in several sources
  - Evaluation of the current state of each barrier is done manually ⇒ outdated/inaccurate information
- Alternative
  - Event processing system that provides the automatic updating of the barriers' states to allow their continuous monitoring
  - Constantly updated view of the integrity of barriers ⇒ actual level of risks to which the operations are

Conceptual block diagram

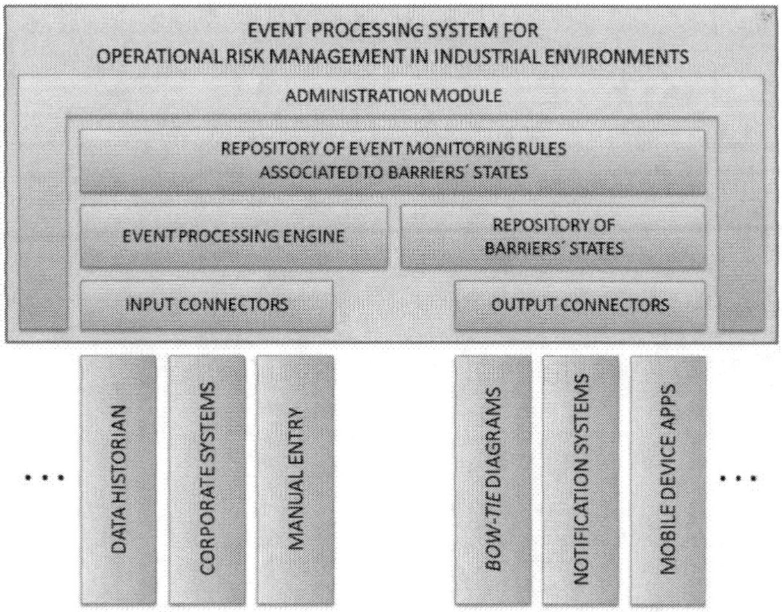

Prototype architecture

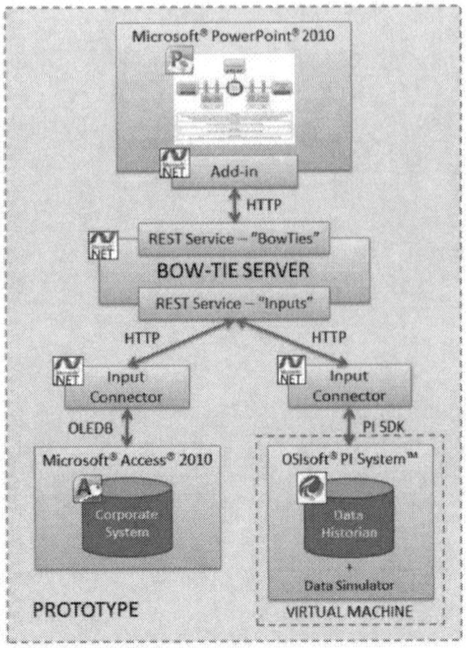

PowerPoint® add-in continuously retrieves bow-tie data from server at user-specified rate and updates barriers' colors

New input values trigger re-evaluation of subscriber barriers' states through re-evaluation of their rules' expressions

Input connectors report value updates to server

Data sources

Business rules and requirements

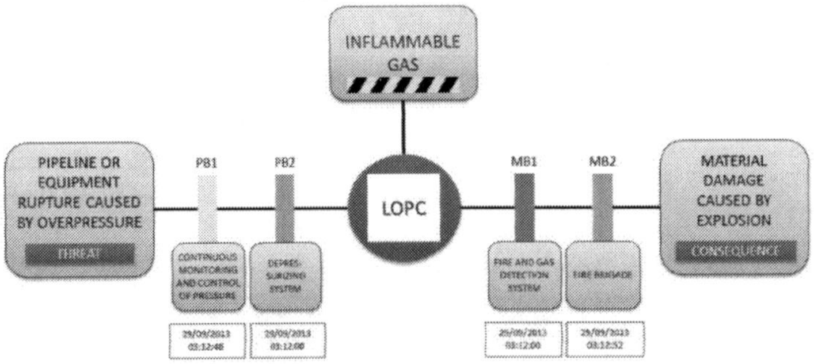

- Barrier states: **Intact**, **Degraded**, **Vulnerable**
- Rules: **boolean expressions + states**
  - Expressions are evaluated according to the order of precedence established by user;
  - Greater precedence true expression = **event** (resulting state);
  - At least one true expression at any time.
- Essential features
  - Rules defined and maintained by end user;
  - Separation between the barriers' states' rules definition/processing and their presentation;
  - Timestamp;
  - Drill-down

Conclusions

- The Event Processing technique is a valid alternative to enable the continuous monitoring of bow-tie diagrams' barriers (successful execution of the prototype system)
- System only is not enough to increase operational safety, it needs to be associated to:
  - A risk management method
  - The engagement of engineers to formulate proper monitoring rules
  - A business process
- Further work
  - Application to other barrier-based risk management methods, either in the oil and gas E&P industry itself or in other industries
  - Security and performance aspects

Thank you!

Ed Silva
Health, Safety and Environment Engineer, MSc
ednevessilva AT gmail.com

Sergio Saad
Computer Engineer, MSc
ssaad33 AT gmail.com

15 AIChE
Spring Meeting, Austin, TX

A Spreadsheet Application for Evaluating Two-Phase Emergency Venting Transients Utilizing an Interface to a Thermodynamic Property Package

Subin Hada
Evonik Corporation
4201 Degussa Road, Mobile, AL 36582
subin.hada@evonik.com

Robert D'Alessandro
Evonik Corporation (Retired)
4201 Degussa Road, Mobile, AL 36582
Cheme1978@gmail.com

J. Andrew Jones
Evonik Corporation
4201 Degussa Road, Mobile, AL 36582
andy.jones@evonik.com

Prepared for Presentation at
American Institute of Chemical Engineers
2015 Spring Meeting
11th Global Congress on Process Safety
Austin, Texas
April 27-30, 2015

UNPUBLISHED

A Spreadsheet Application for Evaluating Two-Phase Emergency Venting Transients Utilizing an Interface to a Thermodynamic Property Package

Subin Hada
Evonik Corporation
4201 Degussa Road, Mobile, AL 36582
subin.hada@evonik.com

Robert D'Alessandro
Evonik Corporation (Retired)
4201 Degussa Road, Mobile, AL 36582
Cheme1978@gmail.com

J. Andrew Jones
Evonik Corporation
4201 Degussa Road, Mobile, AL 36582
andy.jones@evonik.com

Keywords: Two-phase relief sizing, bubbly flow, Churn-turbulent flow, process safety

Abstract

This paper presents a numerical method for determining emergency relief requirements for a top-vented cylindrical vessel or batch-type chemical reactor through an ideal nozzle using a spreadsheet application interfaced to a CAPE-OPEN compliant thermodynamic package. A calculation algorithm based on the DIERS recommended homogeneous equilibrium model are presented for single component system. Fluids exhibiting partial vapor-liquid separation with either churn-turbulent or bubbly flow behavior and disengagement parameters greater than unity were considered. A vessel condition subjected to energy input from an uncontrolled external source such as fire was considered with no chemical reaction. Analytical expressions are used; however, more rigorous solution is achieved by relaxing the assumptions of constant physical properties and constant relieving vent rate of the fluid. The use of a spreadsheet significantly simplifies the tedious unsteady state emergency relief calculations while the use of Aspen Properties® provides reliable thermodynamic data throughout the venting transients.

1. Introduction

Emergency pressure relief is often a dynamic event with rapidly changing physical properties. Having reliable physical and thermodynamic data is vital to a successful safety relief valve design. In addition, this property data must be utilized as a function of temperature, pressure, and composition. As a result, solving emergency relief problems that account for changing physical properties usually require numerical methods that are typically too complex for spreadsheet applications.

The industrial reality involves the design of safety relief valves for two-phase flow [1, 2, 3]. Calculating two-phase venting requirements using most common analytical solutions like Leung [4, 5] and D'Alessandro [6, 7] assume a constant venting rate and constant physical properties for the fluid over the entire venting process. While one assumption may be more severe than the other, they both put some degree of limitations to the applicability of the method for sizing a relief device.

Several computational codes like SAFIRE™ [8] and SuperChems™ [9] have been available for a number of years. These codes can perform thermophysical properties and phase equilibria calculations numerically on pure component as well as mixtures. However, these packages are relatively expensive and difficult to use. Therefore, a practical and easy to use computational tool that computes the two-phase emergency venting requirement under realistic condition is desired.

The contribution presented in this paper introduces a technique to estimate the relief requirement by utilizing:

- a readily available MS-Excel™ [10] spreadsheet application to perform all the differential mass and energy balance calculations, and
- a CAPE-OPEN compliant interface like Aspen Properties® [11] to generate all the necessary thermophysical properties of the fluid based on different thermodynamic descriptions of the phase equilibrium and thermodynamic and transport properties of each phase.

Using the "socket and plug" concept, thermodynamic properties from Aspen Properties® are imported directly into a MS-Excel™ spreadsheet. Here, Aspen Properties® acts as a thermodynamic plug and MS-Excel™ acts as a thermodynamic socket.

The use of a spreadsheet significantly simplifies the tedious unsteady state emergency relief calculations while the use of Aspen Properties® provides reliable thermodynamic data throughout the venting transient.

The novelty of the presented method is that the more rigorous spreadsheet approach relaxes the assumptions of constant physical properties and constant relieving vent rate of the fluid, thereby, relaxing the limits on the applicability of the available analytical methods.

In addition, in industry, spreadsheet calculations are widely used and when combined with a CAPE-OPEN physical properties package, spreadsheets can be an extremely powerful tool for complex calculations. The technique offered is a good engineering practice that ensures the consistency and reliability of the data used and the tools used are readily available.

2. Methodology

The presented paper focuses on the closed form approximate analytical solutions proposed by D'Alessandro [6, 7] for the two-phase emergency relief requirements of top-vented vertical cylindrical containers when the fluids involved exhibit partial disengagement. D'Alessandro's expressions have an advantage over Leung's expression [4, 5] as they allow values for the disengagement parameter other than unity.

The presented work is limited to tempered (vapor pressure) systems where evaporative cooling is available, i.e., the vessel is partially filled with volatile liquid (pure or mixture) that vaporizes or flashes during the depressurization of the vessel. The methods presented in this paper are based on contributions from the Design Institute for Emergency Relief Systems (DIERS) and the American Institute of Chemical Engineers (AIChE) [1].

2.1 Flow Models

Bubbly and churn-turbulent vessel hydrodynamic flow models are considered in this paper to estimate the degree of vapor-liquid disengagement (liquid swell) as a function of vapor throughput. The vent system was modeled as a simple ideal nozzle. A mathematical model to estimate the vent mass flux and volumetric discharge rate as a function of the vessel pressure and the vent entrance quality was modeled using the homogeneous-equilibrium flow model. Figure 1 shows the schematic diagram of the flow regime and flow models that could be considered in the study. The boxes with solid border lines are the cases considered in this work.

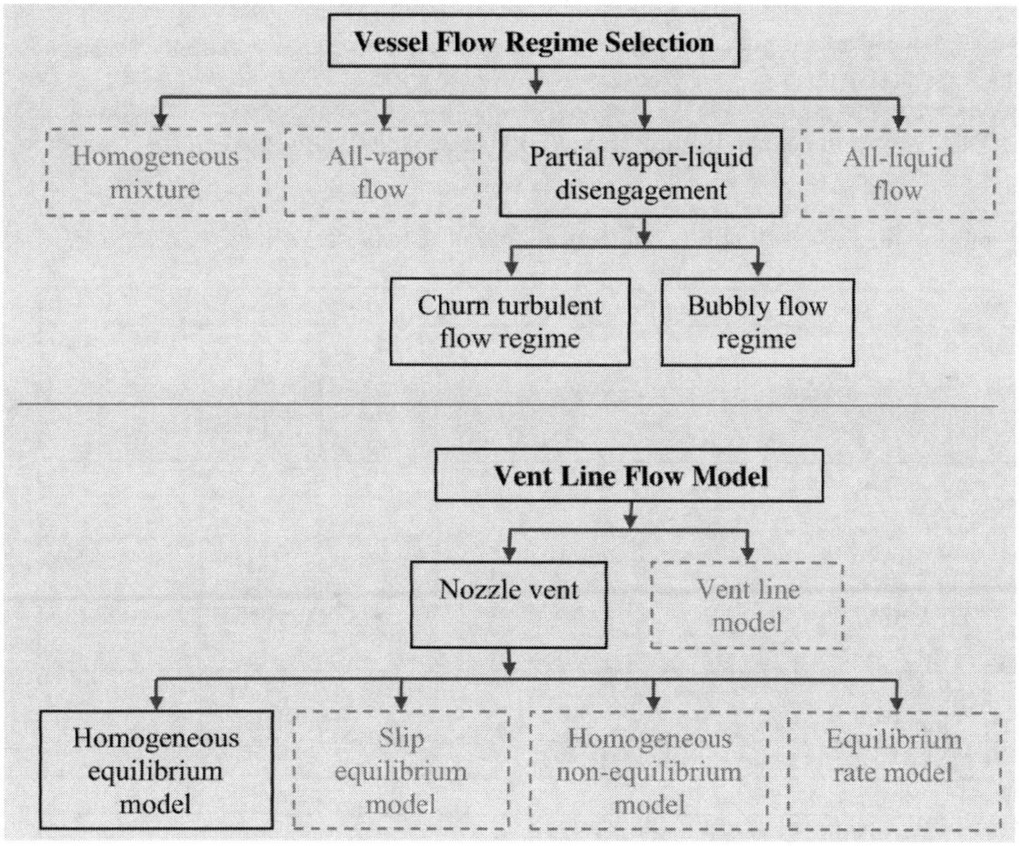

Figure 1: Schematic diagram of flow regime and flow models considered.

Using the DIERS coupling equation, the vessel models and the vent flow models are coupled to provide integrated system models for calculating the volumetric discharge rate from the vessel into the vent [1].

2.1 *Spreadsheet Development*

The rate of change of temperature, pressure, and composition of the vessel contents with time, as a result of external heat loads from fire, were determined by successive heat and mass balances on the vessel contents at small time increments. The calculations were based on ordinary differential and algebraic equations that were integrated numerically forward in time by the explicit Euler method [12].
Once the spreadsheet was set up to do these calculations, Visual Basic for Application (VBA) [10] was used to automate the calculations for every time step that requires solving a series of finite differential energy and mass balances. For every time step, as the temperature and pressure, the physical and thermodynamic properties were obtained from the thermodynamic modeling package Aspen Properties®. The process is repeated until the relief calculation stop criterion (e.g. complete disengagement condition) is reached.

3. Case Study

Pure propane subjected to external fire without undergoing any chemical reaction was used to illustrate the application for one-component system utilizing the methods shown in this paper. The example is taken directly from D'Alessandro's work [6, 7]. The parameters are summarized in Table 1. The churn-turbulent and bubbly flow fluid behaviors with disengagement parameters greater than unity are considered.

Table 1: Propane storage vessel subjected to fire – given parameters.

Vessel			
Diameter	D_c	m	3.49
Height	h_c	m	10.47
Cross Sectional Area	A_{xc}	m^2	9.55
Volume	V_c	m^3	100.0
Vessel Condition at Relief Device Set Point			
Pressure	P_s	bara	4.5
Temperature	T_s	K	271.38
Mass Content	m_s	kg	50483
Void Fraction	α_s	unitless	0.0523
External Heat Input			
Heat Flux	q_F	kW/m^2	26.41
Heat Input	Q_T	kJ/s	3126
Bubbly Fluid Disengagement Parameter			
Bubble Rise Parameter	k	unitless	1.18
Churn Turbulent Fluid Disengagement Parameter			
Bubble Rise Parameter	k	unitless	1.53

Leung and D'Alessandro assumed constant venting rate and physical properties throughout venting transient whereas this work relaxes this assumption by letting the vent rate and the physical properties vary. It was found that the average vent rate (expressed as a geometric mean of vent rate at set condition and at vent rate at turnaround) calculated using short-cut method was close to vent rate at set condition using more rigorous method.

Table 2: Comparison of required vent rates calculated using short-cut method with more rigorous method at fixed set pressure of 450kPa and 10 % overpressure.

		Disengagement Parameter	Vent Area	Average Vessel Void Fraction		Vent Entrance Quality	Required Vent Rate		
			m^2	At Turnaround	At Disengagement	At Turnaround	At Set Condition	At Turnaround	Average
Bubbly Fluid	Leung	**1.00**	-	0.5639	0.9825	0.0491	-	-	117.90
	D'Alessandro	**1.00**	0.0338	0.5675	1.0000	0.0546	125.06	105.65	114.93
		1.18	0.0303	0.4772	0.8275	0.0644	111.40	92.10	101.29
	This Work	**1.00**	0.0297	0.5502	0.9995	0.0646	113.64	97.47	-
		1.18	0.0264	0.4696	0.8132	0.0793	100.53	83.55	-
Churn-Turbulent Fluid	Leung	**1.00**	-	0.1898	0.4731	0.2608	-	-	30.00
	D'Alessandro	**1.00**	0.0117	0.1908	0.4830	0.2592	35.35	25.70	30.14
		1.53	0.0107	0.1743	0.3701	0.2932	31.67	22.75	26.84
	This Work	**1.00**	0.0100	0.1797	0.2751	0.3949	31.15	20.05	-
		1.53	0.0092	0.1652	0.2283	0.4701	27.08	16.96	-

The short-cut results are more conservative than the more rigorous results. These results are consistent with the recommendations given by Leung [4, 5] and D'Alessandro [6, 7]. Figure 2 graphically compares the vent rate estimated using D'Alessandro's method and more rigorous method presented in this paper for Co=1.53 and 10% overpressure.

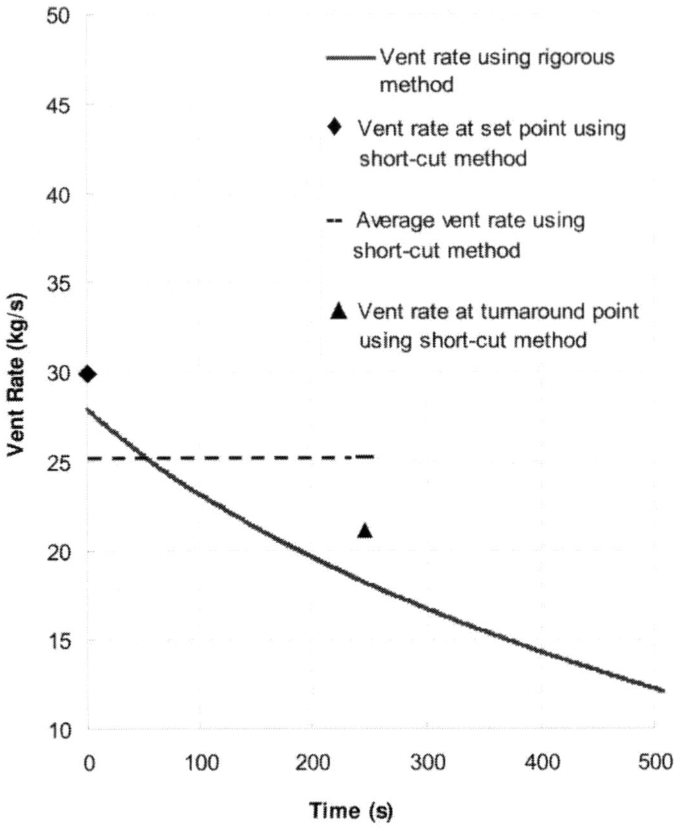

Figure 2: Vent rate using short-cut method and more rigorous method at fixed set pressure of 450kPa and 10 % overpressure with Co=1.5.

3.1 Comparison with SuperChems Solutions

SuperChems^{TM} for DIERS [9] provides a standard tool for designing emergency pressure relief systems which incorporates the latest DIERS pressure relief design techniques and research findings. The software tool has been extensively validated against experimental data to ensure its accuracy.

The transient behaviors for the example of pure propane subjected to external fire are compared to the results generated by SuperChems. The comparisons are plotted in Figure 3. It is clear from the figure that the results are consistent with each other.

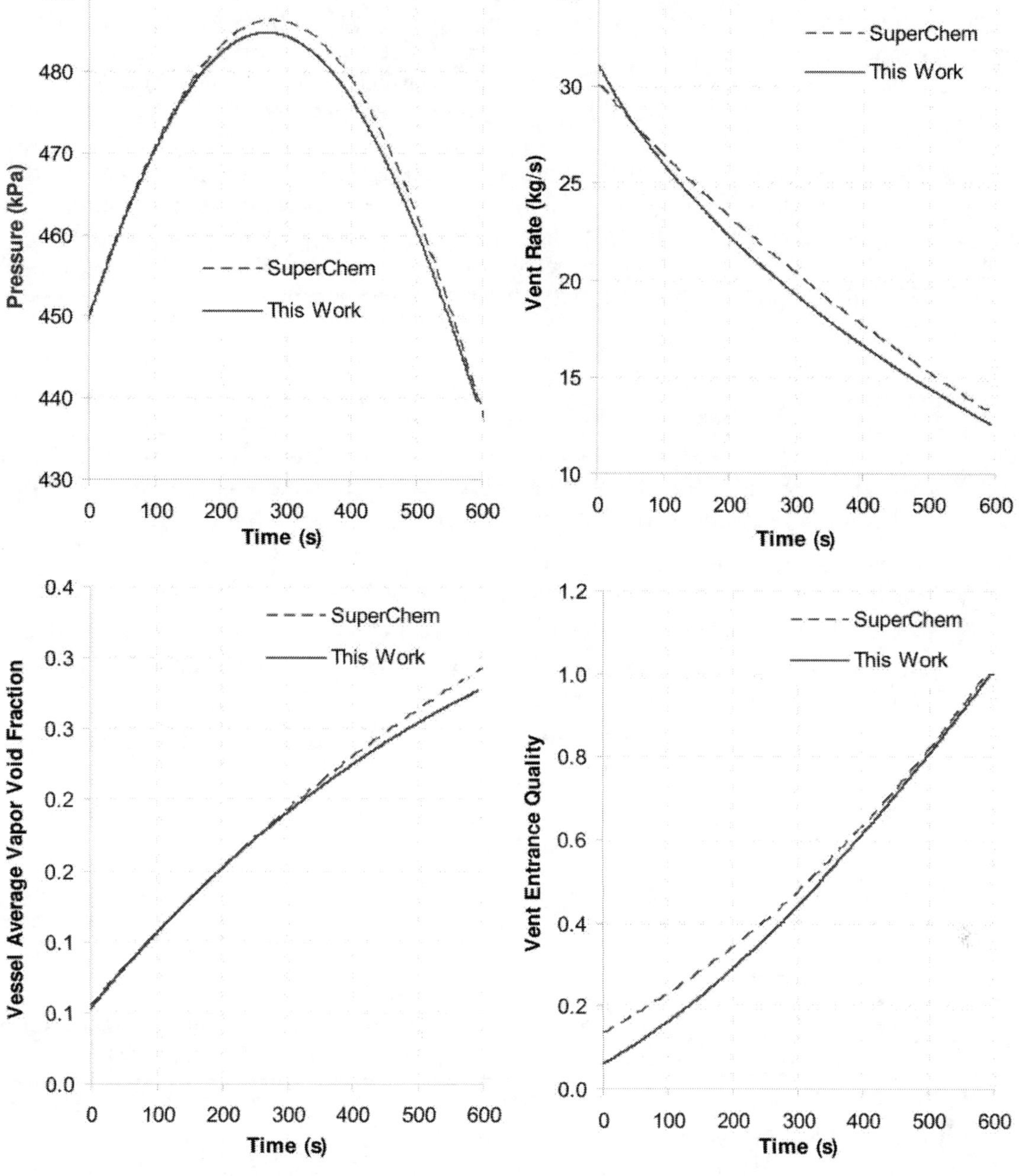

Figure 3: Comparison of more rigorous solutions predicted by method proposed in this paper and solution generated by SuperChems for the example of pure propane with C_o=1.0.

4. Summary and Conclusion

A spreadsheet application for analysis of emergency relief scenarios utilizing an interface to a thermodynamic property package is possible. The spreadsheets were tested for tempered pure and binary systems with relief scenario of external uncontrolled heat input to the fluids exhibiting partial disengagement of bubbly and churn-turbulent flow regime.

The use of a spreadsheet significantly simplifies the tedious coupled differential mass and energy balance calculations while the use of Aspen Properties® provided reliable thermodynamic properties used for numerical evaluation of the temperature and pressure transients. In industry, such a practice ensures the consistency and reliability of the data used. Moreover, in industry, the use of a spreadsheet is practical for essentially all practicing engineers.

5. References

[1] Fisher, H. G., Forrest, H. S., Grossel, S. S., Huff, J. E., Muller, A. R., Noronha, J. A., Shaw, D. A., & Tilley, B. J. *Emergency relief System Design Using DIERS Technology, The DIERS Project Manual*. AIChE, NY (1992).

[2] Boccardi, G., Bubbico, R., Piero Celata, G. & Mazzarotta, B. "Two-Phase Flow through Pressure Safety Valves. Experimental Investigation and Model Prediction." *Chemical Engineering Science*, 60, 5284-5293 (2005).

[3] Darby, R., Self, F. E., & Edwards, V. H. "Properly Size Pressure Relief Valves for Two-Phase Flow. *Chemical Engineering*, 68–74 (2002).

[4] Leung, J. C. "Simplified Vent Sizing Equations for Emergency Relief Requirements in Reactors and Storage Tanks." *AIChE Journal*, 32(10), 1622-1634 (1986).

[5] Leung, J. C. "Overpressure During Emergency Relief Venting in Bubbly and Churn-Turbulent Flow." *AIChE Journal*, 36(6), 952-958 (1987).

[6] D'Alessandro, R. N. "Emergency Venting Requirements for Tempered Systems Considering Partial Vapor-Liquid Separation with Disengagement Parameters Greater than Unity, Part I: Model Development". *Process Safety Progress*, 23(1), 1-15 (2004).

[7] D'Alessandro, R. N. "Emergency Venting Requirements for Tempered Systems Considering Partial Vapor-Liquid Separation with Disengagement Parameters Greater than Unity, Part II: Application". *Process Safety Progress*, 23(2), 86-98 (2004).

[8] Shaw, D.A. "SAFIRE Program for Design of Emergency Pressure Relief Systems." *Chemical Engineering Progress*, 86(7), 14-17 (1990).

[9] Melhem, G. A., and Fisher, H. G. "An Overview of SuperChems™ for DIERS: A Program for Emergency Relief System and Effluent Handling Designs." *Process Safety Progress*, 16, 185-197 (1997).

[10] Excel 2003™, Microsoft Corporation, Redmond, Washington, 2003.

[11] ASPEN PLUS™ Release 9, Volume 2: Physical Property Methods and Models, Volume 3: *Physical Property Data* (1995).

[12] Hoffman, J.D. *Numerical Methods for Engineers and Scientists*. 5th ed., CRC Press, Baca Raton (2001).

402907 Anatomy of a Potential LNG Incident

Abstract:

Recent years have seen the growth in production and handling Liquefied Natural Gas (LNG). While the LNG industry is, in principle, similar to other hydrocarbon industries, the properties of LNG pose some distinct threats for their facilities. Among these are cold embrittlement, liquid metal embrittlement (associated with mercury contamination), rollover with rapid phase transition, and wide temperature swings. Examples of major incidents for these and other related hazards are presented. In addition, design and operational controls to prevent and mitigate the risk arising from these threats are discussed. Various types of LNG releases are described and aspects of emergency response are also reviewed. The paper closes with a comparison of some LNG misperceptions and actual factual information.

Application of Risk Prioritization Matrix of Nonconformities in the Offshore Industry

Tatiana Barboza Coelho Cordeiro
DNV GL / UFRJ
Rua Clapp Filho, 360 – Cachambi – Rio de Janeiro/Brasil
taticord@bol.com.br

Claudia Morgado
Universidade Federal do Rio de Janeiro - UFRJ
Avenida Thrompovisk, s/n – Ilha do fundão – Rio de janeiro – Cep.: 21.941-972, Brasil
claudiamorgado@poli.ufrj.br

Prepared for Presentation at
American Institute of Chemical Engineers
2015 Spring Meeting
11th Global Congress on Process Safety
Austin, Texas
April 27-29, 2015

UNPUBLISHED

Application of Risk Prioritization Matrix of Nonconformities in the Offshore Industry

Tatiana Barboza Coelho Cordeiro
DNV GL / UFRJ
Rua Clapp Filho, 360 – Cachambi – Rio de Janeiro/Brasil
taticord@bol.com.br

Claudia Morgado
Universidade Federal do Rio de Janeiro - UFRJ
Avenida Thrompovisk, s/n – Ilha do fundão – Rio de janeiro – Cep.: 21.941-972, Brasil
claudiamorgado@poli.ufrj.br

Keywords: Audit, nonconformity, ANP3, risk, prioritization and offshore

Abstract

This paper aims to present a methodology for prioritization of nonconformities arising from audits of HSE within the offshore segment. The methodology developed will be an internal management tool of great value in organizations; for, with its implementation the companies will have a more concrete basis to make decisions about the prioritization of corrective actions to, effectively, eliminate the nonconformity, not allowing its recurrence.
For the development of the prioritization methodology it was necessary structuring of some matrices, based on some techniques and methods of prioritizing, available in the literature; such as, Analytic Hierarchy Process, Ishikawa diagram, Pareto diagram and GUT matrix.
Among the developed matrices, we can highlight the array of severity as the cornerstone of the prioritization methodology, according to which, its development was based on legal requirements for offshore segment. Moreover, the methodology assigns a degree of correlation of nonconformities in the Management Practices of Management Operating Safety Management System (SGSO) defined in Resolution N°. 43 of the National Agency of Petroleum, Natural Gas and Biofuels (ANP) of Brazil that this way establishes the proper importance of nonconformities most significant guideline for management systems in the offshore segment.

1. Introduction:

This paper aims to present a prioritization methodology of nonconformities arising from audits in the management system in the offshore segment. This tool is extremely

important because nonconformity which does not receive appropriate treatment can sometimes result in the standstill of operations of the plant, due to the occurrence of an unwanted event, which in the offshore segment can mean a few millions wasted.
According Bird (1974), a major cause of accidents is the lack of control of the administrative elements. The main problems identified are: lack of safety standards, maintaining a loss control program for inappropriate parts of managers and supervisors, inadequate or insufficient rules and failure to comply with standards and regulations. It is necessary to take decisions which concern the organization in cost function and the effectiveness of established actions for the management of nonconformity, it is the key point in a management system. Inspections and evaluations in the present study are related to HSE (health, safety and environment) and will be of great help for organizations in prioritizing corrective actions, therefore, preventing their recurrence.

2 Standards for development of Audits

The offshore industry shows some need for good management of the issues related to the environment, occupational health and safety. In this aspect it is necessary to develop a HSE audit (health, safety and environment) which involves all those subjects. The integrated management system has served as grounds for many researchers. Among these, Ayoade and Gibb (1996) argued that the integration of these systems would be of significant benefit in improving productivity and quality management processes. Today the HSE management in the Brazilian offshore segment is based on technical-scientific entities, standards and national and international legislation. At the start of offshore activities technical regulations were dictated by Petrobras Company. However, with the growth of this industry it was necessary to create environmental laws and designate authorities to monitor the risks related to their operations in an orderly manner

3 Priorization Methodology of Non conformity based on risk

The choice of prioritization technique within a management system is a process that must be analyzed carefully, because each organization has its own characteristics and the technique should adapt to the available environment. The important point is the effective implementation of the action suggested to avoid recurrence of nonconformity. (KEYS, 1997). Normally, the prioritization of corrective actions is made intuitively, taking into account the knowledge of the managers involved in the decision-making process. However, this methodology does not guarantee consistency in the final result. This is a major problem for which there is the need to establish a prioritization methodology. The chosen segment for the application of this methodology was the offshore, which in Brazil is regulated and supervised by the ANP, through Resolution 43, establishing a Safety Management System (SGSO). The practices were developed based on extensive study of the policies adopted in countries like USA, Canada, Norway, UK and Australia. (ANP, 2007)

3.1 Prioritization Methodology

The object of this study is based on an order of magnitude that is allocated according to the percentage distribution of nonconformity found in the audit of Resolution ANP43. The assigned order of magnitude was based on the trial methodology suggested by Saaty (1996/1997) established through a synthesis of orders of magnitude, the judgment of nonconformity according to their degree of importance within the accepted hierarchy. The classification presented in Table 1 is given as follows: "1" for the practice who presented a result below 5% ; "3" for practice that had a score between 5 and 15% and "9" to practice a result above 15%. Table 6 is a summary of the assigned rating. (HADDAD, Galante, Caldas & MORGADO, 2012)

Table 1: Classification of summary management practices
Source: Author

% Non conformity	Order of Grandeur
No nonconformity	"0"
Below 5%;	"1"
Between 5 and 15%	"3"
Above 15%.	"9"

From the award of this order of magnitude is held a joint comparison of the nonconformity of ANP43 management practices and non-conformities found in the audit of HSE. A table model that can be used in this step is shown in Table 2.

Table 2: Comparison Table parity Model
Source: Author

	[illegible]	[illegible]	[illegible]	[illegible]	[illegible]	[illegible]	[illegible]	[illegible]	[illegible]	[illegible]	[illegible]	[illegible]	[illegible]	[illegible]	[illegible]	[illegible]	[illegible]	
PRACTICE OF ANP 43	1	2	3	4	5	6	7	8	9	10	11	12	13	14	15	16	17	
CLASSIFICATION OF NONCONFORMITY / Non conformity [illegible]																		Index of Criticality

The criticality index (Table 2) is a mathematical formulation of the classification of nonconformity and the ANP43 management practices in order to allow prioritization of nonconformities in its preventive aspect. For each nonconformity was awarded an order of magnitude which also was based on the trial methodology suggested by Saaty (1996/1997), in which "0" means that the nonconformity has no correlation with practice management, "1" that the correlation is extremely remote, "3" that the correlation is remote and "9" correlation is likely. (HADDAD, Galante, Caldas, MORGADO, 2012)

3.1.1 Severity Matrix

The methodology aims to prioritize nonconformities related to audit HSE, depending on the area was necessary to develop a categorization of these HSE impacts, synthesized in ascending order of priority, through the following matrix:

Table 3: Severity Matrix
Source: Author

			IMPACT				
			Safety and Health	Environmental	Corporate Image	Fine (UFIR)*	
						Min	Max
CLASSIFICATION OF NONCONFORMITY	9	Very serious	Death Lesion with leave (permanent incapacity)	Catastrophic Environmental impacts due to release of chemical, toxic or flammable substances, reaching areas outside the premises. Causes death or serious injury in the external population or impact on the environment with high recovery time.	International National	3821	6304
	3	Medium	Lesion with leave (temporary incapacity)	Critical Possible damage to the environment due to the release of toxic or flammable chemicals, reaching areas outside the installation. May cause moderate severity lesions on the outside population or environmental impacts with reduced recovery time	National	2472	4418
	1	Low	Lesion without leave	Marginal Irrelevant damage to the environment and the community	Regional	1242	2792
	0	Insignificant	Without lesion	Negligible No damage or damage not measurable	Municipal	630	1935

* Fiscal Reference Unit - UFIR / BRAZIL

The matrix, Table 3, presented the classification of nonconformity divided into 4 categories: 9 (very serious), 3 (Medium) 1 (low) and 0 (insignificant), this order of magnitude is used individually according to the impact to which the nonconformity is connected, as a non conformity "Safety and Health" can be classified as "9" when the aspect associated "Safety and Health" while receiving a rating of "0" when associated with the environmental aspect. HSE impacts assigned to each order of magnitude assigned a rating according to the injury to the worker in case of occurrence of nonconformity, which can be with or without removal or even lead to death. The assignment of the order to the "Safety and Health" and environmental impacts was based on the Guidelines Manual for the development of the State Company's risk analysis studies of basic sanitation and water pollution control technology (CETESB-SP) that is used by most companies to qualify their environmental risks (CETESB, 2003). The values stipulated fines for each category of nonconformity rating are according to the grading and classification of offenses stipulated by NR28. For application to the prioritization methodology was necessary to adapt the data of NR28, in Table 5 shows the fine grading, divided into colors according to the fine intervals depending on the severity of nonconformity (MTE, 2014).

Table 5: Fine Grading

Source: Adaptation NR28

Number of employee	Safety at work							
	I_1		I_2		I_3		I_4	
	Min	Max	Min	Max	Min	Max	Min	Max
01-10	630	729	1129	1393	1691	2091	2252	2792
11-25	730	830	1394	1664	2092	2495	2793	3334
26 - 50	831	936	1665	1935	2496	2898	3335	3876
51-100	964	1104	1936	2200	2899	3302	3877	4418
101-250	1105	1241	2201	2471	3303	3717	[illegible]	[illegible]
251-500	1242	1374	2472	2748	[illegible]	[illegible]	[illegible]	[illegible]
501-1000	1375	1507	2749	3020	[illegible]	[illegible]	[illegible]	[illegible]
More than 1000	1508	1646	[illegible]	[illegible]	[illegible]	[illegible]	[illegible]	[illegible]

The last impact included in the severity matrix is the impact the corporate image, which was based on the understanding Ouédraogo (2011), which explains the importance of assessing the impact of corporate image to the risks and dangers, because in some situations can be the only impact occurs. The Ouédraogo rating for the impact the brand image was the basis for the award of the impact the corporate image stipulated in the displayed severity matrix.

3.1.3 HSE Matrix Prioritization

The HSE Matrix Prioritization is a simple method that, according to an order of magnitude, will identify impacts to which this nonconformity could be related and will help significantly in decision making within an organizational system. In this sense, the HSE Matrix Prioritization should be considered a dynamic tool, which may be completed in order to ensure that nonconformities are compared according to the criticality and severity index for the HSE goals are achieved and actions corrective are conducted in accordance with their actual degree of importance. Table 4, below, shows the model developed the HSE priority matrix, where the non-conformities are sorted by the quantities of orders described above.

Table 4: HSE Matrix Prioritization

Source: Author

	NONCONFORMITY IN [illegible]	CRITICALITY INDEX	SEVERITY INDEX				HSE's PRIORITIZATION OF ACTION
			Safety and Health	Environmental	Corporate Image	Fine	

The allocation methodology of the order will be the same used in the criticality index, where "0" means that the noncompliance has no correlation with the impact of HSE, "1" that the correlation is extremely remote, "3" that the correlation is remote and "9" correlation is likely. Then we present in more detail the calculation of the prioritization of action for nonconformity.

Owning the results obtained prioritization of action of each nonconformity was necessary to use a tool that could identify no longer relevant in accordance prioritization process management of corrective actions resulting from HSE audit. The chosen tool is the Pareto diagram, it is a statistical tool for determining the relative importance of issues or causes and identify thus the most relevant (Frankenfeld, 2003).

4 Conclusion

There is a growing concern for companies of all segments regarding HSE issues. However, the offshore segment companies need greater control of performance of its management system, for its operations offer a great risk both to the environment and worker safety. Due to this scenario, offshore installations are regulated by different standards and resolutions both national and international. Of any magnitude and category companies are susceptible to internal and external elements that directly influence the performance of your management system. Therefore, the organization must have a systematic able to assist in making decisions, especially when it comes to non-conformities related to both internal and external audits. There are some techniques available in collected works for prioritizing action, there is a correlation between their concepts and classifications. In any of the chosen systematic, decision making is fundamentally based on the knowledge of managers and staff involved in this process. Based on the information on prioritization techniques, there was a crossing of data for the development of one of the fundamental pillars of the proposed methodology: the HSE severity matrix. In conjunction with the matrix, to promote decision-making in the classification of nonconformity, it was necessary to adopt the criticality index that is an order of magnitude according to the percentage distribution of non-conformities found in an audit for the model developed was assigned the correlation of nonconformity with the ANP43 management practices. With the results found you can prioritize nonconformity in accordance with their individual criticality index obtained for the audit used as a base. This tool will be of great value to companies in preparing the action plan, because companies often lack the resources to eliminate the no's conformities found simultaneously. Nevertheless, the methodology presented has not eliminated any of the nonconformity, as all have their degree of relevance within the relevant legislation and management system.

5 References

[1] ANP, 2007. Agência Nacional do Petróleo, gás natural e biocombustíveis. Available at: http://nxt.anp.gov.br/nxt/g ateway.dll /leg/resolucoes_anp/ 2007/dezembro/ranp%2043%20-%202007.xml. Accessed on March 22,2013.

[2] AYOADE, Aubrey & GIBB, Alistair G.F. I. Integration of quality, safety and environmental systems in Implementation of Safety and Health on Construction Sites – Proceedings of the first international conference of CIB working commission W99. Lisboa. Portugal. September, 1996.

[3] BIRD Jr, Frank E. Management guide to loss control. Atlanta: Institute Press, 1974.

[4] CETESB - P4621: Manual de Orientação para a Elaboração de Estudos de Análise de Riscos - Mai/2003. Available at: <http://www.cetesb.sp.gov.br/ gerenciamento-de-riscos/Emeg?ncias-Qu?micas/14-Normas-> Accessed on Juny, 2014.

[5] FRANKENFELD, Karoline Pinheiro. Gerenciamento de produtos químicos em unidades de geração de energia em plataformas Offshore: Uma aplicação da metodologia seis sigma. Rio de Janeiro. August, 2003

[6] GIBB, Alistair G.F. Implementation of project management safety manual. in Implementation of Safety and Health on Construction Sites – Proceedings of the first international conference of CIB working commission W99. Lisboa. Portugal. September, 1996

[7] HADDAD, Assed, GALANTE, Erick, CALDAS, Rafaell, MORGADO, Claudia. Hazard Matrix Application in Health, Safety and Environmental Management Risk Evaluation ISBN 978-953-51-0571-8, Published: April 25, 2012 under CC BY 3.0 license.

[8] MTE- Ministério do trabalho e emprego. Available at: <http://portal. mte.gov.br/legislacao/normas-regulamentadoras-1.htm> Accessed on: 02 de maio de 2014.

[9] Ouédraogo, Aristide, Groso, Amela, Meyer, Thierry. Risk analysis in research environment – Part II: Weighting Lab Criticity Index using the Analytic Hierarchy Process. Safety Science. Janeiro, 2011

[10] SAATY, T.L., Decision Making with Dependence and Feedback: The Analytic Network Process (ANP and ECNET Software) Guide, Manual and Examples, Pittsburgh-PA, 1996-1997.

This page intentionally left blank.

396731 Experimental Study on Gas Explosion in an Obstructed Confined Pipe

Abstract:

The acceleration of flame inside a pipe is a complex phenomenon involving several variables including fuel nature, mixture composition and geometrical characteristics of the pipe itself such as length, wall roughness or presence of obstacles. It has been well known that the obstacles in pipes and long ducts would accelerate the flame propagation, and cause the transition from deflagration to detonation. This work sought to evaluate the magnitude of the hydrocarbon-air mixtures explosion using 90-degree pipe and T-junction configuration as a full bore obstacle condition. The experimental work was performed in a confined pipe of 0.42 m3 volume, giving a length to diameter ratio (L/D) of 51. Hydrogen, propane and natural gas-air mixtures were used in this work with equivalence ratio (Φ) ranges between 0.5 to 2.0. The experimental results show that explosion characteristics depended on the carbon content, the layout of obstacles, and the content of argon and nitrogen as an inhibitor. It can be seen from the experimental results that a highest overpressure was observed at a slight above to all fuel-air mixtures as compared to stoichiometric. In addition, records of pressure traces showed a substantial rise to the pressure downstream of the pipe, known as retonation and this phenomenon is discussed.

396332 Uncertainty Analysis for LOPA in the Offshore Industry

Abstract:

According to IEC 61508 and IEC 61511, a variety of methodologies, e.g., risk graph, hazard matrix, LOPA(Layer of Protection Analysis), are introduced to determine target SILs of SISs(Safety Instrumented Systems). Recently, in the offshore industry, LOPA has become the most popular method since it reflects the effects of existing safeguards on a system.

However, even the LOPA method is powerful, there are some uncertainties mainly due to assigned numeric values. With these uncertainties, the target SILs can be affected in terms of credibility. In order to assure that the determined target SILs are credible, uncertainty analysis for the LOPA method can be utilized to show the level of uncertainty of target SILs. To investigate the uncertainty level of target SILs which are determined by the LOPA method, the appropriate uncertainty analysis method should be selected among the numerous methods developed in many applications.

This paper suggests the proper LOPA uncertainty analysis technique based on the natures of the LOPA method. The LOPA method has both qualitative and quantitative attributes. The qualitative attributes come from the C(Consequence or Severity) value which is determined by experts' consensus. The quantitative attributes, on the other hand, exist due to numerical parameters such as the initiating event frequency, PFD(Probability of Failure on Demand) for IPLs(Independent Protection Layers), and intermediate event likelihood. To reflect these qualitative and quantitative characteristics at the same time, the combination of the fuzzy set approach and the Monte Carlo simulation are proposed. By using the proposed method, the case study has been conducted. The results of the analysis will help establish the guidline when applying the LOPA method.

397743 Quantitative Risk Assessment (QRA) with the Inclusion of Human and Organizational Factors and Maintenance

Abstract:

Human and organizational factors (HOF) and improper maintenance have led to number of accidents in chemical process industry, this notion was also strengthen after concluding an analysis of past accidents in chemical process industry between 1983-2013. Therefore, there is a need to include the HOF and maintenance into overall risk calculations. Quantitative Risk Assessment (QRA) study is usually carried out during the design phase of a project, therefore QRA is selected to integrate the effect of HOF and maintenance.

3D Computational Fluid Dynamics (CFD) consequence analysis of potential loss of containments events is widely used in chemical process industry. In this study an analysis will be performed to quantify the HOF with the aid of 3D CFD models which can provide an enhanced detail about consequences of an undesired event.

By considering HOF and maintenance aspects, Location Specific Individual Risk (LSIR) and Individual Risk Per Annum (IRPA) can be calculated in more comprehensive way compare to conventional practices. A case study is also presented with the proposed method to calculate the risk.

396753 Implementing SIS Metrics for Life Cycle Management - Part 1: Proof Testing

Abstract:

Implementing SIS Metrics for Life Cycle Management

By

Keith Lapeyrouse , P.E. **and** **John Harrison**

Reliability Manager | **Senior Solution Specialist, IBU Chemicals**

The Dow Chemical Company | **SAP Labs**

P.O. Box 150, | **4120 Yonge Street, Suite 600**

Plaquemine, LA 70764 U.S.A. | **Toronto, Ontario M2P 2B8 Canada**

kmlapeyrouse@dow.com | **john.harrison@sap.com**

Keywords: Vision 20/20, , Metrics, KPI's, , IT Project Execution, Process Reliability Enterprise Resource Planning (ERP) System,

Abstract

This paper will overview the key elements of implementing SIF Metrics for Life Cycle Management. The key elements of implementation SIS Metrics are:

Defining the business case for what to measure

Defining what measures are most important now

Defining what sources and information needs to be collected to execute the measurement

Data Collection – The new frontier

Executing the Measurement

Driving action with the information

Each of these elements are essential to an effective SIS Metric Implementation … whether for one or for many metrics. These elements can be aggregated into the categories: 1) Business justification, 2) System Implementation and 3) Action Execution. Implementation of SIS Metrics aligns to the CCPS Vision 20/20CITATION AIChE_CCPS01 \l 1033 (American Institute for Chemical Engineers 2014) initiative by supporting the Industrial Tenant of Disciplined Adherence to Standards and the Societal Themes of Harmonization of Standards and Meticulous Verification.

398050 Dust Explosion Kinetics

Abstract:

In this paper we propose a new method for analyzing dust explosion data. The standard approach is to determine K_{st} values from the explosion pressure histories obtained over a dust concentration range using a 20-liter spherical vessel. As an alternative, we treat the 20-liter sphere as a batch reactor and show how global kinetic parameters can be obtained from the same data by using an approach base on integral kinetic data. Global reaction rates are extracted from the pressure history data using chemical equilibrium calculations. Specifically, the method extracts both the concentration dependence and the temperature dependence for the global reaction rates. The results are expressed in terms of the functional dependence of the reaction rate on the concentration of fuel and oxidizer, and the temperature dependence is cast as an Arrhenius activation energy and pre-exponential factor. Although the approach can be used on a combustible dust sample of fixed chemical composition and just one particle size distribution, a more robust set of kinetic parameters can be obtained when samples of several different particle size distributions are tested. The methodology is demonstrated with aluminum dust test data.

398067 Vapor Cloud Hazards of Enclosed Liquid Flammable Processes

Abstract:

Simple approaches to evaluate vapor cloud explosion hazards from releases within enclosed process plants often assume that the entire volume of the building in filled with flammable vapors. This can result high loads that could be overly conservative, particularly for flammable liquid releases. Operations involving solvent processes are often located in such enclosed buildings. Such processes operate at relatively low temperatures and pressures, but with sufficient quantities of solvent to be PSM covered. This paper addresses vapor cloud generation and deflagration from the release of two common solvents, hexane and methanol, within enclosed process buildings using CFD simulation tools. It evaluates the effect of containment curbing on vapor cloud formation and the effect of pressure relief through wall panel failure on deflagration severity. Comparisons are made between the results of the each approach in an effort to aid companies in applying results to their processes and PSM program.

397714 Comparison of Results of the Classified Areas of a Typical Hydrogen Generation Unit (GHU) Using the Standards NFPA 497 and API RP 505

Abstract:

The risk analysis technique called classification of area is widely used in the petroleum industry, being an important tool for the installation of electrical equipment under safety, both in the aspect of equity and the workers. All electrical equipment to be installed in areas with potentially explosive atmospheres need to be certified under the INMETRO, i.e., have a "type of protection". This certification process is costly and therefore reduce this area is synonymous with financial savings for the company. In Brazil the standards used in studies of classification areas are foreign API RP 505 and NFPA 497 and international ABNT NBR IEC 60079-10-1, each with its own peculiarities. This work proposes to compare the implementation of these standards for the classification of areas in the same scenario with the analysis of the differences and similarities between these. The study indicates that it is not possible to use the standards individually. However, a careful review of foreign standard API RP 505 proved possible to reduce the extent of the classified area influencing in a reduced need for equipment with a higher degree of protection.

The Requirements & tools to treat process safety risk as result of Organizational Change

Salvador Ávila Filho
Federal University of Bahia
Salvador – Brazil
avilasalva@gmail.com

Esiron Tárcio Dantas
Federal University of Bahia
Salvador – Brazil
tarciodantas@hotmail.com

Jonas Duarte
Chemtura Corporation
Ontario – Canada
Jonas.Duarte@chemtura.com

Prepared for Presentation at
American Institute of Chemical Engineers
2015 Spring Meeting
11th Global Congress on Process Safety
Austin, Texas
April 27-29, 2015

UNPUBLISHED

The Requirements & tools to treat process safety risk as result of Organizational Change

Salvador Ávila Filho
Federal University of Bahia
Salvador – Brazil
avilasalva@gmail.com

Esiron Tárcio Dantas
Federal University of Bahia
Salvador – Brazil
tarciodantas@hotmail.com

Jonas Duarte
Chemtura Corporation
Ontario – Canada
Jonas.Duarte@chemtura.com

Keywords: organizational change, leadership, checklist, risk, human reliability

Abstract

The organizational changes can trigger events causing accidents, especially when there is substitution of key personnel or functional departments that could stop the initiating events of accident process. The lack of appropriate decision-makers and experts, with the dynamism and knowledge required, creating gaps in functions inside the company. On the other hand, the analysis of long-term or short-term to decisions depends of absent job staff. The speed is not appropriate for the integration between maintenance and operation in the industry.
The organizational change has greater effect on process safety than technological changes. The organizational changes alter the balance of the work and business which, according to most authors of reliability and safety, is the area of the root cause of the major events (accidents). The organizational changes that can lead to greater fragility in process safety at industrial facilities are: working conditions, personnel headcount, functional re-allocation, organizational hierarchy, and, policy changes that affect safety culture. It is essential that analyze the loss arising from organizational change to define what are the main characteristics desired in leadership who will plan, deploy, and monitor these changes.
The proposition is to anticipate the discussion of critical situations caused by organizational changes using a checklist that call a committee composed by internal and external members of the site in study. This committee will propose preventive and corrective actions to avoid events in safety process of this industry. In this paper a checklist was answered by staff involved with a case of organizational change.

Note: Do not add page numbers. Do not refer to page numbers when referencing different portions of the paper

1. Introduction: technical changes

The industry needs physical changes in technologies to meet the challenges of quality, quantity and cost of production. The nature needs a balance in production, having won the positive image of Maintenance Company in the market and society. Thus, governments try with operating licenses stay organized industries in design in our house is that the planet earth.
Economic relations may require urgent need to increase the capacity of the plant, type of product, process parameters, and the production rate. In this situation, the installation of new equipment with new controls cannot cause pleasant surprises for the operational team that is not structured to change. It is important to know what level of knowledge that will be needed for changes in the operation and the principles of the new equipment. The flexibility of the production load, the type of product and quality is a challenge for the sector's sustainability. Changes in process conditions for new quantities, qualities and new products require technical changes that may increase the dangers already installed in the chemical industry. Thus, the CCPS [1] suggests the analysis of changes in process / equipment and the impact on process safety through the Change Management (MOC).

1.1. Change Management Physics and Labor [1]

The technical changes to meet the needs of the market are in the areas of technology, process control and production planning as showed in Table 1. New equipment for increased production in different quality specifications, or radical changes as new technological routes require new control systems to maintain the availability and security processes.

Table 1 - Technical changes that affect safety process

a) Technologies infrastructure equipment & Technological Routes	b) Process Control Quality (Inspection & testing) process security	c) Production Planning Work organization Procedure & Standard maintenance

a) Technological Changes

Technological change requires modification of the basic knowledge and skills available in production. The logistics, the type of product and process, and the type of equipment are considered very important by requiring the installation of key competences to enable the production.
Local infrastructure changes, such as protection for the fire, temporary buildings, permanent buildings, access, and service systems can influence the events leading up to an accident. Changes in technology and operations such as process conditions, process paths, specifications of raw materials, product specifications, new chemical, and promote change in packaging process changes the level of danger demanding new control systems, training, and care to avoid accidents that impact on the men at work, the surrounding environment, and society. The process equipment changes such as building materials, configuration, and design parameters require care in operational control and technological knowledge. Thus, operation and maintenance procedures as well as the tools necessary for its implementation are reviewed for each change of equipment.

b) Process Control Changes

Quality issues of processes and product require changes in process conditions and their changes in instrument tracks. In some cases the increase in production facilities requires a higher level of automation requiring a higher degree of automation in industrial plants, especially in the chemical and petrochemical processing industry has high complexity. With these process control demands, increases the current danger, amount of product processing, temperature and pressure, and often change chemicals unit operations and processes. This requires change in the security project by opening the very current discussion about the possibility of occurrence of major events, the discussion about safety processes.
The need for instrumented systems to reduce the possibility of human error brings a lot of needs audits and calculations of the level of security for the instrumented systems. This an area that the security control changes are not well managed can lead to accidents.
Thus, in simplified form, process changes in control (instrumentation, controls, interlocks and computer control), in quality of material inputs (require changes in inspection and test methods), in process safety automatic systems (SIS), in work rules must be analyzed to avoid accidents.

c) Planning and Production Control Changes

Market demand requires changes in production by changing its resources through the organization of work, review procedures and standards and changing the scheduling of maintenance. In work organization are discussed aspects such as policy change that may cause stress (increased number of overtime), changes in staff support for the control of production and the reduction of the number of operators in turn, change the contracted and the number of days of operation.
The viability of routine activities depends on the definition of standards and task by writing procedures. Changes in procedures, standards, and work practices must be managed to prevent incidents in the operation. The procedure for emergencies when they are changed is also worrying and may increase the impact of accidents. In terms of maintenance, with changes in equipment and technologies, changes in the quantities and process variables, the change of the current product type in the equipment requires modification of maintenance controls for reliability and availability of these. Thus, the plans of preventive and repair requirements to be analyzed by change management to prevent the occurrence of unexpected events such as major leaks from the maintenance activity.

1.2. The need for adjustment in Change Management

Allied physical changes, functional changes can trigger events that cause accidents [2][3][4], especially when there is the replacement of key people or groups that could stop accidents processes. The lack of appropriate people for decision making, managers and specialists, with the dynamism and knowledge necessary, create gaps in the functions within the company. On the other hand, the analysis of short & long-term decisions depends on staff who may be away at work.

1.3. Management of Organizational Change (OCM)

According to the CCPS [5] and CSChE [6][7], aspects related to organizational changes are:

a. Location, communication and person work allocation, and activities of the staff and of turn during inversions, working conditions, emergency changes, changes in job profile, and events involving organizational climate.

b. Dismissal on staff, change in job profiles in production management and security.
c. Team size, level of expertise, integration, time and manner of allocation, liabilities, temporary activities, lack at work and neglected activities.
d. Centralization or decentralization of roles at work, reorganization, control range, type of organization, and changing service providers.
e. Change in corporate policies and procedures in Process Safety.

2. Organizational Structure for decision making

The organizational model found in the oil, chemical and petrochemical companies continues to consider industrial accidents [2] as normal, and after the shutdown they try to maintain operational continuity. Analyzing organizational structure is easy to verify that there are no prerequisites for the formulation of management and thus set is not checked the level of concern with the management changes may impact on your organization's safety culture.
Organizational changes [5] compared to the technical changes [1] have a greater burden on the process safety. The organizational changes alter the balance of work and businesses, according to most authors of reliability and safety [2][3][4] is the area of the root causes of major accidents. It is also known that these organizational changes affect changes in technology when affecting skills in project areas, operation, safety and maintenance. It is important to analyze the motivations that lead to failure resulting from organizational aspects to set the appropriate manager profile.
Looking to the managerial context of an organization, the management of the lower structure aims first care of the upper structure desires, even which may put doubts in the decision-making consequences. It is important to consider that fact that economic modeling of the business and the need for immediate results still prevail in industries management plans.
The knowledge of human reliability & human factors [8] is still subjective and develops only those with a prior concern and study on the subject. In this case there is no way to compare how much a top manager will consider in their decisions impacts on the order of human reliability and their ability to pay attention to undesirable results in management actions.
Understanding human reliability is much more than having technical knowledge expected of man in task execution behaviors, but the practical application of their knowledge in the search for blockages of the factors that enhance human error. In this sense control tools for decision making allowing practical application of human reliability and compulsory analysis of how much each action required in change management, safeguarding weaknesses in the process industry.
It is considered that the lower managers need objective tools to block or overhaul of the wishes of top managers. How have not yet been objective tools for analysis of the consequences within the human reliability, we have invariably noticed that all the requests that impact the topic analyzed, are accepted and implemented, for lack of objective counterarguments.
In the event analysis occurred in petrochemical processes note the high severity in managerial decision-making errors. The discussion on the human factors [8] involved in the events is inhibited by lack of knowledge of the area. A greater understanding of human reliability in conjunction with the technical knowledge to carry out the task allows the identification of the cause of failures that are related to human and organizational errors. The inclusion of human factors analysis in managerial decision making for technical and organizational changes avoids flaws in process safety.

The creation of the Risk management & Human Reliability committee in the organization is an important step to ensure the effective implementation of its tools of control. A survey [7] would be applied in a company to validate the aspects considered to leaders and organizational environment. The representativeness of the committee is presented by an expert in human reliability and risk management. The structure is open and enhanced by hierarchical independence of the plaintiff management unit. Its members would be multidisciplinary and rotary, and may be invited from other units or specialists from academia to analyze more complex decisions.

This structure would be discussed thoroughly the consequences and the minimal protections to ensure control of the process. The fact that a decision to be taken great impact to the committee, already requires greater attention and analysis by leadership, coupled with the tools to take the proposal to bring signatures committee of management. This requests a safer decision.

The severity of incorrect decisions in the management of chemical and petrochemical industry under investigation impacts on internal communities (working body) and external communities (neighbors). The impact of these decisions as well as causing economic losses can cause irreversible damage to human life and nature. Thus, these high-risk decisions cannot be centered on the role of the manager or leader of an organization. These decisions can strongly impact on the community and the nature must be taken in a neutral manner with the participation of human reliability committees to high risk management. The creation of these committees allows you to expand the vision of the consequences to and within the organization, making it possible to lock or review of management actions.

3. Organizational Management Change, risks and tools

High-impact decisions on risk management activities of industries can not be taken believing only in subjective knowledge of a person, even considering that the amount of existing managers in an organization and the dynamic experienced model does not guarantee the proper preparation and training of leadership, considering the factors that alter human reliability.

Recognizing that the hierarchical structure is always respected in conventional administrative models and applied in petrochemical, when a document requests the concurrence of the upper hierarchical structure, this involves an additional attention in the preparation and review process, which is the expectation of this work: enlarge attention and reanalysis of management actions with great impact on safety and risk management of an organization. The figure 1 generally describes the change management process.

The formation of human reliability committee for risk management within organizations is important step to ensure the effective implementation of risk containment in technical changes or organizational changes. A survey [7] should be applied in the company to validate aspects related to human error caused the change processes. This committee should be composed of experts in human reliability, equipment, work processes and management and risk analysis.

The structure must be open and hierarchical independence of the operating units where the changes occur. Its members have multidisciplinary knowledge and are rotatable and can invite experts from other industrial areas or university for the analysis of more complex decisions.

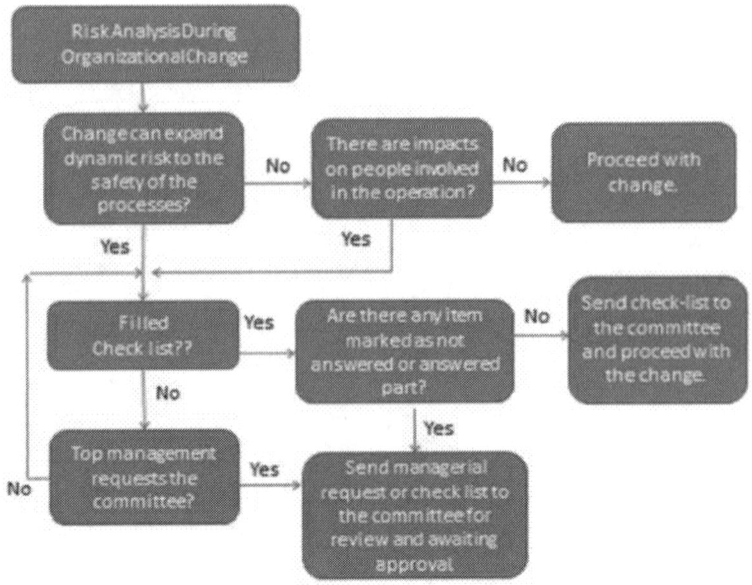

Figure 1. Flow analysis of organizational changes.

The structure of the committee should discuss the changes exhaustively considering the consequences and what the minimum protection to ensure process control and safety. The human reliability committee wants to avoid the failures of the decisions in the management of technical and organizational changes to avoid impacts on process safety. The analysis of biases in management decisions together with the evaluation of technical failures of devices and controls indicates a safer condition for the changes. Leaders have the support and criticism of a multidisciplinary team making the safest and most sustainable decisions.

A checklist is applied before decisions related to technical or organizational change to, according to the identified risks to human error, change decisions making them more secure and sustainable. This list (Figure 2) addresses the following issues that indicate the potential risk of human error: policies in the organization, operation team, team staff, managers, practices for change management, management during the company merger or change of shareholders , changes in the hierarchy, and power level for the decision, monitoring processes and activities, responsibilities, management models adopted, maintenance plans, emergency response plan, operating costs, and changing critical procedures.

The list of questions involve the following topics: History operational occurrences; Leadership; recognition notices about events; experience in operating similar systems; knowledge and technology testing; effect of domino technique for system operation; operational communication; time for decision; proper policy for culture; organizational climate; resistance to change on the team; reliability and task performance; approval of design changes; expansion of impacts in case of accidents; fitness training skills; authorization to perform critical tasks; and following the emergency response plan.

In this list are asked questions that punctuate the need to control actions on the managerial decision demanding opinion of the human reliability committee. If the following service levels are found in small part or missed, the manager must register the control actions taken to mitigate the effects, signed at the end and submitting to the organization's Risk management & Human Reliability committee. The other shipping option for the committee is presented by interest top manager who may find it necessary to renewed discussion of change, requesting the evaluation committee, validating and generating confidence in decision making.

4. Types of analysis of human reliability committee

The main role of the Human Reliability Committee for High Risk Management is to analyze in an impartial and technically the risk potential offered during organizational change. The expectation is that since there is no hierarchical links the committee has the independence of signal risks not previously identified or omitted by an interest changes.
The committee to receive the checklist may request a report to manage with additional data and important pair complete understanding of the proposed information, combined with visits to operating units, saved confidential conditions of the requests.
The Committee shall present its considerations relating proposed changes to historical accident occurrences, and which learning was considered and it should be a point of attention for this condition. In this aspect is taken to end the approval of change, important caveats and reanalysis requests, always bringing the commitment to the safety of the process and people.

5. Study Approach

The current reality of petrochemical plants is extended on its technical and managerial framework, motivated by historical time coincident with the operation time of retirement of its employees. In this context there is a large picture renewal, about 35% in a petrochemical plant in Brazil. With the arrival of new managers and with the mission to operate the unit reducing the workforce by 25%, a sector operations manager proposes a new supervisory model in its plant. The plant began operations in the 70s and is currently undergoing Revamp, in order to adjust technology to maintaining economic viability, improving the quality of products an increase of production speed, at a lower cost.
The proposal drawn up by the sector manager reduces one of the six supervisors, from the most experienced supervisor to account for two operational centers of this plant, and is a renewal of two supervisors after retirement of the above. As always positive and helpful way, the experienced supervisor accepts the proposal, from the sector manager to inform his superior of the implementation of organizational change.
The top manager asks to be carried out the checklist analysis of organizational change and the human reliability committee validates the proposal.
After analysis by specific check list, conducted by sector manager of operations, the proposal is submitted to the committee that analyzes based on information provided by the manager.
The committee to analyze the change of status interview requests with supervisors and involved as a way of expanding information on competence and qualification, at this time, is reported to the committee insecurity in the operation of new equipment and no experience in compressor operations natural gas, the area to be served by the supervisor of two poles. In the study of deepening the committee finds that this unit already had an average proportion of accidents in the start of compressors and that was caused by human error due to ignorance of the task.

5.1 Further Study

Development of leadership profiles to prevent and control events by personnel changes. Analysis of loss due to poorly managed organizational change is essential to define the main features in the leadership to plan, implement and monitor these organizational changes. Then Leaders should:

(a) Keep a balanced decision between the intuitive and cognitive model;

(b) Develop and apply techniques for measurement and control of human behavior and groups in crisis situations;

(c) Know how to develop the perception of risk, commitment, behavior change, common sense and balance in political decisions;

(d) Develop people with: clarity and understanding of the complexity of the tasks and processes, knowing about what are the social risks, capacity to identify human typologies, and cultural trends to avoid fear and failure;

(e) Be aware of impacts and how to act when they come across: questioning staff, needs to establish and achieve organizational objectives, centralization in routine and operational emergency, avoid turnover and develop cooperation in the operation team, develop informal leadership to neutralize the disharmony, and activities of the delegation;

(f) Have ability to recognize and balance the right time to implement changes, achieve expected goals of its management and superior guidelines, knowing dose correctly the speed of execution of management plans; and,

(g) Recognize that each team had a specific improvement model that will have its own response, assessed in association with complexity and criticality of the operations involved

5.2. Checklist Application and opinion of the Committee of Human Reliability

In the checklist completed by the sector manager, the items asking about Hierarchy of Change Processes Technology, Technical Training and equipment of replacement were marked with care in small part, which points to the attention of the Human Reliability committee, but is given attention to review all items in order to validate what is marked as it could have been informed without further reflection.

After meeting of the Human Reliability committee coordinator with the other stakeholders was heard its analysis summarized in the following main considerations: (a) The leadership of renovation exceeds the unit's history, considering not only the new supervision associated with new management; (b) The experienced Supervisor, despite his years with the company, does not have knowledge in natural gas compressor operations; (c) The unit undergoes Revamp with inclusion and replacement of equipment, ordering a new training stage and renewal of knowledge; and (d) The historic plant accidents presents a similar condition that can be configured in the proposal.

In this context the final evaluation of the Human Reliability Committee requests that delayed the change as a way to consolidation of operations and training of new technology, the development of new managerial leadership and supervision and that a compression operation training plan is carried out Natural Gas for the experienced supervisor.

In accordance with the applicants managerial we stipulated period of one year to meet the committee's recommendations and this time a supervisor from another state plant with expertise in natural gas compressor operations will be pursuing its activities in training with the new supervisor.

6. Conclusions

Managing process safety risk as result of organizational changes is always challenging. Further, many technical changes carry on needs for personnel changes that generally is not risk assessed by the main change. Developing criteria and tools for treating process safety risk with focus on organizational changes is the fundamental need in the industry. This paper treated the matter of organizational changed with at plant floor level using current knowledgeable management and supervisory staff working on committees to address the imitating events of personnel; changes inside or outside of technical changes.

7. References

[1] CCPS. Guidelines for the Management of Change for Process Safety. Center for Chemical Process Safety, American Institute of Chemical Engineers, New York, NY, 2008

[2] Michael, Llory, Industrial accidents: The cost of silence, Multiação Editorial, Rio de Janeiro, RJ, 2001.

[3] Carlos de Freitas, Marcelo Porto and Jorge Machado, Expanded Industrial Accidents, FIOCRUZ, Rio de Janeiro,, 2000.

[4] FUNDACENTRO, Prevention of Major Industrial Accidents, Ministry of Labour and Employment, Jorge Duprat Figueiredo Foundation of Safety and Occupational Medicine, São Paulo, SP 2002

[5] CCPS. Guidelines for Managing Process Safety Risks During Organizational Change. Center for Chemical Process Safety, American Institute of Chemical Engineers, New York, NY, 2013.

[6] Rob Cairns. Managing the health and safety aspects of organizational change. CSChE Conference. Vancouver, 2002.

[7] CSChE. Guideline: Managing the Health and Safety Aspects of Organizational Change. 2004

[8] CCPS. Guidelines for Preventing Human Error in Process Safety. Center for Chemical Process Safety, American Institute of Chemical Engineers, New York, NY, 2008

Figure 2 – Check-list - Organizational Changes Management [7][8]

Department:	CHECK-LIST - ORGANIZATIONAL CHANGES MANAGEMENT	Responsible CAHC:
Number:		Result CAHC:
Proposed change for evaluation:		Manager:
		Date___/___/___

	STATUS	Level of service				
	ITEM	Fully	Large Part	Small Part	Not Attended	Not Applicable
	Evaluate whether Organizational Changes are in accordance to the following topics:					
	Organizational Policy					
01	Organizational change is aligned with the strategies of the organization, thus preserving all established pillars?					
	Management Models					
02	Were analyzed the impact on the management model, preserving the Operations security, SMS and respecting the organizational environment?					
	Management Support					
03	It is guaranteed that the key pillars for local management will remain sufficient to organizational efficiency?					
	Management Changes					
04	The management changes were analyzed in adequate conditions of service of way and do not generate management models breaks?					
	Management Merger					
05	In case of merger management, was established technical principles that support the decisions?					
	Hierarchy Changes					
06	Changes hierarchies were well defined and properly porpostas and assimilated by the body technician?					
	Decision-Making Power					
07	The changes ensure that the power to operational decisions meet the need for emergency situations and authorization for critical tasks?					
	Process Technology					
08	The technology involved in industrial operations are known for operating staff and this dominates the technique?					
	Monitoring Engineering					
09	It ensured the continuity of monitoring of Engineering in technical processes guarantee the security operacinal industrial design?					
	Positions of Responsibility					
10	It is widely known positions of responsibility of each participant involved and has confidence in completing the task?					
	Operating Effective					
11	The effective set ensures the correct distribution of alignment activities the safety and control of operations?					
	History of Operational Occurrences					
12	Analyzed the historical reviews occurrences in operacinal unit and respected learning?					
	Maintenance Plans					
13	The proposed maintenance plans comply with the regulatory limits and ensure the preservation of the safety of operations.					
	Operational Communication					
14	Guaranteed is appropriate communication mechanisms and pass services for business continuity, disclosing for all involved?					
	Operating Costs					
15	The approved operating costs are in line with the requirements for proper operation of industrial and rediscuitda unit with the staff?					
	Emergency Response Plan					
16	Organizational change ensures proper compliance with the emergency response plan and no expansion of harmful consequence in case of accident?					
	Technical Training					
17	The offered training plan is appropriate to the level of knowledge required to perform critical tasks?					
	Change of Critical Procedures					
18	Changes in critical procedures are fully aware and we carried out another training with the team involved?					
	Replacement of equipment					
19	The replaced equipment ensure the safety of the project and is fully known to staff?					
	Top manager Signature					
	Blocking actions taken to critical items:					

Using Unrealistically Optimistic Failure Data: Is it Ethical?

Loren L. Stewart, CFSP
exida
64 N. Main St., Sellersville, Pa 18960
lstewart@exida.com

Iwan van Beurden, MSc, CFSE
exida
64 N. Main St., Sellersville, Pa 18960
vanbeurden@exida.com

Prepared for Presentation at
American Institute of Chemical Engineers
2015 Spring Meeting
11th Global Congress on Process Safety
Austin, Texas
April 27-29, 2015

Using Unrealistically Optimistic Failure Data: Is it Ethical?

Loren L. Stewart, CFSP
exida

Iwan van Beurden, M.Sc., CFSE
exida

Keywords: Engineering Ethics, Unrealistic Data, Product Failure Rate, Predictive Analysis, FMEDA, SIL, Functional Safety, SIL Verification

1. Abstract

Performance based functional safety standards such as IEC 61508 [1] and IEC 61511 [2] offer many advantages, including the opportunity to optimize and upgrade Safety Instrumented System (SIS) designs. However, an instrument's performance calculations depend on realistic failure data. While realistic failure data is unsurpassed, unrealistically optimistic failure data can be extremely hazardous. A predictive analysis technique called Failure Modes Effects and Diagnostic Analysis (FMEDA) has been developed along with a component failure rate database that can predict the failure rates of instruments based on their design strength and the expected stress environment. This method has been calibrated with over 150 billion of unit operating hours of field failure data over the last 15 years enabling realistic data to become more attainable. With the knowledge that such data is within a company's reach, using data that is too optimistic is inexcusable (from an engineer's or a purchaser's perspective). This presentation will examine functional safety standards, explore a range of realistic failure data for a simple mechanical device, and discuss the role of engineering ethics in regards to the use of overly optimistic failure data.

2. Introduction

The term "Functional Safety" may be familiar to many people, thus invoking assumed knowledge of the subject. However, as "what is functional safety, and why does it matter?" could be included on several frequently asked questions lists there is a clear need for explanation. The significance and obligation of functional safety in an industry reinforces that this is a topic that is imperative to fully understand. Finding an understanding and complying with functional safety standards play a vital role in everyone's safety, therefore breaking down what it is and why it matters are the first steps to full comprehension of functional safety and a safer world.

A essential aspect of functional safety is performance capability of the safety function. The performance is expressed in failure rates. There are many sources of failure rates, ranging from plant specific data based on years of recorded performance and predictions of failure rates through thorough analysis to simple failure rate calculation based on warrantee information.

This paper

- Provides background information sufficient so that the paper is relatively self-contained
- Presents an explanation of failure rate data and calculations

- Provides an predictive analysis method
- Explains relevant standards and how a company can comply to those standards
- Presents examples of expected failure rates and data along with overly optimistic data
- Provides the importance of data integrity and the ramifications of not complying
- Presents engineering ethics with the level of risk scrutiny

3. Failure Modes Effects and Diagnostic Analysis (FMEDA)

There are various sources for failure rate data. As indicated above, the best source of failure rates would be a company's application of specific field failure data. One significant problem with failure rate data based on field failure information is that quite often a product will become obsolete before enough data is gathered to obtain a failure rate. Using confidence levels of 70% to 90% to will lead to very large confidence intervals if little field failure event information is available, often leading to excessively conservative SIF designs. They are typically outside the realm of realistic and hence lead to skepticism regarding the overall concept of functional safety. Since field failure data is not always available or realistic, a predictive method to an equipment item's performance capability was established in the form of the FMEDA technique. The FMEDA technique was developed by engineers from exida to provide a means to predict not only failure rates but failure rates per failure mode, diagnostics coverage factors, and useful life. The method is based on the complexity and design strength of a product. The FMEDA method accounts for the automatic diagnostics being developed at the time. The technique was first published in 1992 as "Coverage Analysis" in Chapter 6 of [3] and later named FMEDA [2, 3]. The FMEDA method examines each component in a product design. For each component, all failure modes are listed and analysis is done to determine the impact of that component failure mode on the product. An FMEDA is verified by a set of sample fault simulation / injection tests which are done to simulate the component failure mode in the actual product.
Through this thorough evaluation of an equipment item, the FMEDA technique allows for a prediction of the failure rates and failure modes of that equipment item for use in SIL verification.

4. Predictive Analytics

The basis of a FMEDA is formed by the failure rates and modes that are used for the components on which the equipment item is based. This component data comes from a component database that must include failure rates and failure mode distributions of each component as a function of environmental operating profile (expected stress conditions). This core component failure rate database must be compared with actual field failure information to ensure reasonability of the FMEDA predicted failure rates. [6]
The predictive analytics method is an extension of the basic FMEDA technique where the core failure rate database is continuously examined based on collected field failure data [7] and updated to reflect real life experience. This will allow the FMEDA to provide realistic predictions of failure rates for each failure mode of an equipment item. This is also illustrated below in Figure 1.

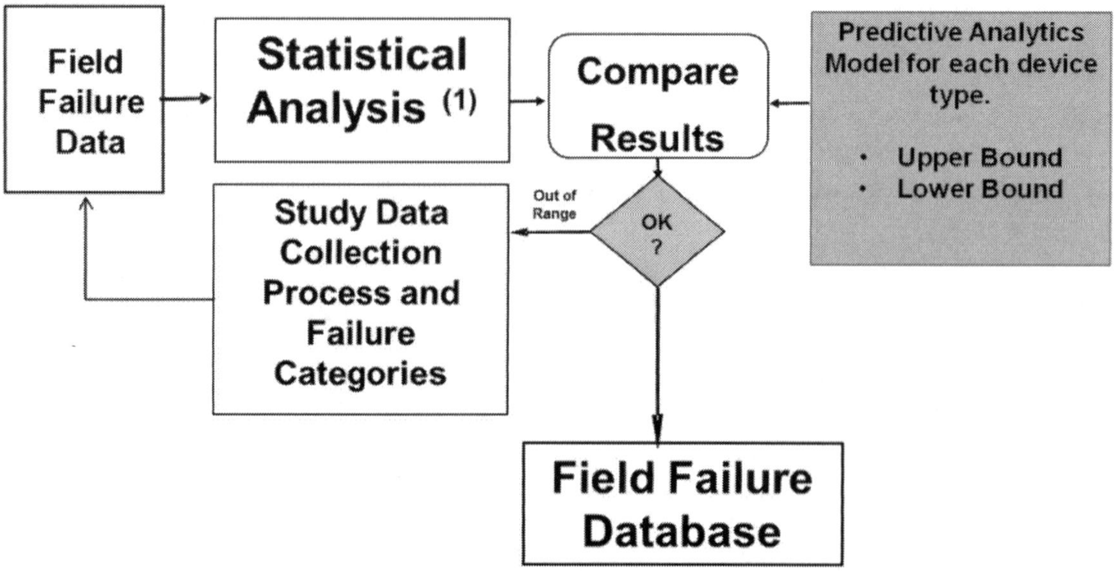

Figure 1 Predictive Analytics

5. Standards/ competency

The goal of Functional Safety is to have an automatic protection function that will perform the system's intended function correctly, or when that cannot happen because of a failure, the system will fail in a predictable (safe) manner.

Working groups and national standards bodies from around the globe have been involved in the process of bringing together the IEC 61508 standard [1]. IEC 61508 is the International Performance Based Standard for All Industries – it applies to suppliers and is known as the "umbrella standard."

IEC 61508, in summary applies to "Automatic Protection Systems" –electrical, electronic, programmable electronic, and mechanical devices. It provides measures of protection against random hardware failures and "systematic" design failures. IEC 61508 can be applied to PROJECT level work – bespoke (turnkey) systems or it can be applied to PRODUCT level work – off the shelf products applied in many applications. The standard targets **Suppliers,** its requirements being for the suppliers of process control and instrumentation for component/element or sub-system safety. It is encouraged for End Users to seek suppliers with products certified to this standard by a reputable, accredited Certification Body.

IEC 61508 is enforced in some countries by governments with the force of law, but in most situations, the standard typically is cited as best practice (accepted engineering practices) and is often required by end-user customers as part of project contracts (such as OSHA or insurance companies). When accidents do happen, the standard can be, and has been cited in civil cases as a commonly accepted standard of performance.

As IEC 61508 is the umbrella, there is a derived standard - IEC 61511: Process Industry Sector [2]. This is essentially the equivalent to ISA 84.00.01-2004 [8] and is used in the US in the same manor. It targets **End Users**, Engineering Contractors, and Integrators in process industries. IEC 61511 covers the entire SIS Lifecycle, including, Risk Analysis, Performance based design, and Operations and Maintenance. It is broken up into 3 sections: requirements, guidelines, and SIL selection, and is a performance NOT prescriptive based standard. This means that it does not evaluate a system on what it is said to do, but rather on its performance. Unlike PED, electrical standards or other standards that have a direct objective, functional safety identifies that every application could be different, so it has a different view for each situation it befalls. An example of this would be NFPA 85, which treats every boiler system the same. If the boiler is in a corn field miles away from any person or building or if the boiler system is in the basement of a children's hospital in the city's center, they have the same required risk reduction. Functional safety and IEC 61511 allows every situation to be evaluated with performance requirements established to match the situation.

6. Equipment Item Performance and Certification

The IEC 61508 performance based standard has established the Safety Integrity Level (SIL). A Safety Integrity Level (SIL) is an order of magnitude set of discrete levels (one out of four) for specifying the safety integrity requirements of the safety instrumented functions to be allocated to the safety instrumented systems. SIL 4 has the highest safety integrity and SIL 1 the lowest.
A Safety Instrumented Function's target SIL is established at a systems level by risk analysis and subsequently documented in a safety requirement specification. Conceptual design engineers then need to define a SIF that will meet the original requirement by a variety of equipment item, redundancy, and testing philosophy selection.
Most equipment items that can be used in safety instrumented function design are certified to a specified SIL. The certification process' end result includes a certificate listing the SIL level for which a product is qualified and the standards that were used for the certification. A good certification assessment will demonstrate high design quality for hardware, software, and high manufacturing quality as well as adequate performance.
The certification of an equipment item addresses the development process as well as the performance of the equipment item expressed through the products failure rates. These failure rates are an input parameter in the overall Safety Instrumented Function probability of failure on demand calculation. Typically the failure rates are predicted through a FMEDA analysis.

7. Questionable IEC 61508 Certificates

Not all certification assessment programs are created equal and unfortunately not all users of these certificates are aware of this. The following table, Table 1, was obtained from [10].

Table 1 Failure Rate Information from Questionable IEC 61508 Certificate

λ_S [1/h]	λ_D [1/h]	λ_{DD} [1/h] (with Diagnostic Test)	SFF (No Diagnostic Test)	SFF (with Diagnostic Test)
1.91E-07	2.10E-08	1.47E-08	90.08%	97.02%

The failure rate information in Table 1 is for a side entry trunnion ball valve. An example trunnion ball valve is shown in Figure 2.

Figure 2 Example Side Entry Trunnion Ball Valve

The certificate (and therefore the data) is for a trunnion ball valve only. No other connected devices are included. Based on that fact if we take a moment to look at the data in Table 1 it is puzzling at best. Looking at the valve we see a couple to components. The ball valve has a valve body, a ball, and a stem that can turn the ball. A simple high level evaluation of potential failures would yield the following:

- Stem stuck, can no longer turn (dangerous)
- Stem break, can no longer turn (dangerous)
- Ball stuck, can no longer turn (dangerous)
- Leakage/Housing fracture (dangerous or no effect based on the material going through the valve)

Now consider the definition of a safe failure. This definition comes roughly down to a failure that causes a spurious trip, i.e. a failure that initiates the safety function without a demand from the process. What are the potential safe failures for a ball valve? Are there any failures that could cause the ball valve to go to the safe state? Reviewing the design with experts in the mechanical FMEDA process, none could come up with a failure that would cause the valve to close by itself in a close to trip application or to open in an open to trip application. In other words, how can the certificate suggest that there are safe failures and actually that the majority of the valve failures are safe (the Safe Failure Fraction is greater than 90% without a diagnostic test)?

In addition to that look at the dangerous failure rate. A well accepted industry reference [11] suggests a dangerous failure rate of 980 FITs for a generic trunnion ball valve. That is 9.80E-7 failures per hour. The generic data in [11] represents a conservative estimate typically twice the average dangerous failure rate for a device like this. However the data in Table 1 suggest that this specific side entry trunnion ball valve dangerous failure rate is only 2.10E-8 failures per hour. That is more than 45 time less than generic data would suggest. This should at a minimum sound the alarms.

Simply looking at the results published on the certificate one would need to conclude that in this case they do not make sense. The analysis that resulted in the certificate does not make sense, how can you have safe failures? Based on this why would anybody trust the data that is presented on the certificate.

8. The Importance of Data Integrity

Merriam-Webster defines BAD as "Failing to reach an acceptable standard: POOR < a bad repair job > [9]." In the functional safety world, we can simplify the definition of BAD data as "Data that leads to unrealistic, often dangerous, designs." The consequences of bad data will have different effects on different applications and the risk varies with use. On one end of the spectrum, using bad data in a marketing brochure, by over selling or exaggeration a statement, the nature of the risk is low. The company's reputation may suffer from the exaggerated claims. On the opposite end of the spectrum, if bad data is used in safety reliability calculations the nature of the risk is very high, where potential loss of life due to under-designed safety systems is possible. Optimistic data leads to unsafe designs, insufficient redundancy, or insufficient testing leading to the required risk reduction not being reached.

One of the premises of IEC 61508 and IEC 61511 is that automated protection systems with diagnostics and periodic testing can provide higher safety reliability than typical control functions. The standards outline the steps that must take place to claim this higher safety reliability. However these steps are only valid if appropriate (or GOOD) data is used.

9. Engineering Ethics

Engineering ethics play a role in everyday life as a safety engineer. Some of the decisions and consequences from those decisions may have relatively low magnitudes while others that one will come across less frequently may have a much higher level of magnitude which could result in catastrophic environmental disasters, legal fees and fines, personnel getting injured or even death.

There are times when a system designer knew the safety reliability data for one device was too optimistic, but still used it because it meant they didn't need to implement online diagnostics or redundancy. They could save money. Naively, it is assumed that when engineers used optimistic data they simply did not know better. The crux of the problem is that there are two distinct world views when it comes to process safety reliability data. One approach is to leave it up to the equipment manufacturers to calculate their own failure rates by looking at field return data. Some of these OEMs believe that all failures are a result of customer misuse and therefore report virtually no failure. This results in data for their valves that appear by inspection to be too good to be true. They report a dangerous failure rate of about 10 FITS (1 FIT is equal to 1 failure per billion hours). This reported failure rate states that the valve will only be unavailable to perform its safety function once every 11,400 years.

This means that if the engineer's plant has 11,400 valves, which is not out of the question, it should be expected that one valve could be in a degraded state any given year. Persons with knowledge of process plants know that a failure rate like that is unrealistic.

Another valve manufacturer supplied data that indicated a dangerous failure rate for its valve of around 450 to 800 FITs depending on the application. Why is this failure rate so much higher? These failure rates are based on a formal analysis of the actual product design (FMEDA) and are driven by a component database with over 60 billion hours of operational data.

Engineers have been told for decades that the responsibility for design and construction to that design is in the hands of the "engineer in responsible charge." If the engineer is a consultant, he or she has liability insurance against an issue of design or construction that causes an accident. If he or she works for a company, the company liability insurance is supposed to provide coverage in the event the design is faulty. But most insurance contracts exclude intentional or deliberate actions.

There is no apparent protection for the engineer who designs the safety system based on "compliance," or the company, regardless of the legal opinions that are offered before the fact. When the failure occurs, the liability is still there and the opposing council will argue for negligence on the part of the engineer and company for self-serving rationalization of engineering principles. Referring back to the example, this is exactly the same issue—reliance on clearly unrealistic test data. The IEC standards require operating plants to have a Functional Safety Management (FSM) process in place. The standards do not encourage compliance, but rather actual increased safety iteratively by following the process the company has established.

10. Notation

FIT	one failure per 10^9 hours of operation
FMEDA	failure modes, effects and diagnostic analysis
IEC	International Electrotechnical Commission
PFDavg	average probability of dangerous failure on demand
SFF	Safe Failure Fraction
SIF	safety instrumented function
SIL	safety integrity level
λ_D	constant dangerous failure rate
λ_{DD}	constant dangerous detected failure rate
λ_S	constant safe failure rate

11. Conclusions

Using the SIL process suppliers can advertise their safety integrity level and attract new business while showing their customers and the industry that they are serious about safety. End users can mandate SIL products to insure they are receiving safer and more reliability products. Unfortunately some third parties are issuing certificates that present unrealistic data and are based on analysis techniques that are not applicable to the process industry. The failure rate data presented is typically at least an order of magnitude lower than average industry data predicts as well as supports based on field failure analysis.

Using Unrealistically Optimistic Failure Data can lead to inadequate Safety Instrumented Function design. This results in exposure of personnel to higher risks than expected. Since procurement expense tends to be lower for instruments with Unrealistically Optimistic Failure Data, the use of that data is questionable at best, and clearly unethical.

12. References

[1] IEC 61508, Functional safety of electrical/electronic/ programmable electronic safety-related systems, Geneva, Switzerland, 2010.

[2] IEC 61511, Functional safety – Safety instrumented systems for the process industry sector, Geneva, Switzerland, 2003.

[3] Goble, W.M., *The Use and Development of Quantitative Reliability and Safety Analysis in New Product Design*, University Press, Eindhoven University of Technology, Netherlands: Eindhoven, 1998.
[4] Goble, W.M., *The Use and Development of Quantitative Reliability and Safety Analysis in New Product Design*, University Press, Eindhoven University of Technology, Netherlands: Eindhoven, 1998.
[5] W. M. Goble and A. C. Brombacher, *Using a Failure Modes, Effects and Diagnostic Analysis (FMEDA) to Measure Diagnostic Coverage in Programmable Electronic Systems*, Reliability Engineering and System Safety, Vol. 66, No. 2, November 1999.
[6] W.M. Goble and J.V. Bukowski, *Development of a Mechanical Component Failure Database*, 2007 Proceedings of the Annual Reliability and Maintainability Symposium, NY: NY, IEEE, 2007.
[7] Collection of life event recording data obtained using the exida SILStat application
[8] ANSI/ISA SP84.00.01 – 2004 (IEC 61511 Mod.), Application of Safety Instrumented Systems for the Process Industries, Raleigh, NC, ISA, 2004.
[9] "Bad." *Merriam-Webster.com*. Merriam-Webster, 2015. Web. 10 Feb 2015.
[10] FS 28710068 Rev. 0., Pibiviesse S.r.l Series E Side Entry Trunnion Mounted Ball Valves, TÜV Rheinland Italia
[11] Safety Equipment Reliability Handbook, third edition, 2007, exida.com LLC, Sellersville, PA, USA

13.Biographies

Loren Stewart, CFSP

Safety Engineer

Loren Stewart graduated from Virginia Tech with a BSME. She has over 7 years of professional experience originating in custom design and manufacturing where she managed and was responsible in supporting key clients. She currently works for exida consulting as a safety engineer, focusing on the mechanical aspects of their customers. Along with assessing the safety of products and certifications, she continually researches and published reports on stiction and is creating a database for the 2H initiative according to IEC 61508.

Iwan van Beurden, CFSE

Chief Technology Officer, Director of Product Development

Iwan van Beurden, CFSE is the Chief Technology Officer, Director of Product Development at exida. He supports exida's End-user services, exida's Manufacturer services, and exida certification activities. Mr. van Beurden is responsible for the development, maintenance, and certification of the exSILentia integrated Safety Lifecycle tool. He also performs IEC 61508 development support, Functional Safety Assessments, a variety of reliability analyses, SIL Selections and SIL Verifications. He also performs process gap analyses to determine the level of effort companies need to undertake for compliance with IEC 61508 or IEC 61511. He gives training for exida and is also an ISA instructor.

Mr. van Beurden previously worked for Yokogawa Industrial Safety Systems in Apeldoorn, the Netherlands. He worked both in the R&D and in the Operations department as a Safety Assessment specialist. He was involved with the implementation of IEC 61508 within Yokogawa Industrial Safety Systems dealing with SIS reliability calculations and the safety education of engineers. Mr. van Beurden is a Certified Functional Safety Expert (CFSE) and a member of ISA. He has published various papers and magazine articles. He holds a Master of Science degree from Eindhoven University of Technology in Eindhoven, the Netherlands, where he majored in reliability engineering and graduated cum laude.

398079 Dynamic Simulation As a Robust Tool for Optimum Flare Systems Design

Abstract:

Having a well designed flare system is critical in today's refining environment. With growing competition and ever increasing costs of material and fabrication coupled with slim margins and tighter environmental regulations, refineries are challenged with walking the tight rope of increasing production yet mainting profitability. Capital costs in designing and expanding a flare disposal system can weigh in significantly in impacting some of these issues.

With growing environmental awareness and statutes, refineries fear falling our of complaiance and as a result run the risk of commissioning flare systems that may be designed for overestimated relief loads. Understanding the nature of routine and non-routine venting over a finite timeline can help home in on superfluous design elements and result in well-tailored systems.

This paper focuses on using dynamic simulation as a tool to achieve an optimum design for a flare system and discuss the various considerations that needs to made and their impact on equipment in the flare network. Insight into variations in relief pressure, temperature, composition and flow rates across various nodes in a flare network through the illustration of some case studies will be offered in this paper. The proper application of dynamic simulation as a tool for flare design can result in significant benefits in terms of equipment availability, plant performance, capital and operating expenditures, and emission reduction.

397220 Best Practice MOC and Pssr 2014

Abstract:

Management of Change (MOC) and Pre Start-up Safety Review (PSSR) are still the most challenging elements of OSHA's Process Safety Management (PSM) standard, the EPA's Risk Management Program (RMP) rule, and now, the US Department of Interior's Bureau of Safety and Environmental Enforcement's Safety and Environmental Management Systems (SEMS). Even though the PSM standard has been around since 1992 and the industry has been managing change for several decades, we still can get it wrong, sometimes with disastrous results. The mindful or diligent efforts of many working in concert are necessary to ensure that change is identified, analyzed and executed in a quality way. Just when we get it right on paper and get the workforce up-to-speed, we have employee turnover, neglect and sometimes regulatory change. New actors and a constantly changing script make it hard to manage change efficiently and effectively. This paper discusses how the chemical process industry has defined the MOC and PSSR best management practices and how they should be automated. In addition, this paper will explore new best practices associated with mobile capabilities for PSSR and MOCs in the field, better defined evaluation processes, leading and lagging indicators and maybe even predictive technologies that will predict higher risk and the need for action to reduce imminent problems.

394645 Application of Risk Analysis in Pilot Plants and Research Labs

Abstract:

Application of Risk Analysis

in Pilot Plants and Research Labs

Abstract

When an accident occurs in a Pilot Plant or Research Laboratory, involving loss of containment, chemical exposure, explosion or fire, people are often seriously and sometimes fatally injured. The subsequent investigation findings often identify the lack of a risk analysis as a root cause factor, before the work was started.

The approach towards the safety of pilot plant and research labs cannot be different from that of the full scale plant. However the uniqueness of challenge is associated with the understanding of the hazards in pilot plants and research labs. Investigations of laboratory-scale incidents reveals that the underlying causes are similar to those found in major chemical plant industries.

The combination of hazard identification and risk analysis is an organized effort to pinpoint weaknesses in the design and operation of facilities that could lead to accidental or unintentional chemical releases, fires or explosions. The studies assist organizations with the goal of improving safety and managing the risk of operations. Investigation findings of accidents in pilot plants and research labs often point to the absence of a risk analysis of the operation that led to the accident.

This presentation discusses the relationship between hazards and consequences of an upset event, the likelihood of the upset happening and the resulting risk to personnel, property and the environment. Some well-established risk procedures have been adapted to the task of hazard identification and risk assessment in the pilot plants and research labs.

The hazard identification process may be accomplished from readily available physical and chemical properties and the health effects of the chemicals to be used in the process. If formal hazard analysis is indicated then more detailed risk analysis techniques, such as Check-List, What-if or HAZOP are available. The choice is driven by the complexity of the planned synthesis and the nature and extent of consequences of an upset. These reviews should be led by an appropriately trained and experienced facilitator.

A powerful advantage of this approach is that it encourages and enables a research laboratory group to consider less obvious ways in which an upset may occur, even though it may seem unlikely at first consideration.

15 AIChE
Spring Meeting, Austin, TX

Integrating Cybersecurity Risk Assessments Into the Process Safety Management Work Process

Harold W Thomas
Exida LLC
64 North Main Street
Sellersville, PA 18960
thomashw@exida.com

John Day
Air Products & Chemicals, Inc.
7201 Hamilton Blvd
Allentown, PA 18195
dayjd@airproducts.com

Prepared for Presentation at
American Institute of Chemical Engineers
2015 Spring Meeting
11th Global Congress on Process Safety
Austin, Texas
April 27-30, 2015

UNPUBLISHED

Integrating Cybersecurity Risk Assessments Into the Process Safety Management Work Process

Harold W Thomas
Exida LLC
64 North Main Street
Sellersville, PA 18960
thomashw@exida.com

John Day
Air Products & Chemicals, Inc.
7201 Hamilton Blvd
Allentown, PA 18195
DAYJD@airproducts.com

Keywords: Cybersecurity, Process Safety Lifecycle, Risk Assessment, Vulnerability Assessment

Abstract

Cybersecurity is rapidly becoming something that process safety can no longer ignore. It is part of the Chemical Facility Anti-Terrorism Standards (CFATS)[8]. In addition, the President's Executive Order 13636[6] – "Improving Critical Infrastructure Cybersecurity," has drawn attention to the need for addressing cybersecurity in our plants as it has been demonstrated that in our new world, they are now a source of potential process safety incident. IEC 61508[2], "Functional Safety of Electrical/Electronic/Programmable Electronic Safety-related Systems (E/E/PE, or E/E/PES)" now has a requirement to address cybersecurity in safety instrumented systems and ANSI/ISA 84.00.01[3], "Functional Safety: Safety Instrumented Systems for the Process Industry Sector" is looking to include this requirement in the next revision. Currently the industry is playing catch up as there tends to be a gap in understanding between information technologists, traditionally responsible for cybersecurity, and the process automation and process safety engineers responsible for keeping our plants safe with help from automated controls and safety instrumented systems. As a result, guidance is being developed, but much of it continues to be a work in progress.

ANSI/ISA-62443-2-1[4], "Establishing an industrial automation and control system security program," requires a high level risk assessment, a detailed level risk assessment and a cyber-security vulnerability assessment to be performed. This paper explains the purpose of these risk assessments and where they fit into the process safety management lifecycle. As the entire cyber-security lifecycle is described, the National Institute of Standards & Technology (NIST) Framework[7] for Improving Critical Infrastructure Cybersecurity is used to better understand the analysis, design/implement, and operate/maintain phases of the lifecycle. As companies have relatively recently begun to understand the need to address the issue of potential cyber-attacks on the process control and protection system, few have a fully implemented work process. Guidance is provided for existing plants to begin the process and how to fully implement the lifecycle as well as how to approach new projects.

1 Introduction

With the advent of cyber incidents that have specifically targeted control and protection systems, it has become clear that the discipline of process safety needs to consider cybersecurity. Compounding the issue is the gap in understanding between information technologists traditionally responsible for cybersecurity and the process automation and process safety engineers whose control and safety systems are at risk. The need to integrate and reduce this knowledge gap is clear. This paper explains how integration between these disciplines can be accomplished.

Management of Functional safety and Cyber for ICS encompasses many aspects, however this can be illustrated in figure 1 below. Beginning with a new grass roots facility or green field site, each of the major phases are shown and how the process safety work process works in conjunction with cybersecurity, showing the relative interaction of the two lifecycles.

Figure 1: Safety/Cybersecurity Integration

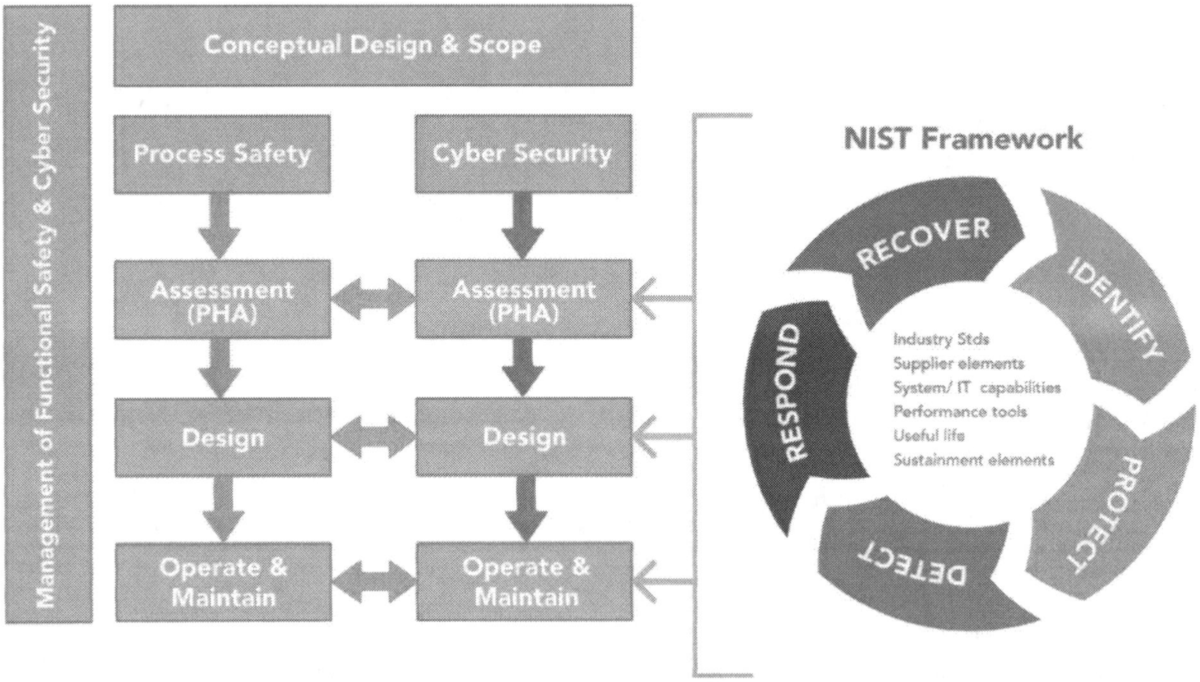

Although this paper will touch upon the full lifecycle, a more in depth focus will be placed on the high level risk assessment, detailed risk assessment and vulnerability assessments. For companies with operating facilities that have previously not considered process safety related cybersecurity, guidance is provided as to how to begin the process of implementing the full lifecycle as a sustainable activity.

Safety and Cyber have similar and co-dependent thought processes throughout the lifecycle. There are specific roles and responsibilities required from Manufactures, Certifying bodies and Owner Operators. Within the roles and responsibilities, specific standards are utilized to manage the lifecycle of products, systems, applications and approaches. Policies, procedures, current industry equipment and software capabilities, level of rigor and useful life attributes all combine to form the basis for cybersecurity management that provides sustainability. This forms a management of functional cybersecurity that is akin to the management of functional safety. The end user is responsible to ensure that the residual risk is commensurate with the level of risk that the business is willing to tolerate.

ANSI/ISA-84.00.01-2004 Part 1 (IEC 61511-1 Mod)[3], Clause 5, *Management of functional Safety and Functional Safety Assessment,* coupled with Clause 6.2, *Safety Life-cycle Structure and Planning* are the governing standards for owner operators with respect to safety instrumented systems. A similar approach to industrial control system (ICS) cybersecurity must be taken.
The National Institute for Standards and Technology (NIST) [7] has developed a framework that is the best practice developed for end users to date to aid in the development of the cybersecurity management systems.

The framework developed by NIST can be used to assist users as they develop the analysis, design/implement, and operate/maintain management policies, standards and procedures for the various phases of the lifecycle. It does not provide guidance on how to accomplish the tasks, rather it is an outline of the things that need to be considered coupled with some level of performance criteria. It is divided into five functions; Identify (I), Detect (D), Protect (P), Respond (R), and Recover (R). Functions then have categories that describe specific aspects of the lifecycle that are relevant to the function. Each category has subcategories that describe numerous activities and performance criteria that are related to the category and function. Finally, the framework provides industry references for each subcategory where more detailed information can be found that an end user might find useful when developing their management systems.

2 Assessment (PHA) Phase

2.1 Cybersecurity Assessment Phase Overview

Figure 2 shows the work process flow of an integrated lifecycle. As part of any new project involving process industry hazards, a scope must be formulated. This is just as true for how to address cybersecurity. One of the requirements in ISA/IEC 62443-3-2[5] is to perform a high level cybersecurity risk assessment. Its purpose is to help define what scope needs to be included. Section 2.2 will discuss the high level risk assessment in more depth.

As part of developing the scope, it is essential that zone and conduit drawings be prepared. To cybersecurity, these are like P&IDs to the process world, and form the basis of what is reviewed during a detailed cybersecurity risk assessment.

The traditional process hazard review, including HAZOP, LOPA or other applicable methodologies should be performed prior to the detailed level cybersecurity risk review as required by ISA/IEC 62443-3-2[5]. This sequence allows the traditional review to identify and analyze the major hazards that need to be considered when conducting the detailed level cybersecurity risk assessment. The high level risk assessment will be discussed in more depth in Section 2.3.

Figure 2: Assessment Phase

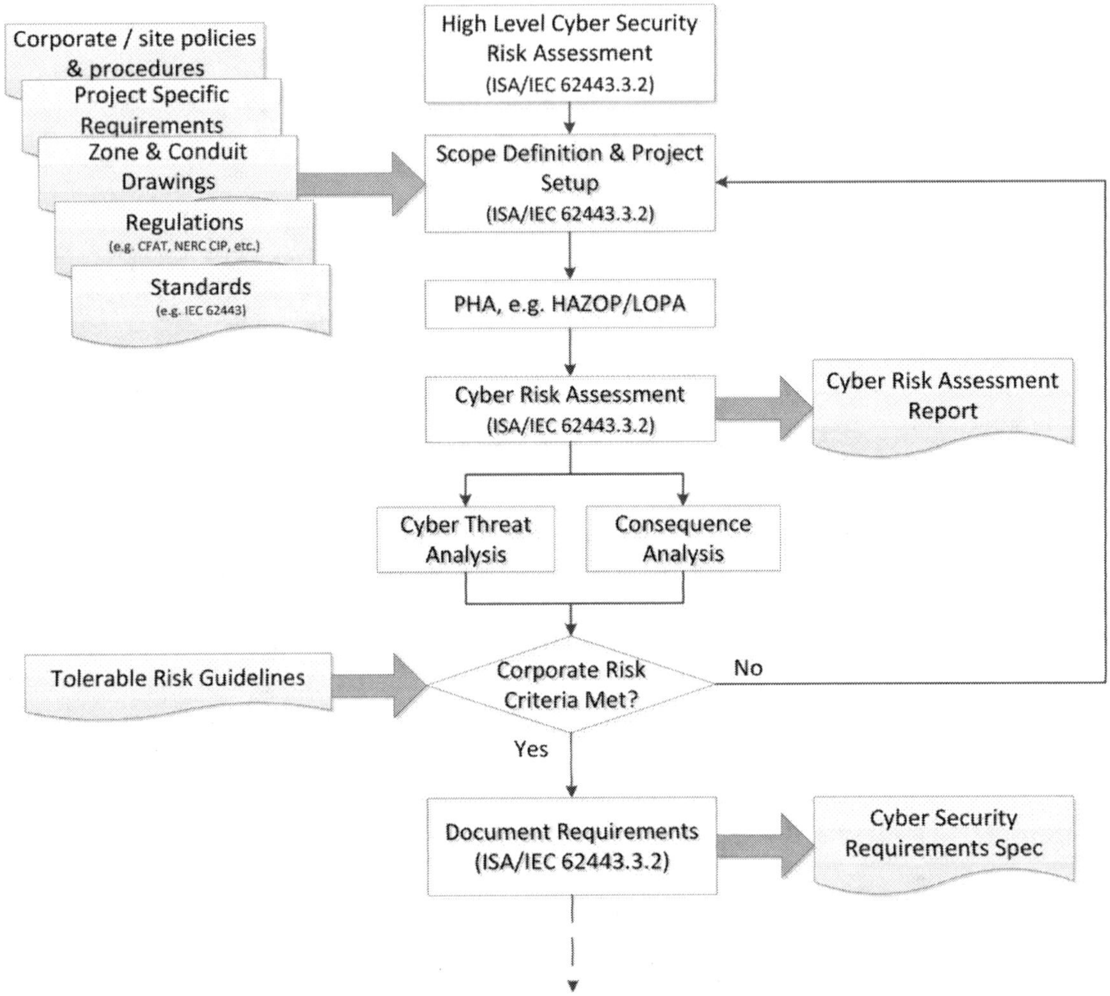

Any major risks identified may need a more rigorous evaluation. This can involve traditional consequence analysis techniques and in the future, there will be a push to understand how to apply more quantitative likelihood analysis techniques, keeping in mind that cybersecurity countermeasures degrade via different mechanisms versus the random and systematic mechanisms that have been addressed by safety instrumented systems and functions. Irrespective, a means to ensure the corporate risk criteria is met is required that addresses both likelihood and consequence. If the risk criteria is not met, then additional measures need to be considered until the risk criteria is met. Once the criteria is determined to be achievable, a cybersecurity requirements specification (CSRS) is documented in a similar manner to the safety requirements specification (SRS) for safety instrumented system (SIS) and functions (SIF) so detailed design can begin.

2.2 High Level Cyber Risk Assessment

2.2.1 Purpose

According to ANSI/ISA-62443-2-1[4] the purpose of a high level cyber risk assessment is to help establish scope. For cybersecurity, a determination of the criticality and consideration of how to respond to various assets being compromised is quite useful.

2.2.2 Procedure

To accomplish, a list of all the types of cyber assets that will be used needs to be developed. Typical cyber assets include but are not limited to:

- Business servers
- Historian
- Personal computers
- Engineering work station
- Operator Consoles for Human Machine Interface (HMI)
- Distributed Control System
- Programmable Logic Controller
- Safety Instrumented System Controller
- Smart field instruments
- Printers
- Etc.

Once the list is complete, the next step is to determine two things; what is the criticality if the device were compromised, and second, if compromised, what type of responses are appropriate. What level of response is required depends upon how much risk may exist if the asset is compromised. Typical responses that can be considered include:

- Keep operating and fix when convenient
- Isolate the asset from other assets & quarantine until threat has been removed
- Shutdown the process

Should something like a Historian server be compromised, it would be more appropriate to isolate while the rest of the process continues to operate. However, should it be determined a SIS is compromised, it would be difficult to justify not shutting down the process. A more interesting question comes about should the basic process control system (BPCS) be compromised. If the SIS is still functional, a case can be made to try and fix on line if it can be determined that the threat is not fatal. This last example raises an important point when selecting a control system at the beginning of a project. As the degree of integration between control systems and safety systems increases, there is less latitude as to the potential responses. In addition, as the level of integration increases, common failure resulting from a cyber-attack can lead to a situation where it is not possible to satisfy corporate risk criteria. Understanding these issues as soon as possible is the best way to minimize the cost and potential schedule impact to a project. When making the determination of what system and level of integration to use, it is advised to reference the 7 core attributes as described in the CCPS Safe & Reliable Instrumented Protection to help make an appropriate decision based on the relative level of risk inherently posed by the process.

With the potential consequences and ease of propagation information documented, each asset should be assigned to a zone and the level of protection recommended to prevent propagation across zones should be established. Using industry recognized and generally accepted good practice as documented in the ANSI/ISA 62443 series of standards should be the starting point. Finally, formal zone and conduit drawings should be developed which can be considered required process safety information input to the detailed cyber risk assessment.

2.2.3 Procedure Summary

- List cyber assets
- Document potential consequences if asset compromised
- Risk rank assets based on potential consequences
- Document ease of propagation with open communication
- Recommend zone location for each asset
- Recommend conduit safeguards for communication between each zone
- Document zone and conduit drawing for use in detailed cyber risk assessment

2.3 Detailed Cyber Risk Assessment

2.3.1 Purpose

A detailed cyber risk assessment is intended to rigorously evaluate the instrument and control system IACS and ensure the proposed design and procedures are capable of satisfying the corporate risk criteria. This paper is recommending a methodology similar to a HAZOP, however, the methodology evaluates cyber nodes that represent the cyber assets that are part of the zone and conduit drawings. Although the concern is with the control and safety systems, the entire plant network including all internet and wireless access points, including 3rd party connections, networks, and devices need to be included as these can be pathways to compromise the IACS.

2.3.2 Methodology Differences

It needs to be understood that the typical process hazard analysis (PHA) has a much greater degree of granularity than a Cyber PHA. In a PHA, individual control loops are considered as potential cause of a hazard whereas in a Cyber PHA, the entire basic process control system would be considered all at once. Non cyber causes are more predictable as the common mode failure of an entire controller would result in all outputs failing the same way, whereas a cyber-attack is insidious as it can cause different outputs to fail so as to result in the worst case consequence.

The final difference is that the PHA will consider safeguards that may or are may not be vulnerable to cyber-attack, whereas a Cyber PHA generally has no practical means to consider safeguards that are not vulnerable to cyber-attack during the hazard identification portion of the Cyber PHA due to the different levels of granularity of what is being reviewed. For instance, if a control failure resulted in high pressure, a relief valve might be listed as a safeguard in a HAZOP. In a Cyber PHA it is not practical to consider this as it does not look at individual loops, but all of the loops, alarms, interlocks contained in a single BPCS. Once the major hazards have been identified, it would be reasonable to consider some safeguards that are not vulnerable to cyber attack on an exception basis.

2.3.3 Leveraging the PHA

After the traditional PHA has been performed, a further step is to consider whether initiating causes and safeguards are potentially vulnerable to cyber-attack. Those that are not vulnerable do not need to be considered in the cyber risk assessment. For those that are vulnerable, the ultimate consequence should be noted. These consequences can be ranked into appropriate categories to be considered in the detailed cyber risk assessment.

The exercise to determine these consequences of concern can be performed by reviewing the HAZOP or by going through the exercise with the layer of protection analyses (LOPA) if performed. If an adequate level of LOPA has been performed, it is somewhat more efficient to use these as they typically cover the hazards. It is not necessary to evaluate every hazard in a detailed cyber risk assessment as once the BPCS and the SIS are protected against major hazards, they would also be protected against the lesser hazards. Table 1 shows a more detailed procedure to leverage the PHA in this manner.

Table 1: Procedure for Leveraging the PHA

Major Procedural Step	Detailed Step	Comments
Perform traditional PHA	1. Use company HAZOP / LOPA procedures	
Excerpt high risk Initiating Events	2. Filter worksheet data to only include high risk consequence severities 3. Export the following data fields: • Initiating event (IE) • Consequence • Severity • Safeguards • PHA Recommendations • Comments 4. Sort the report by descending Consequence Severities	There is no credit for any safeguards at this point
Discard those initiating	5. Review each initiating event.	

Major Procedural Step	Detailed Step	Comments
events not susceptible to cyber influence	a. If it can be caused by a cyber-attack indicate "yes" b. If it is not vulnerable to cyber-attack, indicate "no" 6. Filter the report to only show those cause consequence pairs that are susceptible to cyber-attack.	
Identify safeguards not susceptible to cyber influence	7. Review the safeguards for remaining initiating events as to susceptibility to cyber-attack a. If it can be manipulated by a cyber-attack indicate "yes b. If it is not vulnerable to cyber-attack, indicate "no" 8. Filter the report to only show those safeguards that are not susceptible to cyber-attack	
Perform revised risk ranking	9. Update likelihood of event in the column of the PHA Data Export template. 10. Update risk ranking in the column of the PHA Data Export template. 11. Sort revised worksheet by risk in descending order.	• Only take credit for safeguards not susceptible to cyber-attack • Risk ranking is based on likelihood of cyber threat being perpetrated and failure of the safeguards not susceptible to cyber-attack for the specific risk receptor category.
Provide sorted worksheet to cyber risk assessment team		• The risk ranked worksheet is to be used by the cyber risk assessment team to: o Improve estimate of actual consequences and severity. o Achieve improved granularity of risk when credit for cyber safeguards is assessed.

2.3.4 Risk Assessment Procedure

In preparation for the cyber risk assessment key process safety information needs to be available. This mostly includes what was leveraged from the traditional PHA (Table 1) and the zone and conduit drawings that show:

- Zone location for each cyber assets
- Means of communication protection between zones.

Prior to performing the review, the tool/worksheet being used to conduct the review should be organized to list the zones and assets within each zone. Figure 3 shows a conceptual example taken of a Zone and Conduit drawing using the hierarchy approach from ISA 95.

Figure 3: Zones & Cyber Nodes
Excerpted from
CCPS *Guidelines for Safe Automation of Chemical Processes*[10]
(Peer Review Copy)

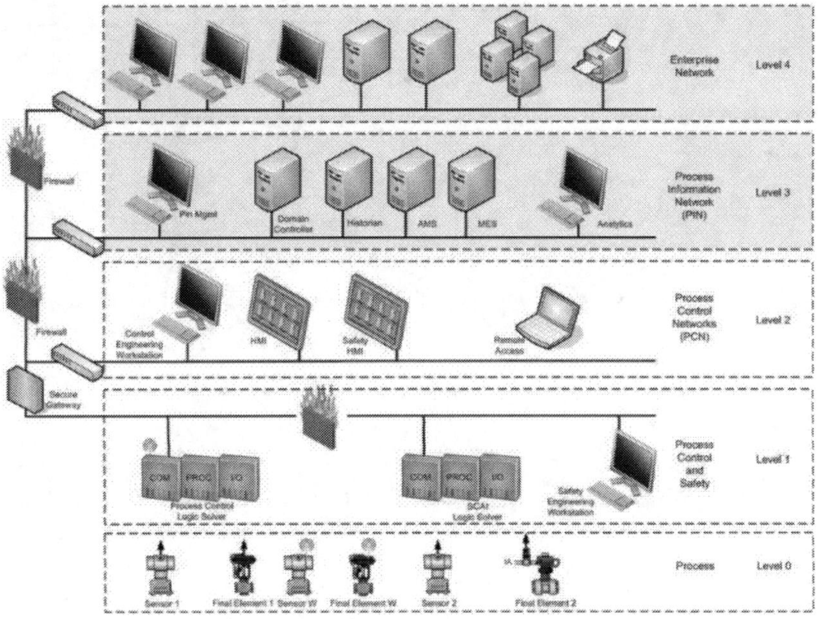

A HAZOP type approach is recommended to perform the cyber risk assessment. This is a qualitative risk assessment method that identifies potential for cyber-attacks that can cause an incident with the potential to result in harm to human life, the environment, property or loss of operating capability. A summary of the method is included below:

- Select zone
- Select cyber node
- Select threat (smart selections based upon type of cyber-attack)
- Identify & record causes

- Identify & record qualitative cause likelihood (without any credit for counter measures)
- Identify & record consequences (without any credit for counter measures)
- Determine & record qualitative severity of consequences
- Identify & record counter measures applicable to the cyber threat and cause
- Determine & Record qualitative likelihood (with existing counter measures)
- Determine if risk is tolerable per risk criteria. If not make recommendation(s)

Figure 4 shows an example worksheet following review.

Figure 4: Example Cyber Risk Review Worksheets

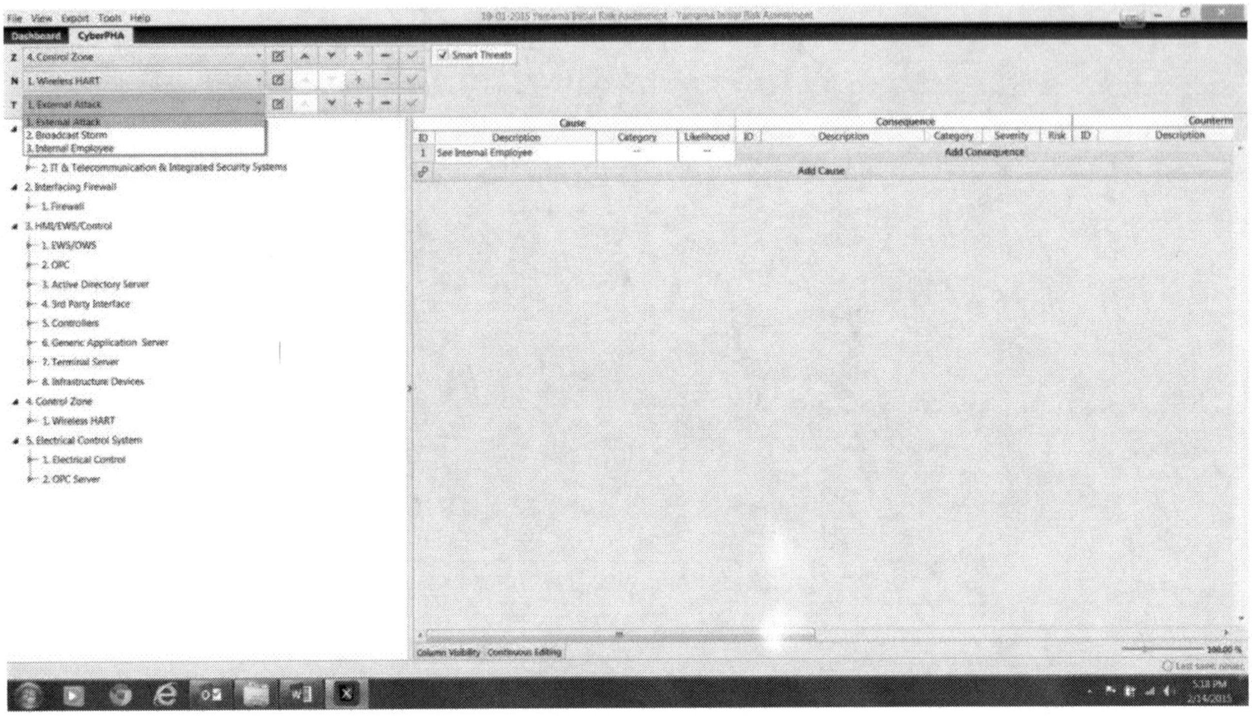

Figure 4 Continued

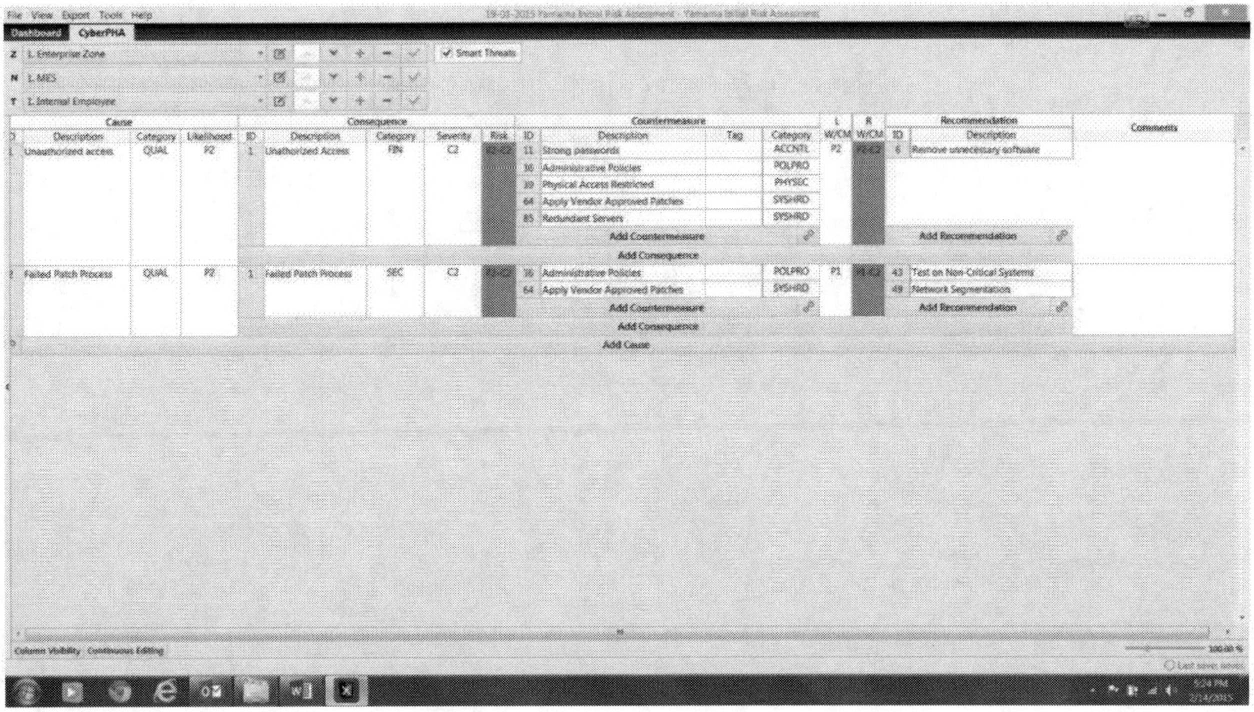

2.3.5 <u>Procedure Summary</u>

In summary, the steps needed to complete the detailed level cyber risk assessment are listed below:

- Determine major hazards vulnerable to cyber-attack from the HAZOP or LOPA
- Document cyber nodes from the zone and conduit drawing
- Perform a Cyber PHA on the zone and conduit drawings
- Make recommendations whenever risk criteria not met

3 Design and Implement Phase

Figure 3 shows the various activities during the design and implement phase. Once the CSRS is provided, an iterative design process between network, IACS, and process safety personnel will take place to ensure that the planned layers of cybersecurity protection including policies, procedures equipment, software capabilities, and level of rigor are appropriate for the cybersecurity performance expectations. Once it has been ascertained that the conceptual design has been validated, detailed design can begin. In parallel with the detailed design, the procedures to test and validate the cyber countermeasures need to be developed.

These test procedures need to support the cyber portion of the factory acceptance test (FAT), the site acceptance test (SAT), pre startup safety review (PSSR) and the initial validation of the cyber countermeasures. This work parallels and should be conducted as part of the SIS validation during its various stages.

Figure 5: Design & Implement Phase

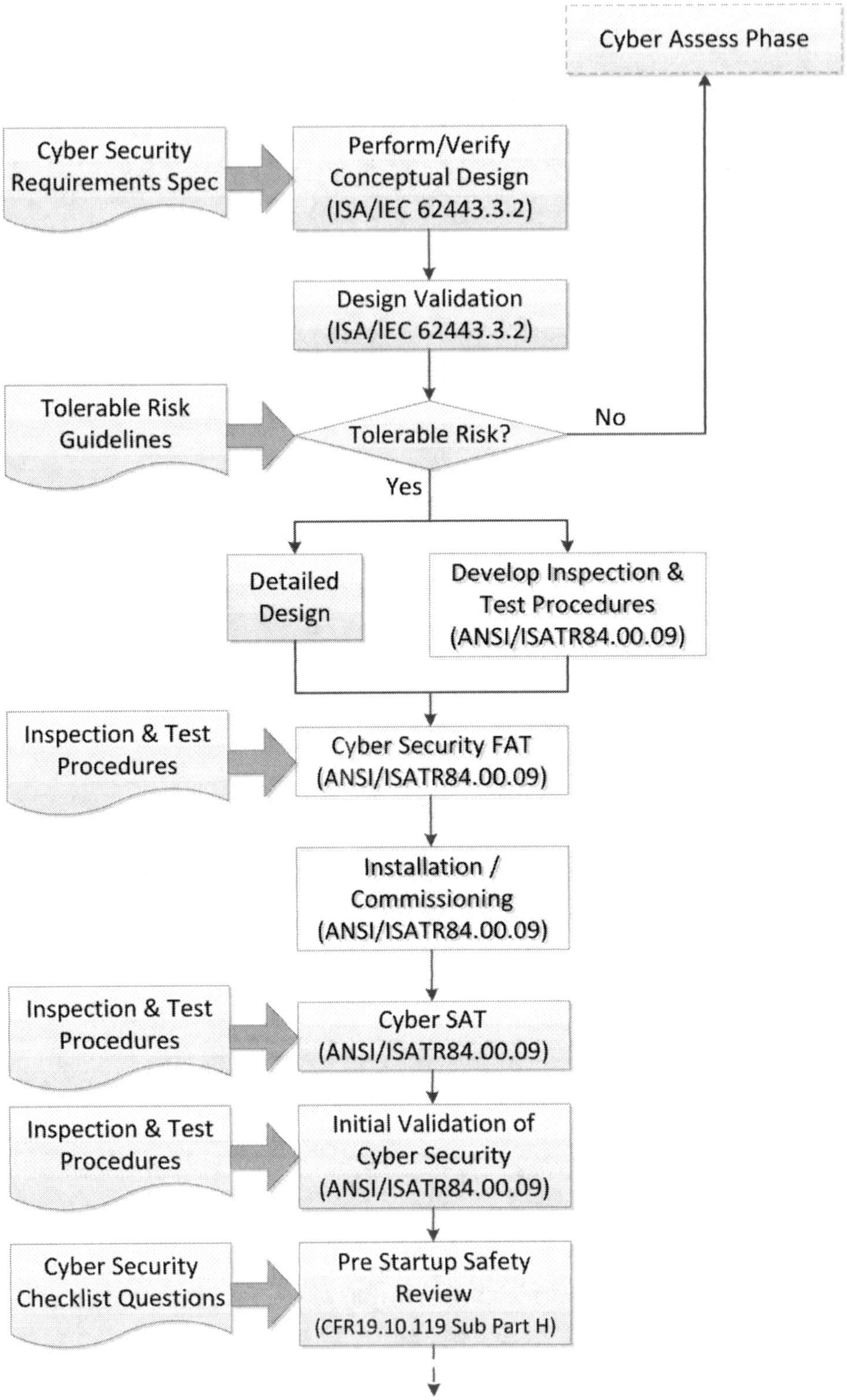

Prior to start-up, it is a process safety management requirement to perform a Pre Startup Safety Review (PSSR). As part of this PSSR, audit questions to assure the quality assurance of the cyber countermeasures being ready needs to be included as a subset of the overall PSSR.

4 Operate and Maintain Phase

4.1 Purpose

Unlike traditional process safety risks, cybersecurity threats continue to evolve, or new vulnerabilities are discovered, potentially making the countermeasures that have been implemented inadequate to maintain an adequate level of safety relative to the corporate risk criteria. As such, it is the end users responsibility to ensure the continued useful life of the cyber protective system, i.e. its technology currency. Procedures need to be in place to measure performance capability and make improvements should they be determined to be necessary.

4.2 Operation and Maintain Phase Overview

Figure 4 shows the various activities during the operation and maintain phase, beginning with startup. Following this, the plant is operated to make product. Periodically, cybersecurity countermeasures need to be revalidated. This is the equivalent of inspecting and testing safety instrumented systems and functions. Vulnerability assessment techniques are the backbone of this exercise and should have been developed and documented as part of FAT, SAT and initial validation.

Any time there is a management of change, potential impacts on cybersecurity needs to be considered. Just as the PHA needs to be periodically revalidated, this revalidation needs to account for cybersecurity as part of the lifecycle work process. The cybersecurity vulnerability assessment is a key part of this revalidation. Finally, if a portion of the facility is being decommissioned, it is recommended that the full lifecycle be employed to ensure that the remaining plant and equipment are not adversely impacted.

Figure 6: Operation & Maintain Phase

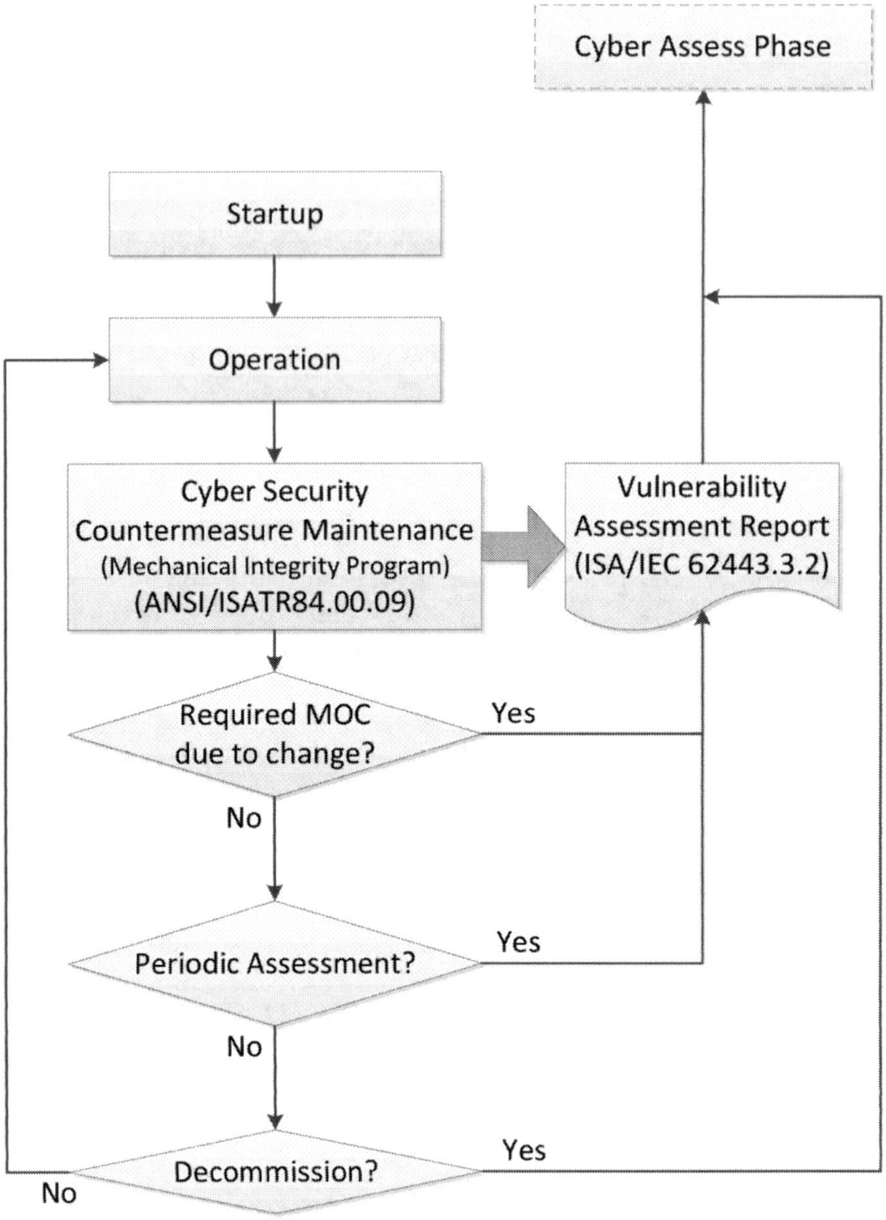

4.3 Vulnerability Assessment

A vulnerability assessment is intended to provide a review of the cybersecurity environment for the control and instrumented safety systems at the facility. It identifies vulnerabilities and then develops recommendations for possible improvements. It is best done in concert with a detailed cyber risk assessment per the lifecycle. During a vulnerability assessment, the following is accomplished:

- Existing facility cybersecurity practices and policies reviewed versus current industry practice, including current industry based alerts;
- Control & instrumented safety system network architecture reviewed and validated versus cybersecurity requirements specification (CSRS);
- Sample configuration data for facility network equipment and control system devices is collected and compared against documented expectations;
- Penetration testing is performed when the plant is offline;
- Findings are documented and compared against CSRS and with current industry best practices;
- Gaps that show corporate risk criteria not being met must be further evaluated in accordance with company policies.

5 Getting Started with an Existing Facility

5.1 Purpose

In today's world, not that many existing facilities have had the benefit of going through the lifecycle as previously described. In this section, the method to implement the lifecycle will be covered.

5.2 Overview

In order to implement the lifecycle, the starting point is during the operation and maintain phase of the lifecycle. A seven step process has been documented in the course titled, *7 Steps to Industrial Control System (ICS) Security*[9]. These steps are as follows:

- Perform Vulnerability Assessment
- Develop & document policies and procedures
- Train personnel & contractors
- Segment the network
- Control access to the system
- Harden system components
- Monitor and maintain system security

Figure 5 shows how these 7 steps ultimately allow a facility to ultimately be fully integrated with the lifecycle described above, achieving a sustainable solution.

Figure 7: Existing Facility Implementation

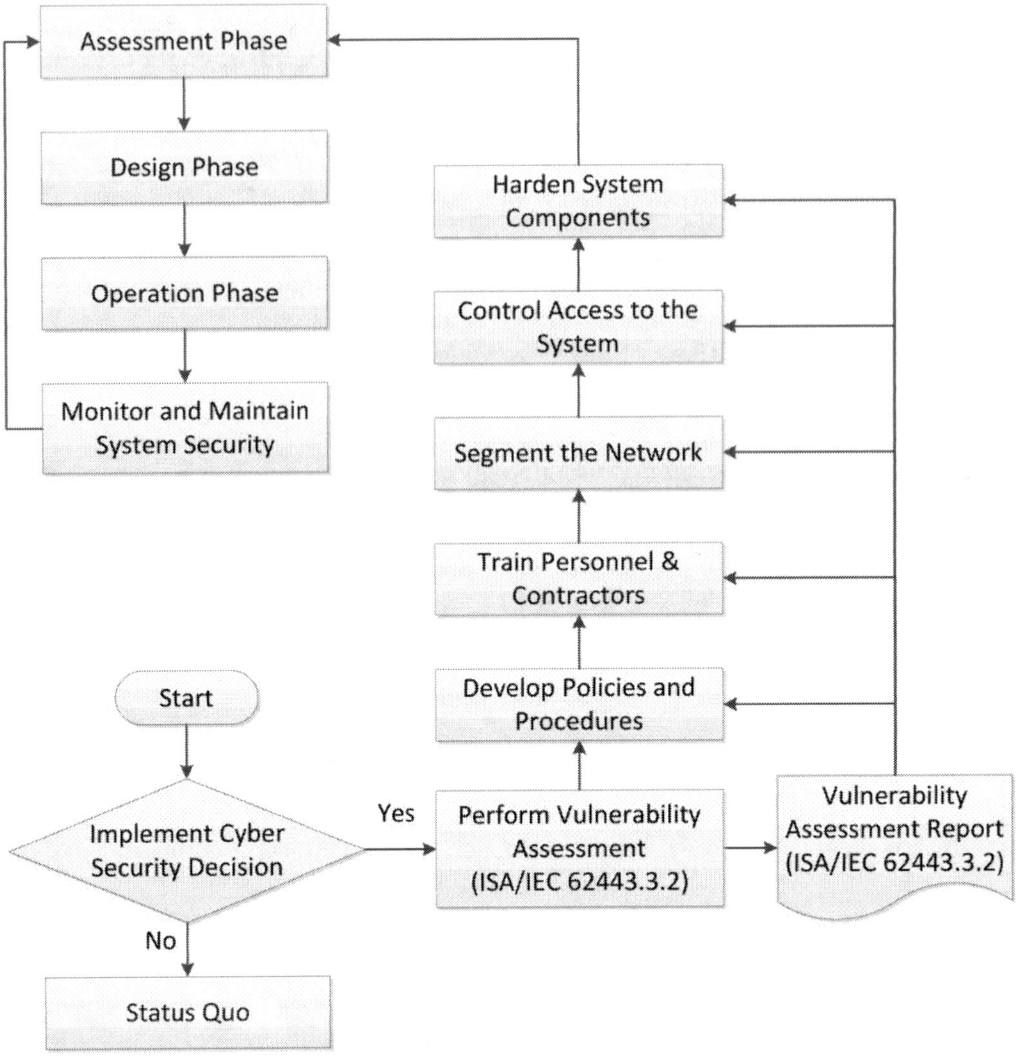

The first step is to perform a vulnerability assessment. When performed for a facility that has not implemented the lifecycle yet, it is somewhat different than what was described in section 4.3. When the first item performed is the vulnerability assessment, all it can do is test and assess the system versus recognized and generally accepted good engineering practices (RAGAGEP).

Once the vulnerability assessment has been performed, it provides valuable feedback to the company, helping them to develop appropriate policies, standards, and procedures. With the policies, standards, and procedures in place, training for employees and contractors may begin. In addition, segmentation of the network, access control and hardening of the assets can also begin based upon recommendations that come out of the vulnerability assessment.

There is still the need to perform the cyber risk assessment, which in turn, will most likely fine tune initial recommendations that emanated from the vulnerability assessment. After the recommendations have been fully implemented, the plant needs to monitor and maintain the system security. At this point, the facility should be fully following the lifecycle as described in sections 2, 3 and 4.

6 Conclusion

As described in this paper, the most effective and efficient means to achieve cybersecurity is to adopt a lifecycle approach, addressing analysis or assessment, design and finally operation and maintenance that is fully integrated with the process safety work process. It is only when the purpose built security countermeasure layers augmented with the performance and lifecycle management (NIST Framework[7]) of functional cybersecurity are followed, does an end user have a comprehensive strategy. As part of this lifecycle, it is important to conduct cyber risk assessments and vulnerability assessments. As the effectiveness of cybersecurity countermeasures may degrade over time due to increasing sophistication of the threat, on-going performance measurements and periodic reassessments are essential to meet the performance level and to judge the effective useful life of any cybersecurity system.

For existing facilities that have not had the benefit of the lifecycle, they can begin implementation via performance of a vulnerability assessment against RAGAGEP and recycle back to the beginning of the lifecycle, performing a cyber-risk assessment and fully integrating the cyber and process safety lifecycle work processes.

7 References

[1] ISA-TR84.00.09-2013, *Security Countermeasures Related to Safety Instrumented Systems (SIS)*; 2013.

[2] IEC 61508, *Functional Safety of Electrical/Electronic/Programmable Electronic Safety-related Systems*, 2010.

[3] ANSI/ISA-84.00.01 Part 1 (IEC 61511-1 Mod), *Functional Safety: Safety Instrumented Systems for the Process Industry Sector – Part 1: Framework, Definitions, System, Hardware and Software Requirements*, 2004.

[4] ANSI/ISA 62443-2-1 (99.02.01)-2009, *Security for Industrial Automation and Control Systems Part 2-1: Establishing an Industrial Automation and Control Systems Security Program*, 2009

[5] ANSI/ISA 62443-3-2, *Security for Industrial Automation and Control Systems Part 3-3: Security Risk Assessment and System Design,* Under Development.

[6] Executive Order 13636, *Improving Critical Infrastructure Cybersecurity*, Feb 12, 2013.

[7] National Institute of Standards and Technology, *Framework for Improving Critical Infrastructure Cybersecurity*, February 12, 2014.

[8] 6 CFR, Part 27, Department of Homeland Security, *Chemical Facility Anti-Terrorism Standards* (CFATs).

[9] exida.com LLC training course; *7 Steps to Industrial Control System (ICS) Security*

[10] CCPS, *Guidelines for Safe Automation of Chemical Processes*, Peer Review Copy.

[11] CFR19.10.119 Sub Part H, *Process safety management of highly hazardous chemicals.*

Additional references not cited:

[12] CCPS, *Guidelines for Safe and Reliable Instrumented Protective Systems*, May 2007

[13] ANSI/ISA 95.00.01, *Enterprise-Control System Integration Part 1: Models and Terminology*, 2000.

[14] ANSI/ISA-62443-1-1 (99.01.01)-2007, *Security for Industrial Automation and Control Systems Part 1: Terminology, Concepts, and Models*, 2007.

[15] ANSI/ISA 62443-2-2, *Security for Industrial Automation and Control Systems Part 2-2: Implementation Guidance for and IACS Security Management System,* Under Development.

[16] ANSI/ISA 62443-2-3, *Security for Industrial Automation and Control Systems Part 2-3: Patch Management in the IACS Environment,* Under Development.

[17] ANSI/ISA 62443-2-4, *Security for Industrial Automation and Control Systems Part 2-4: Requirements for IACS Solution Providers*, Under Development.

[18] ANSI/ISA 62443-3-3 (99.03.03)-2013, *Security for Industrial Automation and Control Systems Part 3-3: System Security Requirements and Security Levels*, Aug 12, 2013.

[19] ISO/IEC 27002, *Information technology — Security techniques — Code of practice for information security management*

396629 Integrating Alarm Management into Your PHA and LOPA Processes

Abstract:

Operator response to alarms is a commonly used layer of protection to reduce risk and to prevent an upset from escalating into an incident. When alarms are used as independent protection layers (IPL), they must follow the applicable standards, guidelines, and regulations for both alarm management (ISA-18.2) and process safety (IEC-61511 and OSHA PSM). This paper will discuss how to integrate the work processes of alarm rationalization, process hazard analysis (PHA), and layer of protection analysis (LOPA) to effectively bring together the alarm management and process safety lifecycles.

Integration of these work processes can deliver significant benefits including enhanced traceability and seamless management of change, as well as a more effective design – ensuring that the operator's response to alarm provides the expected level of risk reduction. Appropriate characterization of the alarm layer of protection also helps to ensure that other layers of protection (such as a Safety Instrumented System) are neither over-designed nor under-designed.

Specific topics include:

- Integration of alarm management into the overall on-going PSM work process
- Properly accounting for the consequences of deviation (mitigated vs. unmitigated)
- Analyzing operator response time in relation to process safety time and safe operating limits
- Analyzing alarm system performance to determine actual demand rate and initiating frequency
- Determining alarm proof test frequency and testing requirements
- Effective operator training
- Setting the alarm priority to indicate the criticality of the alarm in relation to others
- Use of alarm response procedures to help operators respond quickly and effectively
- How tools for PHA, LOPA and alarm rationalization can exchange information seamlessly facilitating the overall PSM work process

395895 High Integrity Pressure Protection Systems: Design, Analysis, Justification and Implementation

Abstract:

HIPPS (High Integrity Pressure Protection Systems) are installed to provide protection to critical equipment ranging from high/low pressure interfaces to equipment overpressure scenarios in upstream and downstream processes. Any of these conditions could result in damage to the plant, the environment, loss of life and financial consequences.

To avert such an event, HIPPS generally shut down, reduce or divert the pressure sources feeding the system that is experiencing the overpressure. In those cases where overpressure in a system causes a huge liquid or vapor relief load being sent to a flare system, the right HIPPS system can help reduce excessive load on an existing flare system and eliminate the high costs associated with purchasing new relief devices, resizing existing flare headers, re-rating flare knock out drums, redesigning the flare stack, etc., in addition to the opportunity costs due to loss of operations.

Currently, the approach is to design HIPPS for flare load reduction as a Safety Integrity Level (SIL) 3 Safety Instrumented Function (SIF) or a SIL 2 SIF (depending on the code or company standard being followed). This work discusses how instead of taking the customary "one size performance fits all approaches", the design could be based on a IEC 61511 Safety Lifecycle view point to determine the required risk reduction and thereby choosing the actual required SIL. This presentation will discuss current practices; review benefits and drawbacks of SIL selection in these scenarios, and describe the impact on total cost of ownership.

393891 Predicting Far Field Blast Overpressures: An Evaluation of Enhanced Input Parameters for FLACS

Abstract:

Vapor cloud explosions (VCEs) are considered one of the main hazards within process safety, considering the destructive potential of blast waves and the possibility of escalation (i.e. subsequent jet fire). In the past decade, the US has experienced severe onshore accidents involving explosion events and BP Texas City (2005) was one of the worst industrial accidents in US history. These types of consequence events have raised the safety concern for the entire onshore safety industry. Predicting potential explosion overpressure is now a recommended practice for new and existing facilities. Until recently, evaluating potential explosion hazards for onshore facilities has mostly been performed utilizing simplistic models such as TNO and BST.

Empirical and phenomenological models (e.g., TNO Multi-Energy Method and Baker-Strehlow-Tang (BST) Method) can to some extent give useful estimation of VCE consequences for simple explosions and crude geometries. Empirical models utilize correlations derived from experimental measurements, whereas phenomenological models employ some methods to represent primary physical processes during the explosion (PHAST).

Alternatively, Computational Fluid Dynamics (CFD) models are able to provide more precise predictions of the energy and resulting overpressure of the blast wave, as well as the ability to consider more realistic physical geometries, gas cloud variations and ignition locations. CFD models can also present detailed dynamic effects in the near and far field, including pressure changes versus time, reflected and redirected blast waves. FLACS is one of the most widely used CFD-tools for explosion modeling. Continually improved from its inception in the 1980s, the primary goal was to predict vapor cloud explosion loads in offshore platforms. As onshore safety restrictions become more stringent, FLACS is becoming an essential tool for onshore safety studies and, in particular, for evaluation of VCE-related overpressure. However, care should be taken when applying FLACS for onshore VCE studies. Investigations into far-field blasts have shown FLACS calculations often underpredict far-field blast pressures when compared to TNO blast curves. Enhanced parameter settings have been developed for the FLACS code and subsequent studies have suggested that more accurate far-field blast predictions are possible. These modified parameters will be utilized, compared, and evaluated for discussion in this paper.

In this study, CFD and TNO models are used to predict onshore far-field blast pressure in under- congested areas. PHAST (Process Hazard Analysis Software Tool) is employed to generate explosion predictions with the simpler models. Based on the available guidelines, FLACS model parameters are set to resolve the blast waves. Several explosion scenarios have been simulated with the aforementioned simulation tools PHAST and FLACS and the results discussed.

402443 The application of the bifunctional environmental deodorizer for sulfur and sour water in SINOPEC TIANJIN

Abstract:

SINOPEC TIANJIN Company Storage and Transportation Department adopted bifunctional environmental deodorizer to clean repugnant substance in the containment, including reaction tower and tank containing sulfur, ammonia and nitrogen during the shutdown process. This action has achieved good deodorization results, and purified the air, protected the environment and the personal safety and facilities, which improve the quality of maintenance and shorten the maintenance duration. After cleaning, waste emissions is allowed.

Safety-Centered Site Master Planning

Co –Authors

Steven Clepper, AIA
Jacobs
Advance Planning Group
10 Tenth Street, Suite 1400, Atlanta, GA 30309

steve.clepper@jacobs.com

Hollie Schmidt, APA
Jacobs
Advance Planning Group
201 N Franklin Street, Tampa, FL 33602

hollie.schmidt@jacobs.com

Prepared for Presentation at
American Institute of Chemical Engineers
2015Spring Meeting
11th Global Congress on Process Safety
Austin, Texas
April 27-29, 2015

UNPUBLISHED

Safety-Centered Site Master Planning

Co-Authors

Steven Clepper, AIA
Jacobs
Advance Planning Group
10 Tenth Street, Atlanta, GA 30309

Hollie Schmidt, APA
Jacobs
Advance Planning Group
201 N Franklin Street
Tampa, FL 33602

Keywords: Risk Assessment, Plot Plan, Facility Siting, Site Selection, Risk Mitigation, Risk Optimization, Plant Layout Optimization, Zoning Plan, Traffic Congestion, Physical Security, Operational Efficiency, Cost Reduction, Logistics Planning

Abstract

Safety-centered site master planning is a holistic, multi-disciplined approach to a common challenge found in many petrochemical refineries and plants. The challenge being, the existence of occupied buildings that are not designed to provide appropriate protection for personnel given the risks presented by their proximity to the core process operation areas and risk generators. The cause of this dangerous condition is often the effect of incremental growth of the plant over time, along with antiquated attitudes about the safe placement of support buildings housing non-essential personnel in close proximity to high risk zones. Safety-centered site master plans improve safety metrics, optimize plant layout, and proactively mitigate risks to personnel and high-value assets.

1. Introduction

Safety-centered site master planning has evolved as a result of a pattern of recent tragic events in the last decade. This approach was developed in response to numerous fatal process-related accidents that have occurred in petrochemical plants in the United States and the desire to avoid future occurrences. Safety-centered site master plans can minimize the risk exposure to employees, contractors, and the public by using the philosophy of "Safety is Good Business", thus reducing the probability for a fatal accident. The vision statement of safety-centered master planning is to *eliminate the risk posed to humans in occupied buildings, including mobile, modular wooden structures, which may not be located in areas that are safe from fires, explosions, overpressures, or exposure to toxic or corrosive materials.*

This comprehensive process requires a core team of technical specialists including petrochemical plant technical experts, land planners and facility programmers who work closely with plant leadership to develop the plan. The plan establishes a framework for future growth and change at the plant. It consolidates all planning data in one digital source and proposes a wide range of strategies to mitigate risk.

Recent refinery and chemical plant events have focused greater attention on the risk of locating non-essential personnel near processing units. Regulatory requirements and industry best practices are currently evolving to respond to the inherent safety challenges at refineries and chemical plants. OSHA and the American Petroleum Institute (API) are leading this evolution by updating key documents related to occupied buildings within petrochemical plants.[1,2]

Safety-centered master site planning is important because employers care about the safety of the people that work at their refineries and chemical plants. The potential financial loses that results from a fatal process incident can be catastrophic. To reduce potential injury and loss of life, it is important to recognize the risks posed by process units and other risk generators to nearby occupied buildings. A thoughtful, well planned site master plan reduces risks to people while improving operational efficiency, thus linking prosperity to safety.

2. The Common Challenge at Most Processing Plants

Effective safety-centered site master plans seek the proper balance between safety and productivity. Often, the perception is that all workers at the plant need to be in the heart of the core process areas to be operationally efficient. This is simply not true. Valid solutions can be identified and quantified to achieve a good balance of these two critical factors. Other, more routine planning factors are still important, but are filtered through the lens of safety. Factors such as mobility, security, utility, environment, and the overall quality of the workplace also contribute to the business of safe and efficient production of chemical and petroleum products.

2.1 How Did We Get Here?

Very few plants are constructed entirely in one phase at one time. In fact, most refineries and chemical plants are the result of cumulative growth through decades of infrastructure investment, migrating operations, changing management or even ownership, surrounding community encroachment and evolving regulations. The initial phase of most refineries and chemical plants consists of a few processing units with nearby support buildings. Typical support buildings include shops, field offices, shipping/receiving facilities, and other buildings to house people that typically do not work directly on the processing units. Incremental petrochemical plant growth often creates problems when existing buildings are not relocated to safer sites when new process units or risk generating infrastructure are constructed. As a result, many support buildings are now "embedded" in high-risk areas due to their proximity to risk. In some cases the people are relocated and the building is repurposed or demolished, but not all. These are the prime candidates for a safety-centered site master plan.

2.2 The Way It's Always Been

The challenge is often exacerbated by the behavioral conditioning of legacy personnel who have been working in these embedded support buildings for a long period of time. Occupants tend to believe that there must be a good reason for the building's location because "it's the way it's always been." As a result, these personnel become somewhat, or entirely, de-sensitized to the risk. A safety-centered approach evaluates the cumulative results of a risk assessment, translates the results into easily understood consequences, then identifies and locates these personnel away from the most severe hazards if their job function does not require them to be immediately in the core of the plant. The cultural changes in behavior that are necessary for successful facility relocation are addressed with the client leadership team as a part of the future plan.

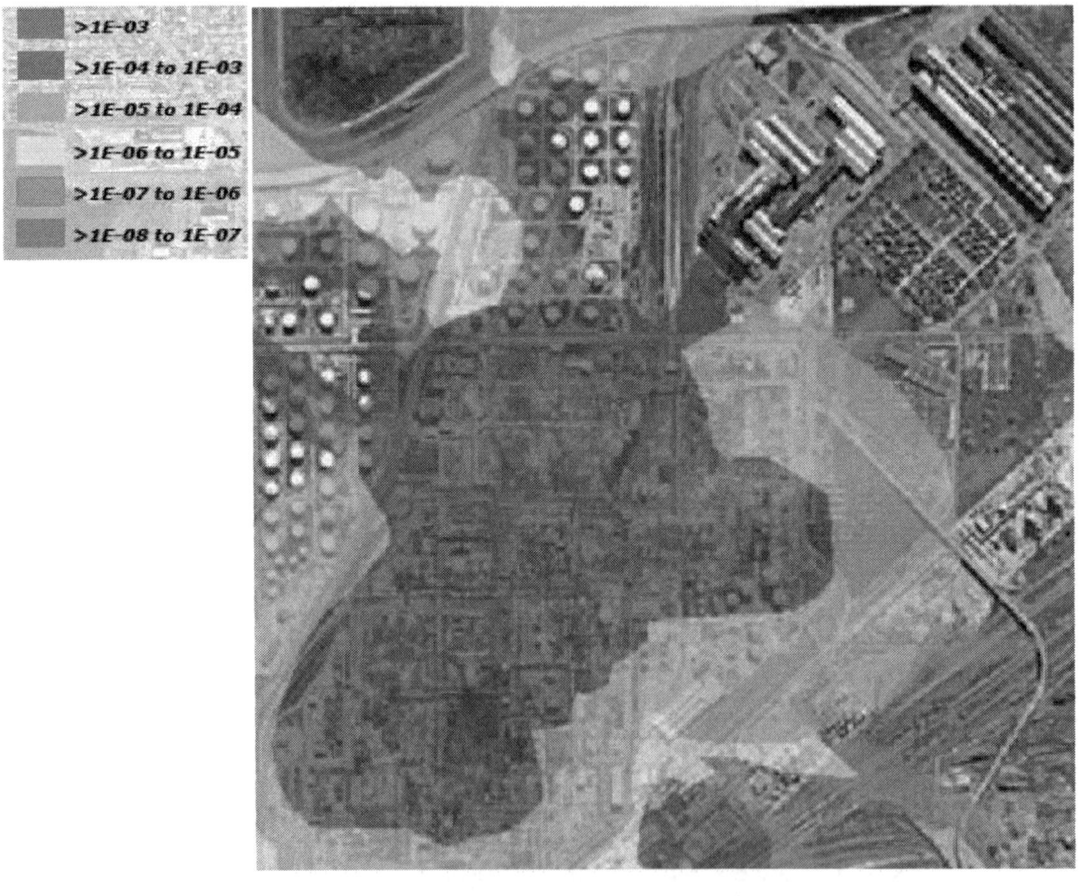

Figure 1 – Example Composite Quantitative Risk Assessment Contours

3.0 Key Goals of a Safety-Centered Site Master Plan

Each safety-centered site master plan should have a clear definition of its vision and goals. The vision and goals are developed with the plant leadership team during a visioning work session, which should be scheduled after the planning team has made a study of existing conditions. Scheduling the visioning and goal setting work session after the existing conditions survey enables the planning team to focus the discussion and ask more targeted questions of the leadership.

The planning team concept works best with a two-level owner team. First, a planning steering committee should be identified that is composed of key plant staff that represent the major departments and will handle day to day involvement with the technical planning team. The top level owner team is the plant leadership team. The planning steering team reports to plant leadership team for issues of policy, direction, and approval of the plan.

The technical planning team is made up of planning disciplines and subject matter experts skilled in planning for these types of facilities. The job of the technical planning team is to lead the owner through the planning process, identify key questions, concepts and other recommendations for the owner's planning steering committee's consideration and action.

3.1 Measurable Results and Desired Outcomes

The science of risk identification and quantification has advanced to include projections of blast overpressure, toxic vapor, thermal radiation, building damage levels, occupant vulnerability, and societal risk. Both consequence-based and quantitative-based risk can be calculated from any number of possible release scenarios. Detailed scientific risk assessments performed by qualified process safety professionals are the standard for assessing the current state of plant safety and is how we measure the potential for improvement. Specifically, the American Petroleum Institute (API) developed two Recommended Practices (RP) API 752 and API 753 which safety-centered site master planning uses as a guideline for recommendations, specifically for facility siting.

API 752 provides guidance for managing the risk from explosions (overpressure), fires (thermal) and toxic material releases to on-site personnel located in new and existing buildings intended for occupancy. API 752 was developed for use at refineries, petrochemical and chemical operations, natural gas liquids extraction plants, natural gas liquefaction plants and other onshore facilities covered by OSHA 29 CFR 1910.119. Figure 1 illustrates an example of composite risk contours using a QRA.

API 753 provides guidance for reducing the risk to personnel located in portable buildings from potential explosion, fire and toxic release hazards. While occupied permanent buildings (e.g. control rooms, operator shelters) located near covered process area are typically constructed to be blast and fire resistant, conventional portable buildings (i.e. light wood trailers) are typically not constructed to be blast and fire resistant. Past explosion accidents have demonstrated that occupants of conventional portable buildings are susceptible to injuries from structural failures, building collapse, and building debris and projectiles.

The ability to generate measurable results for risk reduction, productivity and safety begin with the collection of current and accurate baseline metrics for both risk, which is the purpose of a risk assessment, and production, which helps prove that optimized production or "tool time" does not need to increase to allow for safer work conditions i.e. the desired result. Figure 2 illustrates key safety improvements in terms of number of people, number of buildings, and square footage ranging from high risk on left to low risk on right. This is an example of a typical client:

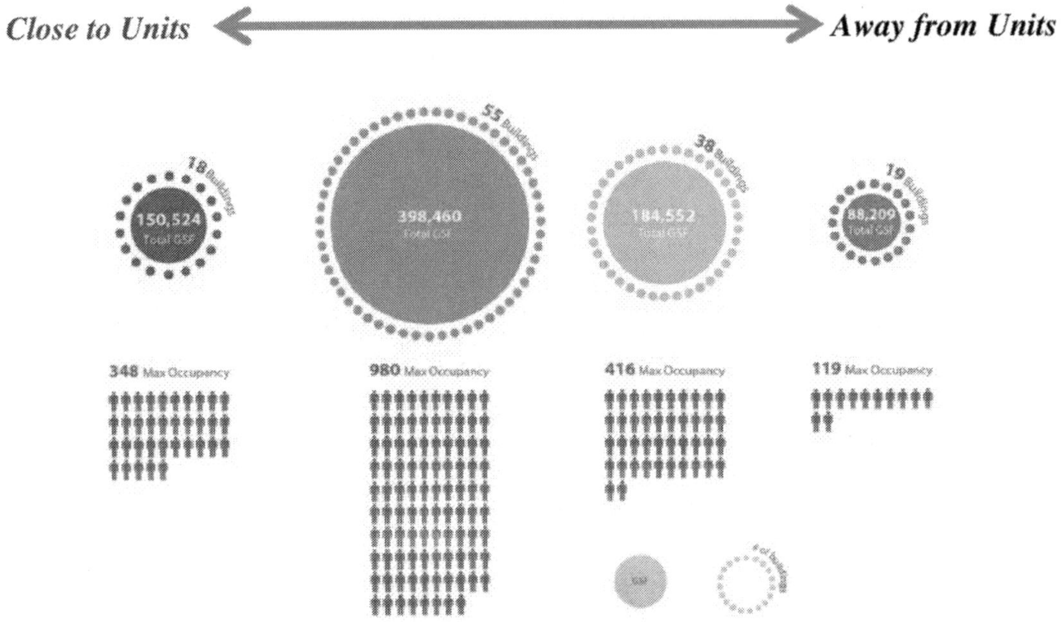

Figure 2 – Existing Conditions before Safety-centered Site Master Planning

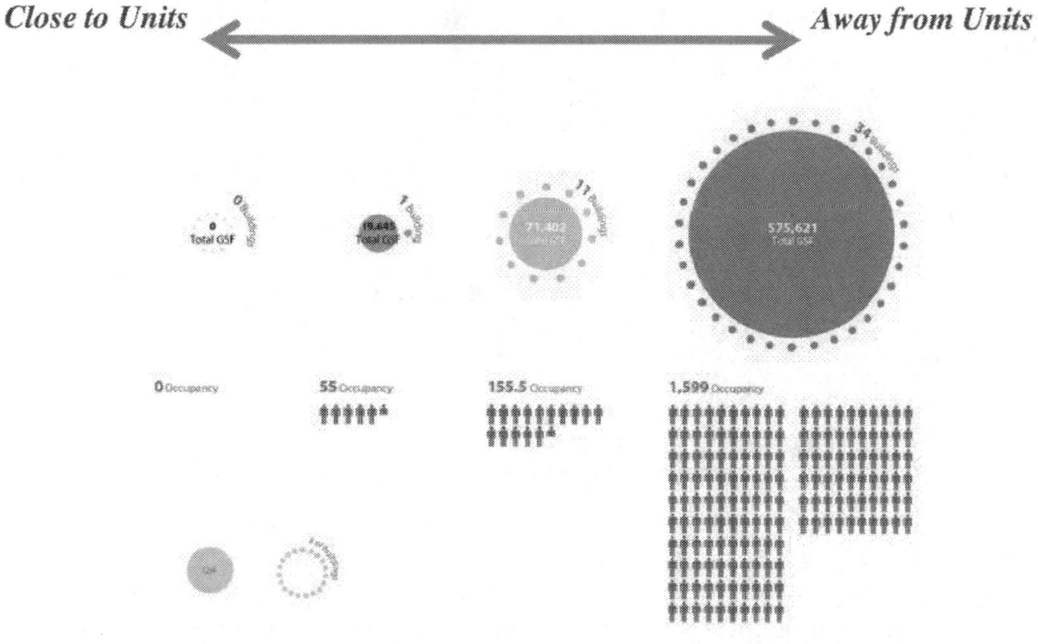

Figure 3 – Effect of Safety-Centered Site Master Planning Implementation

Thresholds and tolerance of risk vary greatly between plants and between corporations. Generally, the highest risk is represented in red and lowest risk is represented in green. Intermediate levels have colors representing a graduated scale of risk between lowest and highest.

3.2 Each Plant is Unique

A key element of the planning process is to address the uniqueness of each plant. Potential differences include feedstock, product, plant size, access to transportation, etc. For the planning team, it's important to quickly determine the unique characteristics of each plant to use as a basis for planning.

Goals of a well-rounded safety-centered site master plan should include:

- Balance Safety with Operations – Balance safety with excellence in productivity, operational efficiency and sustainability
- Improve Operational Efficiency – Increase operational efficiency through optimized functional relationships
- Think Strategically – Envision and aim for an enduring long-term business goals despite short-term priorities of the business environment
- Consolidate Fragmented Operations – Eliminate wasted production time caused by unnecessary movement between fragmented facility locations. Incremental growth in processing capability over time can result in fragmented multiple shops, control rooms, etc.
- Limit Liability – Decrease the potential liability for loss of life, lost production and fines for non-compliance

- Reduce Overall Risk – Reduce risk metrics through good facility siting and other mitigation measures
- Comply With Applicable Regulations –Comply with current facility siting regulations, including OSHA, API 752, and company-specific policies
- Be Open to New Paradigms – Achieve the best outcome by encouraging plant leadership to explore new ways of thinking about the potential of the plant and embrace change
- Be a Good Neighbor and a Good Place to Work - Promote "Industry Best Practices" related to security, environmental protection, and quality of workplace

4.0 Typical Issues To Address

Although there are unique challenges associated with each refinery or chemical plant, there are also a number of typical issues that occur at most locations. It's important to find ways to overcome such issues, which can include: less than ideal management support for the plan, claims of loss of productivity, difficult phasing and incremental (sometimes insufficient) funding. The good news is that a well prepared safety-centered site master plan will suggest ways to overcome these issues and challenges. There are many mitigation strategies available to an experienced planner or subject matter expert who will suggest targeted and feasible solutions. Additional typical issues to overcome can include:

- Lack of a Clear Framework for Growth from Plant Leadership – Lack of a comprehensive plant-wide business/production output plan, designation of land uses and asset management strategy.
- Lack of a Future Vision - Uncertain land acquisition strategy that accommodates growth and creates a buffer zone where it is needed.
- Knee Jerk Reactions – Reactive decision making about facility siting without considering the holistic picture. Often associated with the response to a recent incident on site or within the company.
- Unsafe Embedded Buildings – Existing occupied facilities too close to hazards.
- Emergency Response Equipment at Risk – Many plants have firehouses near processing units where a large event could render firefighting resources unusable.
- Workforce vs. Workplace Imbalance – Modular wooden trailers are often used as permanent facilities to accommodate overflow.
- Security and Traffic Enhancements – A traffic plan is needed to accommodate a large number of vehicles within the secured perimeter due to location of warehouse and shops inside perimeter. Gate traffic congestion sometimes results when vehicles are not segregated by type.
- Inadequate Road and Parking System – Poorly defined roads within the plant and lack of extra parking cannot accommodate a large turnaround event.

- Problems Attracting and Retaining Talent - Lack of investment in employee facilities and workplaces affect recruitment and retention of good employees.
- Community Issues – A community outreach program may be needed to address complains of risk, noise, odors, flares, etc.

5.0 Potential Solutions

Because each plant is different, there is no one solution that fits all. Each safety-centered site master plan incorporates a variety of different strategies and solutions to address the plant's need over time. Solutions can improve traffic access, better perimeter security, appropriate sites for buildings, consolidate operations and other actions specific to that refinery or chemical plant.

Relocating non-essential personnel (those with job functions not requiring close proximity to the units/core risk areas) to safer locations is a primary driver in the master plan. API 752, states that "Consideration should be given to locating non-essential personnel as far as practicable from the hazard and discouraging congregation of personnel in buildings close to process areas."

In order to determine essential versus non-essential personnel, or direct support versus in-direct support of the units, the master plan team utilizes the Circles of Safety tool, a tool unique to Jacobs, that our team has developed specifically for industrial clients with employees who have risk exposure. This tool categorizes personnel by job function in rings of progressive proximity to risk exposure. These Circles of Safety have a direct correlation and influence on where workers sit at their "desk", or perform their job duties, which aims to site future operational support facilities in the safest location possible while still optimizing operations.

In the Circles of Safety shown in Figure 4, employee positions are shown in the circle appropriate for the job's proximity to the process. Those employees working on the process units are located in the center circle. Those that must be near but safe are located in the middle circle and those that do not need to be located near the process are in the outermost circle. Working with plant management to identify the job position placements provides key data on who is considered essential to the process per OSHA and it also provides clues on how to group personnel of similar risks in a facility strategy.

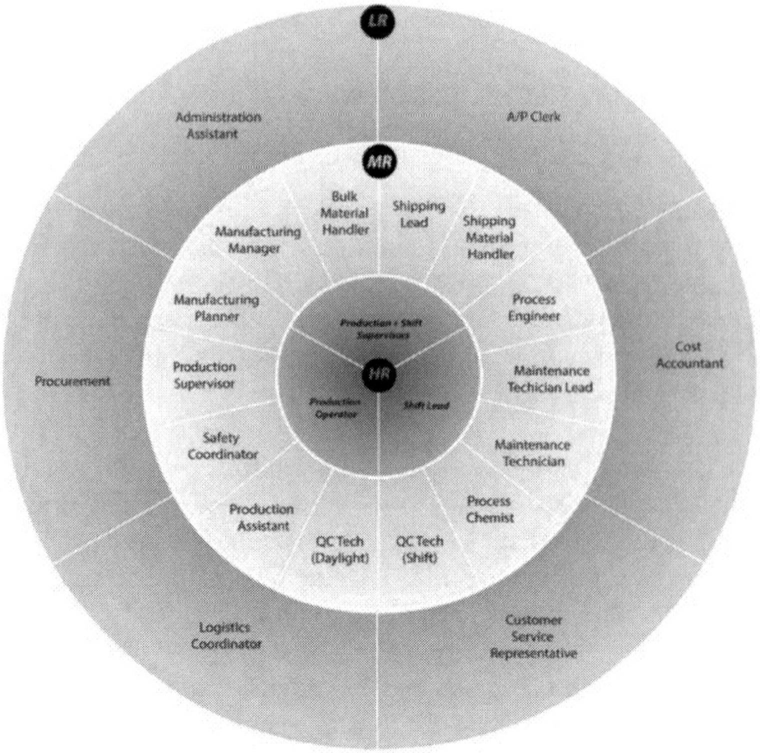

Figure 4 – Completed Circles of Safety Diagram

The results of the Circles of Safety tool, combined with the results of the risk assessment, determine if existing buildings can be examined to potentially harden to be safe or if they need to be replaced or repurposed for non-occupied uses. For replacement buildings and facility re-siting, appropriate building sites are identified and construction types selected based on operational need and lessened risk exposure. Typical non-process facilities that could be impacted include:

- Administration buildings to house office based workers
- Shops to house maintenance personnel
- Maintenance support and their offices
- Warehouses / shipping + receiving to house inventory of materials and shipping /receiving loading and unloading docks for material trucks
- Laboratories to house lab personnel for QC of plant products.
- Changing rooms for plant employees that work in the processing areas of the plant
- Firehouse / Emergency Operations Centers house specialized fire response equipment and provide a central control room for emergency operations

- Guard houses / security buildings house security personnel responsible for gates and perimeter security. These personnel also play a key role in policing the plant and responding to emergencies
- Field offices house office based workers that spend a portion of their time in the process area daily but predominately work in a field office. These personnel typically include contractors and maintenance and operational personnel
- Control Rooms house process control systems and personnel. These facilities can be centralized using contemporary digital control systems and can be located outside the high risk area

Each of these replacement facilities have ideal locations on a well-planned site that includes clearly defined roads, non-process area for location of support facilities, clear entry exit and handling of traffic. Figure 5 illustrates ideal relationships between plant elements.

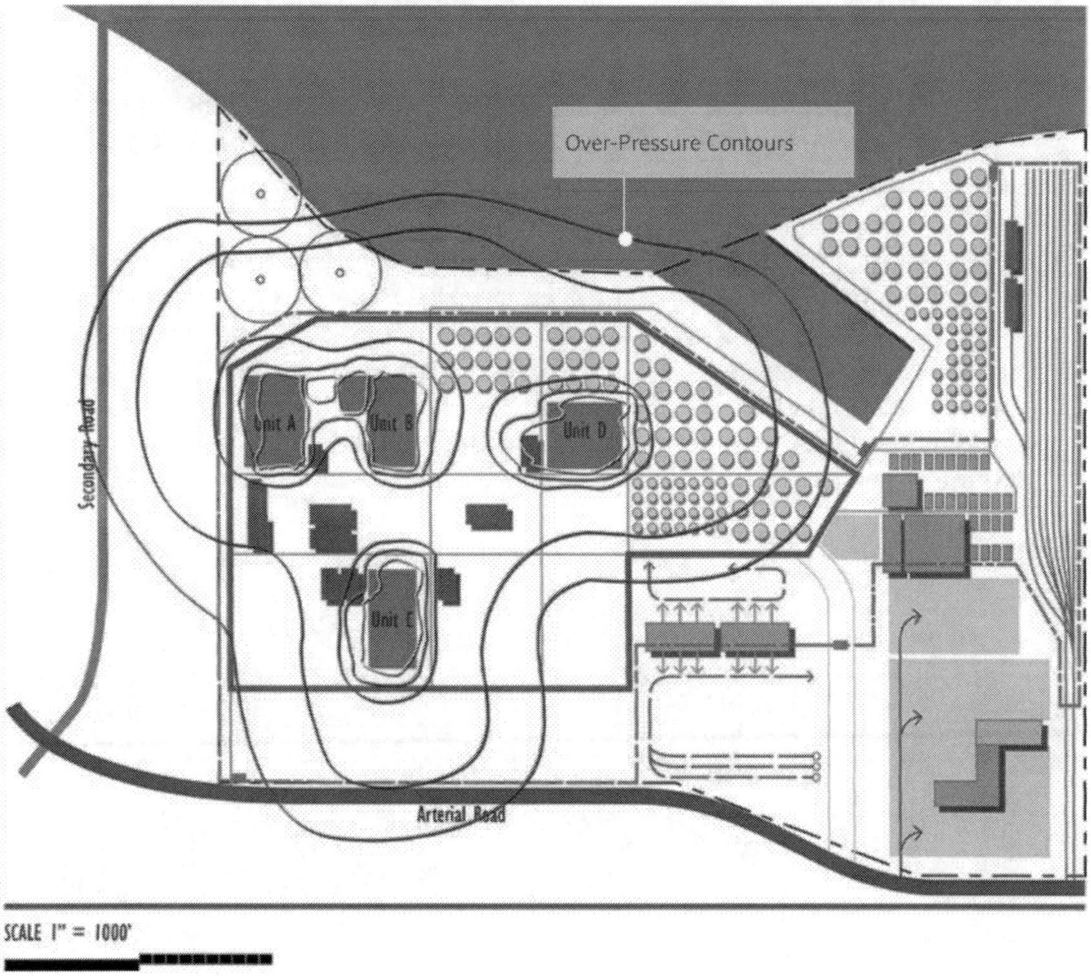

Figure 5 – Ideal Typical Plant Relationships

5.1 Lessons Learned Through Experience

We have learned the following from our experience developing successful safety-centered site master planning solutions:

- Understand That Risk Is the Key – Risk mitigation is the primary planning driver for safety-centered site master planning at petrochemical plants and all other planning considerations are secondary. A safer solution is a better solution.
- Have a Complete Understanding of the Risk – It is important to have a clear and thorough understanding of the major risk factors. The planning team must understand the risk assessment and tabulate the relevant data.
- Identify Situational Safety Issues – Challenge work cultures that place people in unnecessary risk. Often employees are located haphazardly in the plant as a result of individual moves over time, placing process non-essential workers in risk areas. Sometimes employees are moved to a certain location to take advantage of available space, such as an administrative assistant in a shop or an engineer in a control room. Use the Circles of Safety exercise to locate those employees relative to the risk.
- Identify the Most Economical Way To Move People Away From Risk - Relocate people from harm's way wherever possible, especially those non-essential to the process
- Construct Fit-For-Purpose Buildings - For those employees that must work near processing units, construct appropriate facilities that mitigate risk. Almost any situation can be mitigated with high-cost, blast-resistant design and shelter-in-place facilities. The real question is whether it is better to construct buildings outside the risk zone at a lower cost, especially for personnel who are not directly involved in the process. Figure 6 illustrates a well-defined support campus away from the process risk areas.

Figure 6 – Concept of an Ideal Support Campus for Non-essential Personnel at Edge of Plant.

6.0 Conclusion

The benefits of a well performed safety-centered site master plan can be unexpectedly beneficial and comprehensive given the plan uses a holistic, multiple disciplined approach. A safety centered site master plan can be applied to existing sites, brownfield sites and greenfield sites. Three key benefits of a plan to the refinery and chemical plant owners are:

6.1 Improved Safety Metrics

Improved safety metrics, in terms of reduced numbers building damage levels, overpressures, toxic vapors, thermal radiations, and total societal risk will mean:

- Fewer people work in areas of high risk
- Fewer high damage occupied buildings
- Increased use of blast/vapor/thermal resistant buildings built for purpose for essential personnel that must work in areas of risk
- Reduction in risk to the surrounding community due to overpressures, toxic vapor, and radiant heat resulting from an accidental explosion

Reductions in these metrics directly improve the safety of personnel working at the plant and to the surrounding community.

6.2 Optimization of Plant Layout

A well prepared safety-centered site master plan will optimize the layout of the plant and its operations. The comprehensive planning approach will improve the daily operation of the plant and reduce operational expenditures. The plan will do this by providing:

- A facility strategy that is consistent with the business strategy
- A framework for future growth or change that responds to changing business conditions
- Improved security and traffic flows by segregating different types of traffic
- Consolidate fragmented operations
- Close but safe locations for occupied support buildings such as administration, warehouse, laboratory, etc
- A well-designed plan to accommodate additional personnel and vehicles on site during turnarounds and capital projects without affecting ongoing operations
- A land-acquisition strategy to handle future expansion or create buffers fornearby neighbors
- Potential for process expansion area on refinery property by relocating existing buildings to appropriate locations

Several consequential benefits accrue to most plants that perform a safety-centered site master plan by improving efficiency and reducing operational cost.

Figure 6 illustrates the removal of unsafe embedded buildings create opportunities for process expansion inside the security fence.

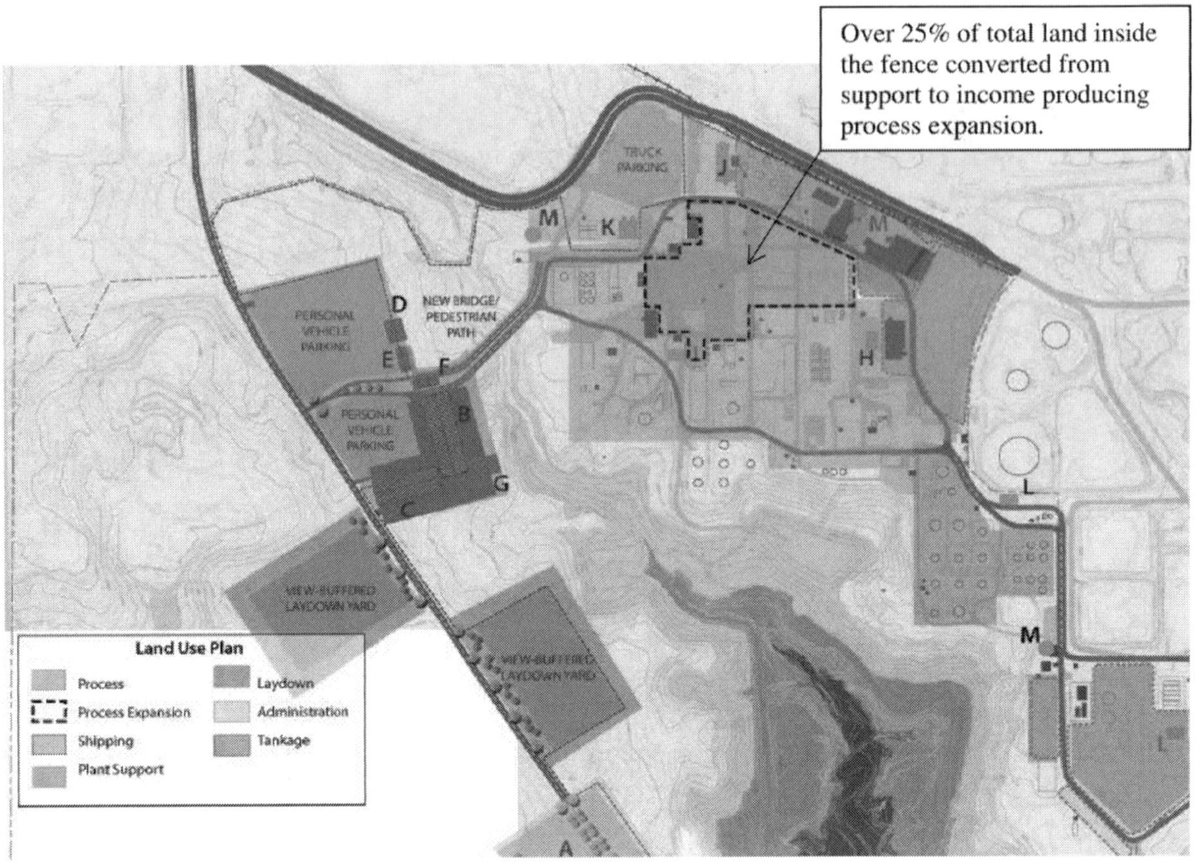

Figure 6 – Removal of Unsafe Embedded Buildings in Process Areas Create Process Expansion Opportunities

6.3 Proactive Loss Mitigation

Petrochemical plant management should seek legal advice on this point. The preparation of a safety-centered site master plan demonstrates leadership and evidence of a good-faith effort to proactively mitigate risk, including:

- Compliance with industry best practices regarding plant layout and non-process facility siting.
- Reduced liability to the company for loss of life, fines, and shut-downs caused by an incident.

The plans are based on a realistic implementation plan. Few plants can fully implement a plan in a 10 year period. Just starting the planning process documents discussions of planned safety improvements. It is far better to avoid costly consequences of an incident by taking active steps to mitigate risk whether that is improving process systems, improving facility siting, through the first step of preparing a safety-centered site master plan.

References

[1]API Recommended Practice 752*: Management of Hazards Associated With Location of Process Plant Buildings, CMA Managers Guide, 1995,* American Petroleum Institute

[2]API Recommended Practice 753: *Management of Hazards Associated with Location of Process Plant Portable Buildings, First Edition,* American Petroleum Institute

Exothermal test for green plastics materials by DSC

C. J. Wang[a], K.Y. Liao[b], J.J. Horng[b]*, J. M. Tseng[a]

[a]Department of Safety, Health and Environmental Engineering, Central Taiwan University of Science and Technology, Taichung, Taiwan
[b]Graduate School of Engineering Science and Technology, National Yunlin University of Science and Technology, Taiwan

Abstract

Food containers made of plastics release harmful gases and nanoparticles due to thermal decomposition, which can pose health risks during use of the containers. Differential scanning calorimetry (DSC) was applied to evaluate the thermal kinetic and basic characteristics of plastic containers, such as polyethylene terephthalate (PET), low density polyethylene (LDPE), polypropylene (PP), polystyrene (PS), and poly lactic acid (PLA). Results indicated that PET, PS, and PLA had the lower decomposition temperature, even less than 100 ℃ compared to other plastics. It is expected that the obtained test results can provide useful safety information for these commonly used plastic containers.

Keywords: containers, plastic, differential scanning calorimetry (DSC), thermal kinetic, safety information

*Corresponding author: *E-mail*: horngjj@yuntech.edu.tw (J. J. Horng)

1. Introduction

Plastic of food containers released harmful gases during thermal decomposition, which can endanger human if these unqualified containers were used. Differential Scanning Calorimetry (DSC) was applied to evaluate the thermal kinetic and basic characteristics for plastic containers in thermal analysis system [1–5]. There are a few paper applied thermal analysis method to described the plastic of food containers.

2. Experimental setup

2.1. Sample

polyethylene terephthalate (PET), low density polyethylene (LDPE), polypropylene (PP), polystyrene (PS), and poly lactic acid (PLA).

2.2. Differential scanning calorimetry (DSC)

Temperature-programmed screening experiments were performed with DSC (TA Q20). The test cell was used to perform the experiment for withstanding relatively high pressure to approximately 10 MPa. ASTM E698 was used to obtain thermal curves for calculating kinetic parameters. For better thermal equilibrium, the heating rate chosen for the temperature-programmed ramp was not to exceed 10 °C min−1. Approximately 3 mg of the sample was used to acquire the experimental data. Non-isothermal tests of the scanning rate selected for the programmed temperature ramp were at 4°C min^{-1}. The range of temperature rise chosen was from 30 to 300 °C for each experiment.

3. Results and discussion

The thermal Characteristics were determined by DSC experimental data at various of 5 figs., 4 °C min^{-1} displayed in figs. 1, 2, 3, 4, and 5 respectively. Fig. 1 shown that the PET onset temperature was about 64.47°C, figs. 4 and 5 shown that PS and PLA onset temperature were 85.72°C and 57.84°C, respectively, according to this phenomenon, harmful gases was released due to thermal decomposition and less than 100°C, these plastic may damage human body during the high temperature.

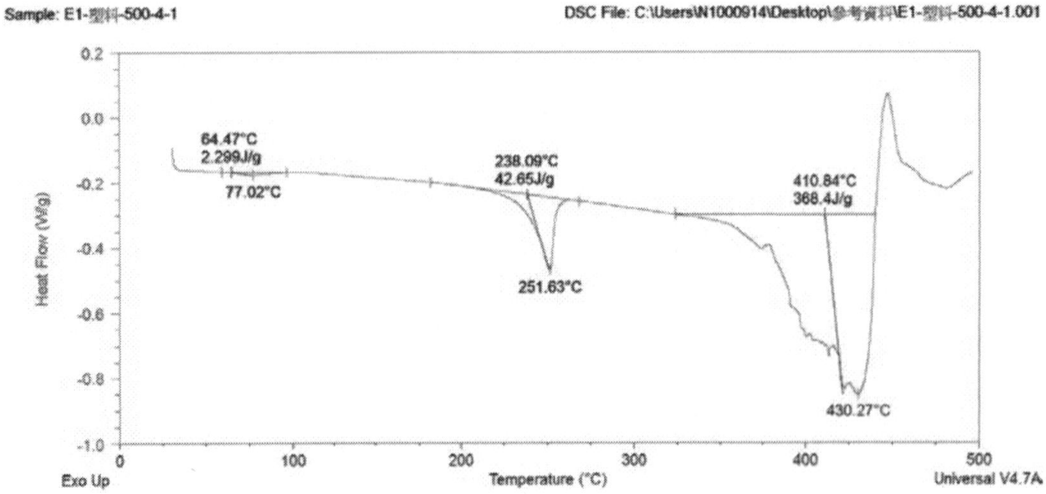

Figure 1. DSC thermal curves of heat flow versus temperature for PET with scanning rates of 4 °C min^{-1}.

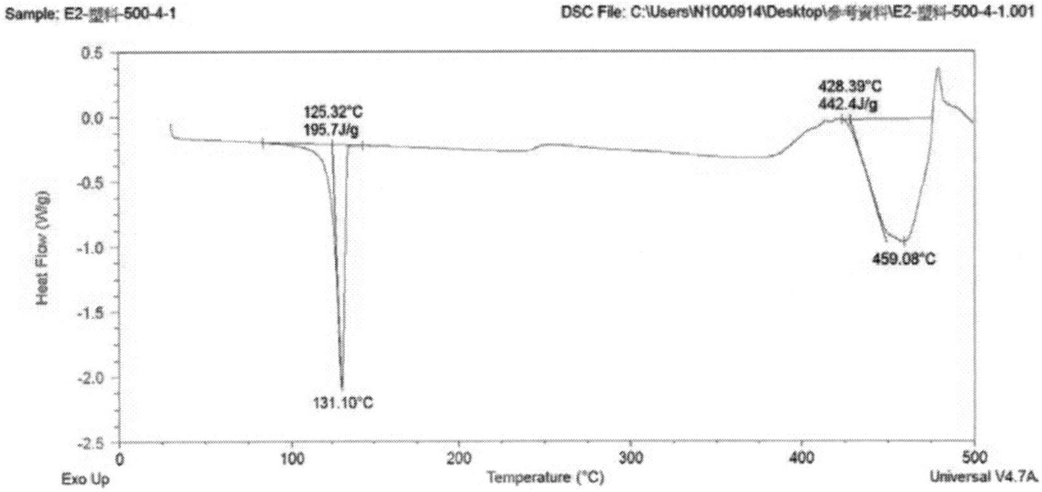

Figure 2. DSC thermal curves of heat flow versus temperature for HDPE with scanning rates of 4 °C min^{-1}.

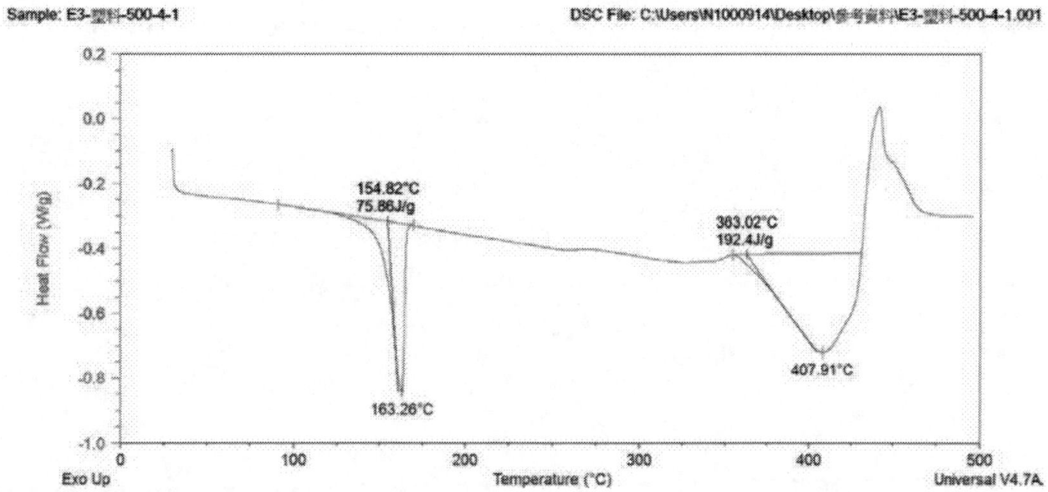

Figure 3. DSC thermal curves of heat flow versus temperature for PP with scanning rates of 4 °C min^{-1}.

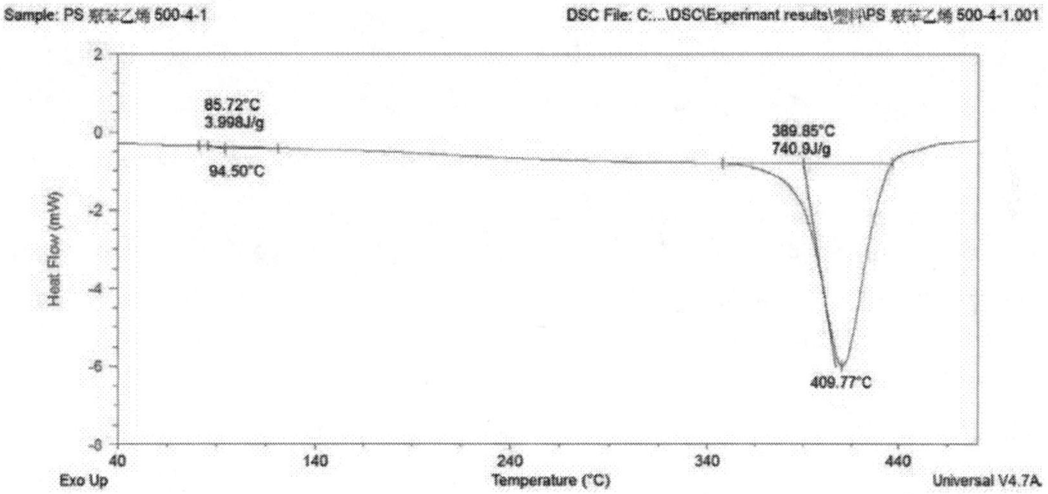

Figure 4. DSC thermal curves of heat flow versus temperature for PS with scanning rates of 4 °C min^{-1}.

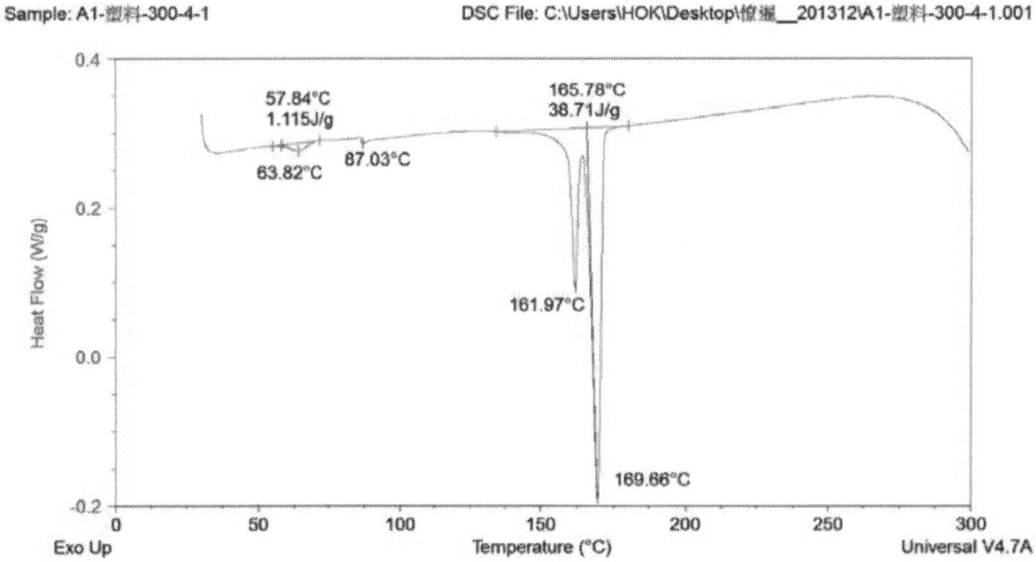

Figure 5. DSC thermal curves of heat flow versus temperature for PLA with scanning rates of 4 °C min^{-1}.

4. Conclusions

Results indicated that food containers have different heat resistance tested by differential scanning calorimetry (DSC). It also shown that harmful gases was induced during the decomposition reactions, in terms of plastic containers, which may damage human body.

References

[1]. I.M. Inuwa, Azman Hassan, De-Yi Wang, S.A. Samsudin, M.K. Mohamad Haafiz, S.L. Wong, M. Jawaid. (2014). Influence of exfoliated graphite nanoplatelets on the flammability and thermal properties of polyethylene terephthalate/polypropylene nanocomposites. Polymer Degradation and Stability,110,137–148.

[2]. Seunggun Yu, Cheolmin Park, Soon Man Hong, Chong Min Koo. (2014). Thermal conduction behaviors of chemically cross-linked high-density polyethylenes. Thermochimica Acta, 583,67–71.

[3]. Lays B. Fitaroni, Juliana A. de Lima , Sandra A. Cruz, Walter R. Waldman. (2015). Thermal stability of polypropylene–montmorillonite clay nanocomposites: Limitation of the thermogravimetric analysis. Polymer Degradation and Stability,111,102–108.

[4]. Saima Shabbir, Sonia Zulfiqar, Issa Sulaiman Al-Amri, Muhammad Ilyas Sarwar. (2013). Enhancement of thermal stability, processability and morphology of polystyrene blends. Thermochimica Acta,573,213–219.

[5]. Shanshan Lv, Jiyou Gu, Jun Cao, Haiyan Tan, Yanhua Zhang. (2015). Effect of annealing on the thermal properties of poly (lacticacid)/starch blends,International Journal of Biological Macromolecules,74,297–303.

The Studies of Flame Propagation Simulation in the Municipal Pipeline

Fenglei Han[1]*; Guangzhe Liu[1] ; Mengmeng Du[1] ;Ren Sun[1]
1 College of chemical engineering, China University of Petroleum (East China)

Abstract

The flame is dangerous, and flame propagation would bring more damage. However, the generation and propagation of the flame is conditional, and it's easier to realize in a closed or half-closed space such as pipeline. The accident of Huangdao '11.22' pipeline explosion showed the harm of the flame in the pipeline. Based on a numerical simulation method, the flame of premixed methane-air in the underground drains was studied, and the results showed the flame propagation state of the premixed methane-air in underground drains in different conditions and the impact on external space. This research could provide the theory instruction for prevention and emergency measures of pipeline explosion accidents, so it has certain practical significance.

Keywords： Numerical Simulation；Flame propagation；Premixed Combustion；Pipeline

* 通讯作者：andyhan0595@gmail.com

1. Introduction

On November 22, 2013, Qingdao economic and technological development zone in Shandong province, the oil in the pipeline spilled and leaked into the municipal drainage culverts, the oil and gas accumulated in a confined space and encountered spark, then caused an explosion, then the heavy casualties and economic losses[1].

The formation of the flame has three factors: fuel, combustion materials and ignition sources. The formation of the flame will causing explosion, and have a devastating impact on the equipment or facilities. Because matter and fuel combustion is inevitable, so how to control the ignition sources became the most important thing.

Figure 1 11 • 22 accident photo on site

2. Numerical model

2.1 Physical model

Physical model was built taking the common pipelines in underground drains as the research object, which simplifying as two-dimension longitudinal section. One of the wellhead was selected as the example.

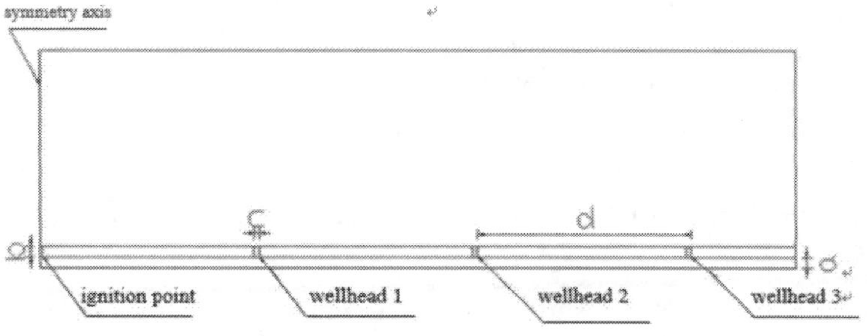

Figure 2 model diagram

The whole layout was axisymmetric and the axis was shown in the Figure 2. The height of the drainage pipeline was a m, distance from the top of the pipeline to the ground was b m, the width of the wellhead was c, and the distance between the two wellhead was d m. The space above the floor was air, and the region of the drainage pipeline was premixed methane-air gas. We choosed two kinds of drainage pipeline to build the models: the specific parameters for model 1: a=1, b=1, c=0.5, d=20; the upper space was 18m, and the pipeline length was 70m; the specific parameters of model 2: a=2.5, b=2, c=0.6, d=20. The upper space was 20m, and the pipeline length was 20m.

2.2 The grid meshing

Considering the flame spread mainly in the interior of the pipeline and the wellhead, the grid meshing of the two simulation models were as shown:

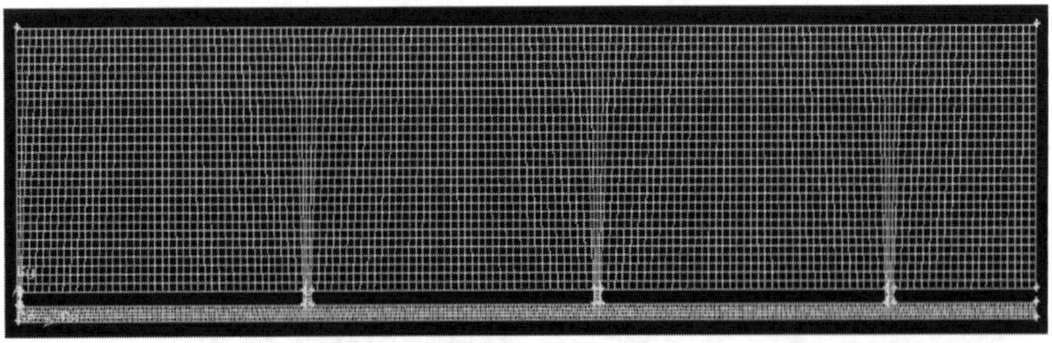

Figure 3 Grid meshing of model 1

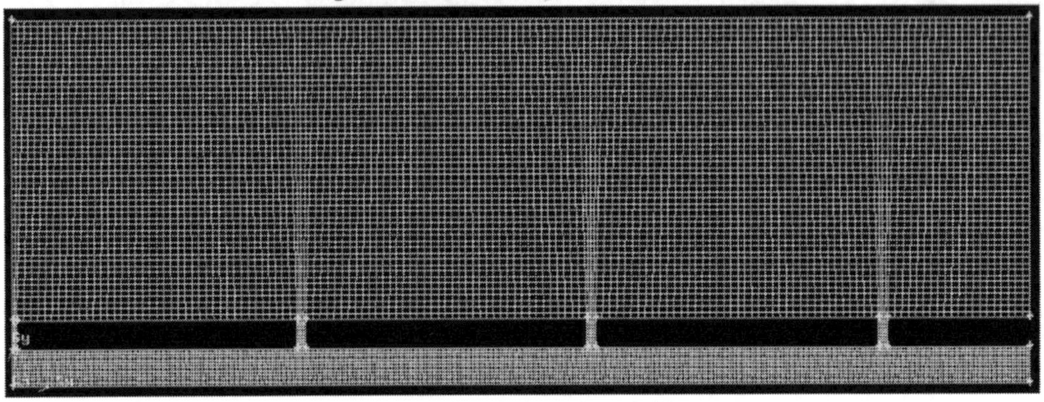

Figure 4 Grid meshing of model 2

2.3 The parameters of the model

When using the Fluent software to solve the simulation model, the model was simplified as:

(1) The air component was simplified as: N_2 was 79% and O_2 was 21%.

(2) The reaction of methane and oxygen is one-step reaction.

(3) The pipe wall is a fully adiabatic wall.

(4) The thermal radiation was not considered.

The 2ddp and unsteady solver of the Fluent were used to solve. The standard k-epsilon double equation model can simulate the methane- air premixed combustion in the sewer pipes. The reaction of methane and air was simplified as one-step reaction.

The patch method was used to initialize the upper space components as the air, and the temperature was 300K, and at the same time, the sewer pipe components were initialized as premixed methane-air gas, which temperature was 300K too. The ignition temperature was 2000K.

3 Results and discussion

3.1 Temperature field

Two different physical models were established according to the description above, the temperature and concentration field was simulated when the methane concentrations is 5%, 10%,

15%, respectively. The figure 5 showed the temperature field as the time changed, when the methane concentration was 5%. The duration time was 10 s, and the generated highest temperature was 2224K. It can be seen in the figure 5 that, the initial temperature was 2000K at the time of ignition. The premixed gas was ignited at the time 0.05s. The flame was in a shape of wedge, and spreaded along the bottom of pipe. When the time is 2.10s, the flame spreaded to the top of the pipe wall, and then speeded up the flame propagation velocity on the pipe wall. At the time of 3s, the flame had spread to the wellhead 1, the premixed methane gas which in the explosion limit range was ignited. And at the time of 3.1s, the flame started to spread around. The flame stopped spreading at the time of 3.3s, because the concentration was below the lower explosive limit. The flame reached the wellhead 2 at the time of 5.1s, where the situation was similar to the wellhead 1. When the flame reached the wellhead 3 at the time of 7.1s, the flame propagation stopped because the methane concentrations cannot reach the explosion limit.

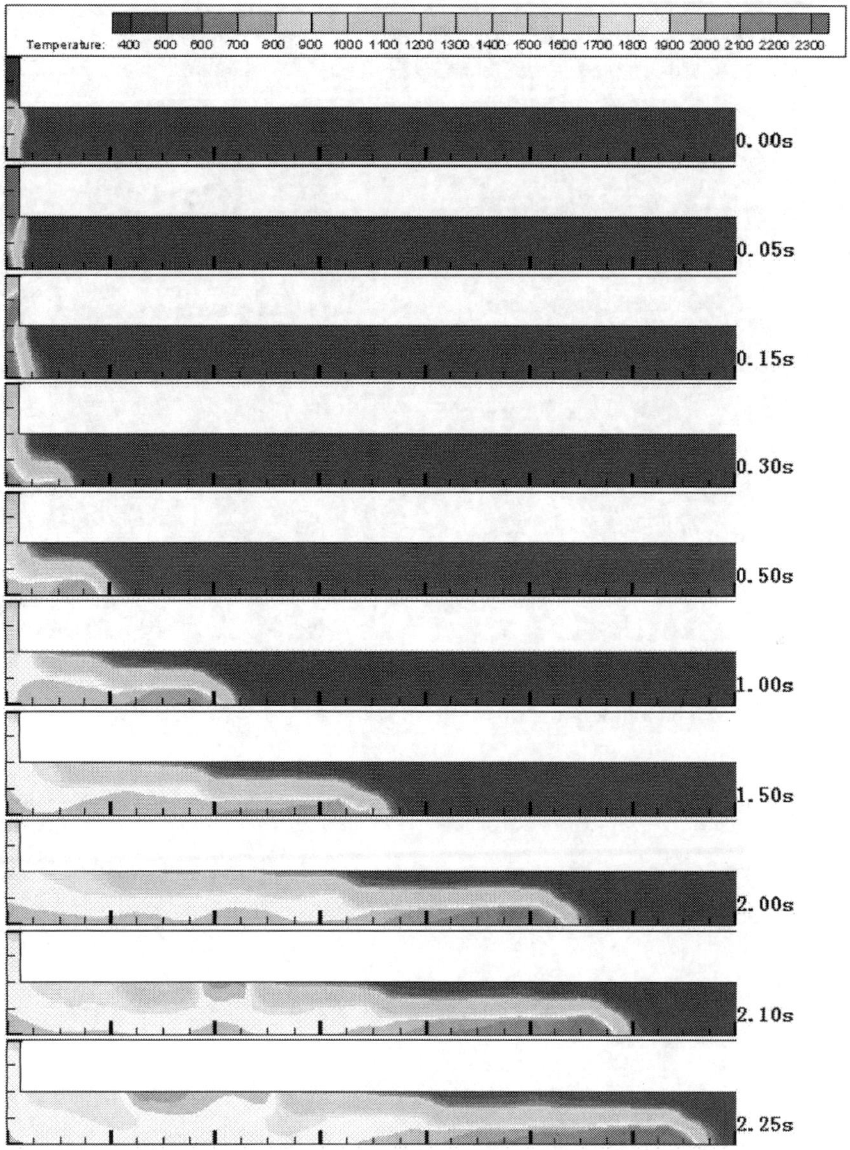

Figure 5 The temperature changed at the x = 0 ~ 14 m, y = 0 ~ 2 m with the time

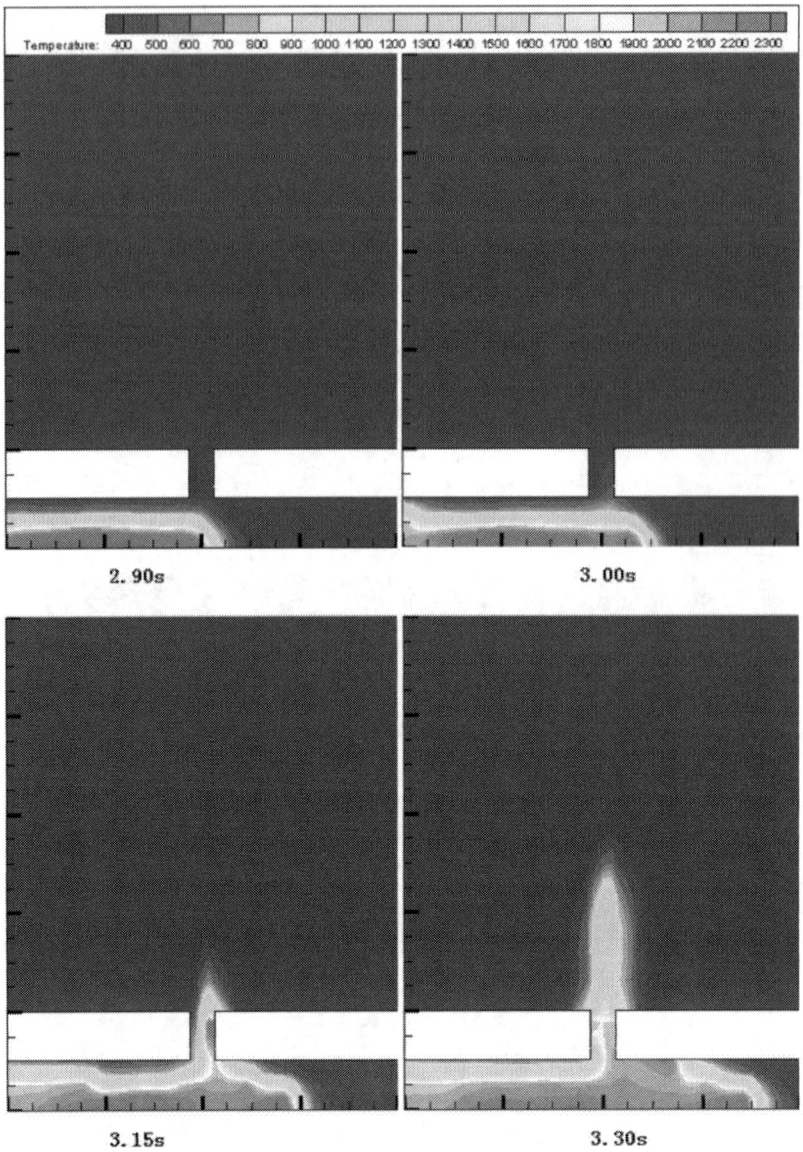

Figure 6 The temperature distribution at the x = 16 ~ 24 m, y = 0 ~ 10 m region

3.2 The flame propagation velocity

In various circumstances, the distance between the ignition point, wellhead 1, wellhead 2, and wellhead 3 was always 20 m, as shown in table 1.

Table 1 The time of the flame reaching each wellhead

Model and the methane concentration	Model 1			Model 2		
	5%	10%	15%	5%	10%	15%
Wellhead 1	2.85s	1.00s	1.10s	9.20s	0.95s	0.95s

Wellhead 2	5.10s	1.40s	2.10s	18.70s	1.50s	1.65s
Wellhead 3	7.15s	∞	3.05s	26.40s	1.95s	2.25s

(Note: methane concentration of model 1 is 10%, the flame is extinguished at the wellhead 2, due to the low concentration of methane gas)

(1) By comparing the flame velocity at each of the wellheads, we can see that the flame was accelerated. According to the results, the pressure generated in the combustion promoted the premixed gas flow in the pipeline, which accelerated the flame propagation.

(2) The order of the effect at different methane concentration was 10% > 15% > 5%, compared from the flame propagation time.

(3) When the concentration of the methane was low, the propagation velocity in model 1 was greater than the velocity in model 2. But it was contrary when the methane concentration was high.

4 Conclusion

The flame characteristics of premixed methane-air was studied in this article. The possible situation after the premixed gas was ignited were simulated based on the Fluent software. And the main conclusions were as follows:

(1) Pipeline was connected with outer space through the wellhead, and gas would flow in the flame propagation process; when the gas in the pipes got out to the outer space, if the concentration was in the explosion limit range, the flame would spread. The temperature diffused quickly when the flame got into the outer space. The external low-temperature air entered into the burning area, reducing the pipe temperature, and diluted the gas concentration in the tube.

(2) The influence of the temperature in the pipeline to the temperature outside depended on amount of methane. At the same concentration, the highest temperature of each wellhead was similar to each other, but the duration was decreased, and besides the ignition had the longest duration.

(3) The flame propagation in the pipeline had effects on the outside at the wellhead. Along with the flame propagation, the effect decreased. And besides the ignition had the greatest impact.

References

[1] The safety committee of the State Council on the Shandong province Qingdao city "11 - 22" Sinopec Dong Huang oil pipeline leakage and explosion accidents in particular Bulletin[J]. Announce of the State Coal Mine Safety Supervision Bureau, State production safety supervision and management administration,2014,02:8-11.

[2] CHEN Zhi-hua* , YE Jing-fang , FAN Bao-chun ,etc. Effects of a wedge obstacle on flame propagation and its structure[J], Explosion and shocks waves, 2006,(03): 208-213.

[3] Li Yuhong, Qi Haiying, Zhang Hongwu.Numerical simulation of pre-mixed flame of methane [J]. Journal of engineering thermophysics, 2002, (01): 119-122.

[4] H E Xue-chao, SUN Jin-hua , CH EN Xian-feng , Richard K .K .Yuen . Experimental and numerical study on flame propagation and structure behaviors of methane-air premixed combustion in tube[J]. Journal of university of science and technology of china, 2009,(04): 419-423.

[5] Fu Qianqian, Zhou Leping, Shi Yuying. Experimental study on Explosion Characteristc of Methane/Air Mixture[J]. Industrial Safety and Environmental Protection, 2009,(02): 27-28.

[6] Yu Jianliang, Zhou Cheng, Liu Renjie. Experimental study of pipeline premixed gas deflagrationp[J]. Petroleum and natural gas chemical industry, 2004,(06): 453-455.

[7] Zhang Lei. Experimental study and numerical simulation of methane/air gas flame acceleration in pipeline[D]. Anhui University Of Science And Technology,2008.

[8] Wang Quan. Study on explosion flame propagation of methane/air pre-mixed gas in the pipeline[D]. Anhui University Of Science And Technology,2006.

[9] M.Gieras,R.Klemens,G.Rarata,P. Wolanski.Determination of explosion parameters of methane-air mixtures in the chamber of 40 dm^3 at normal and elevated temperature.Journal of Loss Prevention in the Process Industries.2006(19): p. 263–270.

[10] Peter Naamansen, D.B.B.H., Solution adaptive CFD simulation of premixed flame propagation over various solid obstructions. Journal of Loss Prevention in the Process Industries, 2002(15): p. 189–197.

402835 The Study on the Vulnerability of Urban Oil and Gas Pipeline Network of the Mechanism of Action

Abstract:

This article contrasted and analyzed the existing analytical methods of vulnerability. The time-space distribution law of the oil and gas pipeline accidents was studied on the basis of the historical accident data and accident scenarios. The damage of the oil and gas pipeline accidents to the disaster bodies of cities was analyzed. The mechanism of action between the oil and gas pipeline accidents and the disaster bodies was discussed. The vulnerability assessment model of urban oil and gas pipeline network was built from the perspectives of the social and economic development of cities, infrastructures, the level of emergency management and so on. Measures were proposed from the aspects of deterring, detecting, delaying, responding and recovering to reduce the vulnerability of urban oil and gas pipeline network so as to reduce the consequences of accidents. Thus, it can give reference for the safe and stable operation of urban oil and gas pipeline network.

Application of HAZOP Study on Some Butyl-Octyl Alcohol Plants

Huang Jiulai, Chen Junjie, Sheng Minbing, Chen Rangqu, Xue Dong
(Xiamen Xibaoyuan Chemical Technology Co., LTD, Xiamen China, 361000)

Abstract: The main raw materials (such as propylene, hydrogen gas, syngas) in the butyl-octyl alcohol production process are of higher explosive and fire hazard. In order to ensure their safety, it is necessary to do safety hazard analysis for these plants. This thesis introduces the application of HAZOP study on some butyl-octyl alcohol plants. The analysis result of the feed gas tank, the waste liquid fuel tank, the syngas feed oxygen injection line and the butyl aldehyde liquid phase hydrogenation reactor shows that HAZOP study is propitious to identify hazard in this process. It is worth promoting the application.

Key words: HAZOP butyl-octyl alcohol plant process safety

1 Introduction

HAZOP study can identify all deviation and potential hazards those are far from the design objective, to provide the basis of eliminate or control them for the design or operating personnel, in most of process hazards analysis methods.[1] In order to solve the problem of process safety issues, the factory usually sets many kinds of independent protection layer in the process that can be extremely easy to become out of control. The safety instrument system (SIS) is considered as the most important line of defense that determines whether a factory is safe or not. Through the layer of protection analysis (LOPA), process safety assessment personnel can further identify whether the security and safety interlock system is sufficient, in order to increase the safety and protective measures during HAZOP analysis. The more effective practice for high degree risk is the layer of protection analysis (LOPA) and safety integrity level (SIL) assessment[2].

2 Overview of HAZOP

The characteristic of HAZOP safety evaluation method is systematic and comprehensive. The basis of HAZOP method uses the data of process flow diagram(PFD) principle and process piping and instrument diagram(P&IDs), unit operation as the nodes, process parameters deviation as the guide, standard regulations as the criterion. Every pipeline, every (Group) equipment, and control circuit must be analysed, identified, to find out the reasons and the consequences, put forward some countermeasures and the improvement suggestion according to the size of the risk for unsafe factors that produced by the deviation of the change of process parameter.

3 Based on the intrinsic safety of butyl-octyl alcohol plants

The security design of butyl-octyl alcohol plants depends on the inherent safety of process design and facilities' basic safety. The process of this kind of unit intrinsic safety can be promoted through the HAZOP analysis that will be able to reduce the risks to an acceptable level.

We found a series of intrinsic safety risk according to the HAZOP analysis of multiple sets of butyl-octyl alcohol plants. This thesis lists two examples.

3.1 The risk of raw gas with liquid condensation

Node intention: The pipeline before the raw gas liquid separation tank uses heat tracing and thermal insulation , and the liquid separation tank is set to can be heated by steam coil. The pipeline after the feed gas liquid separation tank uses thermal insulation treatment in order to prevent the raw gas condensation.

Risk: The following process may lead to security incidents if the water in raw gas can not be effectively separated in the process of liquid separation tank heating.

Suggestions: Setting the pipeline thermal insulation before the raw gas liquid separation tank, cancel the steam coil heating in the liquid separation tank.

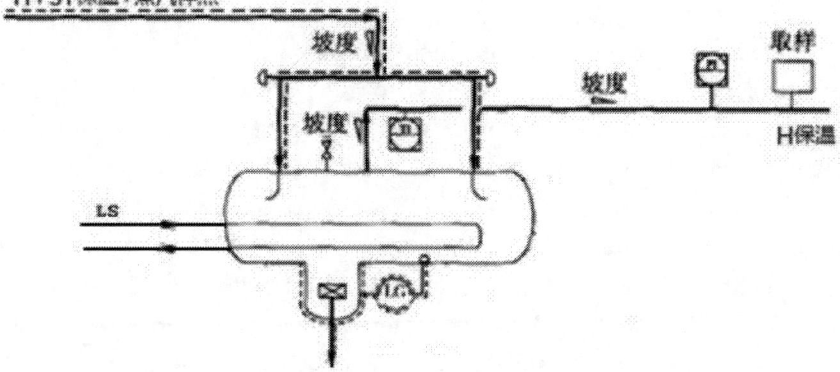

Fig. 1 The schematic diagram of the feed gas liquid separation tank

3.2 The risk of the static electricity generation in the waste liquid fuel tank

Node intention: In some public engineering part of a butyl-octyl alcohol plant, each waste liquid injected into the waste liquid fuel tank through the collection tube, and then distribution processing. The waste liquid fuel tank PID diagram is shown in Fig. 2

Risk: There is a static electricity generation problem of the waste liquid fuel tank. There are some lines of the waste liquid fuel tank can be not satisfied with the requirements of liquid feeding. The problems which caused by too fast flow velocity may produce static electricity generation, and seriously they can lead to fire.

Suggestions: The feed mode of all feed lines of the waste liquid fuel tank should be under the liquid level.

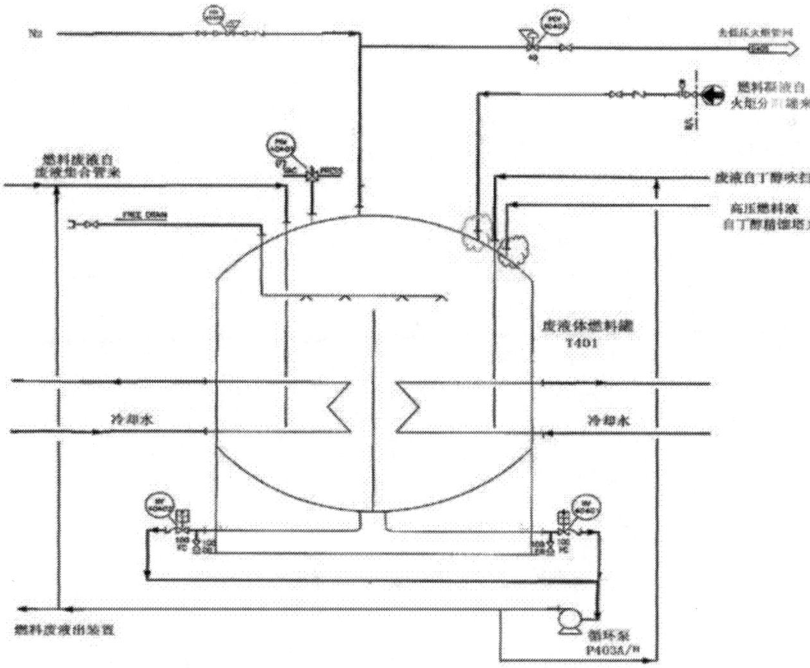

Fig. 2 The PID diagram of the waste liquid fuel tank

4 The instrument safety integrity level requirements of some butyl-octyl alcohol plants

The frequency and consequences of incident scenario from the layer of protection analysis (LOPA) decide whether or not to change the independent protection layer.[3]The 4.1 and 4.2 two node element as examples are elaborated during the HAZOP study on some butyl-octyl alcohol plants.

4.1 The synthesis gas feed oxygen injection line

Node intention: Firstly, the synthesis gas from the border area is heated in the NO.1 synthetic gas heat-exchanger, removed metal carbonyl in the NO.1 synthesis gas purifying tank, added a small amount of oxygen before entering the NO. 1 heat exchanger, assisted with the removal of metal carbonyl. Secondly, the synthesis gas is heated through the NO.2 synthetic gas heat-exchanger and syngas superheater, synthesis gas enters the NO.2 synthetic gas purification tank to clean, in order to remove the oxygen, chloride and sulfide. Then, the synthesis gas goes through the two stage heat exchanger cooling, before entering the filter to remove the catalyst powder and metal precipitates. Most of the synthesis gas enters the stripping steam tower and proposes to steam strip the propylene dissolved in the crude aldehyde product, and then return to the NO.1 carbonyl synthesis reactor. Otherwise another synthesis gas enters the NO.2 carbonyl synthesis reactor.

Risk: This may cause the synthesis gas system overpressure, pipeline rupture, synthesis gas leakage, then may cuase fire , explosion or personnel poisoning.

The part of HAZOP analysis record of this node is in table1, The related PID process is shown in Figure 3.

Table 1 Part of the HAZOP analysis record of the syngas oxygen injection line

Parameter	Guide words	Reason	Consequence	Existing protection measures	level of risk			Suggestions and measures
Pressure	Too high	Pressure control valve(PIC-10303)of the oxygen injection line is failure to open	This may cause the synthesis gas system overpressure, pipeline rupture, synthesis gas leakage, then may cuase fire , explosion or personnel poisoning.	Combustible gas alarm	3	4	12	1. Recommended PIC-10303 adding a high alarm. 2. Adding an independent high pressure oxygen pipeline interlock and a closure of the oxygen inlet valve which should comply with SIL1.

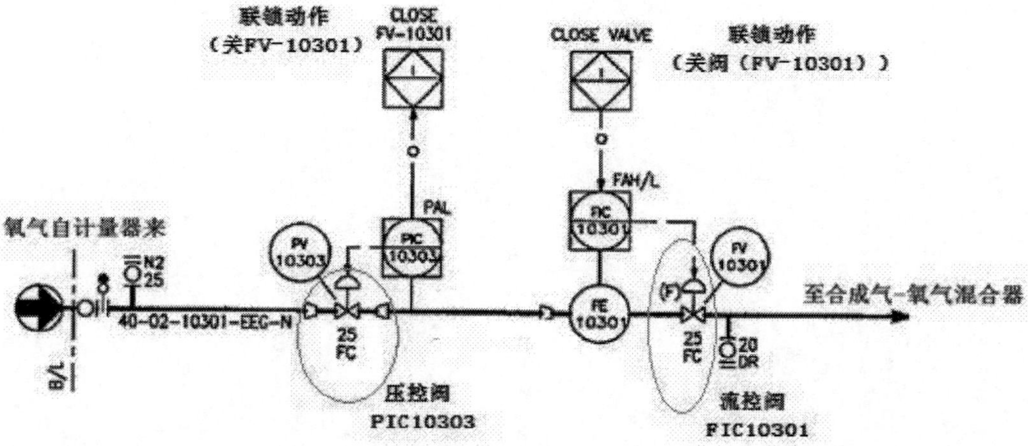

Fig. 3 Part of the syngas feed oxygen injection line PID diagram

In view of this high risk (risk value is 4) , the scene was analyzed by LOPA and SIL grade and put forward some suggestions. The detail is in table 2.

Table 2 Part of the LOPA analysis and SIL evaluation record of the syngas oxygen injection line

NO	The description of the influence event	Severity level	The target frequency	Causation	Times	Protection layer					The possibility of the intermediate event	Safety instrumented function SIF integrity level	Have lessened the likelihood of events	Suggestions
						general process of design	Basic process control system (BPCS)	Alarm And Staff response	Additionalrelief: relief	Additional relief: detection				
1	The synthesis gas system may be overpressure, pipeline rupture, synthesis gas leakage, then may cause fire ,explosion or personnel poisoning	4	10^{-4}	PIC-10303 open	0.1 Times/year		10^{-1}		The NO.1 synthetic gas heat exchanger safety valve(PSV10301)discharge has not been taken into account of 8.7MPa oxygen channeling conditions	Combustible gas alarm	10^{-2}	10^{-2}	10^{-2}	1.suggested PIC-10303 adding a high alarm. 2.Adding an independent high pressure oxygen pipeline interlock and a closure of the oxygen inlet valve which should comply with SIL1.

4.2 The liquid phase butyl aldehyde hydrogenation reactor of A/B (200 unit)

Node intention: The mixed butyl aldehyde from butyl aldehyde isomer tower enters the hydrogenation system reactor to become butyl alcohol. Butyl aldehyde and hydrogen is sent to the first alcohol liquid phase hydrogenation reactor(for liquid phase hydrogenation).This two reactors are the same and in series operation. The liquid phase hydrogenation reactor of butanol includes an internal core tube, external circular bed, and are filled with a hydrogenation catalyst in pieces.

Butyl aldehyde through the external cylindrical catalyst bed of two series connected converters, with the addition of hydrogen, then becomes butanol. Because of the reaction exothermic, therefore it needs to use the circulating pump of butanol liquid phase hydrogenation, extract a cycle logistics from the external bed bottom, mixed with the fresh feed into the first converter after cooling. Recycling stream is cooled with the steam produced by side of the liquid phase hydrogenation of circular cooler. A part of circulation stream is extracted before it enters the cooler, mixed with fresh hydrogen , then enters the internal core tube of the series of reactors. In some of these residual unreacted butyl aldehyde will react, after that, crude butanol products are sent to butanol liquid phase hydrogenation deaerator.

The part of HAZOP analysis record of this note is in table3. The related PID process is shown in Figure 4.

Table 3 Part of the HAZOP analysis record of the butyl aldehyde liquid phase hydrogenation reactor A/B (200 units)

Parameter	Guide words	Reason	Consequence	Existing protection measures	level of risk			Suggestions and measures
flow	too high	The liquid phase hydrogenation reactor of butyl aldehyde flow control valve A (FIC20106) is failure,then it opens	It may cause hydrogen loss, the butanol degassing tank (V-208) to be overpressure, rupture, and hydrogen, methane, butyraldehyde etc. may be leakage, then may cause fire and explosion.	PIC20406 (the pressure control valve of V-208) control, PSV20401, hydrogen alarm.	3	4	12	Suggested setting an independent pressure high alarm for V-208.
		The liquid phase hydrogenation converter of butyraldehyde flow control valve B(FIC20206) is Failure,then it opens	It may cause hydrogen loss, the butanol degassing tank (V-208) to be overpressure, rupture, and hydrogen, methane, butyraldehyde etc. may be leakage, then may cause fire and explosion.	PIC20406 (The pressure control valve of V-208) control, PSV20401, hydrogen alarm.	3	4	12	Suggested setting an independent pressure high alarm for V-208.

In view of the above risk scenario, HAZOP analysis group has carried on the LOPA analysis. The detail of LOPA analysis result is in table4.

Table 4 FIC20106 fault open LOPA plot analysis record

NO	The description of the influence event	Severity level	The target frequency	Causation	Times	Protection layer					The possibility of the intermediate event	Safety instrumented function SIF integrity level	Have lessened the likelihood of events	Suggestions
						general process of design	Basic process control system (BPCS)	Alarm And Staff response	Additional relief: relief	Additional relief: detection				
1	The tank of V-208 is overpressure, rupture,then butanol may be degassed. Hydrogen, methane and butyl aldehyde etc. may be leakage, then may cause fire or explosion.	4	10^{-4}	FIC20106 or FIC20206 fault open	2×0.1 Times/year		10^{-1}		10^{-1} PSV20401 (the safe valve of butanol degassing tank)	Set a combustible gas alarm	2×10^{-3}		2×10^{-3}	Suggested setting an independent pressure high alarm for butanol degassing tank V-208.

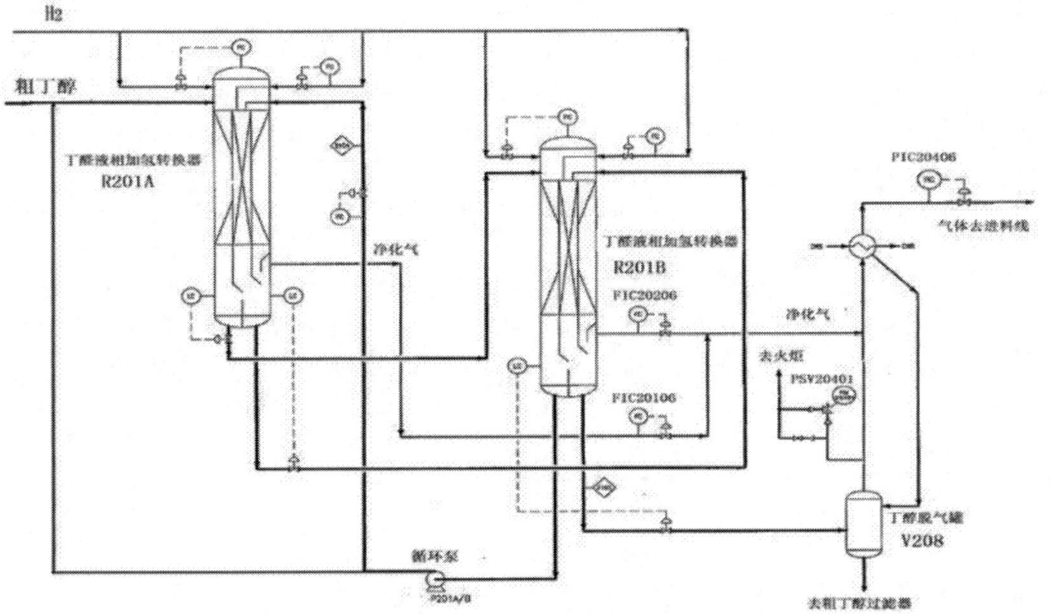

Fig. 4 The PID diagram of the Butyl aldehyde liquid phase hydrogenation reactor A/B (200 units)

5 Conclusion

Through the application of HAZOP study on some butyl octyl alcohol plants, three main aspects, such as the impact on personnel safety, environmental pollution and damage to property were considered. The HAZOP team analyzed 63 pieces of PID figures, 20 analysis nodes, 380 discussion and analysis of problems, and identified 7 greater risk, put forward 48 recommendations in some butyl octyl alcohol uint.

The hazard and operability problems in these butyl octanol units which were identified by HAZOP study mainly list in the following aspects:

(1) Some individual high risk places did not set safety instrumented interlock system. For example, the pressure of the oxygen feed is too high, and there is a larger pressure difference, but it is only controlled by a pressure control valve. If the pressure control valve is failure, the pipeline and equipment after it may be overpressure and rupture, then may cause the synthesis gas and oxygen leakage, and cause the risk of fire and explosion. Suggestion: the oxygen line should be set an overpressure safety instrumented interlock when the oxygen feed is shut-off.

(2) The early alert measures of high temperature and high pressure are inadequate. Such as octanol liquid-phase hydrogen degassing tank (V305) may be overpressure, rupture, then hydrogen, methane, butyl aldehyde etc.may be leakage, then may cause fire or explosion. Suggestion: An independent pressure high alarm should be set on V305.

(3) Some pipelines and equipments exist adverse currents, which may lead material channeling. Especially the different materials from different devices may lead excessive temperature and overpressure, material entrainment and material channeling, so these safety problems must need to focus on.

(4)The key valve should be set to LO or CSO: people need to start from the two aspects of design and management, and make sure that the key valve is in correct state. The accident condition should be the important points to consider.

(5) A lot of risk assessments need further confirmation and accounting, because it is the

design phase HAZOP analysis . For example, it should be confirmed the valve XV11603 trigger condition at stability of the feed line of butyraldehyde tank T102 emergency whether or not shut-off interlock. Also it should be confirmed the highest temperature of synthesis gas superheater steam valve E105 when TV10402 is fully opened , whether or does not exceed the design temperature of 250℃ DEG.

HAZOP analysis that combines with LOPA technology provides an important basis for the improvement of process safety management of butyl-octyl alcohol units. The evaluated analysis problems have also been the owners, design institutes and foreign patent owners recognized, it is significantly important to identify the potential defects and errors and effectively improve device to an arrival essential safety level in the design . It is also an important way to let enterprises deepen the understanding of the risk of this unit, strengthen the safety operation and improve its safety level.

Reference

[1] The People's Republic of China State Administration of Production Safety Supervision and Management.AQ/T:3049-2013,Hazard and operability analysis (HAZOP analysis) Application Guide[S].2013:4, Beijing, China.

[2] The People's Republic of China State Administration of Quality Supervision Inspection and Quarantine Chinese International Standardization Management Committee. GB/T 21109-2007.1/IEC 61511-3:2003, The functional safety of safety instrumented systems for the process industry sector , first part: framework, definitions, system, hardware and software requirements[S].2007:14 , Beijing, China.

[3] Chongguang WU, Zhenyu Su, Xiansheng Song, Chenglong Sun, Guangwen Zhang, Jiulai Huang, Yi Lu etc. The Hazard and Operability Analysis (HAZOP) Application Guide[M]. Petrochemical Press.2012:18, Beijing:China.

Critical Success Factors for Effective PHAs

Arlyn H. Poppen, CSP
Practical Process Safety, LLC
3576 Lemieux Circle
Eagan, MN 55122
ahpoppen@comcast.net

John W. Herber
Process Hazard Management Services
11 Moonlight Bay
Stillwater, MN 55082
john@PHMS.us.com

Prepared for Presentation at
American Institute of Chemical Engineers
2015 Spring Meeting
11th Global Congress on Process Safety
Austin, Texas
April 27-29, 2015

UNPUBLISHED

Critical Success Factors for Effective PHAs

Arlyn H. Poppen, CSP
Practical Process Safety, LLC
3576 Lemieux Circle
Eagan, MN 55122
ahpoppen@comcast.net

John W. Herber
Process Hazard Management Services
11 Moonlight Bay
Stillwater, MN 55082
john@phms.us.com

Keywords: PHA scope, PHA team, leadership skills, PHA techniques, information, safeguards, PHA traps

Abstract

Process Hazard Analysis (PHA's) have been an industry practice since the early techniques were develop at Imperial Chemical Industries (ICI) in the 1970's and have been a part of the OSHA PSM rule since its inception in 1992. While PHAs are consistently completed on processes, they are also frequently cited in incident investigations as ineffective in identifying critical hazards, missing required safeguards, being inadequate for the complexity of the process, or being ineffective in other ways.

Much has been learned over the years on what makes a PHA successful in preventing incidents. Substantial effort is required in the preparation for a comprehensive PHA. Appropriate methodologies must be used. PHA teams must be comprised of people with the right knowledge and skill sets. Information available to the PHA team must be complete and comprehensive. The time spent doing the PHA must be adequate for thorough analysis of the process. Recommendations from the PHA must be carried though to closure.

This paper will review the success factors for effective PHAs, list some practices that should be employed for improving the quality and effectiveness of any hazard analysis, and provide the user with an outline for conducting high quality PHAs.

1. Introduction

Hazard identification and analysis are one of the most important parts of any safety program. In process safety, the term Process Hazards Analysis (PHA) covers any of the generally recognized and accepted techniques. The intent of this article is to discuss success factors common for all of the techniques.

1.1 History

Development of the process hazard analysis is not well documented, however, development of the Hazard and Operability Study (HAZOP) is usually attributed to Imperial Chemical Industries (ICI). PHA's were conducted as a matter of internal practice by some companies and in various forms and degrees of detail for many years prior to the OSHA PSM regulation. HAZOP is probably the first published technique dating back to the early 1970's. PHA's were required by OSHA beginning in 1992, with the Process Safety Management Standard.[(1)] That standard lists six techniques plus the option for "equivalent methodologies". Since then, several other techniques have been implemented and recognized to enhance the PHA toolbox.

1.2 PHA Requirements

The PSM rule has seven requirements for PHAs. This paper will discuss four of them. Ignoring the phase-in over the first five years of the rule, the retention of PHA documentation and the requirements for PHA revalidation. The remaining four requirements are:

- Use specified methodologies
- Address the hazards of the process
- Include proper expertise on the PHA team
- Address recommendation of the PHA team

Additionally, there are other practices that are not "required" by the OSHA regulation, but are practical necessities for a successful PHA. Some of these necessities are preparation of PHA documentation, collection of required PSI, organization of the process under analysis, selection of the team members and physical arrangements. These are the areas where the success factors can be most readily applied.

1.3 Benefits of Thorough PHAs

If a PHA is done well and the recommendations are properly addressed, the process should operate safely across its lifetime. We all wish safe operation was that simple. But we are well aware that many other issues such as unknown technical factors of the process, operating procedures/practices, integrity inspections, and change management are also critical to sustained safe operation. However, without an accurate analysis of both the hazards and safeguards, the process has less chance of operating safely and in fact, many of the other safety activities are highly dependent on the quality of the PHA.

1.4 Going Beyond the Requirements of the PSM Standard - General Duty

Regulatory requirements are the minimum expectations for managing safe processes. Regulations alone cannot provide all the guidance needed to understand the chemistry, process equipment, siting, and many other factors necessary for safe operations. That is the responsibility of the owner/operator; they also need the process to operate reliably and to meet customer demands for quality and timely delivery. The personnel at the site have a right to a safe workplace and the adjoining community has an expectation that the process will have as little impact on them as possible. All of these factors support a requirement for more diligence in the PHA process – from planning through action item completion.

Because the PHA has been so widely accepted as an effective tool at incident prevention and mitigation, it has come to be recognized as an industry standard for assessing chemical processes for hazards and hazard management. If an incident does occur on a chemical process which is not specifically included under the PSM rule because of the chemicals used is not listed or the threshold quantity (TQ) has not been met, the owner/operator may still be held in account under the General Duty Clause(2). The US Environmental Protection Agency (US EPA) has even more clearly and succinctly defined this expectation in the General Duty Cause of the Clean Air Act of 1990(3).

> "Purpose of General Duty - ...The owners and operators of stationary sources producing, processing, handling or storing such substances have a general duty, in the same manner and to the same extent as Section 654, Title 29 of the United States Code [OSHA General Duty Clause], to identify hazards which may result from such releases using appropriate hazard assessment techniques, to design and maintain a safe facility taking such steps as are necessary to prevent releases, and to minimize the consequences of accidental releases which do occur."

Because the other PSM elements are based on the assumption that a thorough PHA has been completed, the PHA is the linchpin that holds it all together.

2. PHA Application

There are two basic approaches to conducing a PHA. If the nature of the operation is such that a collection of equipment and facilities are used in various configurations to carry out multiple, diverse types of process activities (i.e.: mixing, reactive chemistry, extraction, separation, etc.) it is usually most effective to conduct the PHA in four separate phases addressing respectively: 1) the site; 2) the process buildings/facility; 3) the respective equipment available to the process; and 4) each of the respective chemical process operations expected to be conducted on the equipment.

If the nature of the process is a collection of equipment and facilities that are essentially static in their configuration and used to carry out similar process operations (i.e.: mixture chemistry using similar chemicals but perhaps in different combinations or concentrations) it is typical to conduct a PHA based around the equipment and facilities while considering the hazards presented by the family of chemicals anticipated to be used in the process.

2.1 Required Processes and Threshold Quantities

Which processes need a PHA? The simple answer is; all of them! Each process needs a hazard review commensurate with the hazards and risks the process presents. That risk may be more than the properties of the chemicals or energy release (safety focus), but also may reflect the criticality of that process to the financial viability of the operation or company (business focus).

The initial PSM rule included a requirement for a five year plan to perform initial PHAs with applicability of PSM on the process based on exceeding a Threshold Quantity of one of the chemicals included on a list in Appendix A of the rule. The expectation was that the plan for conducting the PHAs be based on the risk presented by the process and other factors. The risk may be based on hazards of the materials, the amount used or stored on-site, the proximity of the inventory to the property line, the number of people potentially impacted by an incident, or other risk factors based on company culture or risk tolerance.

2.2 Connected Processes

Clearly, processes that contain a listed chemical, over the threshold quantity (TQ), must, by regulation, have a PHA with a few exceptions such as atmospheric storage of flammables. One of the confusing clauses in the PSM rule is the inclusion of processes that are 'connected to or could be impacted by a PSM-covered process'. These criteria may expand the number of PHAs that are needed.

An approach to consider when determining PHA applicability is: "one layer beyond the covered process". In addition, utility and support services for the process may introduce hazards leading to an initiating event for an incident. Going the one layer beyond and addressing utilities and support activities ensures that any carry-on effects that might impact the process are evaluated.

2.3 Risk Based PHA Applicability

The criteria specified in the PSM rule (listed above) are useful, but there are other processes that may need or deserve a PHA to protect employees, the environment, business assets, or continuity of the business. All areas where chemicals are stored, processed or handled, including waste streams, should have a hazard review at some level. A drum storage warehouse may have a limited quantity of materials, but may have a more diverse inventory, have fewer procedures or lack engineered safeguards. Atmospheric tank farms may be excluded from coverage under the PSM rule but they often present some of the highest risks at a facility dues to the sheer volume of materials stored.

PHA applicability needs to be reviewed for overall process design, operation, proper storage of potentially incompatible materials and how to manage a spill or other loss of containment event. The same may apply to a waste collection or storage area, especially since waste streams may not be as stable or well understood as raw materials or finished products. There are many areas such sewers or waste treatment operations where chemical reactions are possible without the controls normally present on vessel designed for chemistry. Waste water treatment is a chemical process in and of itself and must be safely managed. Often companies write these off as, "just dirty water" and overlook the potential for catastrophic events.

If a process has chemical listed in the PSM rule or another regulation it should be a candidate for a PHA even though it may be below a TQ. It may not be the highest risk, but should be properly prioritized into the PHA list. A reactive chemical review and combustible dust hazard assessment should be part of the overall hazard assessment process as well.

2.4 Role of PHA Guidance in the Facility PSM Program

Guidance on how to conduct a PHA can be found in many books and courses; just type 'PHA' into a web browser and several hundred hits will appear. But that is generic guidance, not specific to your company or plant. The roles and responsibilities for the PHA process are typically found in the PHA section of the facility's PSM program. While the PSM rule provides the requirements, the facility PSM program should explain the "how & why" of PHAs. Typical issues related to PHAs that should be covered by the facility PSM program are:

- Prioritization of the initial PHAs
- Roles and responsibilities for the persons to be involved with PHAs
- Required PSI and where it is retained
- Selection of the proper PHA methodology
- PHA team requirements including facilitator qualifications
- Handling of PHA logistics
- Risk ranking and prioritizing the recommendations / action items
- Documentation requirements and procedures for the PHA
- Methodology and practices for PHA revalidation

Without this guidance document, PHAs will be performed inconsistently and will lack the quality needed to properly protect employees and the business enterprise.

2.5 PHA Effectiveness and Efficiency

What do the terms efficiency and effectiveness mean with regard to PHAs? Effectiveness could be defined as assessing all applicable process deviations, their impacts, and safeguards to address the issues that have the potential for a catastrophic event (incident prevention). PHA efficiency is getting the most effective PHA completed as expediently as possible. These two attributes can be somewhat mutually exclusive. The challenge is to balance them to get the best possible review of process hazards with reasonable effort.

What makes a PHA effective? The real effectiveness of a PHA may not be known for many years after the study. If no events occur, the PHA should be considered a success. However, that's a long time to wait for feedback. If an incident does occur, that also does not necessarily mean that the PHA was a failure. However, the investigation needs to search for gaps in effectiveness of the PHA together with other physical evidence and root causes of the incident.

2.6 PHA Focus

The first step in setting out to do a PHA is to define the focus of the PHA. Conducting a PHA is like taking a journey. When setting out it is highly desirable to have a clear understanding of where the journey will take us, how we will get there, the expected time involved, and issues of importance along the way. So it is with conducting a PHA. Some of the most critical issues involved with the PHA include a clear understanding of:

- Why is the PHA being done?
- What are the consequences of concern?
- What process / part of process is to be analyzed?
- What operating modes will be considered?
- What level of detail is required?
- What kind of assumptions will be used?
- Are all the significant stake holders in agreement with the above

2.7 PHA Timing

So, when is a good time to do "the" PHA? Actually, PHA is really more of a continuum than an event. For new processes just coming out of the design stage, the PHA should begin before the design is completely finalized to make sure that the process is as inherently safe as possible. For existing processes that were included in the initial implementation of the PSM rule or for processes that become a candidate for a PHA after they have been launched, the initial PHA will identify and allow correction of deficiencies that may have been pre-existing. From there, the PHA becomes a continuum of updates and revalidations and action item closure for the life of the process.

2.7.1 Pre-PHA

When a new chemical process is being designed, it is wise to apply the PHA analysis techniques as the design progresses, most likely starting when a firm preliminary design has been completed. Applying PHA techniques at this point allows the design team to comprehensively analyze the new design for possible flaws or weaknesses that could result in unsafe operation. Because the PSM rule does not require that the PHA be started at this point in a process life cycle, documentation can be less formal and need not necessarily be retained for the life of the process. However, the lessons learned and recommendations identified will allow problems to be designed out while the process is still mostly lines on paper and not a mass of welded steel and complicated controls which are much more difficult and expensive to modify.

2.7.2 Final Design PHA

At some point between when the final design is reached and construction is complete and prior to commissioning of the process, a formal initial PHA must be completed. This is the time at which all of the points made under Section 2.6, PHA Focus (above) must be employed and the formal PHA documentation process begun. Also, remember that prior to "the introduction of hazardous materials into the process" all applicable recommendations and action items from the PHA must be completed.

2.7.3 Update / Revalidation / Redo Following Construction

Once the formal initial PHA is completed, the requirements of Management of Change (MOC) must also be implemented. This requires that any change made to the design, any substitution of process components, really any change that is not a like-for-like must be subject to the MOC process. Recognizing that even with a generous sized staff and the best of intentions, applying flawless MOC to construction work is a daunting if not impossible task. To assure that all of the "as built" changes to the process are accounted for and evaluated for hazards they might present, it is a good practice to update the PHA once final construction, commissioning and startup of the process is complete. This approach will help assure that no unaddressed hazards have sneaked past MOC and gotten into the process during construction and startup, and that the PSM documentation package is accurate and up-to-date. Also, any new hazards seen during checkout / start-up or during early operation can be inserted into the PHA.

2.7.4 5-year revalidation

Whether the process is a new construction sporting a shiny new PHA, or a veteran of many years of service with a history of PHAs and revalidations dating back many years, each PHA must be "revalidated" at least every five years. The issue of "How to Revalidate?" brings with it a whole new series of questions and evaluations to select the proper methods and approach for the revalidation. These issues are beyond the scope of this paper. Guidance on PHA revalidation can be found in the CCPS books *Guidelines for Hazard Evaluation Procedures*(4) and *Revalidating Process Hazard Analysis.*(5)

2.8 Recommendations and Action Items

The primary outputs from a PHA are recommendations or action items intended to improve the safety of the process. Documentation of these recommendations or action items and follow-up to assure that each is fully addressed and brought to closure is vitally important to the effectiveness of the PHA.

2.8.1 Writing well

To be effective and properly implemented, recommendations and action items identified by the PHA team must be clearly communicated. In the context of the PHA team meeting, the team members are engaged in the process and are focused on the discussion at hand. As such, they are well aware of what kind of improvements are needed to improve the safety of the process. In reality, this often leads to sketchy documentation of the recommendations. Days, weeks or even months later when it comes time to execute the recommendation, it can be difficult to remember or identify exactly what the PHA team's concern was. Therefore, it is critically important that each PHA recommendation / action item be accurately and concisely documented. The best rule to follow is that the recommendation as written can be taken completely out of the context of the PHA and be readily understood as to what needs to be done and why. This requirement falls as a heavy burden on the PHA leader and scribe.

2.8.2 Risk Ranking

At the conclusion of a PHA, especially the initial PHA or a complete redo of a PHA, it is not unusual for recommendations to number in the dozens, even into the hundreds for a large process. In order to productively address and resolve this number of activities it is imperative to have a plan. Often it is tempting to start by reaping the low-hanging fruit – complete the easy recommendations first and put off the hard, expensive or time-consuming recommendations until later. In reality, the recommendations should be addressed in the order of risk with the highest risk issues addressed first and the lower risk issues later on. Therefore, a structured method of risk ranking should be part of the overall PHA process.

Many companies have established risk ranking tools that are used in conjunction with other safety or environmental issues. It may be possible to use or adapt one of these existing tools for use with the PHA recommendations. In addition, there are many publications that document a broad variety of risk ranking tools, from simple to complex, that can be employed to help prioritize PHA recommendations.

2.8.3 Completing Recommendations – Continuous Progress Toward Closure

In an ideal world there would adequate resources to begin resolution of many PHA recommendations simultaneously and carry them to completion in short order. In reality, however, such is rarely the case. Nonetheless, there is an expectation that continuous progress toward closure be maintained until all of the recommendations have been resolved. Sometimes it is tempting to manage the workload based on the established deadline and not start on corrective actions until as close as possible to the required completion date. With PHA recommendations there is an expectation in good industry practice that work progress continuously toward closure. As part of the PHA follow-up and documentation process, it is important to periodically document the progress-toward-closure of PHA recommendations.

2.8.4 Documentation of Closure

As with the documentation of the recommendation itself, the documentation of closure should be complete, concise and able to stand on its own outside of the context of the PHA. Obviously, "Done" or "Completed" are woefully lacking.

There are several tests to determine if closure is adequately documented.

- Will this description be useable in six months or more?
- Are the specifics of the closure action well documented (equipment specifications, equipment drawing numbers, document identification, etc.?)
- Could someone such as an auditor read the documentation and understand what was done?

2.8.5 The Four Acceptable Reasons Not to Complete a PHA Recommendation.

There will be times when even the best ideas that occur to the PHA team to improve the safety of a process will not be able to be accomplished. There are four generally accepted reasons that will allow a PHA recommendation or action item to be set aside:

a. The recommendation is based on analysis which contains factual errors.
b. The recommendation is not necessary to protect the health and safety of the employees and contractors.
c. An alternative measure would provide a sufficient level of protection.
d. The recommendation is infeasible.

This is an important point for the PHA team to understand. Once a recommendation has been made and documented, it becomes quite difficult to retract. If a recommendation is later decided to be unnecessary and meets one of the above four criteria, it is very important to thoroughly document the rationale that supports the decision to not implement against the day when an incident may occur and be found to have root in a recommendation that was not completed.

3. Top Four Success Factors – #1 - Planning and Preparation

The cold hard fact is that executing a successful PHA means doing everything that is required and doing them all completely and doing them all right. That is an onerous task to say the least! A PHA can be compared to a complex medical procedure. You simply can't leave out any of the steps and expect the procedure to be successful.

The critical success factors for a PHA will be broken down into four categories for the purposes of this paper.

1. Planning and preparation
2. Time investment in the PHA
3. The PHA team
4. Information available and used for the PHA

Planning and preparation for a thorough PHA is nearly as big a task as the actual analysis of the process by the PHA team. Depending on the amount, quality and availability of information, the size of the process, the number of ancillary and support systems to be included, the total number of PHA nodes and other factors, it is not unusual to have the PHA facilitator / PHA coordinator and perhaps various support teams to invest 2-4 hours of advance preparation for every hour spent by the PHA team in actually doing the analysis.

3.1 PHA Scope

One of the most important parts of any successful project is preparing a comprehensive master plan. Whether the project is the design and construction of new equipment or facilities, development of a proposal for services to be provided for a client, or preparing to execute a PHA, it is of utmost importance to define the details. These include what is to be included or excluded, what are the milestones, what is the budget, who needs to be involved, what are the physical boundaries (where will the project start and stop), and what approvals are required.

Adequately defining the scope of the PHA is one of the top four success factors for a PHA.

3.1.1 Physical scope

The first step in the master plan for a PHA typically involves defining the scope and objectives. Why is the PHA being done? Common reasons include regulatory compliance, compliance with company policies, incident investigation, and new process design review.

Next the boundaries of the PHA need to be clearly defined. Typical issues include:

- Where will the PHA start and stop? It is very important to define the specific physical points in the process where the PHA will start and stop.
- What equipment / facilities will be included in the PHA?
- Will utility and support systems be included?
- Will ancillary activities such as material receiving, delivery, storage and shipping be included?
- Will hazardous waste generation and handling be reviewed?
- What parts of the process under study may also be part of other processes, and what kinds of issues might reflect back or forth between these linked processes?

3.1.2 Operating Modes

What operating modes for the process will be included in the PHA and how will they be addressed by the PHA team? The PSM rule requires twenty different issues to be addressed in Operating Procedures(6) including normal operation, process startup, process shutdown, emergency shutdown startup after emergency shutdown, service / utility failures and maintenance / changeover activities. Which of these will be reviewed by the PHA team?

3.1.3 Assumptions

Each member of the PHA team will enter the PHA process with their own set of ideas and experience. It is important to get everyone on the same page by clarifying the assumptions that will be used (or not used) by the PHA team. Examples of assumptions include:

- The process work as designed (Especially important for new processes under design.)
- The equipment is fit for its intended use
- Only trained operators will be used to operate the process
- Written procedures accurate and will be followed precisely at all times
- Facility policies (i.e.: PPE, security access, cardinal safety rules) are well defined and consistently followed and enforced

3.1.4 Consequences of Concern

When the PHA deliberations begin many, many issues are likely to be identified. Which of them are to be of concern to the PHA team? Fire, explosion and toxic release are required to be considered under the PSM rule. However, other topics may also be of concern including environmental damage, property damage, employee injury, on-site and off-site impacts, operability, waste, energy efficiency and many others. The broader the scope of consequences that are to be of concern to the PHA team, the more time the PHA will take to accomplish and, most likely, the longer the list of recommendations that will need to be addressed. When these "other concerns" are included in the PHA, the level of the environmental incident or injury needs to be defines and documented. (flammable liquid spill over X gallons).

Remember that PHA documentation is required to be maintained for the life of the process. Keeping the list of the consequences of concern to a limited, important few will reduce time, effort and documentation size.

It may be advisable to consult with legal counsel about the possible impact of long-term documentation of items not required to be maintained under the PSM rule. Items of interest that are identified by the PHA team that are outside the defined consequences of concern should be tabled to a "parking lot" for further consideration outside the PHA meetings.

3.1.5 Approval by process owner

Finally, as the scope of the PHA is defined, it is important to assure that all of the critical stake-holders in the process are in agreement with the scope. At a minimum, the critical stake-holders include:

- Unit management
- The EHS organization
- Engineering (especially for new processes under design and construction)

If all of the critical stake-holders are not on board with the scope and approach of the PHA, there may be a lack of cooperation in providing the resources necessary to execute the PHA or to carry out the recommendations of the PHA team.

3.2 PHA Technique(s)

One of the most important decisions affecting the success of a PHA is the methodology that is selected for analysis of each node of the process. The PSM rule lists a series of acceptable PHA methodologies and also allows for use of "an appropriate equivalent methodology"[7]. The following PHA methods are listed in the PSM rule and are presented in the order of least to most complex:

- What-If
- Checklist
- What-if / Checklist
- Hazard and Operability Study (HAZOP)
- Failure Mode and Effects Analysis (FMEA)
- Fault Tree Analysis

Additional analysis methods that are commonly used for special applications and very high hazard processes include:

- Layers of Protection Analysis (LOPA)
- Quantitative Risk Analysis (QRA)

What-If and Checklist methods are typically considered appropriate for site, facility, and building assessments as well as procedures, preliminary design reviews, and simple process operations such as basic mixing / blending and packaging activities. The HAZOP method is generally accepted as the "gold standard" for portions of a process where fluids (liquids or gases) are used and which are interconnected by piping and/or ductwork systems. FMEA is sometimes used to focus in on control systems. Fault Tree Analysis is often employed where multiple failure modes may be anticipated that could lead to a catastrophic event. LOPA is a supplemental technique that can be used to determine if additional layers of protection (safeguards) may be appropriate.

FMEA, Fault Tree Analysis, LOPA and full QRA methods all require some degree of quantitative data on the probability of failure on demand of safety systems or various discrete components to achieve the most benefits.

Therefore, a complete PHA on a process is likely to employ multiple PHA methodologies in order to assure efficient and thorough analysis of all of the various issues and components related to the process under study. If the proper study methods are not employed, the likelihood of a successful PHA is diminished.

3.3 Risk Tolerance

As issues that fall within the consequences of concern are identified by the PHA team, it is important that there be a common consensus of what constitutes an acceptable risk before recommendations / action items are generated. Each person on the PHA team will have a personal standard for risk tolerance; what is acceptably safe and what is not. The same will hold true for management as the recommendations from the PHA are carried forward for authorization to complete.

It is critically important that a single standard be employed to define what constitutes tolerable risk from an organizational perspective. Typically, the members of the PHA team are not at the level of management that determines what risks are or are not acceptable to the company. Likewise, many senior managers may not have the technical background to fully understand the impact of the consequences of an incident. Therefore, it is paramount that the organization's risk tolerance be defined before the PHA is undertaken. If the risk tolerance is too high, the PHA may lose effectiveness because potential hazards may go unaddressed. If the risk tolerance is too low, the number of recommendations will increase and the resources needed to address all the recommendations may become unduly burdensome.

Risk tolerance is usually most easily communicated using a risk matrix where probability of occurrence and severity can be estimated in semi-quantified terms and categories defined where improvements are required, potentially required and/or not required. This can help the PHA team define when recommendations for improvement are warranted.

4. Top Four Success Factors – #2 - Time Investment in the PHA

Spending an adequate amount of time on each node to thoroughly identify and analyze all relevant issues is one of the top four most important success factors for a PHA.

How much time spent on the PHA is enough? Unfortunately, there is no universal answer. A thorough PHA needs to take as long as it must take. Factors that impact the time length include:

- Size and complexity of the process
- Availability, accuracy and level of detail of the Process Safety Information including:
    - Process and Instrumentation Diagrams (P&IDs)
    - Chemical property and behavior information
    - Detail on safe operating limits for the process
    - Alarm and response schedules
    - Process control computer logic
- Level of knowledge of the PHA team members
- Ability of the PHA team to work without interruption
- PHA methodology used
- Level of preparation for the PHA (advance screening for reactives, combustible dust, etc.)
- The level of focus on the scope and objectives set up for the PHA

For planning purposes for a PHA on a typical chemical process, the first few HAZOP nodes may take 4-8 hours each to accomplish, although as the PHA team becomes familiar with the PHA method and develops a discipline for focusing only on the consequences of interest, a HAZOP node may be reasonably accomplished in 2-4 hours. Checklists may allow the group to move along more quickly. What-If methodology can be quite time consuming if lists of questions must be developed from scratch. The What-If / Checklist method may provide some structure and reduce the time spent on a node.

5. Top Four Success Factors – #3 - The PHA Team

Another of the four critical success factors for a PHA is the composition of the PHA team. The PHA team must be made up of people with the proper technical knowledge of the process including operational practices, equipment design and performance, properties of the materials associated with the process, applicable codes and standards, human factors, maintenance procedures and practices and many other subjects.

5.1 PHA Leadership Competency

To be successful, the PHA team needs a strong leader. The PHA team leader is typically the PHA facilitator, however, in cases where the facilitator is from outside the immediate organization such as a consultant or specialist resource from a corporate group, it may also be necessary to identify a PHA coordinator who is resident in the facility where the process is located and who can manage the overall execution of the PHA from PHA scope definition through final documentation and closure of the PHA recommendations.

The person identified to be the PHA facilitator must have some foundational skills to successfully and productively lead the PHA team. These foundation skills include:

- Technical qualifications and understanding such as:
    - A basic understanding of the process under study
    - Good communication skills
    - Excellent organizational skills
    - Ability to manage group dynamics
    - Good language and writing skills
    - Computer skills and knowledge of software that may be used during the PHA
    - A general understanding of the hazards of materials (flammables, reactives, combustible dust, toxics, etc.)
    - Risk assessment skills
    - Consequence modeling (release modeling) understanding
    - Basic understanding of codes, standards and industry practices applicable to the process
- Experience leading PHAs using the methods that will be employed during the PHA (HAZOP, What-If, Checklists, LOPA, etc.)

5.2 Define Team Composition

The PSM rule states that, at a minimum, the PHA team must be made up of a group "...with expertise in engineering and process operations, and... shall include at least one employee who has knowledge specific to the process being evaluated...[and] one member... must be knowledgeable in the specific PHA methodology being used."[(8)] In a later letter of interpretation, OSHA has stated that one PHA team member must be a process operator.

While one could argue that a PHA could be carried out by a single person that is a trained operator who is also an engineer and trained facilitator, this would be a recipe for disaster. A typical PHA team might consist of five to ten members, with some subject matter experts rotating through the PHA team as various pieces of the process are studied. It is also important that the PHA team not become too large so as to be difficult to keep focused on the analysis that is taking place.

Potential makeup of a solid PHA team should have members selected form at least the following possible resources:

- Facilitator with appropriate training and experience
- Process Operator
- Knowledgeable technical person – process or manufacturing engineer
- Scribe - familiar with PHA documentation practices
- Capital equipment / design engineer(s)
- Process control engineer(s)
- Maintenance personnel (mechanic, instrument tech and electrician)
- Process / maintenance supervisors and technologists
- EHS specialists – industrial hygiene, general safety, chemical safety, etc.
- Laboratory / product development personnel
- Code specialists
- Vendor representatives for chemicals and/or equipment used with the process
- Process owner representatives (for licensed processes)
- Consultants / special experts

Factors determining team makeup include the complexity of the process and the equipment used with the process, availability and experience of in-house process and equipment engineers, unique hazards of some materials used with the process, whether or not the process has been licensed for use from a third party, empirical risk presented by the process, level of experience with operating the process, and pervious incident experience to name just a few.

5.2.1 Required PHA Team Members

As noted above, the minimum required members of the PHA team must include:

- A process operator
- A person with engineering knowledge of the process
- A person with knowledge of the process under study
- A person trained in the PHA methodologies being used (a facilitator)

5.1.2 Core PHA Team Members

To maintain continuity of the PHA process a nucleus of persons should participate in the PHA study sessions from start to finish. At a minimum this most likely would include:

- A person with engineering knowledge of the process (senior process engineer)
- A person with knowledge of the process under study (process engineer or process technologist)
- The PHA facilitator
- The facility PSM coordinator (may be the process engineer)

5.2.3 Ad Hoc PHA Team Members

Depending on the size of the process and the number of ancillary process components included in the PHA some members of the PHA team may rotate through the team as various parts of the process are studied. In addition, based on the depth and technical acumen of local personnel, it may be necessary to engage the services of consultants in various process specialties (electrostatics, flammable liquids, reactive processes, etc.), or persons with specialized knowledge of chemical hazards and behaviors, or code specialists with in-depth understanding of hazardous occupancy design requirements.

Ad hoc members of the PHA team may be drawn from the following:

- Maintenance personnel
- Process / maintenance supervisors and technologists
- EHS specialists – industrial hygiene, general safety, chemical safety, etc.
- Laboratory / product development personnel
- Code specialists
- Vendor representatives for chemicals used with the process
- Process owner representatives (for licensed processes)
- Consultants / special experts
- Persons with experience with similar processes at other company facilities

5.2.4 PHA Team Training

Typically, some degree of training is needed for virtually every member of the PHA team. Normally, the facilitator will provide an overview training to the members on the PHA team on the PHA methods that will be used. This ensures that the entire PHA team understands how the PHA will be conducted and will reduce confusion caused by past PHA experience. A person with a solid understanding of the process itself should provide training to the PHA team on how the process is expected to work and the hazards that are understood to be present. Often this is done on a node-by-node basis by a maintenance or process engineer or specialist who is very familiar with the specific part of the process under study.

If the facilitator is coming from outside the facility, training will also need to be provided to the facilitator in the process operation, documentation systems used to maintain the PSI, facility safety requirements and practices, previous incident history and other topics.

6. Top Four Success Factors – #4 - Information Available and Used for the PHA

The last of the top four success factors, but by no means the least, is the importance of having complete, detailed and accurate information for use by the PHA team. Ideally, virtually all of the Process Safety Information (PSI) listed in the Process Safety Information paragraph of the PSM rule(9) should be available for the PHA team during their analysis. In reality, all of the PSI will probably not be available in the level of detail that is desirable.

Assembling all of this information and assuring that it is current and accurate will require considerable effort in advance of the PHA team meetings.

6.1 PSI – The Bare Essentials

There is some information that it is so important that the PHA should not be started without it. This information includes:

- Complete and accurate Process and Instrumentation Diagrams (P&IDs)
- Description of all process chemistry (including side reactions and by products)
- Material and energy balances
- Relief system settings, design and design basis
- Materials of construction of process equipment and piping and ventilation systems
- Details of mechanical integrity practices in place
- Understanding of potential non-process reactive chemical issues (inadvertent mixing, contamination, etc.)
- Understanding of potential combustible dust issues

- Special or unique hazards (i.e.: extremely high / low temperatures or pressures)

If any of this information is missing, it needs to be developed before the PHA is begun. If information is out of date, it needs to be updated before the PHA is begun. Therefore, one of the first activities to be undertaken after the scope of the PHA is defined is assembly and review of the available PSI.

Reactive chemical and combustible dust hazards are frequently addressed as independent hazard assessments done as part of the pre-PHA activities. Drawings, especially P&IDs, layouts, and piping diagrams need to be verified for accuracy and updated or redone as appropriate. Relief system designs need to be verified against equipment actually in place. Mechanical integrity information needs to be reviewed, gaps identified, as well as any prominent findings identified to be carried forward into the PHA meetings.

If the PSI is not verified as available and carefully scrutinized for accuracy before the PHA is begun, the PHA team will end up generating action items to confirm, verify or study issues related to missing information. This can lead to the PHA moving forward using assumptions rather than facts. The assumptions can lead to erroneous conclusions on process safety, as well as a need to revisit all parts of the PHA that were done based on assumptions: A considerable waste of effort by the PHA team and a reasonable likelihood that some critical issues may be missed.

6.2 PSI – Additional Information

Some information used by the PHA team may need to be generated or presented anecdotally in the course of the PHA meetings. For example, it may not be practical to develop a written narrative of the process control computer logic, therefore, this information may need to be made available in the form of a knowledgeable engineer or technologist who can interpret the control logic directly from the process control computer system. In addition, operating procedures may need to be verified during the PHA to assure that practices in place on the process floor actually reflect the detail and order set forth in the written operating procedures.

7. Additional Success Factors

7.1 Pre-PHA Activities

Depending on the nature of the process to be evaluated by the PHA, it is often necessary to engage in some pre-PHA activities such that the PHA team can work as productively as possible. Examples of pre-PHA activities might include:

- Reactive chemical review

- Combustible dust study
- Operating procedure review
- Job Hazard Analysis (JHAs) or Job Safety Analysis (JSAs)
- Qualification and training requirements
- Mechanical integrity program review
- Detailed review of past incidents

7.2 Process Familiarization

It is important that everyone working on the PHA be familiar with the process under study. For some PHA team members this will not be an issue because they regularly work with the process as operators, maintenance personnel, or engineers. For other PHA team members such as consultants, special experts, vendor representatives, outside facilitators, and corporate support personnel, however, it may be necessary to conduct more detailed orientation and/or training such that these persons have a good grasp of the process and are capable of recognizing possible hazards and areas where improvements may be needed.

It is also necessary that everyone be familiar with the PHA method(s) being used. This can be done by conducting a brief training session on the PHA methods, and a discussion of the process under review at the start of the PHA meetings. When possible, it is advisable to tour of the covered process areas with the PHA team. If the PHA meetings are being held off-site, it is helpful to have photographs of the process and included equipment, control panels, utility services and the like that can be used during the PHA meetings.

Also, as the PHA progresses, it is helpful to have a knowledgeable person such as a process engineer, supervisor or lead operator describe the intended function of each node of the process including operating practices and all safeguards prior to detailed discussion of the respective nodes.

7.3 Similar Processes

Frequently, there may be multiple process operations that are similar or virtually identical. In these circumstances, it is tempting to do the PHA based on a "representative" operation and assume that all of the other similar operations have the same hazards, safeguards and needs for improvement as the sample process. In reality, although the processes may at first glance seem to be virtually identical, they rarely are. Piping runs may differ in length and configuration, controls and instruments may not all be identical, siting influences may be different, minor modifications to one or more of the similar units may have altered the equipment or changed operating characteristics.

To assure that all of the hazards and safeguards are properly analyzed, it is possible to do a full and complete PHA analysis on one unit and then "clone" the other units based on the sample, but looking at each deviation and issue for possible differences from the sample.

7.4 Safeguards

There are many and varied types of safeguards used to protect chemical processes. Some safeguards are obvious and others less so. In general safeguards fall into one of three groups: 1) design features; 2) protective (preventive) devices/systems; and 3) mitigating devices/systems. As with the overall process operation, it is important for the PHA team to receive an explanation of the intended function / operation of each safeguard on the process.

To assure that no safeguards are overlooked, it is often helpful to create a list of safeguards for reference during the PHA meetings. Note issues and safeguards that may not be found specifically listed in PSI – containment systems, traffic nearby, electrical classification, equipment vulnerable to damage, lighting, sampling locations, catwalks and access, and distances to control room or emergency equipment.

7.5 PHA Traps

Up to this point, this paper has presented things that contribute to the success of a PHA. There is also a long list of traps, actions or inactions that can as easily detract from the success of a PHA. Some of the more common traps / detractors include:

7.5.1 Assumptions

Previously, reference was made to clarifying assumptions that will be accepted by the PHA team as they complete the analysis. In the context of this paragraph, the assumptions refer to assumptions that are made during the PHA deliberations when needed specific information is not available. ("I think that pressure relief valve is set to relieve at 250 psi.") Whenever such issues are identified, it is important to confirm the validity of any assumption made with an action item to verify that the assumption is correct, and if the assumption is not accurate, to revisit the issue again in the full context of the PHA.

7.5.2 Fails Safe

One common and very important safeguard used with chemical processes is how controls or equipment fails. But "fails safe" needs to be clearly defined. Does the valve fail open or closed? Does the cooling water system fail on or fail off? Does loss of motive power (i.e.: compressed air or hydraulics) still allow the safeguard to move to the safe condition? These important descriptors need to be included in the PHA documentation.

7.5.2 Bluster Factor

Frequently, there is a member of the PHA team who assumes (rightfully or wrongfully) the role of "process expert". A person who's knowledge, understanding and/or opinion is accepted without question. Frequently, this expert may have a significantly dominating personality that other members of the PHA team may not attempt to challenge the expert even though they have factual information or experience that contradicts or is significantly different from of the expert's position. While it is one of the jobs of the facilitator to manage group dynamics, each member of the PHA team needs to understand that they have valuable knowledge and contributions that need to be brought before the PHA team.

7.5.3 Silence Isn't Golden

As with the "Bluster Factor" described above, there is also a concern that important issues may not be adequately brought before the PHA team because a person may be concerned about ridicule or possible disciplinary action if they were to admit that things are not perfect. For example, on the third shift the operators frequently skip certain steps in the operating procedure or do things out of sequence for whatever reason. Again, the facilitator must play an important role in drawing out input from each member of the PHA team. However, it is equally important that the environment of the PHA team meetings are non-intimidating. (i.e.: Like Las Vegas – What is said here stays here with no fear of reprisal.)

7.5.4 Too Many/Too Few Sessions

Like it or not, doing a PHA is hard, time-consuming work. Consequently it is frequently the case to try to cram as much as possible into long days or, conversely, schedule PHA team meetings for only short durations of one or two hours per meeting. Both have potential negative impact on the effectiveness of the PHA. In addition, almost everyone involved in the PHA team continues to shoulder other responsibilities of their "regular job" while also serving as a member of the PHA team.

If meetings are too long or too frequent, team members may lose focus on the discussion because of the pressure of other pending work, or simply because of fatigue from the high level of concentration required in the PHA meetings. If the meetings are scheduled for too short a time or with too long an interval between sessions, the PHA team may not be able to adequately address issues because of time constraints and the thread of discussions may not survive the breaks between the meetings.

Typically, a good balance is to schedule PHA meetings for half day sessions such that the PHA team members have some time to keep current with other responsibilities and to perhaps do a bit of follow-up on questions from earlier PHA meetings such as to avoid the need for an action item. In some cases, where major participants on the PHA team are traveling away from home or are otherwise not able to function productively in the off periods, it may then be necessary to conduct the PHA as full-day sessions on consecutive days. When such is the case, an off-site venue is advisable to allow everyone on the team to minimize interruptions and distractions.

7.5.5 Problem identification vs. Problem Solving

One of the most common traps that a PHA team can fall into with a significant impact on both the PHA efficiency and effectiveness is to become engaged in developing solutions to the problems identified during the PHA sessions. Since many of the likely candidates for membership on the PHA team are technical people, they are trained to solve problems and enjoy solving problems. However, the mission of the PHA is to identify hazards of the process and the safeguards provided to manage those hazards and to make recommendations on where changes are needed to improve the safety of the process.

If the PHA team also takes on the role of solving the problems that it identifies, there is a serious risk of overlooking possible solutions because inadequate time is spent reviewing possible alternatives. Specific, detailed recommendations may later be found to have a better alternative available. Also, some members of the PHA team may not have the expertise or a role in the development of solutions and may lose focus while solutions are being developed.

It is important to bear in mind that the PHA team's role is to review the process to identify hazards, the related safeguards and to make recommendations on where improvements should be considered. Leave developing solutions for the recommendations and action items to a separate team.

8. Summary

If a PHA is done well and the recommendations generated from the PHA are properly resolved, the process should operate safely across its lifetime. There are many success factors that can be used to help evaluate if the PHA has been done effectively. The top four success factors for a PHA as discussed in this paper include:

- Planning and preparation for the PHA
- The time investment in the PHA
- The makeup of the PHA team
- The information available and used by the PHA team

In order to be successful in preventing unwanted incidents, the PHA must be executed in all of the necessary detail and using all of the necessary resources. The PHA is an expansive task, but when all the analysis is properly completed and all of the recommendations and action items are resolved, a major step will have been taken to help ensure the safety of the process.

9. References

1. Title 29 US Code of Federal Regulations, Part 1910 – Occupational Health and Safety, Subpart H – Hazardous Materials, § 1910.119 – Process Safety Management of Highly Hazardous Chemicals

2. Title 29 US Code, Part 654(5)

3. Title 40 US Code, Part 68 § 112(r)(1)

4. CCPS. *Guidelines for Hazard Evaluation Procedures, Third Edition.* Center for Chemical Process Safety, American Institute of Chemical Engineers, New York, NY, 2008.

5. W.L. Frank and D.K. Whittle. *Revalidating Process Hazard Analysis.* Center for Chemical Process Safety, American Institute of Chemical Engineers, New York, NY, 2001.

6. Title 29 US Code of Federal Regulations, Part 1910.119(f)

7. Title 29 US Code of Federal Regulations, Part 1910.119(e)(2)

8. Title 29 US Code of Federal Regulations, Part 1910.119(e)(4)

9. Title 29 US Code of Federal Regulations, Part 1910.119(d)

402435 Security Risks and Countermeasures of Pesticide Formulations Production Process

Abstract:

This paper investigates the actual production of pesticide enterprise, combined with the major features of the flammable and explosive process of intermittent and non closed operation, the low level of automation equipment and the operator of the strong seasonality and liquidity. A comprehensive identification of the unreasonable of analysis process, maladjustment of the equipment and facilities, defects of safety management. Put forward relevant countermeasures for the three areas of process engineering, equipment and facilities and safety management. Pointed out the direction for pesticide enterprise to rectification of the hidden dangers and improve the safety management.

394181 Incorporate Safety during Process Design

Abstract:

Process facilities are generally designed in accordance with good engineering practice. Safety features like inherently safer design, Basic Process Control Systems (BPCS), Alarms, Relief valves, Shutdown valves, etc. are incorporated to take care of upsets during Normal/ Emergency operation, as well as Startup.

Equipment and Piping are designed to withstand the maximum expected pressures, which can be foreseen during the design and engineering phases of a Project. This however should not lull the operators into a sense of complacency and lead them to believe that the facility will be operated with unlimited flexibility. Problems can arise if due attention is not given to the criteria adopted for the design of various systems. This could happen if critical parameters are not monitored, good operating practices are not followed and abrupt changes are made in feed composition or operating conditions. Inherent safer practice has also been challenged with the introduction of electric submersible pumps (ESPs) in new and existing wells as the maximum discharge pressure under block-in conditions exceeds the maximum allowable working pressure (MAWP) of existing piping. A safe alternative to replacing the piping is the use of a high integrity pressure protection system (HIPPS). Though HIPPS protects the downstream piping, but can result in significant ESP damage and production losses due to blocked ESP discharge.

It is understandably difficult to incorporate multiple features to guard against mal operation or human failure, but is desirable to try and anticipate problems and take pre-emptive action at the design stage rather than carry out corrective actions following a failure. Some of these aspects are illustrated with a description of the fall out on account of not taking into consideration the limitations in the design of the existing facilities.

Study on Application of LOPA in HAZOP on China Refinery Design

CAI Mingfeng WANG Zhigang ZHOU Hao WANG Wenjuan

(1 *CPECC East-china Design Branch, Qingdao,* 266071, *China*

2 *CNPC East-China Design Institute, Qingdao,* 266071, *China)*

Abstract: Recommending the demand of HAZOP and LOPA evaluation method in the design stage from China department of work safety and large scale petroleum enterprises, summarized the feature and analysis process of HAZOP and LOPA, obtained the complementary by comparing the advantages and disadvantages, put forward the feasibility about comprehensive assessment of the two tools. With the example of refinery RDS unit which is in basic design stage, briefly stated the process about comprehensive assessment of the two tools, and verified the feasibility and advantages of comprehensive assessment of the two tools. Affirmed the advantages and put forward shortages of comprehensive evaluation of the two tools through analysis theory and project case. Suggested to continuously improving the analysis process in design stage to make it more adapted to the design of domestic project, improve safety design integrity of the domestic design institutes, protect the device to more steady and safely。

Key words: HAZOP, LOPA, refinery design, accident scenario, risk matrix, failure frequency

HAZOP (Hazard and Operability Studies) analysis method is one of the most widely used international petrochemical industry in the process of qualitative risk assessment method. The world's advanced industrial countries and emerging industrialized countries have been or are being vigorously promoted and applied to petrochemical industry HAZOP analysis method.

HAZOP analysis combined with the LOPA (Layer of Protection Analysis) joint analysis is the latest application of recent domestic risk assessment methods. LOPA is organic supplement and deepening of HAZOP method [1]. It can greatly improve the accuracy of the analysis and it is one of the encouraging recommended method of the domestic government, enterprises and institutions and other departments.

1 Policy for China and National Petroleum Companies

State Administration of Work Safety, the Ministry of Industry and Information Technology published "About dangerous chemical enterprises implement the implementation opinion which is ' state council on further strengthening notification of enterprise safety in production work' "(National security supervision bureau 186th 3 〔2010〕). It required the device of large and dangerous chemical process must take HAZOP analysis after completing the preliminary design.

State administration of work safety, ministry of housing and urban-rural published "on further strengthening the dangerous chemicals safety design of construction project management"(State Administration of Work Safety 76th 3 〔2010〕). It required the Project Party Company should take the initiative in the construction project design contract to require Design Company taking HAZOP review to design and involving two focus one major construction project, HAZOP analysis must be carried out in basic design stage. (Two focus one major construction project: the key regulatory dangerous chemical process, supervision of dangerous chemicals, major hazard installations)

One of State administration of work safety leaders said at all levels conference many times, " design safety is eugenic and superior nurture of petrochemical projects" and design company played an important role in the safe operation of processing equipment, and encouraged design company carrying out the HAZOP analysis in basic design stage.

"China national petroleum corporation hazard and operability analysis work regulations" asked the projects which did not take HAZOP analysis in the preliminary design stage could not conduct preliminary design review.

2 HAZOP and LOPA Application Guides from China and National Petroleum Companies

State Administration of Work Safety issued "Hazard and operability studies (HAZOP studies) -Application guide" on June 8, 2013 which put into action on October 1, 2013.

China National Petroleum Corporation(CNPC) issued " Hazard and operability analysis guide " on March 30, 2011(Q/SY 1364—2011)which put into action on May 1, 2011. This guide is one of the domestic earliest technical documents based on the HAZOP analysis. This standard stipulates the hazard and operability analysis preparation, procedures and measures of tracking and other technical requirements, and creatively puts forward the first domestic HAZOP risk analysis matrix, and provides the technical support to set guidelines and standards for government departments and other business.

State Administration of Work Safety has organized to draw up "chemical companies covering layer analysis application guideline". The guideline will be implemented in the near future.

3 Overviews of HAZOP and LOPA

3.1 HAZOP

3.1.1 Summary

HAZOP is a systematic and structured qualitative risk assessment method, mainly used to determine existing risk and operation problems in the engineering design in the design stage. It uses the design of guide words in advance, through the determination of deviation analysis to review the design safety and hazard of causality. HAZOP analysis determines whether the risk level is acceptable according to the degree of event frequency and consequences, and risk matrix. HAZOP analysis not only pays attention to the danger of the device itself, but also involves the unit operation, etc. Figure 1 is HAZOP analysis process [2].

3.1.2 Limitation

HAZOP analysis is the most widely used domestic analysis method in the process hazard analysis (PHA). It has distinct characteristics which advantage do not relate once again and mainly simply overview its limitation.

Firstly, the accuracy of risk level depends on the participants' level of technology, experience, and different analysis people will give different risk grade. Especially the risk frequency, it is got according to the experience, completely a kind of qualitative analysis, and the accuracy is not high, so it needs further adopt the method of semi-quantitative or quantitative analysis [3].

Secondly, it is difficult to identify the existing safety protective measures or suggestions whether they can have the effect of the reduction factor for all deviation, consequences and reasons. It cannot guarantee that proposed measures can meet the requirements of the existing consequences severity.

Moreover, the causes and consequences that HAZOP analysis getting is often not enough. That lacking the analysis of some intermediate events causes that suggestion measures cannot fully cover accident chain.

3.2 LOPA

3.2.1 Summary

LOPA is a semi-quantitative analysis tool of the risk assessment. It usually characterizes the risk of the scene using probability of initial ramifications, severity of consequence and the size of the order of magnitude independent layer failure frequency. LOPA risk calculation is usually specific to the single initial reasons and the adverse consequences dual dangerous drama type of risk. And the risk is for dangerous plot. LOPA is the analysis tool that based on the information of qualitative risk assessment (such as HAZOP analysis). Figure 2 is the protective layer of accident and figure 3 is the analysis process of LOPA [4].

3.2.2 Advantage and limitation

Relative to the qualitative and quantitative risk analysis method, the LOPA has obvious advantages:

a) Compared with qualitative analysis, such as HAZOP, LOPA provides more reliable risk identification, and gives numerical plot of the specific frequency and consequences.

b) Compared with the quantitative analysis (such as fault tree analysis), it spends less time, simple analysis, and can improve the efficiency of risk analysis meeting.

Of course, LOPA analysis also has its own limitations:

a) LOPA itself is not a complete process of risk analysis tool, relying on the qualitative analysis methods (such as HAZOP), and it can maximize the use of their own advantages only standing on the shoulders of the qualitative analysis.

b) The foreign petroleum companies have their own libraries of risk analysis of the failure frequency, which data root in all the factory equipment failure frequency statistics in the system, and the data is real and effective. But there is not a China Petroleum company owning equipment failure frequency statistics, so domestic LOPA analysis only can turn to foreign databases, generating that analysis will inevitably exist error.

By comparison two methods of analysis, two methods of analysis have strong complementarity, through the organic combination can complement each other, make the analysis more complete and comprehensive, can better identify intermediate events, existing safety measures, etc., and the conclusions are also more accurate and objective [1].

Table 1 Combination of HAZOP output information and LOPA input information

Requested information of LOPA	Provided information of HAZOP
Influential incidents	Consequence
Severity level	Severity of consequence
Cause	Factor
Trigger a possibility	Reasons for the frequency
Protection course	The existing security measures
The additional requirement is reduced	Recommended practice

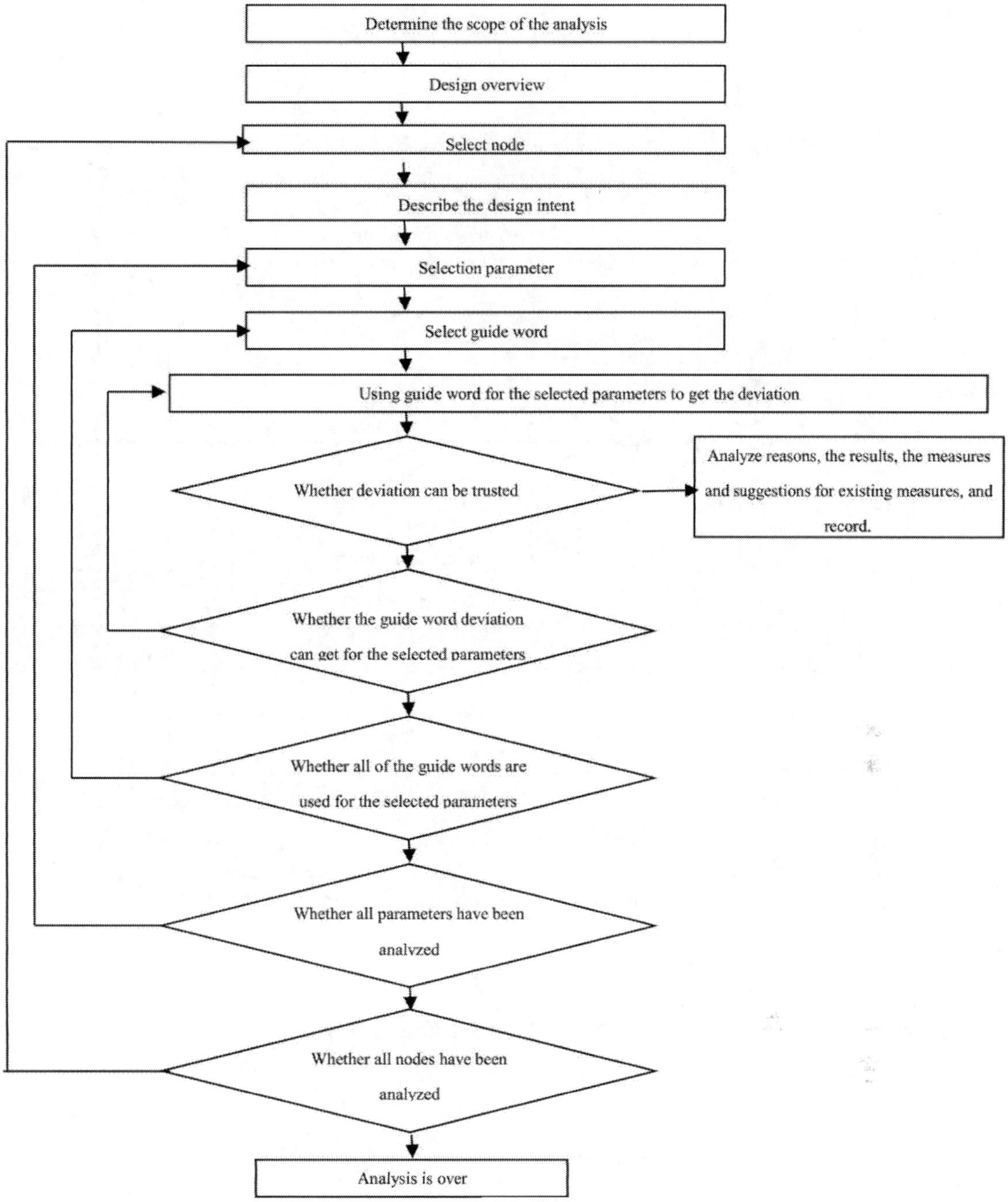

Determine the scope of the analysis
Design overview
Select node
Describe the design intent
Selection parameter
Select guide word
Using guide word for the selected parameters to get the deviation
Whether deviation can be trusted
Analyze reasons, the results, the measures and suggestions for existing measures, and record.
Whether the guide word deviation can get for the selected parameters
Whether all of the guide words are used for the selected parameters
Whether all parameters have been analyzed
Whether all nodes have been analyzed
Analysis is over

Figure 1 HAZOP analysis process

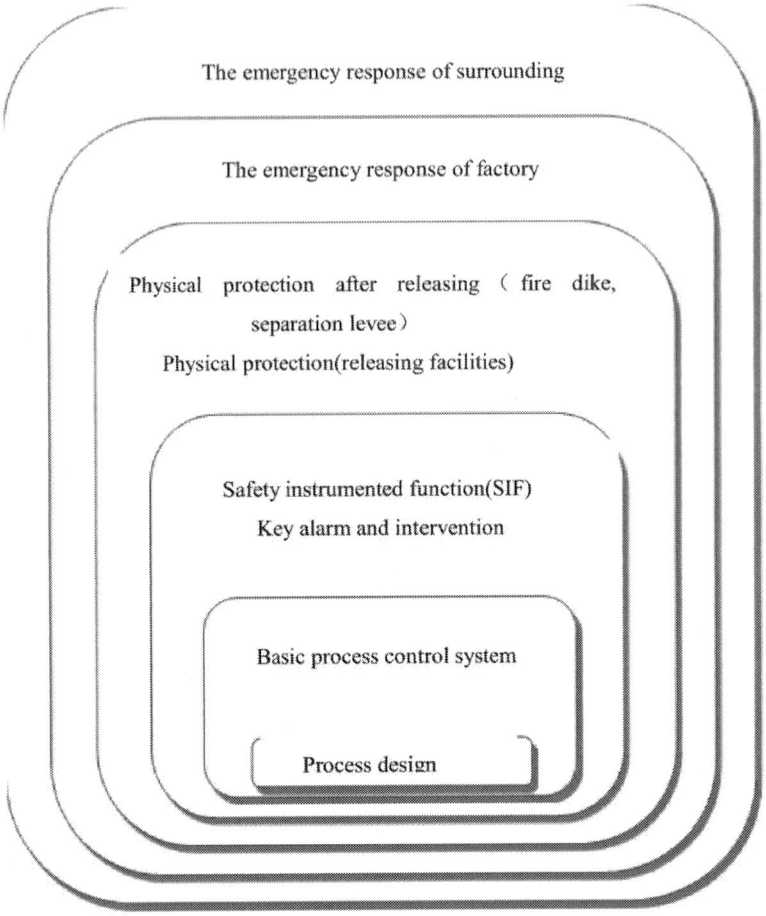

Figure 2 Layer in accident

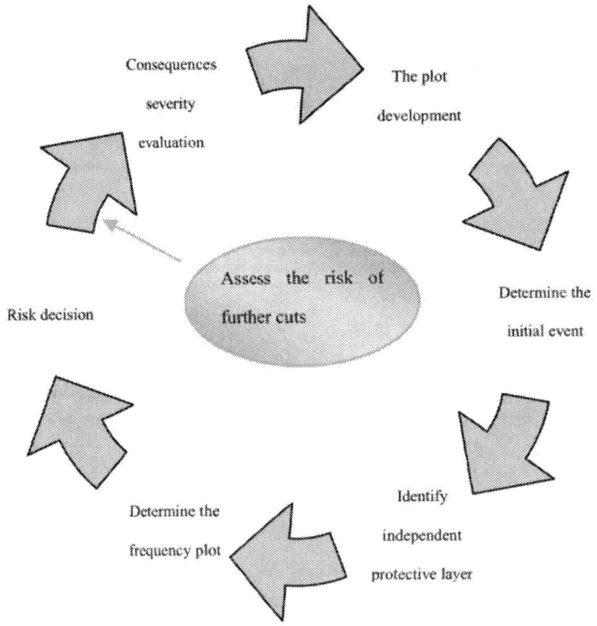

Figure 3 LOPA analysis process

4 LOPA Applies in Design Phase of the RDS (residual oil hydrogenation) Unit based on HAZOP

One petroleum refining project lies currently in the stage of basic design. According to requirements of the State Administration of Work Safety, the RDS unit belonging to the "two key one significant project", HAZOP must be conducted. In order to more accurately identify problems that existing in the basic design, the analysis uses HAZOP and LOPA comprehensive evaluation method.

According to " Fire prevention code of petrochemical enterprise design"(GB50160-2008) and "Code for design installations in explosive atmospheres"(GB50058-2014), residue hydrodesulfurization unit belongs to the top device of fire risk, and the most zones of device pertain to explosion danger area 2. The considerable part of raw materials, products, by-products in the production are inflammable, explosive medium, and a lot of materials in the process of production are in environment of high temperature and high pressure. From the process of raw material input to product output, once there are equipment, facilities and pipeline sealing, operator wrong operation, high temperature and high pressure equipment, especially the pressure vessel technology parameters abnormal fluctuations and other reasons, all can lead to process medium leaking, and flame is easy to cause fire explosion accident.

4.1 Process description

a) Reaction part adopts hydrogen furnace after mixing, heat low technological process. Fractionation part adopts the typical towers stripping process. Fractionating column uses the fractionating column to feed heating furnace heating, and diesel side stripper adopts reboiler stripping.

b) The device sets two heating furnace, heating furnace of reaction feeding furnace and fractionating column feeding, and the two stoves share a burning tank furnace.

c) Back flush filter is set, which filters the solid particles larger than 25 microns in diameter, to prevent oil as raw material of solid impurities into bed reactor, and make the faster pace of increase in the reactor bed pressure drop and affect the operating cycle.

d) Water injection is set at the entrance of cold low pressure separator and the reaction product air cooler, to avoid ammonium salt deposition plugging equipment and piping.

e)Raw material buffer tank and filtered oil raw material oil buffer tank use nitrogen protection, to prevent they contact with air to generate polymer and gelatin, effectively prevent scaling.

4.2 Device hazard analysis

4.2.1 Fire explosion

The chemicals of device are mainly involved in naphtha, diesel, hydrogen, hydrogen sulfide, light hydrocarbons, and fire, explosion are the main risk factor for this project. There are most flammable and explosive substances involved in the process of production for, and the material leakage, improper operation, external factors will cause fire explosion.

4.2.2 Poisoning

There is a large number of poisonous hydrogen sulfide gas in this device, in addition there are toxic and harmful substances such as dimethyl disulfide, MDEA, nitrogen, etc. Leakage, gathering of this kind of material can cause the operator suffocation or poisoning.

4.2.3 Corrosion

Each equipment has the characteristics such as high operating temperature and operating pressure, medium (such as hydrogen, hydrocarbon and hydrogen sulfide) inflammable, explosive, corrosive in this device, and the device corrosion is one of the most important reason for equipment corrosion causing material leakage, fire, explosion and poisoning. Device corrosion is the major risk factors.

4.2.4 Others

This device also has the danger such as noise, burning hot, high-altitude falling, electric shock, etc.

Residual oil hydrogenation unit main distribution of dangerous and harmful factors is shown in table 2.

Table 2 Dangerous and harmful factors distribution table of device

Hazardous and harmful factors / Evaluation unit	Fire explosion	Poisoning and stifling	Scalding	Falling accidents	Mechanical injury	Noise	Electric shock
Reaction zone	√	√	√	√			
Fractional area	√	√	√	√			
District heating furnace	√	√	√				
Compressor area	√	√			√	√	√
Desulfurization area	√	√		√			
Pump housing area	√	√	√		√	√	√
Heat exchanger	√	√	√	√			

4.2 HAZOP

Two parts of reaction and desulfurization, fractionation exist in residual oil hydrogenation unit. 25 parameters, 39 deviation, 29 nodes are used in this HAZOP analysis.

The analysis of device technology has carried on the comprehensive and systematic HAZOP analysis, dividing 29 nodes, the nodes involved deviations, causes, consequences, and protective measures are detailed analyzed. Recommended practice is divided into two categories: general advice and suggestion, and total of 113. Whereby suggestion is divided into five: equipment (including instrument, pipeline) related measures, the update operation procedures related measures, DCS related measures (high and low and its interlocking), update drawings and other relevant measures.

Residue hydrodesulphurization unit belongs to high temperature and high pressure device, high temperature and high pressure parts mainly concentrated in the reaction, and the oil, gas, hydrogen, hydrogen sulfide and other fire explosion, poisonous and harmful material exist in reaction, therefore this article mainly analyze the part.

This article selects the sixth node to explain. The sixth node mainly takes in high pressure separator, hot low pressure separator and other major equipment. There are high temperature and high pressure material (oil and gas, hydrogen, hydrogen sulfide), and hot high pressure separator operating temperature is 350 ℃, operating pressure is 16.4 MPa. Operating temperature of thermal low pressure separator is 350 ℃, and the operating pressure is 2.8 MPa within the scope of this node. Due to high pressure and hot low pressure separator pressure difference is very big, there is the risk of high pressure running low pressure. So it selects high pressure deviation of thermal low pressure separator to analyze.

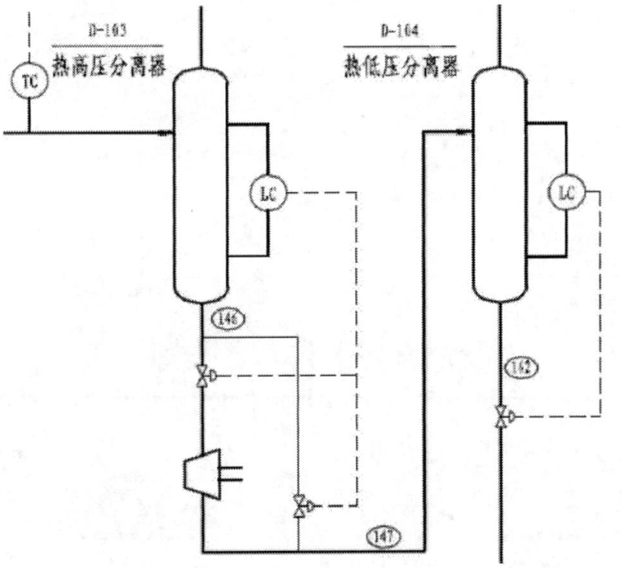

Figure 4 process flow diagram (interlocking and relief valve not showing)

Table 3 can be seen that the major reason of high thermal low pressure is hot high liquid level control circuit fault, leading to no hot high level, high pressure gas into the low grade heat. By comparison with high marks and hot low pressure parameters, thermal low pressure ability is difficult to bear the high pressure of 16.4 MPa gas, physical explosion.

Through qualitative analysis, referring to the risk matrix, the initial risk grade of the deviation is 11, belonging to the unacceptable risk. It need evaluate existing security measures whether can lower the initial risk to an acceptable level of risk.

It can be seen that the existing security measures are setting a high level low interlocking and hot low relief valve in view of the selected deviation, by P&I. If only HAZOP analysis, at present it can get the initial risk to an acceptable range experience through the expert qualitative analysis of the existing security measures, and the results of the analysis greatly influenced by expert experience, but it is difficult to guarantee the accuracy of analysis.

Table 3 HAZOP record chart

Parameter	Deviation	Reason	Consequence	Frequency	Severity	Level of risk	Safety measures
Thermal low pressure separator pressure	Hot high pressure low pressure separator high stress	Hot high level control circuit fault	The thermal low overpressure explosion	1E-1～1E-2	6. distensible	11	1、Set a hot high level low interlocking (2 of 3)2、Heat setting low relief valve

Risk assessment	Possibility							
	1	2	3	4	5	6	7	8

Risk assessment		Possibility							
		1	2	3	4	5	6	7	8
Severity	1	1	2	3	4	5	6	7	8
	2	2	3	4	5	6	7	8	9
	3	3	4	5	6	7	8	9	10
	4	4	5	6	7	8	9	10	11
	5	5	6	7	8	9	10	11	12
	6	6	7	8	9	10	11	12	13
	7	7	8	9	10	11	12	13	14
	8	8	9	10	11	12	13	14	15

Figure 6 risk matrix

4.3 LOPA evaluation

On the basis of the existing HAZOP analysis and using the LOPA analysis can solve this problem, seeing table 4. Assume that the initial event is control circuit fault, and the valve largely opens, according to the failure frequency library, the initial frequency is 0.1. And suppose that there is inspection worker, through the analysis, researchers exposed probability is 0.3 (according to the inspection frequency to determine), the BPCS is failure (initial event is control circuit fault), so the BPCS, operation and reporting to the police can't work. Existing safety valve and interlock protective layer can be used as an independent to reduce accident frequency, and the final incident frequency can reduce to 3e-7.

Table 4 LOPA evaluation table

SIF	Set a hot high level low interlocking (2 of 3)																	
		Initial Event		Enable event condition		Condition amendment				Protective layer								Mitigated event probability
Influence events (personel)	Severity	Description	Frequency	Description	Probability	The ignition probability	Researchers exposed probability	Mitigated event probability	Other	The general process Design	BPCS	Operation and reporting to the police	Safety relief	Other	Additional ease	Intermediate event probability	SIF integrity level	
Thermal explosion low	6	Control circuit fault, lead to the valve largely open	0.1				0.3						0.001			3e-5	2	3e-7

It can be seen in table 4 and table 5 that through LOPA it can reduce the probability of 3e-7, less than 10-6.And it has cut the risk from red to green, belonging to the acceptable risk according to risk matrix, so there is no need to improve the SIL level interlock.

Table 5 LOPA residual risk assessment based on the HAZOP

Parameter	Deviation	Reason	Consequence	Frequency	Severity	Level of risk	Safety measures	residual risk	Suggestion
Thermal low pressure separator pressure	Hot high pressure low pressure separator high stress	Hot high level control circuit fault	The thermal low overpressure explosion	1E-1～1E-2	6. distensible	11	1、Set a hot high level low interlocking (2 of 3)2、Heat setting low relief valve	7	

5 Conclusions and recommendations

5.1 Conclusions

a) It can be seen form the application instance, HAZOP analysis can deviate from the complete analysis of causes and consequences, and can be targeted from the reason to the consequences of each link analysis, make sure whether is equipped with safety protection measures. But only through the HAZOP qualitative analysis, it is difficult to analyze the existing protection measures whether can provide adequate protection.

b) Introducing LOPA assessment on the basis of HAZOP analysis. It is able to calculate half quantitative analysis of the existing security measures after cutting the risk of frequency, through risk matrix can be known whether residual risk is acceptable risk, whether you need increase new security protection facilities, and whether it can provide design basis for process instrumentation design.

c) Overseas project which the current domestic large refining design institutes participate in are LOPA and SIL evaluation method to determine the safety level of instrument, and other design institute or experience to carry on the design, leading to unnecessary redundant design, contributing to overprotect and waste of investment.

d) There is no oil company owning equipment failure frequency statistics in home, and domestic LOPA analysis equipment failure frequency library draws lessons from foreign companies. Whether it is fully applicable to domestic programs is still controversial, causing analysis will inevitably exist error.

5.2 Recommendations

a) At present domestic government and some oil companies make force to HAZOP analysis to greater danger of craft device, suggest whether LOPA evaluation method can prevail in China. It can improve the safety of design more precise under the condition of the mature technology of forced LOPA assessment.

b) Suggested that the domestic government agencies dominant the statistics of refinery equipment failure frequency. Of course, each big petroleum companies also need to change thinking, actively cooperate with the statistical work. Only does the safe foundation work fine, can we fundamentally improve the petroleum chemical engineering design and security level of operation, benefit the community.

References

[1] Zhou Rongyi, Li Shilin, etc. Study on Application of LOPA in HAZOP[J]. China Safety Science Journal, 2010,20(7):76-81

[2] China National Petroleum Corporation Enterprise Standard. Hazard and Operability Analysis Technology Guide[S].Q/SY1364-2011

[3] Wu Chongguang, etc. The Application Guide of HAZOP and Operability Analysis (HAZOP)[M]. China Petrochemical Press,2012

[4] Bai Yongzhong, Dang Wuyi, etc. Protective Layer Analysis-simplify the Process of Risk Assessment[M]. China Petrochemical Press,2010

[5] Li Qi, Bai Yongzhong, etc. Study on Coating Technology Based on the HAZOP Analysis[J]. Health Safety and Environment,2010,10(9):34-37

398136 Field Application of Risk-Based Approach for Fatality Prevention

Abstract:

Industry data has shown that fatality rate does not necessarily correlate with traditional safety metrics, such as incident rates or injury frequencies. Even organizations with world class safety performance can exhibit fatality trend that is not supported by their otherwise good safety records. This presentation proposes that the application of risk-based approach to daily tasks and conditions can help prevent fatalities. The idea hinges on equipping personnel with the necessary tools to recognize situations with high potential for serious injuries or fatalities, and ensuring that all protection mechanisms are in place and will function as intended when required.

402433 PROBLEMS AND RESOLUTIONS OF SIS ENGINEERING APPLICATION

Abstract:

This article systematically introduced the basic conceptions of safety instrumented system (SIS), safety instrumented function (SIF), safety instrumented levels(SIL) as well as SIS specific methods in the practical application, pointed out problems and resolutions of SIS in domestic engineering application, thereby it can help operators to analyze, judge and solve SIS problems in production and design.

396941 Developing Process Engineers into Effective HAZOP/LOPA Facilitators

Abstract:

There are many factors driving the increased application of HAZOP/LOPA, especially for facilities under design (reference "Assimilating Design Formulation and Design Review into a HAZOP," GCPS 2012), such as:
- Evolving regulatory requirements
- Management's prioritization of safety
- Management's recognition that HAZOP/LOPA can move the design project forward and drive resolution of key design issues

With this increased application, there is a growing need for effective HAZOP/LOPA facilitators and very organized approaches to implementation of HAZOP/LOPA, as well as organizing for effective follow-through as the design matures; however, having the individuals with the correct skills, and retaining them can have many associated challenges:
- Recognizing talent
- Training
- Supporting HAZOP/LOPA facilitation being embraced as a desirable career path
- Providing a work-balance that encourages retention

Many oil & gas companies have worked hard to develop an in-house team, only to find that individual skills may have been poorly matched with the need, or individuals are anxious to be re-assigned to other activities. Those companies that have moved towards the use of consultants for facilitation of HAZOP/LOPA may have encountered related problems. Whether the HAZOP/LOPA Facilitator Team is from a consultancy or from the Owner/Operator, similar challenges exist of finding the individuals with the correct skills, and finding ways to keep them focused on your project/facility needs.

The objective of this paper is to describe the challenge for Management and to provide key tips for things to look for in an employee or consultant:
- Interpersonal skills
- Critical thinking
- "Thinking on your feet"
- Technical prowess
- Problem-solving
- Attention-to-detail
- Application of best HAZOP/LOPA practices

In addition, for the Owner/Operator developing a team of in-house support for HAZOP/LOPA Facilitation, this paper will help Management understand what individuals with these skill sets are looking for and provide key tips for:
- Retention
- Training
- Professional development

396522 Investigation of Self-Ignition Risks of Iron Sulfides in the Sour Crude Oil Storage

Abstract:

The sour crude oil and its evaporation react with corrosion and bio-corrosion products of the storage equipment to form iron sulfides. The pyrophoric nature of iron sulfides leads them to ignite spontaneously and ignite oil and flammable substances. Self-ignition of iron sulfides can cause serious problems during storage of the sour crude oil. To investigate the cause of the thermal reaction and the thermal hazard of self-ignition, different kinds of iron sulfides were heated at different heating rates and in different atmosphere. The reaction heat, onset and peak temperatures, other thermodynamics and kinetics characteristics of iron sulfides in different conditions are analyzed. It is demonstrated that the presence of oxygen atmosphere seems to be the most important factor to induce the self-ignition. Moreover, the results showed that the smaller the diameter of iron sulfide is, the lower the onset temperature of self-ignition is. The injection or supplying with nitrogen into the oil storage equipment should be a feasible approach to prevent the self-ignition.

397764 Optimal Sectioning of Hydrocarbon Transport Pipeline By Volume Minimization and Environmental Vulnerability Assessment

Abstract:

Pipelines are usually used to transport liquid hydrocarbons due to the fact that they are considered one of the safest ways to achieve this task. Despite this fact the US DOT have reported 149 incidents until May 2014, totaling 6,296 net lost bbl. Some of these spills represent serious environmental damage such as the one at Michigan on August 2010, leaving 16 miles of Kalamazoo river polluted; and the one that took place at North Dakota on September 2013 affecting 30,000 m^2 of soil.

In countries characterized by a reach biodiversity, not just locally but globally important ecosystems can be lost or seriously damaged. The spill at Monagas Venezuela on February 2012 affecting more than 200 species, 4 of them endemic is a clear example.

In order to minimize the consequences and thus the risk, pipelines have shutdown systems involving sectionalizing valves for mitigation; making the number and location of equipment imperative in the design and management of the pipeline; especially for areas where topography characteristics are a main variables due to mountain conditions and ecological importance.

To do so some researches have proposed optimizations based on the minimization of spilled volume or the minimization of costs associated to environmental impact by optimizing the number or the location of valves. This poses an alternative way of sectioning, which supports the standards and recommended practices on industry; for example 16 km is the maximum distance between valves proposed by the National Association of Pipeline Safety

Representatives on the United States.

This project has as output the optimal sectioning; both number and location of valves for liquid hydrocarbon transport pipelines, the optimization is divided on two stages: volume minimization and the incorporation of environmental vulnerability of the area. Both stages are programmed on Java generating a computational tool.

Stage one (volume minimization) include the operational variables of the pipeline and the characteristics of the topography, so the volume can be divided on: static and dynamic volume. The calculation associated to the static spilled volume is achieved by the hydrostatic forces on the pipeline associated to topography characteristics coupled with a source model. The selection of the number and location of valves is based on the minimal spilled volume after the location of the valves by a linear optimization algorithm using flow on networks. The algorithm consist on the division of the pipeline by i nodes, each one associated with a maximum spilled volume ($max\ V_i$), the algorithm founds the network composed by the nodes with the minimum sum of maximum volumes.

Stage two (environmental vulnerability) takes into account not just the spilled volume but the vulnerability of the area. To assess the last one, a number of factors divided on hydric, biotic and abiotic are selected and ranked on five levels, each one associated with the level of damage given a specific characteristic of the area describing an specific factor, where the highest level represent the highest damage and thus a highly vulnerable area. To include this vulnerability characterization on the optimization algorithm a weight a_i is assigned to each of the five levels and then included with the spilled volume on the objective function. The optimization algorithm includes for this case the weight on the characterization of the node by its multiplication with the spilled volume.

For both stages the maximum distance between valves is included as a restriction and it is defined by the Class location on CSA Z662 and an additional weight is summed at the end per valve installed by the algorithm.

The results obtained support the decision making process for the location of sectionalizing valves, already stablished on standards recommendations and representing the minimum volume and environmental damage configuration. Fundamental optimization variables are identified such as the pre stablished distance between nodes on the optimization algorithm, as smaller the distance less uncertainties, bigger resolution times and well distributed the valves.

All the distances between nodes represent a minimization on spilled volume, but it also can be observed that configurations with higher number of valves not necessarily implies less spilled volume, indicating the importance of the location of the equipment. An advantage of the methodology and the tool is the use of factors, since its number can vary depending on the available information to characterize all the segments of the pipeline.

Regarding the weight per valve installed, as it increases, the number of valves and its location stays constant, meaning the algorithm at a certain point takes into account the weights associated to the network (composed by the volume and the environmental vulnerability) over the importance of the weight associated to each installed valve.

Finally through the results it can be concluded that the location of the valves is an important variable of decision and that the optimal configuration not only depends on the number of valves. The more detailed the information on the pipeline (less distance between nodes) better resolution, and higher resolution times. Including the cost of the valve on the network implies a reduction near to the 50% of the number of the valves for the higher costs, and reaches a constant value at this point.

Assessment of Reliability and Risk at Dynamic Systems: A Discussion of Social-Technical Factors in the Exploration and Production of Oil at Pre-Salt Layer - Brazil

Monique Nunes Gomes de Abreu
Federal University of Bahia
Salvador - Brazil
mona_ssa@hotmail.com

Salvador Ávila Filho
Federal University of Bahia
Salvador – Brazil
avilasalva@gmail.com

Prepared for Presentation at
American Institute of Chemical Engineers
2015Spring Meeting
11thGlobal Congress on Process Safety
Austin, Texas
April 27-29, 2015

UNPUBLISHED

Assessment of Reliability and Risk at Dynamic Systems: A Discussion of Social-Technical Factors in the Exploration and Production of Oil at Pre-Salt Layer - Brazil

Monique Nunes Gomes de Abreu
Federal University of Bahia
Salvador - Brazil
mona_ssa@hotmail.com

Salvador Ávila Filho
Federal University of Bahia
Salvador – Brazil
avilasalva@gmail.com

Keywords: Dynamic systems; Pre-Salt layer, Risk assessment; Reliability analysis

Abstract

Economically and historically, the world and its economy have been dependent on different commodities. The existence of one of the most used nowadays, oil, precedes the modern times. In its constitution are mainly organic compounds, as hydrocarbons of various molecular weights, accumulated miles and miles beneath the earth. Currently, the extraction of oil is one of the largest and most important activities in the world. This is due to its high use in several applications, such as LPG (Liquefied Petroleum Gas) or cooking gas, gasoline, kerosene and thousands of other plastics products we may find in our daily activities. However, this is a non-renewable resource, so more and more efforts are being made towards finding oil and its extraction from underground. More recently, a large reservoir was discovered, situated at distance from the coast and depth never explored before, as it is referred to, the Brazilian Pre-salt layer. Undifferentiated of any exploration activity, extraction and transportation of oil from seabed offer risks. There are tools available for qualitative and quantitative risk assessment. In order to be effective and successful, the analysis of these risks must include the understanding of dynamic socio-technical factors, since it is necessary to deal with enormous challenges when it comes to extracting the oil. Some challenging issues are: extreme conditions of temperature and pressure at high depths, abrupt climate changes, salt layer drilling technologies, personnel logistics, as in moving staff in and out of the platforms.

In this paper, we initially try to understand the importance to investigate future scenarios in risk and reliability analyzes to make correct decisions about the investment of controls and appropriate assets and technologies to oil exploration. In clear words we show these steps: development of risk analysis; quality tools to risk and reliability; failure mode analysis and reliability; quantitative analysis; and dynamic factors (that include human factors). These stories about risk & reliability become more comfortable the discussion about a case of study entitled the Brazilian pre-salt oil.

1. Introduction

Economically and historically, the world and its economy have been dependent on different commodities. The existence of one of the most used nowadays, oil, precedes the modern times. In its constitution are mainly organic compounds, as hydrocarbons of various molecular weights, accumulated miles and miles beneath the earth.

Currently, the extraction of oil is one of the largest and most important activities in the world. This is due to its high use in several applications, such as LPG (Liquefied Petroleum Gas) or cooking gas, gasoline, kerosene and thousands of other plastics products we may find in our daily activities.

However, this is a non-renewable resource, so more and more efforts are being made towards finding oil and its extraction from underground. More recently, a large reservoir was discovered, situated at distance from the coast and depth never explored before, as it is referred to, the Brazilian Pre-salt layer.

Undifferentiated of any exploration activity, extraction and transportation of oil from seabed offer risks. There are tools available for qualitative and quantitative risk assessment. In order to be effective and successful, the analysis of these risks must include the understanding of dynamic socio-technical factors, since it is necessary to deal with enormous challenges when it comes to extracting the oil. Some challenging issues are: extreme conditions of temperature and pressure at high depths, abrupt climate changes, salt layer drilling technologies, personnel logistics, as in moving staff in and out of the platforms.

2. The development of Risk Analysis

The need for analyzing the risk of operations and its consequences came about through an evolution of thought. Originally, long ago, the one and only factor that guided the production of a good or service was intuition. Products were made by hand, in such way that imperfections and defects were accepted, since they do not prevent the product, from fulfilling their ultimate purpose.

Over the course of time, it was noticed that improvements could be made by adding knowledge to the process. This knowledge provided the creation of design patterns. Thus, it was possible to map the failures and trace projections of expected impacts. However, the association of intuition with knowledge did not, by itself, put an end to/mitigate the consequences of general accident, from that moment on, another variable was included in the matter, risk analysis. With the evidence of this new need, design techniques and risk analysis were developed, highlighting techniques to quantify the risk (Figure 1) based on frequency of failures and severity of impacts, based on expert opinions. Also, the category of risk of each activity was evaluated and estimated so that protections could be provided for the most dangerous operations.

Figure 1. The need for qualitative risk analysis

2.1 The qualitative tools

Tools such as Process Hazard Analysis (PHA), Hazards and Operability Study (HAZOP) and Qualitative Fault Tree are used to evaluate and estimate the impacts of production. However, it was noted that such estimated impacts differ, in some circumstances, for the actual impacts. This information was archived in databases. The gathering of this information enabled the introduction to reliability studies. In other words, reliability means the ability of items or systems of maintaining their expected function, in a given period of time. Progresses in this direction have been made through the development of tools such as Vulnerability Study, Reliability Study, Failure Mode and Effects Analysis (FMEA), Quantitative Fault Tree and Layer of Protection Analysis (LOPA). Regarding that direction, adjustments to the risk analyzes were incorporated, confirming or readjusting frequency and severity of events evaluated in previous qualitative assessments.

2.2 *Failure mode analysis and reliability*

Nowadays, the matter of reliability is widely discussed and mentioned. The economical global situation implies, more and more, quick changes and effective adaptations in order for a company to maintain a strong competitive status inside its market. The maintainability of this competitiveness, by nature, brings obvious and rigorous consequences: the need to generate more revenue by using less and less resources. Regarding that train of thought, enters the concept and use of reliability, which, according to Moubray [1], is the ability of an item or system to maintain its original purpose intact, for a given period of time, under given circumstances, improving process safety, durability and productivity. Raposo [2] says that the definition of reliability has a close relationship with the security in an enterprise, to the extent that a poor performance or a non-performance of the required function results in consequences that can cause harms, loss of lives, or violation of environmental laws. Also, situations in which people and machinery do not perform satisfactorily tend to represent extra costs that may, or may not, fit the project's budget, which jeopardizes the company's image and performance before the competitors and clients that, evidently, are becoming more and more exigent.

2.3 *Quantitative analysis (Figure 2)*

On the other hand, despite the development in the field of quantitative risk assessment, it has been noted that the estimated impacts differ from the real ones, which drove the studies to the conclusion that the problem was in making the assumption that all economic systems are controlled and constant, but, they are definitely not. According to Rasmussen [3], the further understanding of how risk assessment works involves the comprehension of the dynamic-environment complexity, in such a way that human, organizational and environmental factors are considered in the analysis.

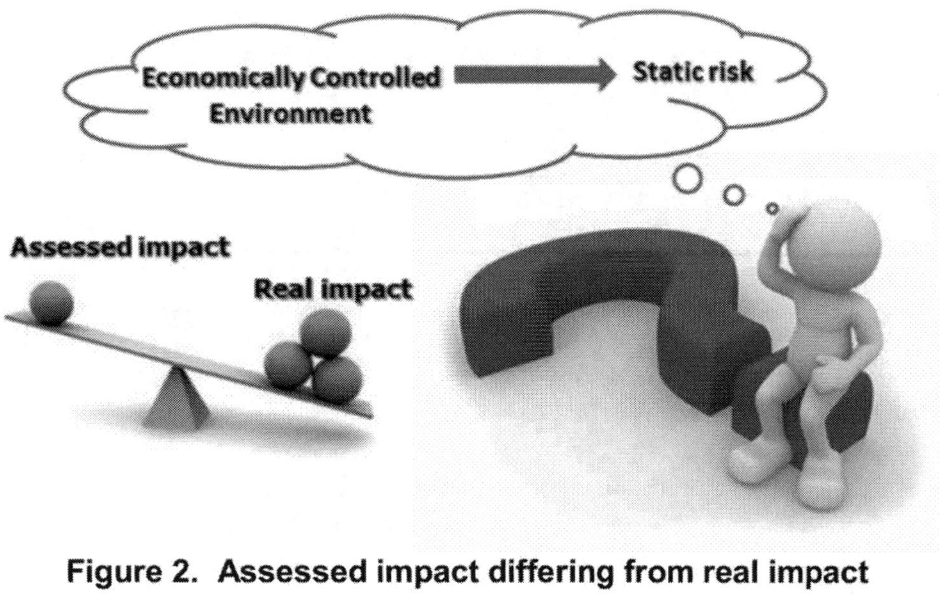

Figure 2. Assessed impact differing from real impact

2.4 *The dynamic factors*

Dynamic factors (Figure 3) represent a large influence on reliability analysis, which means they also impact on the probability of a failure event. Conforming to Rasmussen [3], it is extremely important to take into account the dynamic risk, in numerous occasions. In that previously mentioned unstable scenario is where almost - or absolute all -operations take place, and, by considering it, one is able to deal more efficiently with uncertainties and undesirable situations.

Climate changes and human factors are common examples of the dynamism faced by companies, on their operational activities. Barroso [4] says that the human behavior and his perception of the risks he is exposed must not be excluded of the operational risk management, as the analysis of human factors in the risk assessment helps preventing errors that might lead to process accidents. The more one comprehends how these elements affect the system, the more it is possible to minimize the difference between the estimated and real impacts. Therefore, this is a matter of unquestionable importance for companies that wish to improve their performance.

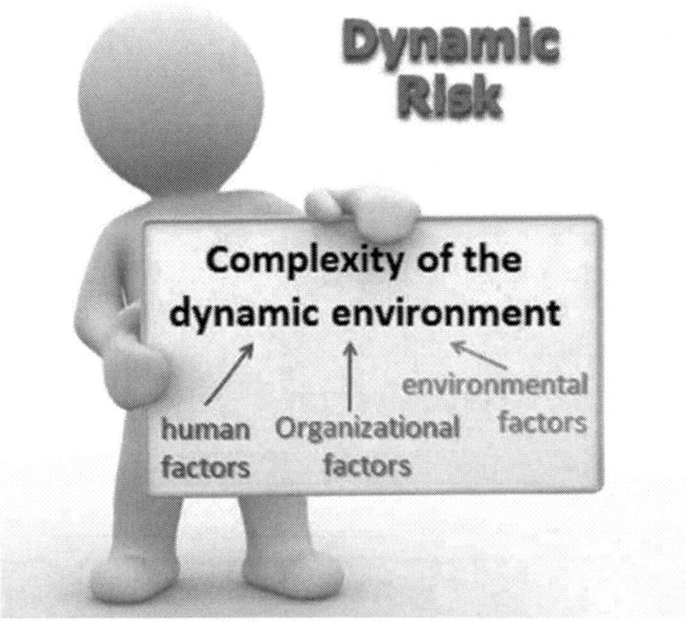

Figure 3. Dynamic factors

According to Ávila Et Al [5], there are tools for assessment of risk, including dynamic factors, such as the Social HAZOP (SH). The SH understands all the agents inside the failure, through a systemic view, including equipment, instruments, and people, who develop, operate, modify, and improve organizations. Also, Ávila [6], shows a tool named LODA (Dependent Layers of Operation Decision Assessment) which provides a dynamic means of analyzing the behavior of operators in terms of risk, seeking responses from the inventory of knowledge and based on social relations at teamwork to examine whether the knowledge inventory in the team is sufficient to treat an emergency.

3. The case study, the Brazilian Pre-Salt Oil

During this study case, it is pretended to discuss the dynamic aspects of reliability and risk associated to the petroleum extraction and exploration, in Brazilian Pre-salt. This region is referred to as the biggest oil reservoir discovered in the last ten years, and it is located three hundred kilometers from the coast, reaching an eight-kilometer depth, in the middle of the ocean. The exploitation of this oil is of a great challenge to those who make plans and efforts to do so, because it involves extreme circumstances never dealt with before, not to mention that it demands the newest and state-of-the-art technologies. Renó [7] points that the drilling and exploration of wells in the pre-salt-layer is a process that requires a critical analysis of various factors due to the geological complex profile of the region, high levels of temperature, pressure, presence of corrosive gases and difficulty of access due to great deeps. In addition, there is the fact that this type of operation is relatively recent, so methods and standards used for drilling wells in conventional locations need to be readjusted or even fully developed to attend the conditions of the pre-salt-layer.

Firstly, when it comes to offshore exploration, there are certain known, but not less complicated, difficulties, such as: the choice of what material to use in ducts that transport the oil up to the platform, in order to hold against severe conditions; the oil upstream control system; the complete logistics of people and necessities in and out of the platform and, evidently, personnel training to avoid accidents.

Advanced and modern knowledge are needed for the risk analysis of the enterprise, due to a more stressful environment and more demanding structures, regarding seas operations. It is important to discuss the environmental conditions in the salt layer after its penetration to assess the possibility of a plastic deformation due to higher temperatures and pressure levels.

Another important discussion is the one concerning oil-leakage containment technologies in high seas and the assessment of the consequences to avoid another disaster event, like Deepwater Horizon (BP). Macondo Prospect oil field, where the drilling for BP was occurring, is located about 60 km southeast of the Louisiana coast and the accidental leakage took place in a 1.5 km depth. According to the National Commission on the BP Deepwater Horizon Oil Spill and Offshore Drilling [8], the Deepwater Horizon blowout produced the largest accidental marine oil spill in U.S. history, an acute human and environmental tragedy. The explosion killed 11 workers and injured several others and efforts to stop the oil spilling took over 80 days. The pre-salt-layer exploration deals with distance from coast 5 times greater than BP's and depths over 5 times greater. The emergency planning is a key point in this enterprise, to avoid disaster events even greater than BP's. In order to treat and overcome emergency situations with that kind of distance from the coast, it is necessary to establish the right transportation mode and more suitable local rescue facilities to mitigate the effects of possible future events. Specialists say that it takes one day and a half for the leakage-containment support team to get to the ships.

The necessary training and emergency simulation, in high seas, are highly difficult. They demand high simulation technologies and approximation to catastrophic situations (with the discussion of evasion of installation). The exploration team faces the challenge of working with high temperature and pressure levels variation in the depths of the ocean, which certainly impacts the environment.

This kind of process deals with a larger amount of flammable products, which, definitely, increases the risk of explosion, especially because it comes to ships and platforms. This risk can be amplified by socio-technical dynamic factors, such as: vastly multicultural staff with a great possibility of communication problems; onboard workers performing activities that cause extra stress creating an environment susceptible to human failure or management errors. According to Ávila [9], the environment of the critical task, must take into account, technical, social and human aspects, so that the social-human risk analyzed helps reaching intangible results such as improved communication, and achieves tangible results as the reduction of accidents and failures.

In this way, the sustainability of organizations in such complex context of production can be achieved taking on account an integrated reliability, which analyzes dynamic aspects. Ávila [10] says that integrated reliability (Figure 4) is a system where management and organizational environments are proposed to become people and equipment more reliable. In this way, technical and social aspects are analyzed considering the interrelation of process, equipment, operational and human reliability, which involves automation, man-machine integration, development of tasks, process control as well as knowledge and skills.

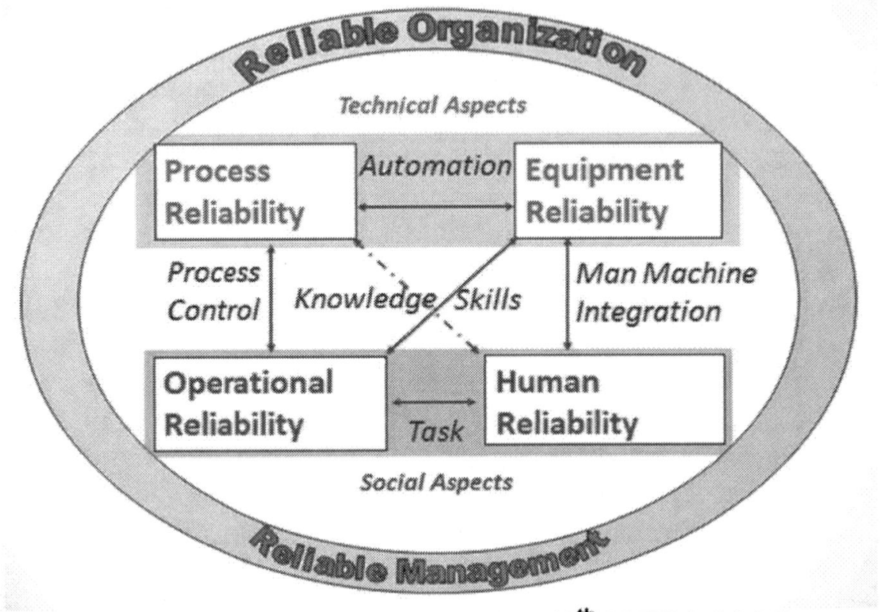

Figure 4.Reliable organization (5th ICEFA 2014)

As presented in Figure 5, the risk assessment, which analyzes dynamic aspects of the exploration and production of oil at pre-salt layer, incorporates the integrated reliability study and must be executed in a cyclic way. As the exploration of the pre-salt-layer is a new enterprise, there is no database available, so the risk assessment begins with the study of Physical Failure Phenomenology (PFP) and Social Failure Phenomenology (SFP). PFP aims to understand technical factors that might lead to interruption of operation due to structural failure, like corrosion, turbulent flow, two-phase flow, fatigue of materials, fracture by low temperature etc. SFP aims to understand social factors that might lead to interruption of operation due to low efficiency in task, like social failure due to paternalism, excessive flexibility, egocentrism and communication problems aggravated by multiculturalism which leads to social failures, as omission and skipping steps of tasks. The analysis of the interruption of operations provides social and technical failure data that feeds the study of integrated reliability, which is essential to the assessment of risk itself. The results of risk assessment re-feed the study of PFP and SFP by pointing the most critical points that need to be studied more in depth. This cyclic way of analyzing and managing the risk is crucial to the success of this enterprise that demands the newest and state-of-the-art technologies since it is involved in extreme environmental conditions never dealt with before.

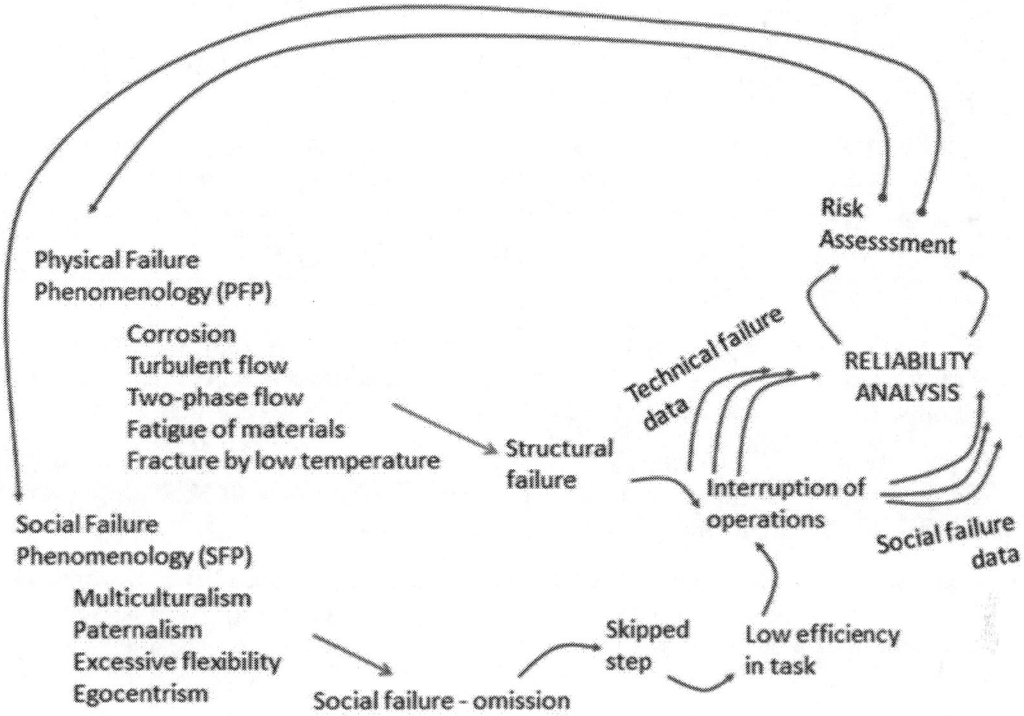

Figure 5. Risk analysis cycle for pre-salt layer exploration

4. Conclusion

A discussion regarding the development of the risk analysis was presented addressing the creation of qualitative tools of assessment, the classic idea of reliability, and the evolution towards a quantitative analysis that. It was concluded that this evolution, in one hand, improved the risk analysis, but, in the other hand, still made estimated impacts differ from the real ones due to the assumption that all economic systems are controlled and constant.
In this way, it was introduced the importance of risk assessment works which involves the comprehension of the dynamic-environment complexity, in such a way that human, organizational and environmental factors are considered in the analysis.
Dynamic aspects of reliability and risk associated to the petroleum extraction and exploration, in Brazilian Pre-salt were discussed. It was shown that the sustainability of organizations in such complex context of production can be achieved taking on account an integrated reliability, which analyzes dynamic factors. Also it was proposed a cyclic way to analyze the dynamic risk in the exploration of the pre-salt-layer.

5. References

[1] J. Moubray, Manutenção Centrada em Confiabilidade (Reliability-centred Maintenance), Aladon Ltd, Lutterworth, 2000.

[2] J.L.O. Raposo, Manutenção centrada em confiabilidade aplicada a sistemas elétricos: Uma proposta para o uso de análise de risco no diagrama de decisão, Electrical engineering master's thesis, Federal University of Bahia, Salvador, Bahia, 2004.

[3] J. Rasmussen, Risk management in a dynamic society: a modelling problem, Safety Science, Vol. 27, No.2/3 (1997), pp.183-213.

[4] M.P. Barroso, A influência dos fatores humanos nas técnicas de análise de risco APP e APR: estudo de caso de uma plataforma de perfuração de petróleo no Nordeste, Industrial engineering master's thesis, Federal University of Bahia, Salvador, Bahia 2013.

[5] S.F. Ávila, F.L.P. Pessoa and J.C.S Andrade, Social HAZOP at an Oil Refinery, Process Safety Progress, Vol. 32, Issue 1, (2013), pp.17-21.

[6] S.F. Ávila, Dependent Layer of Operation decision analyzes (LODA) to calculate Human Factor, a simulated case with PLG event, 7th Global Congress on Process Safety, Chicago, Illinois, 2011.

[7] J.A.S. Renó, A. Nascimento, P.A. Duarte and J.L. Gonçalves, Avaliação técnica dos fatores críticos para otimização de perfuração nas camadas do Pré-Sal, Rio Oil & Gas 2014 Expo and Conference, Rio de Janeiro, RJ, 2014.

[8] National Commission on the BP Deepwater Horizon Oil Spill and Offshore Drilling, Deepwater: The Gulf Oil Disaster and the Future of Offshore Drilling, Report to the President, 2011.

[9] S.F. Ávila, Review of Risk Analysis and Accident on the Routine Operations in the Oil Industry, 5th CCPS Latin American Conference on Process Safety, Cartagena de Indias, 2013.

[10] S.F. Avila and D. Santos, Integrated reliability: failure assessment to process, operation, equipment and human integration (POEHint), a case of propene pump operation, Sixth International Conference on Engineering Failure Analysis (5th ICEFA), Lisbon, 2014.

398600 Experiment Study on Acoustic Emission Detection of Internal Leakage of Gas Ball Valves

Abstract:

Valve internal leakage experiment using acoustic emission (AE) method was conducted to build the relationship between AE signals and various detect location, pressure, leakage interval and valve size. Empirical formula for ball valve internal leakage was obtained based on AE amplitude. First, the (AE)ASL value and inner leakage rate of different pressure, valve size, failure mode and the damage degree were obtained by develop the small opening and sealing damaged failure models, which are two common failure modes in ball valve internal leakage. Stepwise regression and partial least square method were further applied. Results show that fitting formula of the average prediction error is less than 10% in certain conditions (valve fault type and pressure conditions), and is promising in rapidly predict valve internal leakage rate.

397142 Why Do We Still Have Blast Walls on Offshore Platforms?

Abstract:

The objective of this paper is to step back and think why we still have stand-alone blast walls in the design of offshore platforms. Blast walls are reactive mitigation protection required if there is an explosion over pressure concern that can jeopardize the escape and evacuation of personnel off the platform. It is interesting to see that it has become common practice to install this mitigating protective system during conceptual phases of platform development rather than putting effort in reducing the potential and magnitude of potential hydrocarbon accidental releases as per international guidelines such as the UKOOA/UK HSE "Fire and explosion guidance – Part 1: Avoidance and mitigation of explosions" - Issue 1 (2003) Table 2.1 or API RP 2FB – "Recommended Practice for the Design of Offshore Facilities Against Fire and Blast Loading" and ISO 13702 –" Petroleum and natural gas industries – Control and mitigation of fires and explosion on offshore production installations – requirements and guidelines". This paper will be based on the authors experience and lesson learned during the design of offshore installations utilizing Computation Fluid Dynamics (CFD) explosion modeling, at the pre-conceptual stages of a platform development project, as a design tool, in the decision for the final layout of the facility. The concept of an inherent safe approach to a design will be explained in parallel to how this approach affect the inputs to the final quantitative risk analysis study that is performed during most of the current offshore facilities design. This process or approach is utilized as a means of demonstrating "As Low As Reasonably Practicable – ALARP) in some regulatory regime. The intent of this paper is to have practitioners step-back and review how blast walls are justified during a conceptual design phase of an installation development and realize that a focus on a better understanding of this explosion overpressure hazard mechanism, early in a project life cycle will provide a more cost effective & safer design and operations of a platform. It must be reminded that the key is in minimizing the loss of containment to improve the production availability of a producing facility and thus the use of an inherent safe design process.

Dynamic simulation of discharge process safety and optimization for multi-stage ground flare

ID# 400905

Anfeng Yu[1], Peng Wang[1], Jiedong Zhang[1], Zhongshu Han, Yongzhong Bai, Wenyi Dang*
SINOPEC Safety Engineering Institute, Yan'an 3rd Road, Shinan District, Qingdao, Shandong
[1] These authors contributed equally to this work and should be considered co-first authors
* Corresponding author. Email: Dangwy.qday@sinopec.com

Keywords: Ground flare; multi-stage; dynamic simulation; pressure drop.

Abstract

Multi-stage discharge system is a remarkable feature for ground flare, through which the stages could open or close according to the discharge amount and the high efficiency could be realized. If the multi-stage discharge parameters could not be fixed properly, many problems may occur, such as frequency open/close action of stage valves, inefficiency of the combustors, and the high backpressure of related facility. In order to solve this problem, the dynamic simulation model was established based on a real ground flare for a lubricating oil hydrogenation unit. The pressure drop through the whole pipe network was analyzed. The main pipeline pressure, treatment amount of each stage/combustor was calculated. Meanwhile, the action of stage valves were tracked. Results shows that the multi-stage discharge system is complicated and could be impacted by stage pressure setting. The dynamic model could help to the proper design of such kind of system.

1. Introduction

Ground flares have attracted more and more attention for petrochemical plants and oil & gas fields in China, especially for the intense land resource area [1]. Compared with the elevated flares, ground flares have the benefits of high efficiency, flame invisibility and low heat radiation, which are important for the environment protection and community relationships [2].

Multi-stage discharge system is a remarkable feature for ground flare, through which the stages could open or close according to the discharge amount and the high efficiency could be realized. Therefore, it's an important issue to determine the stage pressure, number of stages and the treatment capacity of each stage. If this issue could not be solved properly, many problems may occur, such as frequency open/close action of stage valves, inefficiency of the combustors, and the high backpressure of related facility [3]. Due to the complexity of discharge process and the structure of discharge pipelines including numerous combustors, it's difficult to determine the proper parameters of discharge process with traditional hydraulic calculation.

In order to solve this problem, the dynamic simulation model was established including the main pipeline, water seal tank, arrester, and eight stages with the combustors based on a real ground flare for a lubricating oil hydrogenation unit. The pressure drop through the whole pipe network was analyzed. The main pipeline pressure, treatment amount of each stage/combustor was calculated, and the action of stage valves were tracked. Based on the model, the effect of key parameters were discussed, and the optimization were carried out.

2. Modelling and Methodology

2.1 Discharge parameters

2.1.1 Discharge gas and curve

The discharge gas composition is shown in Table 1, and the discharge curve is shown in Figure 1. It should be noted that one facility may have several discharge scenario with different discharge curves which should be considered thoroughly during the design of multi-stage process. In this research, the typical discharge process of the aiming facility was chosen.

Table 1. Discharge gas composition of the ground flare

Component	mole fractions
Methane	0.025103
Ethane	0.030703
Propane	0.070707
I-Butane	0.047605
N-Butane	0.035503
N-Pentane	0.016902
Hydrogen	0.651266
Nitrogen	0.122212

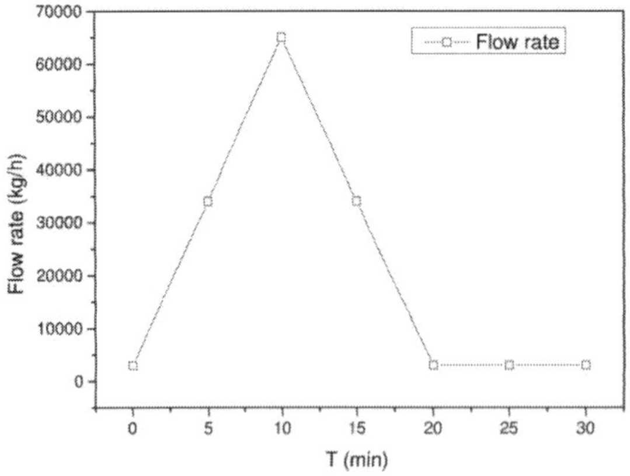

Figure 1. Discharge curve of the facility

2.1.2 Discharge pipelines

Pressure drop is the key parameter during the simulation, thus it's very important to construct the simulation model accurately following the real structure. The pipeline between the discharge point to the ground flare inlet is shown in Table 2, and each stage's detail parameters are shown in Table 3.

Table 2. Pipeline parameters between the discharge point to the ground flare inlet

Position	Diameter (m)	Length(m)	Elbow types	Number of Elbows
Discharge point to KO tank	0.4	100	--	--
KO tank	2.4	7.0	Standard Elbow 90 sudden expansion/contraction	2 1/2 0.2
KO tank to water-seal tank	0.5	26.2	Standard Elbow 90	4
Water-seal tank	2.4	7.0	Standard Elbow 90	2
Water-seal tank to flare inlet pipeline	0.9	31.64	Standard Elbow 90	5

Table 3. Each stage's detail parameters

Stage	Diameter (m)	Length (m)	Valve types	Valve numbers	Combustor numbers
1st stage	0.20	6.605	Arrestor	1	3
			Ball valve	1	
2nd stage	0.20	6.605	Ball valve	2	3
			Butterfly valve	1	
3rd and 4th stage	0.25	6.605	Ball valve	2	6
			Butterfly valve	1	
5th and 6th stage	0.35	6.605	Ball valve	2	12
			Butterfly valve	1	
7th and 8th stage	0.40	6.605	Ball valve	2	23
			Butterfly valve	1	

2.2 Model construction

Through the whole system, the pressure loss is caused when the discharge gas flow through the pipeline, elbows, valves, inlet and outlet. It's simple to calculate the pressure loss induced by the pipeline with elbows and valves. Nevertheless, the porous structure of combustors makes the combustor's outlet pressure drop calculation complex since each combustor has 127 orifices and each stage has more than one combustor. Meanwhile, the dynamic simulation also requires the model concise and easy to converge. Considering this situation, the equivalent curve of pressure drop and flow rate is utilized successively for the single combustor and each stage [4].

2.2.1 Pressure loss

First, based on the relationship between orifice size and Cv number, the Cv number of 8 mm's orifices in the combustors is set to 2.6049. Then the 7 branches of one combustor is simulated separately using ChemCAD to get the equivalent Cv number. After that, one combustor is simplified to one item with Cv value 329.6 instead of 7 branches with 127 orifices. The equivalent curve is shown in Figure 2.

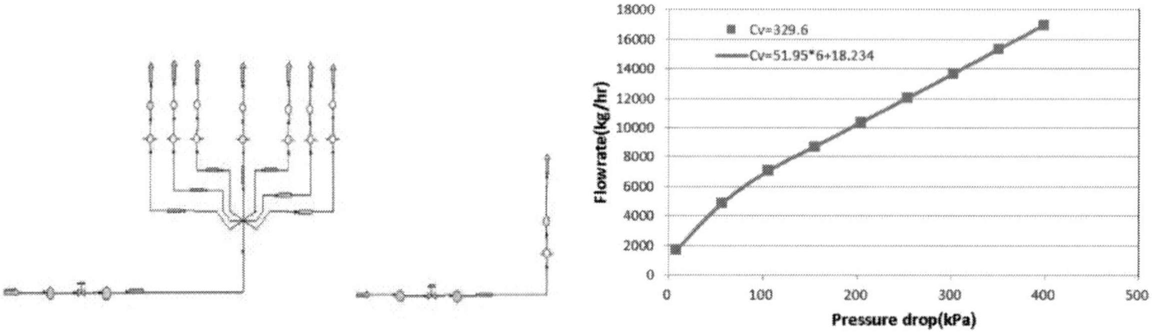

Figure 2. Equivalent curve of one combustor compared with 7 branches with 127 orifices

Furthermore, each stage is also replaced with one item based on the equivalent Cv in ChemCAD in order to simplify the calculation. As illustrated in Figure 3, the 1st and 2nd stage's three combustors was predigested to one item.

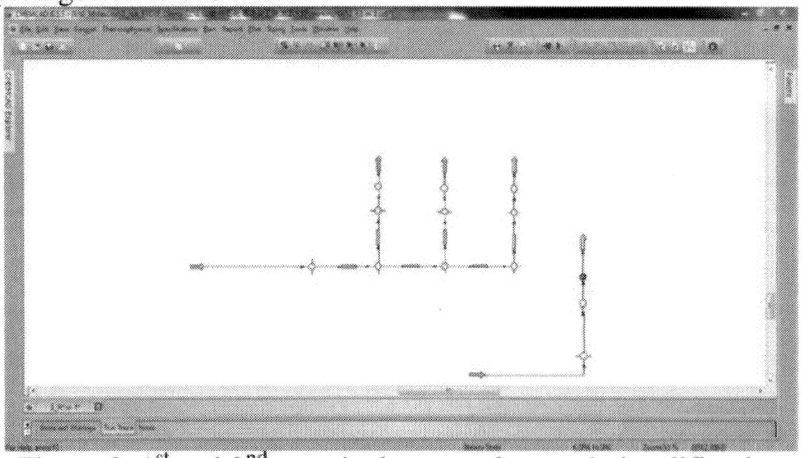
Figure 3. 1st and 2nd stage's three combustors' simplification

In the same way, each stage's equivalent Cv was calculated as shown in Table 4.

Table 4. Each stage's equivalent Cv

Stage	Combustors' number	Equivalent Cv
1st and 2nd	3	813
3rd and 4th	6	1620

5th and 6th	12	3320
7th and 8th	23	5670

In addition, the 1st stage has a flame arrester to avoid back fire which should also be taken into consideration. With the relationship between air flow rates and pressure drop, the equation is fit as shown in Equa.1.

$$y=3.79412E\text{-}12x^2+8.24319E\text{-}07x$$

In this equation, x is flow rate(ft3/hr) and y is pressure drop(psi), the unit is consistent with the ChemCAD's internal setting.

2.2.2 Multi-stage control

Multi-stage control strategy is realized in ChemCAD with User Added Modules (UAM) and programmed with Visual C++ language [5]. The front control panel is constructed with Symbol Builder and Screen Builder module in ChemCAD as shown in Figure 4. The open/close trigger pressure could be changed conveniently through this panel instead of through the programming.

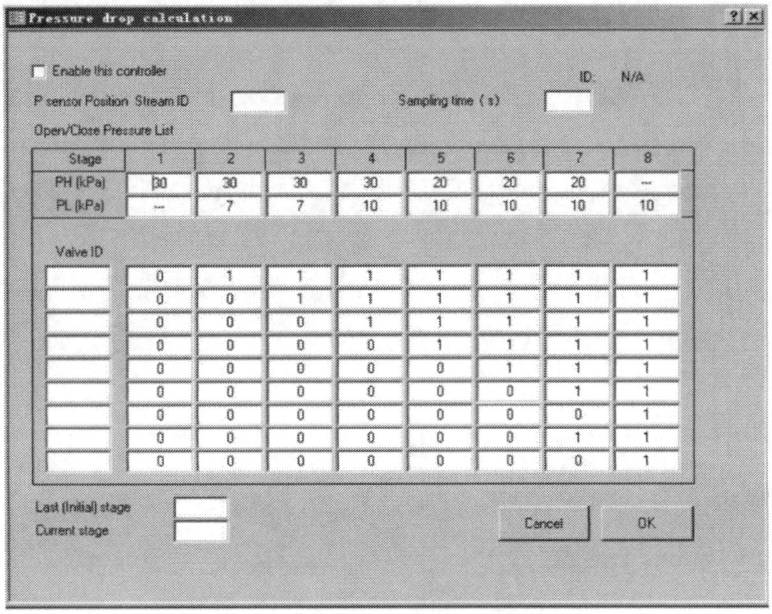

Figure 4. Multi-stage control pane

2.2.3 The mainstay model

Based on the above pre-work, the mainstay model could be concisely constructed as shown in Figure 5. The tank was used to simulate the pressure drop through the main discharge pipeline and the buffering effect due to the volume capacity of the whole discharge system.

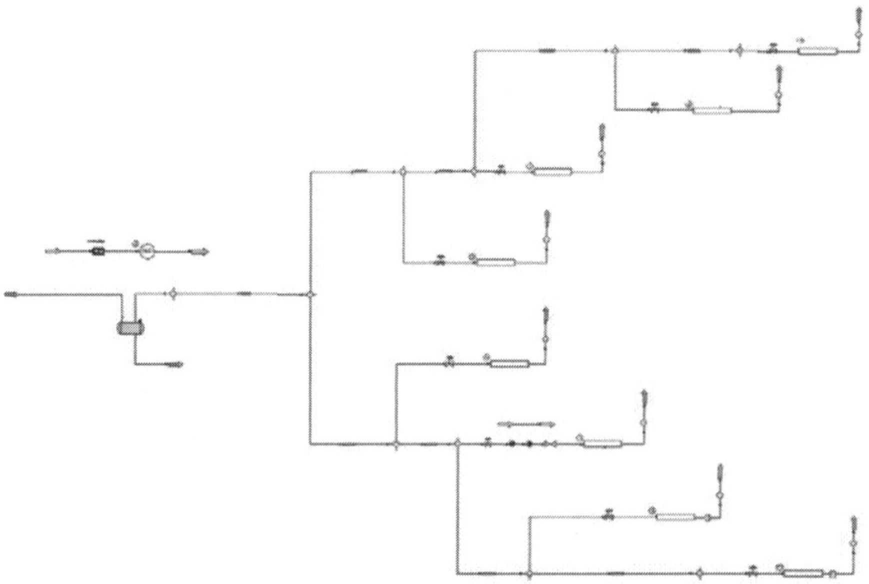

Figure 5. The mainstay model of the ground flare with 8 stages

2.3 Key parameters discussed

During the simulation, the key parameters were changed including the stage open/shutdown trigger pressure and combustors' number of each stage as shown in Table 5 to Table 6.

Table 5. Changing of valve open pressure

Stage	1	2	3	4	5	6	7	8
Valve open pressure Original(kPa)	-	30.0	30	30	20	20	20	20
Valve open pressure Control 1(kPa)	-	25	25	25	20	20	20	20
Valve open pressure Control 2(kPa)	-	35	35	35	20	20	20	20

Table 6. Changing of valve shutdown pressure

Stage	1	2	3	4	5	6	7	8
Valve open pressure(kPa)	-	30	30	30	20	20	20	20
Valve shutdown pressure Original (kPa)	-	7	7	10	10	10	10	10
Valve shutdown pressure Control 1 (kPa)		12	12	15	15	15	15	15
Valve shutdown pressure Control 2 (kPa)		10	10	12	12	12	12	12

Table 7. Changing of combustor's number of each stage

Stage	1	2	3	4	5	6	7	8
Combustors' number Original	3	3	6	6	12	12	23	23
Combustors' number Control	3	5	8	8	10	10	21	23

3. Results and Discussion

3.1 The effect of stage-open pressure on main pipeline pressure

The main function of ground flare is to decrease the main pipe pressure and avoid high back pressure for discharge point. Therefore, the main pipe's pressure should be tracked. As shown in Figure 6. Due to the large diameter of main pipe and the vessel, the pressure of the main pipe is almost the same with ground flare inlet, indicating the pressure drop in main pipe is very low. When the open pressure changed from 25.0 kPa to 35.0 kPa, the main pipe pressure also decreased which was easy to understand. However, the maximum pressure in the main pipe during the discharge process are always larger than the setting points. For example, when the open pressure was set to be 25.0 kPa shown in Figure 6(b), the maximum pressure of main pipe could reach up to 29.9 kPa. This phenomena could be explained by the steep discharge curve, delay of control system as well as the valve action duration. So when the stage-open pressure was determined, it should not be directly set to the maximum allowance back pressure of the discharge point.

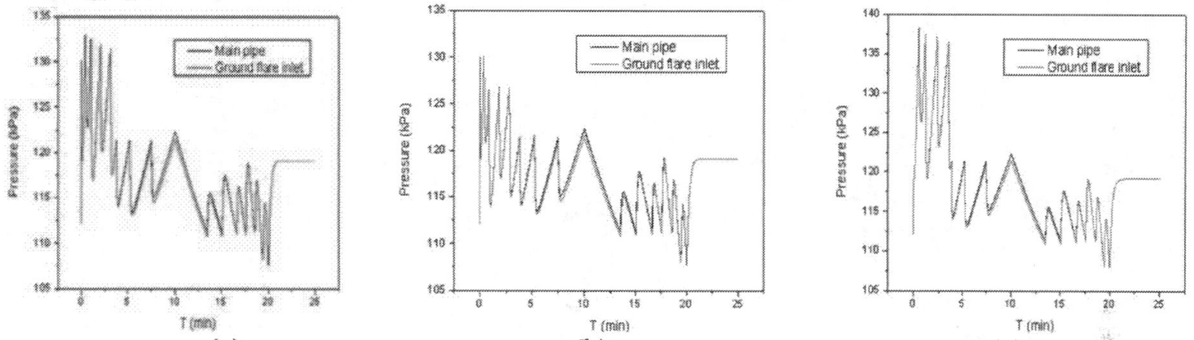

(a) (b) (c)

Figure 6. Changing of pressure of main pipe and ground flare inlet with time: (a) Original with the open pressure of first 4 stage 30 kPa; (b) Control 1with the open pressure of first 4 stage 25 kPa; (c) Control 2 with the open pressure of first 4 stage 35 kPa.

3.2 The effect of stage-open pressure on each stage's combustor treatment

The open pressure could also influent each stage's combustor treatment as shown in Figure 7.As the open pressure decreases, the treatment load of the first 4 stages' single combustor also decreases and is more gradual which helps the combustor keep a high and stable combustion efficiency. Furthermore, the treatment load of the first 4 stages is always larger than the other stages, so it's better to distribute the first 4 stages' combustors to ensure a good air supply [6].

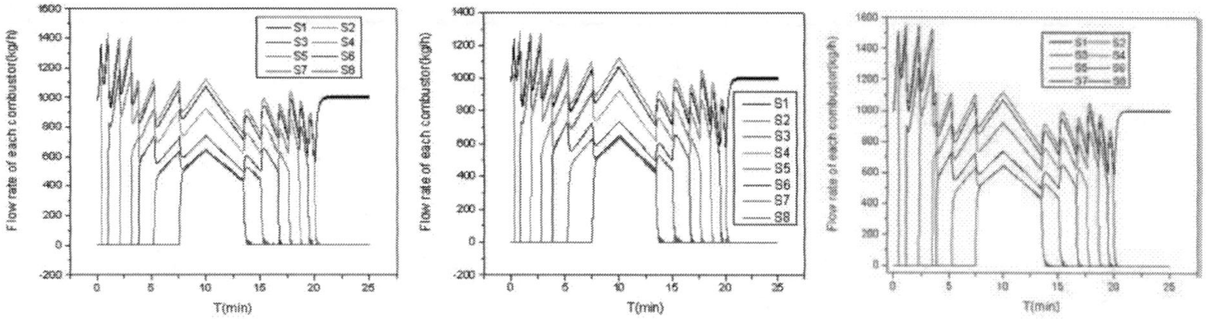

(a) (b) (c)

Figure 7. Each stage's combustor treatment load under different open pressure: (a) Original with the open pressure of first 4 stage 30 kPa; (b) Control 1with the open pressure of first 4 stage 25 kPa; (c) Control 2 with the open pressure of first 4 stage 35 kPa.

3.3 The effect of stage-shutdown pressure on discharge process

The shutdown pressure also plays a very important role during the multi-stage discharge process. As shown in Figure 8 when the shutdown pressure significantly increased, the stage valve may open and shut down repeatedly due to the pressure fluctuation after the acting of stage valve.

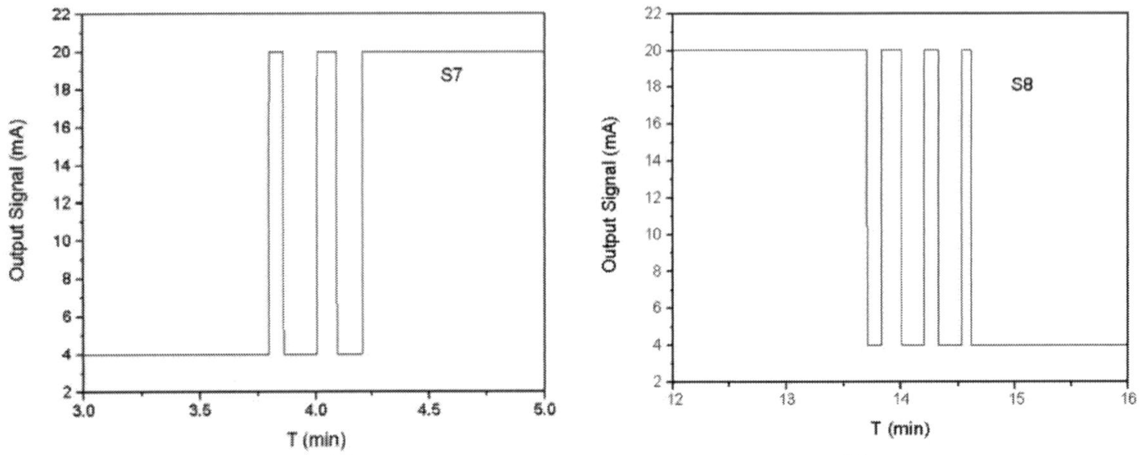

(a) (b)

Figure 8. Valve action of Stage 7 and Stage 8 in Control 1 during the discharge process: (a) Valve of stage 7; (b) Valve of stage 8.

Compare with Control 1, the shutdown pressure in Control 2 only has a slight increase. The simulation results show that the repeated open/shutdown action could be avoided. Meanwhile, the first stage treatment load show a more gradual tendency due to the timely shutdown of senior stage during the later discharge process as shown in Figure 9.

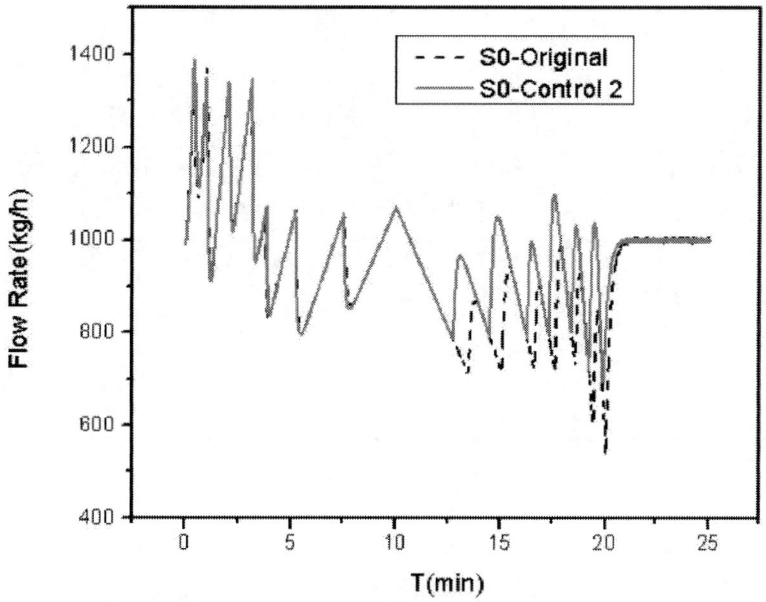

Figure 9. Comparing of single combustor's treatment load after the slight increase of shutdown pressure.

4 Conclusion

Through the equivalent Cv number and multi-stage control stretagy programming, the ground flare model was successfully constructed with which the key parameters effect on multi-stage process were investigated. The main pipe pressure may exceed the setting open pressure during the discharge process, so the open trigger pressure should be calculated with the dynamic model. Proper decreasing of open pressure could mitigate the first few stages' treatment load and contribute to the high and stable combustion efficiency. Improper shutdown pressure setting may cause the repeatedly open/shutdown action of stage valve, while the proper increasing of shutdown pressure may help to the gradual treatment load for the first few stage.

This research show that the multi-stage discharge system is complicated and could be impacted by stage pressure setting and combustors' number of each stage. The dynamic model could help to the proper design of such kind of system.

5. References

[1] API 521-2007. Petroleum and natural gas industries—Pressure-relieving and depressuring. American Petroleum Institute, Washington, DC.

[2] API 537-2008. Flare Details for General Refinery and Petrochemical Service. American Petroleum Institute, Washington, DC.

[3] H. G. Fisher, A. R. Muller, et al, *Emergency Relief System Design Using DIERS Technology: The Design Institute for Emergency Relief Systems (DIERS) Project*

Manual. Design Institute for Emergency Relief Systems, American Institute of Chemical Engineers, New York, NY, 1993.

[4] Xiangjuan Qi, Shiyu Li, and Lingdong Sun, Applied of ChemCAD in distillation column dynamic simulation. Comp Appl. Chem. 25(2008), 752-756.

[5] ChemCAD user's guide, Available at http://www.chemstations.com/content/documents/chemcad_6_user_guide_-_online.pdf Accessed on June 20, 2014.

[6] Schwartz R, White J and Bussman W, *The John Zink Combustion Handbook*. Boca Raton, FL: CRC Press; 2001.

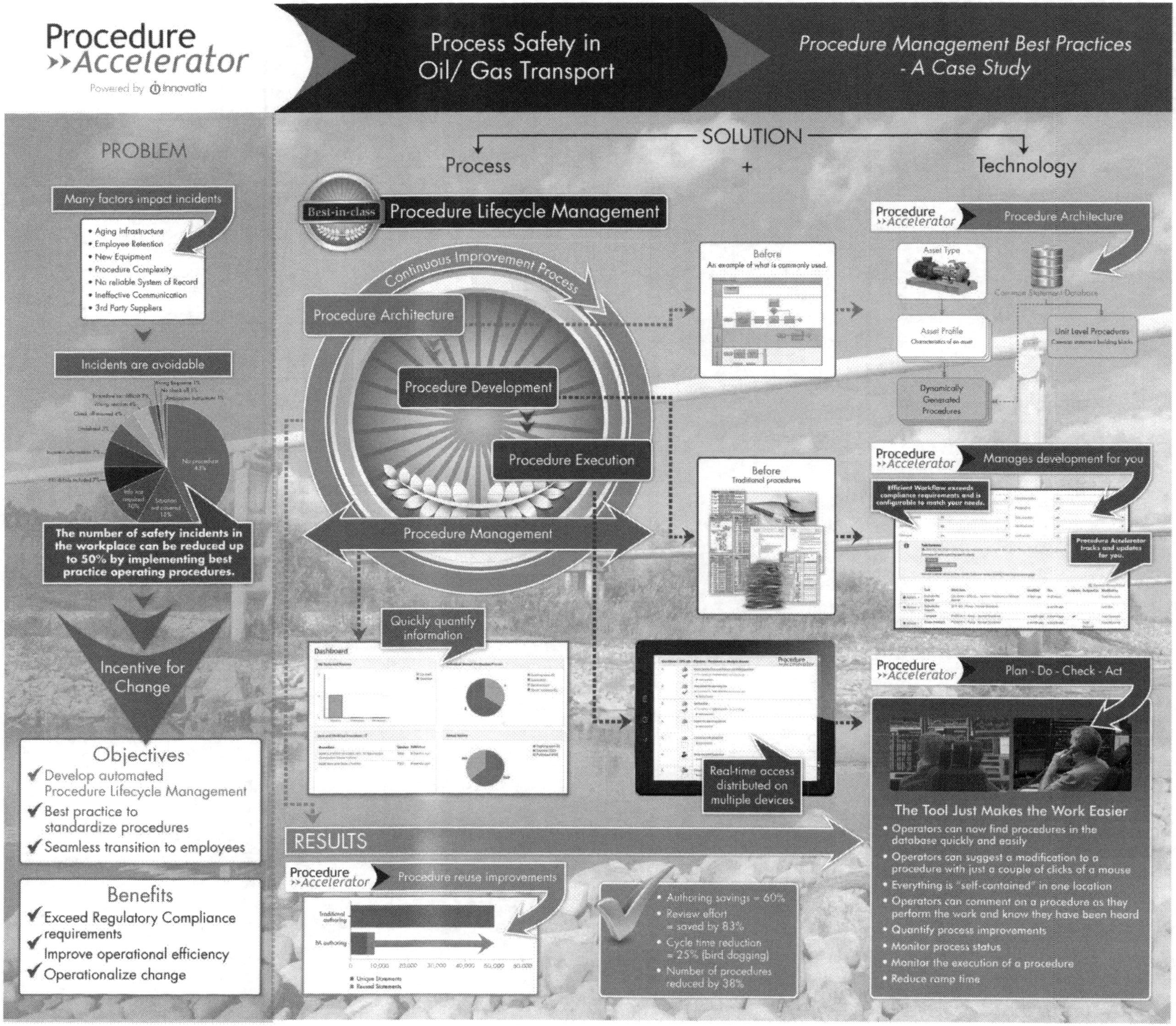

Procedure
>>Accelerator
Powered by innovatia
Process Safety in
Oil/ Gas Transport
Procedure Management Best Practices
- A Case Study
PROBLEM
Many factors impact incidents
• Aging infrastructure
• Employee Retention
• New Equipment
• Procedure Complexity
• No reliable System of Record
• Ineffective Communication
• 3rd Party Suppliers
Incidents are avoidable
The number of safety incidents in the workplace can be reduced up to 50% by implementing best practice operating procedures.
Incentive for Change
Objectives
Develop automated Procedure Lifecycle Management
Best practice to standardize procedures
Seamless transition to employees
Benefits
Exceed Regulatory Compliance requirements
Improve operational efficiency
Operationalize change
SOLUTION
Process
+
Technology
Best-in-class
Procedure Lifecycle Management
Continuous Improvement Process
Procedure Architecture
Procedure Development
Procedure Execution
Procedure Management
Before
An example of what is commonly used.
Before
Traditional procedures
Quickly quantify information
Dashboard
Real-time access distributed on multiple devices
Procedure Accelerator
Procedure Architecture
Asset Type
Common Statement Database
Asset Profile
Characteristics of an asset
Unit Level Procedures
Common statement building blocks
Dynamically Generated Procedures
Manages development for you
Efficient Workflow exceeds compliance requirements and is configurable to match your needs.
Procedure Accelerator tracks and updates for you.
Plan - Do - Check - Act
The Tool Just Makes the Work Easier
• Operators can now find procedures in the database quickly and easily
• Operators can suggest a modification to a procedure with just a couple of clicks of a mouse
• Everything is "self-contained" in one location
• Operators can comment on a procedure as they perform the work and know they have been heard
• Quantify process improvements
• Monitor process status
• Monitor the execution of a procedure
• Reduce ramp time
RESULTS
Procedure reuse improvements
Traditional authoring
PA authoring
0
10,000
20,000
30,000
40,000
50,000
60,000
Unique Statements
Reused Statements
• Authoring savings = 60%
• Review effort = saved by 83%
• Cycle time reduction = 25% (bird dogging)
• Number of procedures reduced by 38%

Predicting Thermal Behavior of Hazardous Chemicals

*Nadia Baati, Annik Nanchen[1], Francis Stoessel[1], Thierry Meyer**

Ecole Polytechnique Fédérale de Lausanne, Institute of Chemical Sciences and Engineering, Group of Physical and Chemical Safety, ISIC-GSCP, Station 6, 1015 Lausanne, Switzerland, tel: +41 21 6933614.

[1]Swissi Process Safety, Mattenstrasse 24, WRO-1055.5.27, 4002 Basel, Switzerland.

**for correspondence: thierry.meyer@epfl.ch*

Abstract

This Quantitative Structure-Property Relationship (QSPR) study aims at developing predictive models capable of reconstructing thermal signatures of 96 structurally diverse chemicals. By extracting only a few key characteristics of the thermograms issued by Differential Scanning Calorimetry, they are entirely reconstructed, thus conserving the overall compound's thermal behavior. Initially, global models are developed for the entire data collection, then a subset is examined apart for building more accurate local models.

Keywords: Quantitative Structure-Property Relationship (QSPR), Differential Scanning Calorimetry (DSC), thermal behavior.

1. Introduction

When handling chemicals, safety is a primary concern for production, storage, transportation and use. It is important to characterize the given chemicals and the processes in which they are involved. For that purpose, various analytical techniques, simulations and small scale experiments can be deployed in order to identify and quantify the faced hazards. Without a proper characterization of the compounds and the process conditions, risk analysis would not be complete and potential hazards would not be correctly confined. This would leave the process with design flaws and any deviation from the intended behavior could have severe consequences.

Obviously, these considerations are taken into account when designing any process, however, accidents still occur. According to the EU Major Accidents Hazards Bureau, in most cases, the underlying causes are flaws in the process and safety design [1]. Thus, this crucial preliminary phase should be conducted with major care and attention.

One particular aspect, for both chemical industry and R&D field in chemistry is the thermal stability of the compounds. Indeed, many of them are thermally unstable, and can undergo spontaneous and undesired secondary reactions such as decompositions for instance. Even the desired reactions can be highly exothermic, and when performed, the consequent heat release should be carefully anticipated. Moreover, all the intermediate compounds should be considered as their presence at some point of the process will also pose additional related risks. Then, in order to implement the adequate preventive and protective measures regarding thermal risks, the thermal behavior of the chemicals is to be very well defined: would the chemical decompose? and if so, at which temperature? how much heat would be released and at which rate? These questions should be answered and the process designed in order to enclose those eventualities.

Differential Scanning Calorimetry (DSC) is a technique that has been widely used to tackle these considerations. During a DSC test, a small sample of the studied chemical and reference sample are submitted to a temperature profile. In power-compensated DSC, they are both kept at the programmed temperature by controlling their respective temperatures and adjusting the heating power to the sample. Heat-flux DSC is slightly different as the heating power supplied to both the sample and reference is the same and temperature difference is allowed to occur [2]. The result from both DSC modes is the same: a thermogram which represents the heat release rate or consumption by the sample as a function of temperature. The thermogram is thus a representation of the compound's stability regarding temperature as it reveals in which temperature range an eventual decomposition would be triggered, the energy that would be released and the rate of the heat released which are the key data mentioned above as necessary for a safe process design.

Nonetheless, for a newly developed product that is unavailable for testing, an intermediate that is practically impossible to isolate, or when in need for screening several products at once, laboratory tests are inapplicable or become time and resource demanding; simulations would then be a preferred mean to estimate the data usually experimentally acquired. Given this context where predictive models could play an interesting role, the aim of the present study is to propose Quantitative Structure-Property Relationships (QSPR) to estimate DSC-derived data from the chemical's molecular structure.

2. Materials and Methods

2.1. Model development

There are mainly two major methodologies of modelling chemical and physical properties of compounds based on their molecular structures. On one hand there are the various methods relying on group contributions [3, 4] and on the other hand, the Quantitative Structure-Property Relationships (QSPR). The latter were initially developed as Quantitative Structure-Activity Relationships in the 70's for studies on biochemical activities. Considering the successful models developed, their application domain expanded and reached various fields and eventually applied to hazardous properties of chemicals and safety related data (i.e. flash points [5], flammability limits [6], auto-ignition temperatures [7], etc...). The procedure to develop QSPR models is not unique as there are several calculations methods to generate descriptors and the models' complexity depends on either they are based on simple regressions or complex neural networks, but the basic principles and the actual procedure applied here are as follows.

QSPR studies rely on the molecular descriptors ability to define the molecular structures to the highest extent by quantifying or associating numerical values to most aspects of the structure. Here, the Codessa Pro software [8] was used on 2D molecular representations of the studied compounds in order to produce optimized 3D geometries and derive numerical descriptors of their constitution, geometry, topology, and electronic, quantum chemical and thermodynamic properties. Over thousand descriptors are generated. A first selection eliminates the descriptors with the lowest variances or with missing values resulting in a working set of approximately 350 descriptors per structure.

The models are built as multi-linear regressions following the stepwise regression method. This feature selective technique is performed in order to select among the structural descriptors the ones that correlate the best the modelled properties. The descriptors that do not enhance the models are excluded. The more parameters included in the models, the better the models fit the targeted data; however, there is a risk of obtaining overfitting models: when too many parameters are included, the models lose their generalizability and fail to describe data external to the training set. Therefore, before the modelling phase begins, the data are divided into a training and a validation set. The training set is the basis for developing the models and the validation set serves to evaluate the models predictive power by verifying how accurately they estimate property values of compounds external to the training set.

2.2. Experimental Data Set and Data Treatment

Based on the results from previous studies conducted to develop predictive QSPR models of key characteristics of DSC thermograms of nitro aromatic compounds [9-11] , this work intends to cover a larger range of chemicals, therefore the set is composed of 96 compounds, with diverse structures and without a common particular functionalization. Nevertheless they were selected based on their thermal reactivity and medium to highly exothermic decompositions.

Previous studies proposed models of thermal behavior described by the onset temperature, i.e. the temperature at which the decompositions are triggered, and the enthalpy [10, 11], but these two-value characterizations do not enclose the global thermal behavior. In order to capture the entire thermogram as a fingerprint of each chemical, we apply an effective method in which the curves are considered as being successions of several asymmetric Gaussian curves [9]. Five key characteristics are extracted from each peak of the curve: the position in terms of temperature (T [°C]), amplitude (A [W/mol]), width (W [°C]), asymmetry (AS [-]) and partial area which corresponds to the enthalpy (PA [kJ/mol]) of each peak. All thermograms were treated using the AKTS software [12] for extracting these data that are then treated sequentially for building the predictive models. Recovering the entire predicted thermogram is achieved by back processing the estimated data from the predictive models.

3. Results and Discussion

3.1. Global models

Among the study set, not all thermograms presented the same number of peaks, thus only the first appearing exothermic peak of each thermogram is considered. The models for each of the key characteristics are built independently from the others and the performance of the models measured as the goodness of fit of the model to the training set and the generalizability as the goodness of predicting the validation set properties. The results are summarized in Table 1.

Table 1: Global models summary

	PA	A	T	H	AS
Training Set R^2	0.68	0.76	0.84	0.73	0.88
Validation Set R_{EV}^2	0.64	0.69	0.76	0.59	0.63
Number of parameters	20	23	17	16	22
Training Set Size : 88 observations					
Validation Set Size : 8 observations					

A graphical representation of the model for the peak's position in terms of temperature T is presented in **Figure 1**

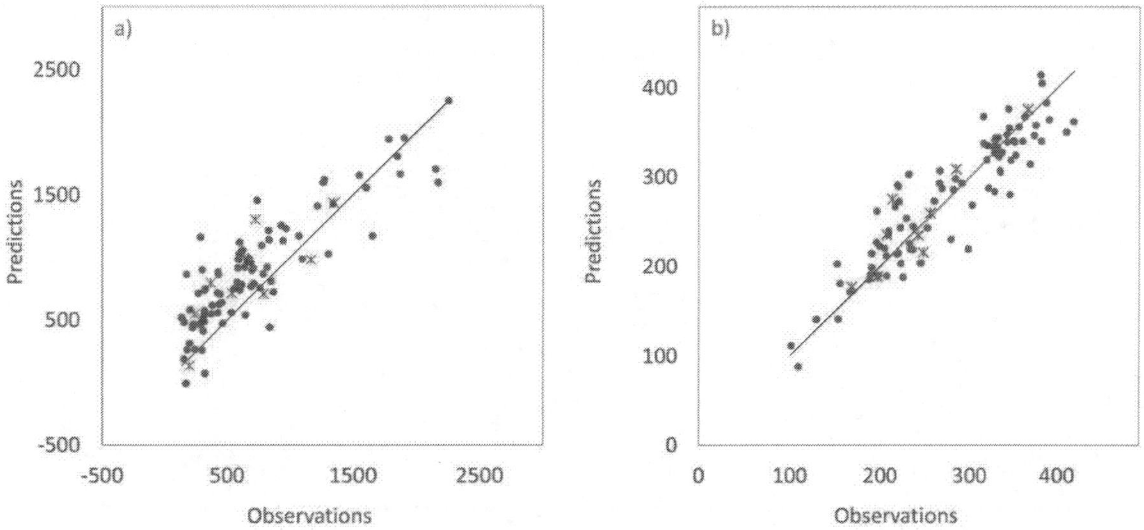

Figure 1: Graphical representation of models for a) amplitude A, both axes in [W/mol] and b) peak position T, both axes in [°c] – training set in blue, validation set in red, line represents perfect match of simulations to observations.

Table 1 indicates that the model performances on the validation set are lower than on the training set which indicates a probable overfitting. The temperature model shows the best performance when compared to the other models regarding the fit to the training set, and suffers the least from the overfitting issue as it also performs best on the validation set. This is mainly due to the fact that the model for T includes less parameters than the other models: 17 descriptors were selected to describe the data out of the 351 descriptors available. As an indicative rule, it is accepted that the number of parameters within a models should not exceed 20-25% of the number of observations used to train the model in order to avoid overfitting. In this case, the models were trained on 88 observations and 8 were kept as the validation set. Thus it would have been preferable not to exceed 17 parameters per model; however this lead to weaker models.

Nevertheless, the thermograms were still recovered from the estimated data, and the overall result is, despite the limited models obtained, a strong indication of the chemicals thermal behavior. Figure 2 exhibits the observed and simulated thermograms for 4-methyl-3-nitro-anilin and dicumyl peroxide. The latter is one of the compounds from the validation set, and therefore all the key characteristics used for recovering its corresponding simulated thermogram are predictions from the models. The amplitude is overestimated by the corresponding model in most cases as the simulated values for A either for training and validation set are higher than the expected values, and this is confirmed by the thermograms shown in Figure 2. On the other hand, the peak position and overall shape are relatively well described and thus the simulations give an effective information regarding the thermograms aspect, especially the onset and increasing heat flow at the triggering of the reaction.

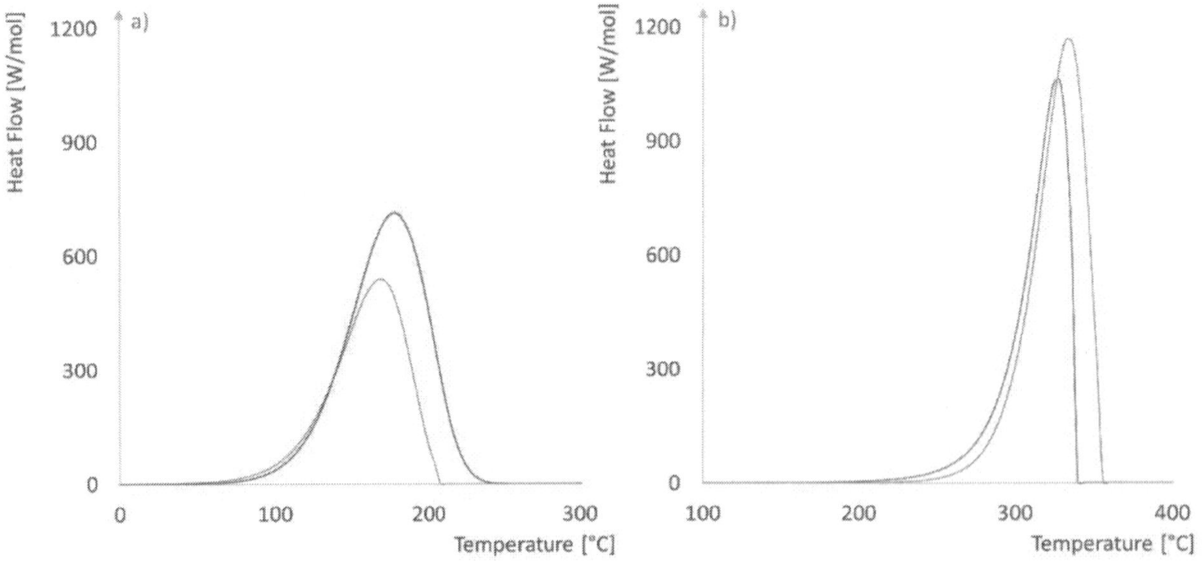

Figure 2: Measured and simulated thermogram for a) dicumyl peroxide (from validation set - measurement in blue, prediction in red and b) 4-methyl-3-nitro-anilin (from training set - measurement in blue, fitting in green)).

3.2. Local Models

From the literature and our previous trials [9-11], QSPR models could be expected to achieve much higher performances when the observed chemicals belong to the same chemical family. The data set is composed of a miscellaneous collection of chemicals without any specific functionalization. However, several data have common specific chemical characterization. For instance among the 96 compounds, there are five peroxides, five nitrites, six diazocarboxylates, and seven tetrazoles. As these subgroups contain too few data to be appropriately analyzed with the statistical tools, it is necessary to merge them and form larger sub-sets. Considering the data space of the thermograms' five key characteristics, the tetrazoles and diazocarboxylates were the closest subsets. Thus, models were developed for these particular compounds in order to see if better results could be achieved. The term 'local models' refers here to the fact that these chemicals are localized within a given region of the data space and are, from this perspective, similar.

The adjusted results to this narrow set are summarized in Table 2. The reduced number of observations contributes to easing the good fittings from simpler models, however the improvement on the models performance is also due to the observations vicinity in the data space.

Table 2 : Local models summary

	PA	A	T	H	AS
Training Set R^2	0.927	0.948	0.965	0.968	0.983
Cross-Validation R^2_{CVOO}	0.847	0.885	0.92	0.942	0.967
Number of parameters	3	3	3	3	3
Training Set Size : 12 observations					
Validation Set Size : 1 observation					

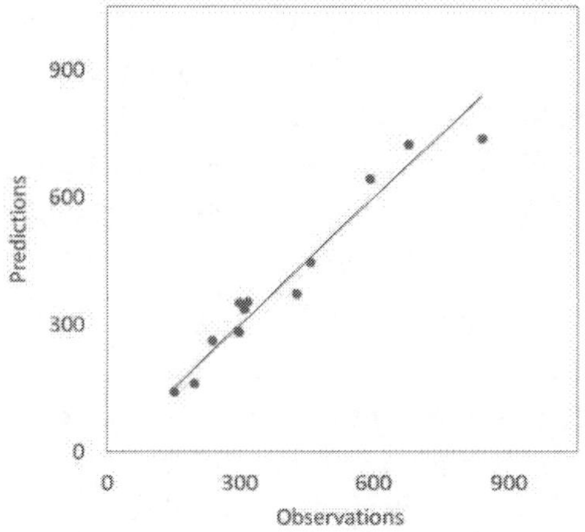

Figure 3: Graphical representation of model for amplitude A, both axes in [W/mol], line represents perfect match of simulations to observations.

Table 3: Example of detailed local model

Property Amplitude A [W/mol]		
Number of parameters	3	
Model Evaluation	R^2= 0.948	R^2_{CVOO}=0.885
Model		
$A = -3067 + 675 N_{C,R} + 3420\ MNO_{BOmin,H} + 35.6\ HACA\text{-}1$		

Descriptors	**Abbreviations**	**Coefficients**
Intercept		-3067
Relative number of C atoms	$N_{C,R}$	675
Min. bond order for atom H	$MNO_{BOmin,H}$	3420
H-Acceptor Charged Area	HACA-1	35.6

Besides, due to the restricted amount of data within the local sub-set, no observations were withdrawn for performing an external validation with data that were not used to train any of the models. Instead, "leave-one-out" cross-validation was performed: this method leaves one of the observations out of the training set and predicts its property value based on the regression performed on all other data, iterates this process and gives an averaged result.

As for the global models, the thermograms were reconstructed from the simulated data and 2 examples are presented in Figure 4.

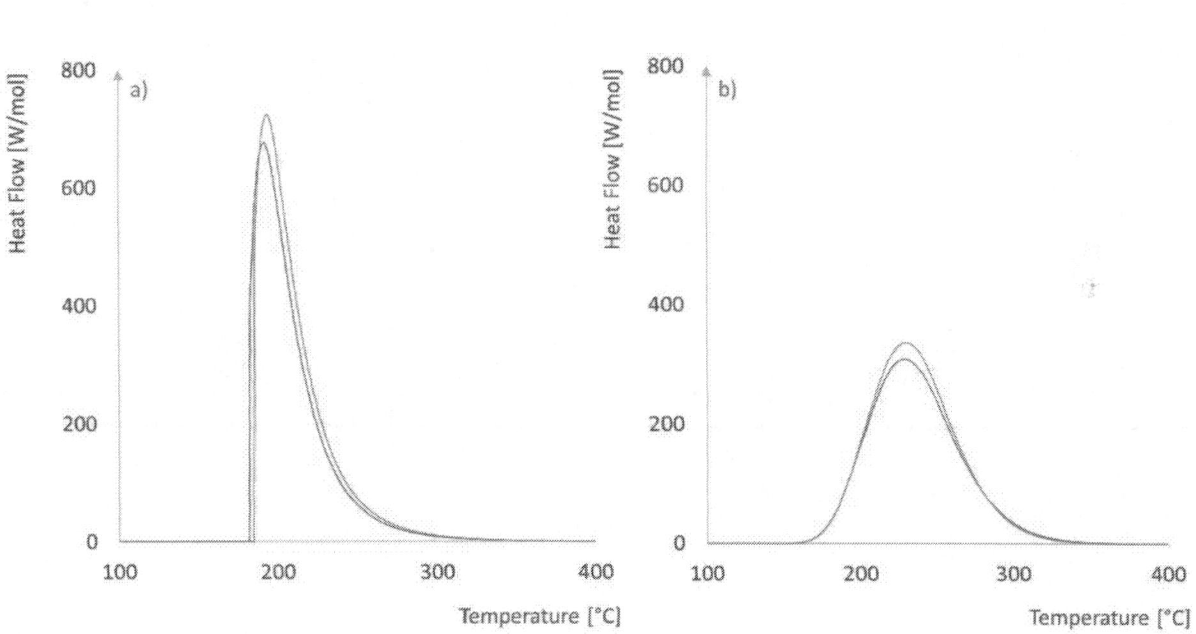

Figure 4: Measured and simulated thermogram for a) 3,4-Dihydro-3-hydroxy-4-oxo-1,2,3-benzotriazine and b) Bis(2,2,2-trichloroethyl)azodicarboxylate (measurements in blue, fittings in green).

The examples exhibited in Figure 4 illustrate well the main reason that underlies the attempt to recover fully the DSC thermograms rather than describing them with two values, i.e. the onset and the reaction enthalpy. In these cases, both samples initiate a decomposition around T_o =180°C and generate an enthalpy of ΔH_r=-186kJ/mol, however the peak shapes are very different and imply different reactivity and rates of decomposition. Moreover, the models were successful at accurately predicting these different thermal behaviors.

4. Conclusion

The QSPR method for producing predictive models was applied to DSC thermograms of a large data set of 96 miscellaneous chemicals. Moreover, the entire thermograms shapes have been modelled in order to obtain an estimated curve, which gives strong indications on the reaction kinetics. Initially, the data set was studied entirely, and then a local sub-set, composed of chemicals with similarities from the point of view of the data space was extracted and studied apart. The local models performed better than the global models and achieved higher accuracy for all property values. If applied sequentially, global and local models could help having a first estimation which is then refined into a more accurate prediction. Finally, it is important to note here that, when applied to sensitive data such as safety related information, predictive models should be handled with precaution. They should not replace a proper experimental investigation, but rather be a helpful tool to determine where to focus resources and to consider the inherently safer procedure.

Acknowledgments

This project is supported by the Swiss Commission for Technology and Innovation (project no. 14711.1 PFIW-IW). Assistance provided by AKTS and Novartis was greatly appreciated.

References

[1] A. Jacobson, J. Sales and F. Mushtaq, Underlying Causes and level of learning from accidents reported to the MARS database, J Loss Prevent Proc 23 (2010) 39-45

[2] G. Höhne, W. Hemminger and H. J. Flammersheim, Differential Scanning Calorimetry, Springer, Berlin 1995.

[3] T. Grewer, D. J. Frurip and B. K. Harrison, Prediction of thermal hazards of chemical reactions, J Loss Prevent Proc., 5, (1999), 391-398.

[4] J. Marrero and R. Gani, Group-contribution based estimation of pure component properties, Fluid Phase Equilibr., 183-184, (2001), 183-208.

[5] A. R. Katritzky, I. B. Stoyanova-Slavova, D. A. Dobcheva and M. Karelson, QSPR modeling of flash points: An update, J Mol Graph Model 26-2, (2007), 529-536.

[6] T. A. Albahri, Prediction of the lower flammability limit percent in air of pure compounds from their molecular structures, Fire Safety J 59, (2013), 188-201.

[7] Y. Pan , J. Jiang, R. Wang , H. Cao and J. Zhao, Prediction of auto-ignition temperatures of hydrocarbons by neural network based on atom-type electrotopological-state indices, J Hazard Mater, 157, (2008), 510-517

[8] CODESSA, Codessa Software (V. 1.0 RC2), CODESSA Pro, University of Florida, 2002.

[9] N. Baati, A. Nanchen, F. Stoessel and T. Meyer, Predictive Models for Thermal Behavior of Chemicals with Quantitative Structure-Property Relationships, Chem Eng Technol, 2015 (38)- *in press*

[10] S. R. Saraf, W. J. Rogers and M. S. Mannan, Prediction of reactive hazards based on molecular structure, J Hazard Mater 98 (2003), 15–29.

[11] G. Fayet, P. Rotureau, L. Joubert and C. Adamo, On the prediction of thermal stability of nitroaromatic compounds using quantum chemical calculations, J Hazard Mater, 171 (2009), 845–850

[12] AKTS, AKTS-Thermokinetics (V. 3.8), Advanced Kinetics and Technology Solutions, www.akts.com

Ignorance is No Defense...Audit Management Best Practice

Mike Bearrow and Thelma Kipe
Principal Consultant, Process Safety Management
Rolls-Royce Controls and Data Services, Inc.
8303 McHard Rd #200
Houston, TX 77053
Michael.Bearrow@controlsdata.com
Thelma.Kipe@controlsdata.com

Disclaimer - All opinions expressed herein are those of the author and not of Rolls-Royce CDS Inc.

Prepared for Presentation at
American Institute of Chemical Engineers
2015 Spring Meeting
11th Global Congress on Process Safety
Austin, Texas
April 27-29, 2015

UNPUBLISHED

Ignorance is No Defense...Audit Management Best Practice

Mike Bearrow and Thelma Kipe

Principal Consultant, Process Safety Management

Rolls-Royce Controls and Data Services, Inc.

8303 McHard Rd #200

Houston, TX 77053

Michael.Bearrow@controlsdata.com

Keywords: Audit, best practices, continuous improvement, corrective action, preventive action, MOC, CAPA, process safety management, PSM, compliance, SEMS

Abstract

No one wants to be ignorant and everyone wants to do the right thing. Why do prominent companies make the headlines for compliance failures, fines and worse? Are they unlucky or not paying attention? Are they using the wrong tools for the job? Do they have inadequate processes or are they using the wrong people?

Whatever the reason, being ignorant of how well you are complying with the PSM standard (RMP, BSSE requirements, OHSAS...) means you are rolling the dice. Rolls-Royce, our parent company, conducts 2000 audits per year and manages the results to ensure thousands of gas turbine engines provide thrust when needed, safely transporting people and cargo to their destinations. Getting that wrong is not an option. Driving compliance to high standards and continuous improvement must be achieved. I have infrequently observed the same level of excellence in other industries. Helping the petrochemical industry take auditing to a new level using best practices is the purpose of this paper.

Contents

- These People are very smart – He meant Process Safety
- What is hard about auditing?
- What industry does today…
- PSM, Auditing & Lawyers
- Auditor High Level Process - Do Loop
- What does the Audit Process Look Like?
- Who should be part of the audit team?
- Audit Management Nitty Gritty
- Audit Best Practice Means
- Managing Results from an Audit - Best Practice Results Management
- The Most Successful Companies…
- What are the Best Practice Audit Metrics
- Baker Panel Report - CSB Chairman to Sr. Managers

Ignorance of the law is no defense.

Ignorantia juris non excusat is a legal principle holding that a person who is unaware of a law may not escape liability for violating that law merely because he or she was unaware of its content. Of course we are all aware of the laws that govern safety and process safety. We just might not be aware of how well we are complying with those laws or if we are complying in a world-class way.

My background as an auditor

In 1992, I worked as an apprentice for Science Application International Corporation (SAIC) with several PSM "blue bloods" including Ray Brandes, Ian Sutton and Geoffrey Kaiser and others, conducting audits, HazOps, and creating operating procedures for chemical plants and refineries. When we were contracted to conduct a PSM audit, we took days to prepare and would show up with protocols in hand and an excel spreadsheet for documentation. We would always meet with site management for a kickoff meeting to explain why we were there and what we were going to do. Sometimes the management team would dare us to find something. I thought that was odd, since that was why we were there in the first place.

Our job was to assess their compliance, to look for weakness in their process safety management systems and provide opportunities for improvement. The recommended actions would be areas where processes could be changed usually resulting in a safer environment, higher quality and sometimes even increased performance. Why would a management team not want us to successfully identify these improvement opportunities?

In an attempt to be thorough we would take one night out of the week to show up at the plant on night shifts to sample the management systems when the management team was not there. This approach always produced results in the form of audit findings and recommendations.

No one wants to be ignorant

No one wants to be ignorant and everyone wants to do the right thing. Why do prominent companies make the headlines for compliance failures, fines and even process safety events? Are they unlucky or not paying attention? Are they using the wrong tools for the job? Do they have inadequate processes or are they using the wrong people?

Whatever the reason, being ignorant of how well you are complying with the PSM standard (or RMP, BSSE requirements, OHSAS…) means you are rolling the dice. Rolls-Royce, our parent company, conducts 2000 audits per year and manages the results to ensure thousands of gas turbine engines prod thrust when needed, safely transporting people and cargo to their destinations. Getting that wrong is not an option. Driving compliance to high standards and continuous improvement must be achieved. I have infrequently observed the same level of excellence in other industries. Helping the petrochemical industry take auditing to a new level using best practices is the purpose of this paper.

Overview

"What gets measured gets managed" *Peter Drucker*

The saying *"What gets measured gets managed"* implies what gets measured is what is important to your organization. Whether it's for process safety management, risk management, ISO, OHSAS or equivalent, the audit element is essential to make sure you are keeping your eye on the ball. Catching risks and mitigating them before something goes wrong is always preferable to reacting to an incident. A robust audit program can help your organization be proactive and continually improve.

The audit management element is meant to compare what should be with what is. The results of that analysis come in the form of major and minor findings as well as opportunities for improvement and best practice. In order to make full use of these results, a best practice audit management system is required. This paper explores how your company can leverage the audit element to your advantage and facilitate continuous improvement in all areas of your business.

Many times the audit group is small and uses archaic methods to record and measure performance. Taking the time to build a best practice process is wasted unless you ensure usability, effective training and key measurements to validate the process. Failure to measure, and leverage the results, will ensure that findings of bad practice and also good practice will not be useful.

What is a Best Practice?

A **best practice** is a method, process or technique that has consistently shown results superior to those achieved by other means, and that is used as a benchmark.

A best practice is a process or technique that:

- Gives a company or facility a tangible competitive advantage.

- Usually takes advantage of technology, but not always.
- Improves EHS performance, product quality, and can also lower cost.
- Gives management more visibility, control and influence over outcomes.
- Must be a working system and not just a theoretical best practice.

How do you really know you have a Best Practice? You know because you are beating the competition with a unique process or technique or most of the companies in a market have adopted the same practice because it is yielding superior results. Superior results can be measured in lack of downtime, less customer complaints and fewer injuries. How do you know you are leveraging audit best practice: by the smoothness of the business, implemented improvement opportunities and the lack incidents that are tied to audit findings.

Why Do We Audit?

You audit because you want to know if you are living up to expectation! Expectation is defined by standards, procedures, rules, legislation, or business goals. Not knowing whether you are meeting your commitments should keep managers up at night. Knowing where you stand and what is being well managed is the best you can do. Around the world there are many standards, guidance and legislation that require you to assess if you are complying and if so, how well.

These standards include:

- US EPA's Risk Management Programs for Chemical Accident Prevention (40 CFR PART 68) or EPA RMP,
- US OSHA's Process Safety Management of Highly Hazardous Chemicals standard (29 CFR OSHA 1910.119) or OSHA PSM,
- US Department of Interior's Bureau or Safety and Environmental Enforcement (BSSE) - 30 CFR Part 250 Subpart S - Safety and Environmental Management Systems (SEMS),
- OHSAS 18001, 2007 revision - ANSI/AIHA Z10-2005.
- European Union Law Seveso II Directive (see Annex III of EU Council Directive 96/82/EC).

The 14 Elements of Process Safety

In the US, the chemical process industry is governed by the PSM standard which has 14 elements (and EPA's RMP rule, with most of the same elements). They are:

1. Employee Involvement	**8. Mechanical Integrity**
2. Process Safety Information	9. Hot Work
3. Process Hazard Analysis	10. Management of Change
4. Operating Procedures	11. Incident Investigation
5. Training	12. Emergency Planning & Response
6. Contractors	**13. Compliance Audits**
7. Pre-Startup Safety Review	14. Trade Secrets

This paper focuses on number 13. It is one of the continuous improvement aspects of the PSM standard. Like incident investigation, it is intended that you leverage it for learning and improvement. In the case of incidents, you are intended to learn from unwanted events and create corrective action. In the case of audits, you are intended to learn from identifying system weaknesses before something goes wrong and take preventive action.

The PSM Standard has similar elements to other standards. One audit management system can handle them all.

Program Element	OSHA PSM	EPA RMP	BSSE	OHSAS 18001	SO9000/ 14000	CMA Responsible Care
Commitment & Accountability		•	•	•	•	•
Employee Participation	•	•				•
Management System/Planning	•	•		•	•	•
Process Safety Information	•	•	•	•	•	•
Hazard Analysis/Risk Assessment	•	•	•	•	•	•
Corrective Action/Task Management	•	•	•	•	•	•
Operating/Safe Work Procedures	•	•	•	•	•	•
Training	•	•	•	•	•	•
Contractor Safety	•	•	•	•		•
Pre-Startup Safety Review	•	•	•	•		•
Maintenance/Mechanical Integrity	•	•	•	•		•
Safe Work Permits / Control of Work	•	•	•	•		•
Management of Change/Control	•	•	•	•	•	•
Incident/Accident Investigation	•	•	•	•	•	•
Emergency Planning & Response	•	•	•	•	•	•
Audit	•	•	•	•	•	•
Performance Measurement & Reporting				•	•	•
Registration/Certification		•			•	•

What industry does today…

Many companies still consider the audit element to be a test, quiz or quest to be passed. That was not the intent that the original authors of the PSM standard. An audit should be considered an opportunity to search for weakness in your management system like looking for buried treasure. Audit findings should be treated like an opportunity to get better without loss.

Unfortunately in some companies, it is considered a game of hide and seek. Trying to fool the auditors is the game and the wanted result of the audit is no findings, no weaknesses, sound the all clear siren. No findings means that there is no room for improvement and your management systems are all working perfectly. In my experience, this result means that the audit was conducted poorly and that the time was wasted. You have documented that you have accomplished the audit as required, but you have missed the mark on why and not improved or shared best practices.

A famous company hired another famous company to conduct an ISO 9003 recertification audit. The auditor was neither curious nor experienced in the business processes he was auditing. He produced a final audit report with no findings and a recommendation for recertification. On the surface, it looked like a success for the hiring company because they would be recertified without having follow-on recommendations to address. After a review of the audit report and discussions with the staff being audited, it was clear that accepting the certification credit from an inept auditor was not ethical or a method for driving them to best practice. The audit was rejected, another auditor was sent, and the company was then able to get recertification (along with findings to address), knowing that the audit was done accurately.

Many companies construct a small team of part time auditors who infrequently conduct audits. It is common to use inexperienced engineers for this function to give them some experience. They use spreadsheets and email to conduct, track and report on audit results. They submit reports to senior managers once per quarter and they assign the recommendations to the site PSM coordinator. Sometimes, the draft audit report is quarantined until it gets a legal review. That is a wasted step and can sometimes delay progress and even color the truth so that the weaknesses are not addressed as intended, or worse exposed through an incident that could have been prevented. Sometimes, these audits are "owned" by the legal department within the organization. Just remember, if the audit documentation is client-attorney privileged, privilege does not exist as far as OSHA or EPA are concerned, since employers are required to maintain (and produce on demand) the last two compliance audit reports. If the audit is not enhancing the compliance with standards and improving the process safety culture, then it is really worthless.

PSM, Auditing & Lawyers

Honesty, integrity, and diligently trying to do the right thing is your only defense in a court of law or public opinion. Playing hide and seek with people who do that for a living is not an effective strategy. In the event that you get an external audit "for cause" or because of a complaint, having a great internal audit program and recommendation management system is your best defense.

If there is a bang, the media will put OSHA, EPA, DOT, USCG and any regulatory authorities in the headlines, along the families of the injured and their representatives. Attorneys are not PSM experts…ever! They are concerned with protecting a company from legal harm, placing blame on others and/or getting money from another company regardless of the facts. If you make something client-attorney" privileged, that means it does not exist as far as proof you tried to the "right thing". If OSHA cannot see your PSM/RMP Compliance Audit, that document does not exist and you are definitely out of compliance…on purpose. Operators and mechanics have great and long memories. Even if you have a "Retention Policy", the truth will likely come out if push comes to shove.

Making a regulating authority wait or telling them they have no right to see what they are looking for is a losing strategy. Do not think that exculpatory signage or language like "Not available in all areas, must be 21…" or "We make them sign a waiver to work here!" will excuse you from negligence. Engineers, Operators, Mechanics and Managers are not lawyers and should try not to think or act like one. Unless you are in the magic business, forget the smoke and mirrors.

What does the Audit Process Look Like?

They all work like this. You plan the audit (Phase 1), you prepare for the audit (Phase 2), you conduct the audit (Phase 3), you create and assign corrective action (Phase 4) and finally you validate that the corrective actions were successful (Phase 5).

Figure 1: Audit Management Process

Besides the 5 phases, there is a need for constant communication, analysis and reporting. That is nearly impossible using spreadsheets and manual email. More on this later…

Conducting the Audit (Phase 1) - Do Loop

Conducting an audit is pretty simple and is actually reminiscent of the process hazard analysis requirements of the PSM standard. You ask questions. You get answers. If you don't like the answers you make recommendations for getting better, safer, or smarter. You continue asking questions until you run out of questions. The iterative nature of asking questions makes this audit phase look like a do loop.

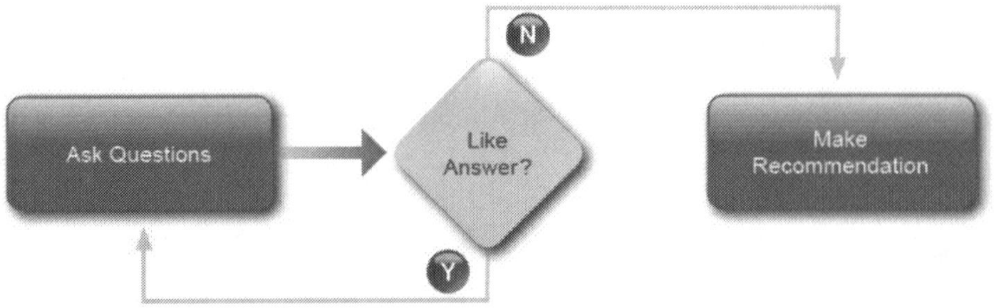

Figure 2: Conducting the Audit

Who should be part of the audit team?

That's a great question! Should the audit team be made up of operations experts, PSM experts, long tenured operators, engineers, mechanics or managers? Should they be all from your company or all external? What's important about the auditor? The most important characteristic of an auditor is someone who is curious. Other characteristics that are important are experience with your industry, chemical processing experience, maintenance background, detail oriented, easy to visit with and one that has the heart of a servant. After all they are looking for weakness in your management systems that might result in someone getting hurt or worse.

Some new standards like BSSE's Safety and Environmental Management Systems (SEMS) require an independent 3rd party (I3P) to audit your operations. I believe that is an essential element to a best practice audit management system. Companies need industry perspective and it is hard to audit yourself in terms of industry best practice if you do not know what that means. A friend of mine used to say, I3P is important because a fish does not know it is wet.

On the other hand, internal auditors will sometimes go much deeper on an audit because they have deep understanding of the operations. No outsider can do that. Therefore, having a mix of internal and external domain experts conduct your audits is the best approach.

Audit Management Nitty-Gritty

A best practice audit management system begins with an understanding of the requirements. In the case of PSM there are 14 elements, but there is so much more than just PSM that you can measure while you are at it. You can measure operational excellence, RMP, BSSE, OHSAS... What else needs to be measured? Once you know all of the requirements then you need protocols or checklists to help you assess the level of compliance. The entire corporation and future audits should use the same set of questions for repeatability and to ensure that there is a minimum of depth of observation.

How many audits can you conduct in a year? What if you have an audit wish list of hundreds and you can't get them all done? How do you decide which ones get done and which ones don't? Once you decide what can be done and the audits are prioritized, then how do you assign audit leaders and get approval to use these resources?

What tools can you use to make the job standard, repeatable, visible and valuable? I recommend a built for purpose audit management software, PC, laptops, big chief notepads, iPADs/tablets. Use all the technology you can get your hands on. It's a big job and using old fashion tools is like finding the fountain of youth while stranded on a deserted island; there could be jewels of wisdom but no one will ever know.

A high level best practice audit management system looks like the continuous improvement management system depicted below. Improvement ideas are fed into the system. These ideas for improvement are vetted and assigned as CAPA (corrective action, preventive actions). Each of these ideas for improvement are addressed and documented in a database management system. Real-time dashboards and other reports illustrate progress.

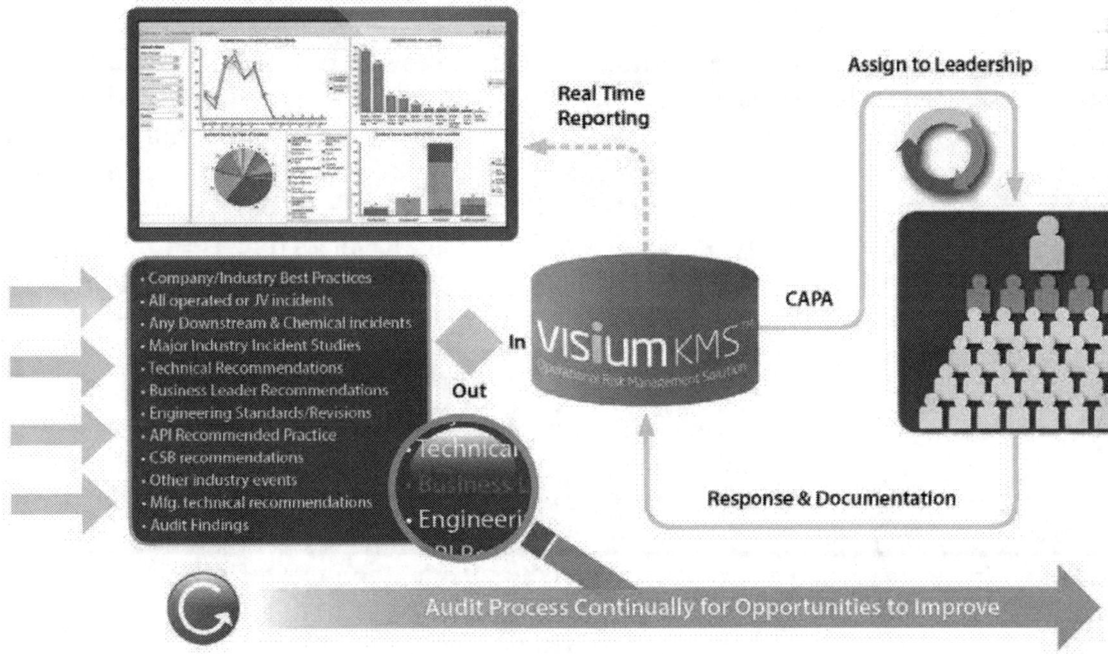

Figure 3: Continuous Improvement Loop

Audit Best Practice Means

Ensuring that you have an audit management solution that works for all requirements (all inclusive) is step 1. There is no need for separate audit solutions for environmental, PSM, operational excellence, RMP, OE, SMS, EMS, EHS, Quality, or security. If you have all of the audit needs collected in one system, you will surely need an automated process for proposing and prioritizing audits that need to be accomplished (planning). You will need a means to approve the audit schedule and resourcing (planning). Then you need an efficient and effective way to assign and remind lead auditors of their assignments (notification and reminder). You will need experienced internal auditors in combination with independent third party consultants that can give you industry perspective (resourcing).

You will need an enterprise audit management software that can be used anywhere, preferably on laptops or tablets and can be used off-line and synchronized with the enterprise audit solution. You need the ability to categorize findings as Major, Minor, opportunity for improvement or best practice. In addition, you need the ability associate findings with a company specific root cause.

Once your final report is created, you need to be able to review audit results on-line with managers and then be able to push out these results to an enterprise dashboard and reporting system for all to see. A best practice audit management system will enable you to see a prioritized list of findings and their associated recommendation; including the "as-is" and "to be" risk. Senior Managers who are constantly measuring audit program results as well as the tracking of change required to address findings will need to know how the risk will be reduced by the change.

Managing Results from an Audit - Best Practice Results Management

So the results are twofold. First, there are the findings and recommendations for the site that has been audited. These need to be assigned and tracked to closure. These recommendations can then also be evaluated for their applicability at other units, sites, or facilities so that those lessons can be applied across similar functions before they get cited; or worse yet, experience a process safety event. In further implementation of the PSM standard, there is an expectation of a performance safety culture cultivated by the sharing of ideas and openness of correcting findings. The best practices that have been identified need to be shared and actioned at the other sites. This is where most companies fall short. There is no requirement in the PSM standard to identify best practices while auditing.

There is no requirement to share these best practices and there is certainly no requirement to make sure these best practices are implemented at applicable sites. But wouldn't that make sense? Of course it would, and that is a best practice management system. Sharing good practice can be equally important for improving the safety culture and encouraging the right behaviors. The ability to learn at a site level and leverage the learning globally is a rare quality in a company.

Managing results from audits begins with omni-present reporting. Once you have planned your audits and are conducting them, you need a real-time view of audit progress and results. Are you meeting the planned audit schedules? Reporting dashboards are the best way to view real-time graphs which provide non-invasive, right now, information that can be used by everyone to manage continuous improvement. Transparent reporting of results is essential to ensure everyone is being held accountable. Careful thought should be applied when designing dashboards so it is clear whether there is action needed or issues to resolve as well as conveying good results and behaviors.

The Most Successful Companies…

The most successful companies will have executive leadership that believes PSM and the audit management element are important. You can do without this, but it makes it so much more difficult. It is important that the management team embrace PSM as a good way to do business. PSM is not an add-on, with no payback. The most successful companies leverage technology by automating processes where possible and continuously improve the audit process and the information system.

Having an audit management system that is best practice is a moving target, so if you are not getting better, you are losing ground. You need to be auditing the audit management system and assessing its effectiveness. Are you squeezing all the goodness out of the audit effort?

You need to be able to measure the risk of the findings ("As Is") and the "To Be" Risk after corrective actions are completed.

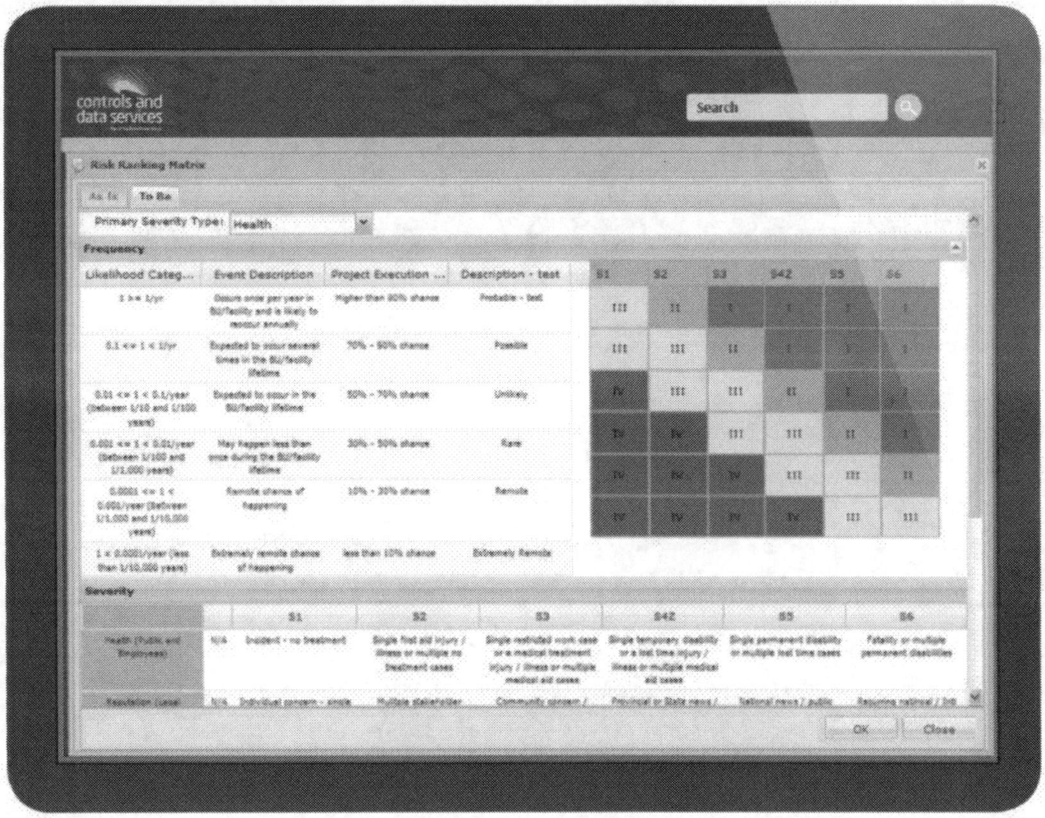

Figure 4: Risk Ranking the Weakness

What are the Best Practice Audit Metrics?

You need to measure the audit plan, the progress to plan, the types of audits that you do, the recommendations (Major, Minor, OFI and Best Practice), Root Causes, Best Practices discovered and actioned. What in the world does that mean?
In order to take full advantage of the audit management system, you should make sure that practices are not only shared but acted upon by the other sites in the company that might benefit from them. How do we do that? You assign senior managers at each facility best practice actions that have been discovered in audits and they must either implement the best practice at their site or explain how they have already addressed that best practice in another manner.

The dashboard below is an example of audit management metrics. Quadrant 1 (top right) - Auditor Item Timelag – This shows how long does it takes to get a draft report out, the final report out (once all has been accepted) and the time needed to close the audit. These metrics are needed to show the number of days to Audit Closure. Quadrant 2 (top left) - Auditor Items Schedule by Month - This chart is important because it is showing how many audits are planned, assigned, approved, open and closed. Quadrant 3 (bottom left) reveals the kind of audits that are being accomplished (ISO, PSM, RMP, OE…). It is important to know where you are spending your audit budget. Quadrant 4 (bottom right) - Auditor Items Open REC/MOCs by location. This is an "aging" chart. All standards require that you implement corrective actions without undue delay.

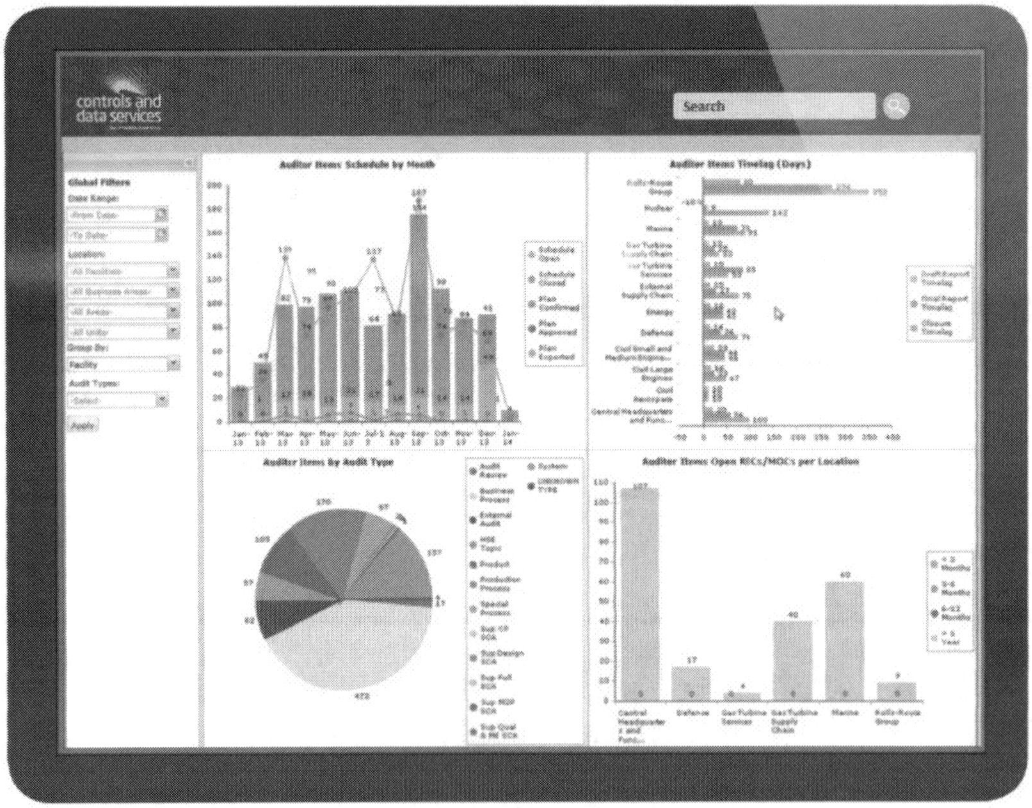

Figure 5: Measuring Audit Management System Performance

Identifying the "Finding" Root Cause

Another metric measured by world class companies is the "finding" root cause (basic and root). Basic and root cause analysis reveals an interesting story concerning why you have system weaknesses. The "why" is important so you can address the underlying (root) cause. A root cause is a cause that once removed from the problem fault sequence, prevents the final undesirable event from recurring. In conducting incident investigations, various methodologies are employed to identify the root causes for the incidents. In conducting audits, there should be the same focus on finding the true cause of the failing so that appropriate measures can be taken. The table below illustrates examples of root causes that could be associated and then reported upon is dashboards to drive corrections. This will add additional insight into the kinds of work you need to do as a company to get better.

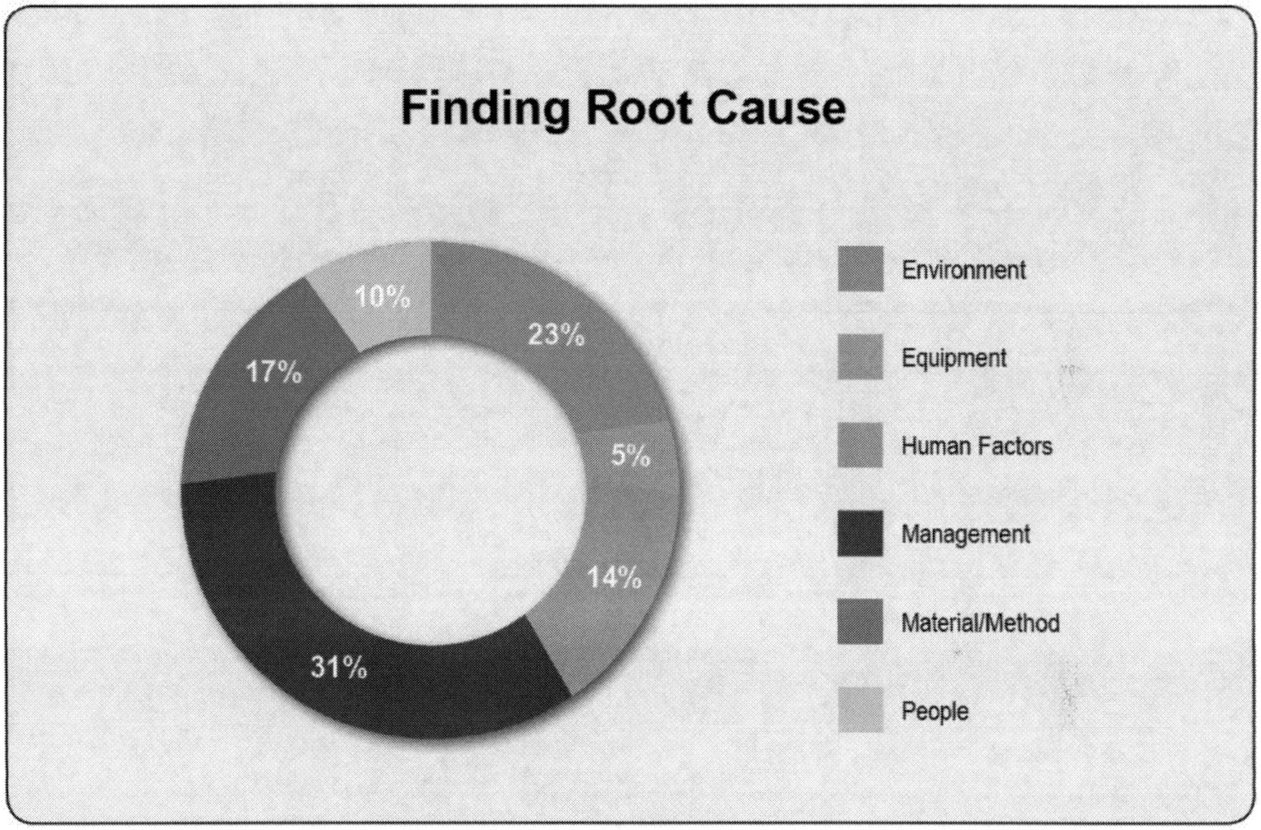

Figure 6: Identifying Basis Causes

Even more understanding can be gleaned if you identify the detailed root causes like the ones presented below.

Environment - Inadequate Conditions
Environment - Lack of Workspace
Environment - Poor Ergonomics
Environment - Poor Housekeeping regime
Environment - Unpredictable event
Equipment – Availability

Equipment – Control
Equipment – Failure
Equipment - Inadequate/incapable
Equipment – Maintenance
Equipment – Setting
Human Factors – Compliancy
Human Factors – Distractions
Human Factors – Fatigue
Human Factors - Lack of assertiveness
Human Factors - Lack of awareness
Human Factors - Lack of communication
Human Factors - Lack of resource
Human Factors - Lack of teamwork
Human Factors - Negative norms
Human Factors – Pressure
Human Factors - Repetitive/monotonous
Human Factors – Stress
Management - Inadequate Communication
Management - Lack of resource
Management - Poor governance
Management - Responsibilities not defined/understood
Management - Training insufficient or inadequate
Material/Method - Common cause process failure
Material/Method - Document - No method of recording
Material/Method - Document - Record incorrect
Material/Method - Documentation lost/not available
Material/Method – Identification
Material/Method - Inadequate Process Capability
Material/Method - Inappropriate /unclear instructions
Material/Method - Instruction not available
Material/Method - Insufficient Planning and control
Material/Method - Insufficient process/procedural definition
Material/Method - Insufficient process Validation
Material/Method - Lack of process tools
Material/Method - Material not to required standard
Material/Method - Parts Time expired
Material/Method - Poor material handling
Material/Method - Poor Source and Method control
Material/Method - Process/procedure not defined
Material/Method - Requirements not captured
People - Incorrect decision taken
People - Incorrect interpretation of instructions
People - Lack of awareness/knowledge
People - Lack of competency

Figure 7: Identifying Root Causes

Best Practice Action Tracking Metrics

Once you have identified what needs to be done (corrective actions, preventive actions and possibly MOCs), it is critical that all required activities are tracked to closure. If continuous improvement ideas don't get managed, your company has gained nothing but a check in the box that the audit was completed. No progress; just an embarrassing paper trail…yikes! What is being measured below are the number recommendations (corrective actions, preventive actions),

MOCs, emergency MOCs, past due recommendations, implementation status and changes in target dates. These are best practice action tracking metrics.

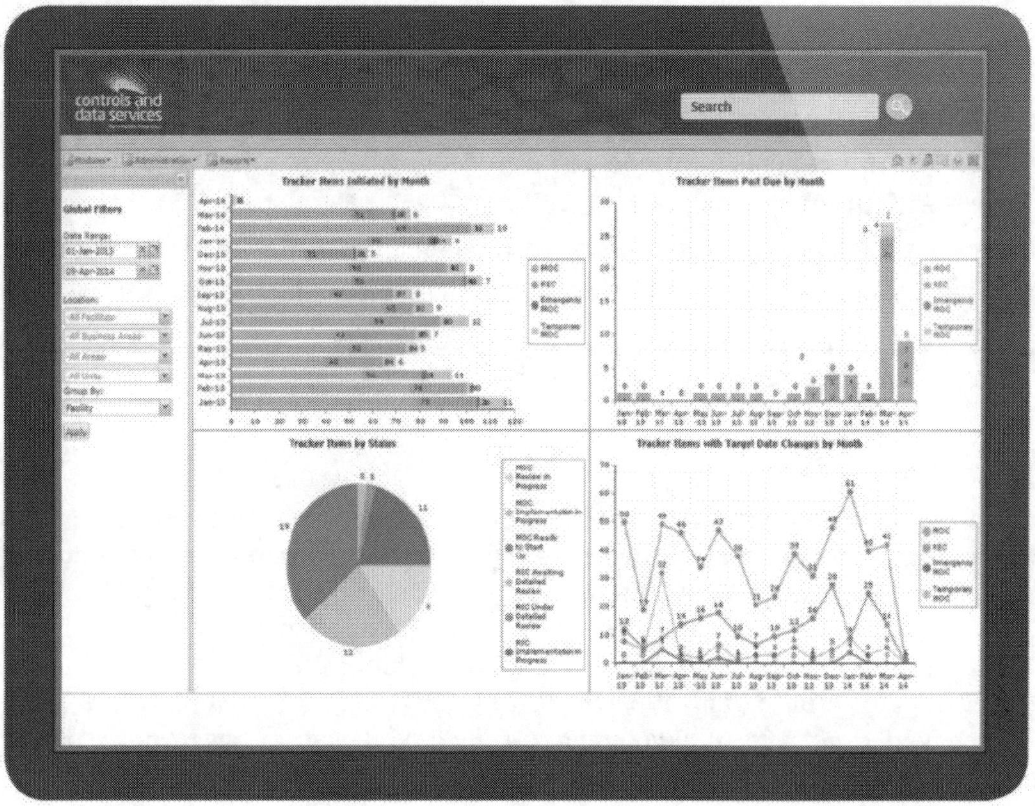

Figure 8: Managing the Change

And don't forget the Baker Panel Report - CSB Chairman to Sr. Managers

"Process safety programs protect the lives of workers and the public deserve the same level of attention, investment, and scrutiny as companies now dedicate to maintaining their financial controls."

"The boards of directors of oil and chemical companies should **examine every detail of their process safety programs** to ensure that no other terrible tragedy like the one at BP occurs."

Best Practice Summary

- **Automate** your entire audit management system; planning, preparation, conduct, recommendation tracking, verification, communication and reporting.
- Ensure your **audit management system covers all of the audit types** (PSM, RMP, BSSE, ISO…) your organization must comply with
- Create a **full-time audit group** that is staffed with experienced (operations) and domain (PSM, RMP…) members who are curious and energetic about finding system weakness and best practices.

- Augment that audit team with **independent 3rd party auditors** that can bring industry perspective.
- Ensure that each finding has a **basic and near root cause** identified so you can see where to concentrate your efforts in the future.
- Ensure that all recommendations have an **"As-is" and "To Be" risk ranking** so it is easy to see how big the opportunity is
- Ensure that recommendations concerning system weakness are assigned and **tracked to closure with urgency**.
- Ensure that **best practices are assigned to senior managers** at all relevant locations so they can take advantage of the learnings. These should be managed with the same urgency as system weakness.
- Use **dashboards** to highlight the health of each audit phase.

Conclusion

The strongest fear known to mankind is the fear of the unknown! Don't be afraid of Audits. Be afraid of being unaware (ignorant). Adopt a strategy of being aware of your circumstance and identifying where your management systems have weaknesses. Take solace in knowing that the weaknesses have been identified, analysed, communicated, and most importantly, that corrective and preventive actions are being managed in a timely fashion. Leverage audit management best practice to lift up the entire organization by ensuring best practices identified are best practices utilised. Then you won't care if ignorance is a defence; because you're not ignorant, you're informed, aware, and taking effective action.

References

[1] US EPA, Risk Management Programs for Chemical Accident Prevention (40 CFR PART 68) or EPA RMP, 2015

[2] US OSHA, Process Safety Management of Highly Hazardous Chemicals standard (29 CFR OSHA 1910.119) or OSHA PSM, 2013

[3] US Department of Interior's Bureau or Safety and Environmental Enforcement (BSSE) , 30 CFR Part 250 Subpart S - Safety and Environmental Management Systems (SEMS), 2014

[4] OHSAS 18001, ANSI/AIHA Z10-2005, 2007

[5] European Union Law Seveso II Directive, Annex III, EU Council Directive 96/82/EC, 1996

396777 Screening Tool for Runaway Reaction Characterization. Optimizing Efforts for Emergency Relief Systems Design

Abstract:

Not all reactive systems present the same thermal risk level, and accordingly, not all of them require the same level of detail for Emergency Relief System design. The present paper illustrates and easy-to-use and effective screening tool for characterizing any reactive system based on information extracted from basic calorimetric tests. The proposed tool is able to characterize two key parameters of both desired and potential secondary reactions in case of a runaway: (1) Adiabatic Temperature Rise; and (2) Time to Maximum Rate under adiabatic conditions. The screening tool is able to provide four key temperature levels which contain the thermodynamic and kinetics information of the runaway reaction. Based on the criticality class estimated, it is reliable to technically justify the optimum efforts required for ERS design. The methodology is an added value tool for the decision-making process able to save time and money when questioning if complex, time-consuming and laborious two-phase dynamics simulations and kinetic model developments are required or not.

396910 Facilty Siting Challenges

Abstract:

Advancing in construction methodology and minimizing footprint of process area in the recent years requires more focus on process safety and facility siting. Previously, most of the engineering companies used spacing table to define the distances between occupied building and source of hazard.

Conducting performance and risk based studies to place the occupied buildings in a safe location is of a great importance. In addition, identifying buildings containing safety critical element (SCE) requires further investigation to ensure the process will not be interrupted in any event that might impact the building. Consequence and Risk analyses ensure that the risk to these buildings is mitigated to ALARP (As Low As Reasonably Practicable).

Atkins Consulting Canada Ltd. has developed a method to apply process safety studies (HAZID, Consequence Modeling, QRA ,etc) in facility siting and equipment arrangement.

401416 Process Safety in the Midstream Sector Under the Shale and Tight Oil and Gas Boom

Abstract:

Keywords: Midstream, accidents, fire, loss of containment, Process Safety Management.

Abstract

During the last decade, the "shale revolution" has stimulated production of oil and natural gas in the United States. Technical advances, combined with high oil and gas prices have enabled increased production of the abundant oil and natural gas resources in the United States.

The extraordinary growth of this segment of the oil and gas industry has resulted in challenging conditions for safe construction and operation of green-field facilities.

We examine the contributing factors on a set of incidents and accidents related to construction and operation of facilities associated with the U.S. shale boom. Specifically, we look at factors common to accidents on new natural gas processing plants and associated pipelines.

397901 Lessons Learned during Air Tightness Testing of Shelter-in-Place Buildings at Chemical Facilities

Abstract:

When properly implemented, the shelter-in-place building concept can protect personnel during a toxic gas release at a chemical manufacturing facility sufficiently and long enough for the event to pass, for occupants to don appropriate PPE and evacuate, or for a rescue to be coordinated by emergency responders. There are many issues to consider when assessing if such a building is suitable as a shelter-in-place for toxic hazards. While quantitative physical characteristics, such as the air tightness of the building envelope and the HVAC design, are key parameters to consider, more qualitative aspects such as written procedures, personnel training, and functional performance of the HVAC systems are also critical for the success of a shelter-in-place building.

Testing the air tightness of such buildings can provide insight to more than just the quantitative results of air change rates or leakage areas of the buildings. This testing has logistical issues that can require interaction with site safety personnel, interaction with the building occupants including operations personnel, and the shutting down of HVAC equipment, all of which may provide an opportunity for the test engineer to gain a deeper understanding into how successful the buildings shelter-in-place strategy might be. BakerRisk has recently conducted air tightness testing at roughly 50 shelter-in-place buildings at chemical facilities. This testing has consisted of tracer gas tests, blower door tests, and infrared thermography. Additionally, BakerRisk has used methods such as visual inspections, building design review, review of procedure documentation, shelter-in-place drills, and formal and informal interviews with site safety personnel, operations personnel, building occupants, and HVAC technicians to better assess the suitability of a shelter-in-place buildings.

The quantitative testing and these more qualitative methods have provided insight into shelter-in-place procedures, personnel training, building design and building envelopes, HVAC systems, maintenance procedures, expedient sheltering, and locating major leaks. Many of these insights can be problematic, while other insights can be enlightening. This presentation will discuss a variety of observations regarding these subjects relating to sheltering-in place. While buildings can be tested to quantify how air tight they are, there are many other things to consider when assessing if shelter-in-place building is suitable, or if a shelter-in-place strategy will be successful. Several methods can be employed to gain insights into this assessment, some of which may have a greater impact on the risk that the test results themselves.

Lloyd's Register Energy

A simplified PSM guidance for smaller facilities using a Risk Based Approach

Swarup Bade, P.E, Process Safety Consultant
Lakshmana Tangirala, Project Manager
Lloyd's Register Energy Americas, Inc.
1330 Enclave Parkway, Suite 200 Houston, TX 77077
W http://www.lr.org/energy
E swarup.bade@lr.org; lakshmana.tangirala@lr.org

Process Safety Management (29 CFR 1910.119) is the application of management principles and systems to the identification, understanding and control of process hazards to protect employees, facility assets and the environment.

14 Elements of PSM (Process Safety Management)

Employee Participation
- Develop a written Employee Participation Plan for Employee Involvement
- Consult with employees on the conduct of the development of PSM Elements
- Provide access to PSM information

Process Safety Information (PSI)
OSHA groups PSI in 3 categories:
- Hazards of the chemicals and flammables in the process
- Information related to the technology of the process
- Information pertaining to the equipment in the process

Process Hazard Analysis (PHA)
- What if / Check list
- HAZOP (Hazard and Operability study)
- FMEA (Failure Mode Effect Analysis)
- FTA (Fault Tree Analysis)

Operating Procedures
- Initial startup: Normal, temporary and emergency operations.
- Normal and emergency shutdown.
- Operating limits; Consequences and corrections of deviation
- Engineering & administrative controls and personal protective equipment
- Safety systems and their functions

Training
- Process Overview and operating procedures
- Safety & health hazards, emergency procedures
- Refresher Training
- Record of the training

Contractor Safety
- Requirements include:
  - Obtain and evaluate safety performance and programs
  - provisions of the emergency action plan
  - Develop and implement safe work practices
  - contract employee log

Pre-Startup Safety Review (PSSR)
- Construction and equipment is in accordance with design specifications
- Safety, operating, maintenance, and emergency procedures
- Modified facilities meet the requirements contained in Management of Change
- Training of each employee involved in operating a process

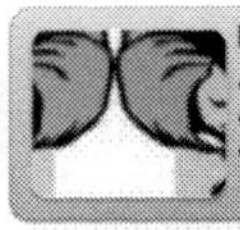

Mechanical Integrity
- Test & Inspections (T&Is) and documentation of T&Is
- Equipment deficiencies
- New Equipment and material Control

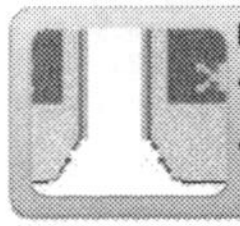

Hot Work Permit
- Fire protection and prevention to be implemented (29 CFR 1910.252(a))
- Dates and identity of the object on which hot work is performed

Management of Change (MOC)
- Establish and implement written procedures to manage changes (except for "replacements in kind") to process chemicals, technology, equipment, and procedures; and, changes to facilities that affect a covered process.

Incident Investigation
- Investigate each incident
- Establish an incident investigation team
- Incident Reports and Corrective Actions
- Report Review

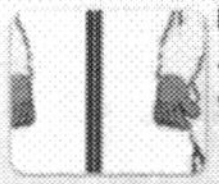

Emergency Planning and Response
- Emergency action plan for the entire plant
- Procedures for handling small releases

Compliance Audits
- Certify compliance with the provisions of the PSM Standard at least every three years to verify that the procedures and practices developed under the standard are adequate and are being followed.

Trade Secrets
- Standard 1910.1200, employees and their designated representatives shall have access to trade secret information contained within the process hazard analysis and other documents required to be developed by this standard.

Implementation of a good PSM program is critical for the safety of personnel and for the assets of process facilities. It helps prevent the loss of revenue and productivity due to accidents. Moreover, it will satisfy the NEP program and reduce or eliminate workplace hazards associated with the release of highly hazardous chemicals and help prevent the facilities from being fined by OSHA audits.

National Emphasis Programs (NEP):
- Provide for planned inspections in high hazard industries
- Focus efforts on specific hazards
- **Refinery NEP** focuses on implementation of PSM in Refineries
- **Chemical NEP** focuses on implementation of PSM in all other PSM covered facilities

Comparison of Refinery and Chemical NEP's
(as per OSHA 2012 report)

Type of PSM Inspection Program	Citation Per Inspection	Penalty Per Inspection
Refinery NEP	11.2	$ 76,800
Chemical NEP	8.4	$ 31,600

Top 1910 Standards Violated
(as per OSHA 2012 report)

Standard	Description	Number	%
1910.119	Process Safety Management	891	59.9 %
1910.147	Lockout / Tagout	55	3.7 %
1910.120	Hazardous Waste & Emergency Response	47	3.2 %
1910.134	Respiratory Protection	36	2.4 %

Top Refinery NEP Cited PSM Elements
(as per OSHA 2012 report)

Element	Description	Number	%
j	Mechanical Integrity	198	19.5 %
d	Process Safety Information	177	17.4 %
f	Operating Procedures	174	17.1 %
e	Process Hazard Analysis	168	16.5 %
l	Management of Change	92	9 %

Top Chemical NEP Citations by PSM Element
(as per OSHA 2012 report)

Element	Description	Number	%
j	Mechanical Integrity	207	23.2 %
d	Process Safety Information	186	20.9 %
e	Process Hazard Analysis	141	15.8 %
f	Operating Procedures	125	14 %
l	Management of Change	49	5.5 %

CCPS Risk Based Process Safety Elements

1. Safety Commitment
   - Process Safety Process Safety Culture
   - Compliance with Standards
   - Process Safety Competency
   - Workforce Involvement
   - Stakeholder Outreach
2. Understand Hazards and Risk
   - Process Knowledge Management
   - Hazard Identification and Risk Analysis
3. Manage Risk
   - Operating Procedures
   - Safe Work Practices
   - Asset Integrity and Reliability
   - Contractor Management
   - Training and Performance Assurance
   - Management of Change
   - Operational Readiness
   - Conduct of Operations
   - Emergency Management
4. Learn from Experience
   - Incident Investigation
   - Measurement and Metrics
   - Auditing
   - Management Review and Continuous Improvement

Matrix for execution based on Risk Elements

	Phase 1: Safety Commitment & Understanding				Phase 2: Managing Risk & Learning from Experience				
PSM Elements / Factors	1	2	3	4	1	2	3	4	5
Time taken									
Total Cost									
Resources									
Current Status									
Safety Impact Rank									
Overall Priority									

Summary: This poster aims to present a simplified guidance for PSM implementation for smaller facilities, using a Risk Based Approach. A matrix presented intends to assist in developing risk based scores for PSM elements, combining Risk Based Process Safety Elements from CCPS and Time, Cost, Resources, Safety etc. considerations for a facility. The approach can be used to create a new PSM program or develop an existing program.

397593 Unlock the Value of Operational Data

Abstract:

Many forward-thinking industry executives now embrace the operational excellence mindset and approach and hold the belief that the best run organizations must operate not only productively, but safely and reliably. With more and more companies adopting process-based enterprise data platforms, predictive analytics offer a very promising approach to help organizations prioritize their investments and focus their efforts in risk mitigation. It may also allow the discovery of new solutions that are not likely to emerge from conventional observational programs or safety perceptions surveys.

This study aims to answer the research question that whether a combination of metrics that represents an asset's incident management process behavior can be found to effectively reflect the asset's risk reduction capability and therefore predict its safety performance. It stresses the importance of data quality and integrity for predictive analytics to avoid misleading analytical exercises. One of its key objectives is to identify proven benchmark indicators that are actionable, objective, routinely measurable from day-to-day activities, and are believable and predictive.

The presentation will highlight the rigorous process used in the predictive analytics as well as the systematic steps and change management approach to operationalize the benchmark indicators at local operational units. It features a case study of an energy company's effort in applying predictive analytics to gain insights from the operational data that they collected from an enterprise management system. The case study covers more than 900 million work hours with over 30,000 incidents and identifies benchmark indicators such as team competency, risk-driven actions, root-cause based mitigation, and learning-mindedness for estimating the corresponding injury rates (adjusted $R^2 > 0.8$). The predictive model was further validated using a new set of data.

Findings from this study showcase how organizations may leverage a process-based information system to drive performance by capturing day-to-day reporting behavior, measuring responsiveness and diligence in process execution, capturing the lessons from successes and errors, and allowing "learning" to take place systemically. While this study provides statistical evidence to support the strong correlation between process behaviors and safety performance, it also emphasizes the organizational factors, such as engagement, that are necessary to enable and sustain effective risk mitigation efforts.

How to Play a Better Central Role by Blending PSM System into HSE System

Liu Bin
CNOOC Safety Technology Service Co. Ltd
1 Xingang Yongtai Road Binhai District Tianjin, China
Liubin72@cnooc.com.cn

Sun Shu
CNOOC Safety Technology Service Co. Ltd
1 Xingang Yongtai Road Binhai District Tianjin, China
sunshu@cnooc.com.cn

Prepared for Presentation at
American Institute of Chemical Engineers
2015Spring Meeting
11thGlobal Congress on Process Safety
Austin, Texas
April 27-29, 2015

UNPUBLISHED

How to Play a Better Central Role by Blending PSM System into HSE System

Liu Bin
CNOOC Safety Technology Service Co. Ltd
1 Xingang Yongtai Road Binhai District Tianjin，China
Liubin72@cnooc.com.cn

Sun Shu
CNOOC Safety Technology Service Co. Ltd

Abstract

Abstract: It is known that PSM system enjoys a unique advantage in preventing and avoiding major process safety accident encountered by petrochemical and chemical industry. In recent years, PSM system has been being accepted and applied in China, firstly implemented by several leading companies, like CNOOC and CNPC, where had witnessed remarkable achievements. However, there are still lots of problems need to be solved in terms of blending PSM system into HSE system and fully making use of advantages of PSM system and HSE system. On the basis of its original HSE system, one company which belong to CNOOC had blended PSM system into HSE system and had fully made use of management advantages of PSM system and HSE system. As for the blending between PSM and HSE, CNOOC had blended some elements (like management of contractors, management of work permit) set up in PSM and HSE, by which a "Three Directions" "Four Lines" feasible and security management mode had been formed. The newly formed management mode is more suitable for the basic production performed in a company. More importantly, it is beneficial for a company to improve safety management performance.

Key words: PSM (process safety management), HSE, blending, "Three Directions", "Four Lines"

1. Introduction

Nowadays, the safety management used by China Petrochemical Corporation is mainly HSE system which will control different kinds of risks. However, with the development of production technology and production science, the requirement of safety production management gets stricter and higher. In recent years, we had witnessed some accidents, like the blowout in Kaixian County Chongqing and explosion in Jilin Petrochemical Company. Under this background, we have to rethink that whether there is something wrong with our HSE systems. Obviously, we can see that even though we relied on HSE systems, we did not pay much attention to process security. It is known that PSM (process safety management) enjoys some exclusive advantages

in preventing and avoiding process safety accidents happened in petrochemical enterprises and chemical enterprises. On September 6th 2010, our government issued Implementation Guideline for Process Safety Management of Chemical Enterprise (AQ/T 3034—2010) which was put into practice on May 1st 2011. Since then, process safety has been generally brought into China and firstly executed in CNOOC and CNPC where great achievements had been obtained.
In China, there are 12 elements involved in PSM, some of which are the same as that involved in HSE, including contractor management, emergency response, hot work etc. In this case, it is a key and emergent problem that how to combine HSE systems with PSM systems and how to make good use of their advantages. In this document, it mainly talks about the combination between HSE and PSM and how to well use their advantages in CNOOC. This document gives several examples, like how to combine the elements in HSE and PSM, by these examples, a practical and feasible "Three Directions" "Four Lines"safety management system which will better comply with the production facts in different enterprises and improve the performance of safety management might be obtained and formed.

2. HSE Management

In recent years, HSE has been considered as a common management system in international petroleum and natural gas industries. HSE system, combining experience of various countries and reflecting the standard operation of petroleum and natural gas enterprises located in big cities, sticks to a scientific management philosophy of "prevention first, promise by leaders, participation by all staff and continuous improvement". So HSE is a guarantee for petroleum and natural gas enterprise to realize modern management and go international. Formation and development of HSE system is based on many years' experience of petroleum exploration, which reflects an integrated management philosophy. However, there are some weaknesses of HSE, like it pays too much attention on document management and relevant procedures are too complicated, the practical application of safety technology is not emphasized, especially that safety guarantee for process system is too weak. Luckily, PSM is able to well work in preventing and managing process accidents.

2.1 Pyramid of Accident Management

In 1931, Heinrich, an American safety engineer published a book related to prevention of safety accidents which is called *One Scientific Method.* In this book, he proposed a famous principle "Accident Pyramid" which is based on an analysis of occurrence probability of 0.55million work accidents. And this principle was to be used for the operation of insurance companies. By this principle, it thinks that if there is one fatal and severe accident, then there will be 29 minor accidents; if there are 29 minor accidents, then there will be 300 non-injured accident and attempted accidents and large amounts of unsafe behaviors and insecure situations. The relationship among fatal accident, minor accident, non-injured accident, attempt accident, unsafe behavior and insecure situations is shown by a pyramid, which is called "Accident Pyramid".
With the development of modern production technology, the requirement of management gets stricter and higher. Today, based on the current management, we have a deeper understanding on Heinrich's "Accident Pyramid" (refer to Fig.1 for the new accident pyramid"). From the figure here below, we can see that the accidents are divided into accidents, attempted accidents and

near-miss. In the pyramid, the upper four tiers are accidents, the fifth tier is attempted accidents and the sixth tier is near-miss.

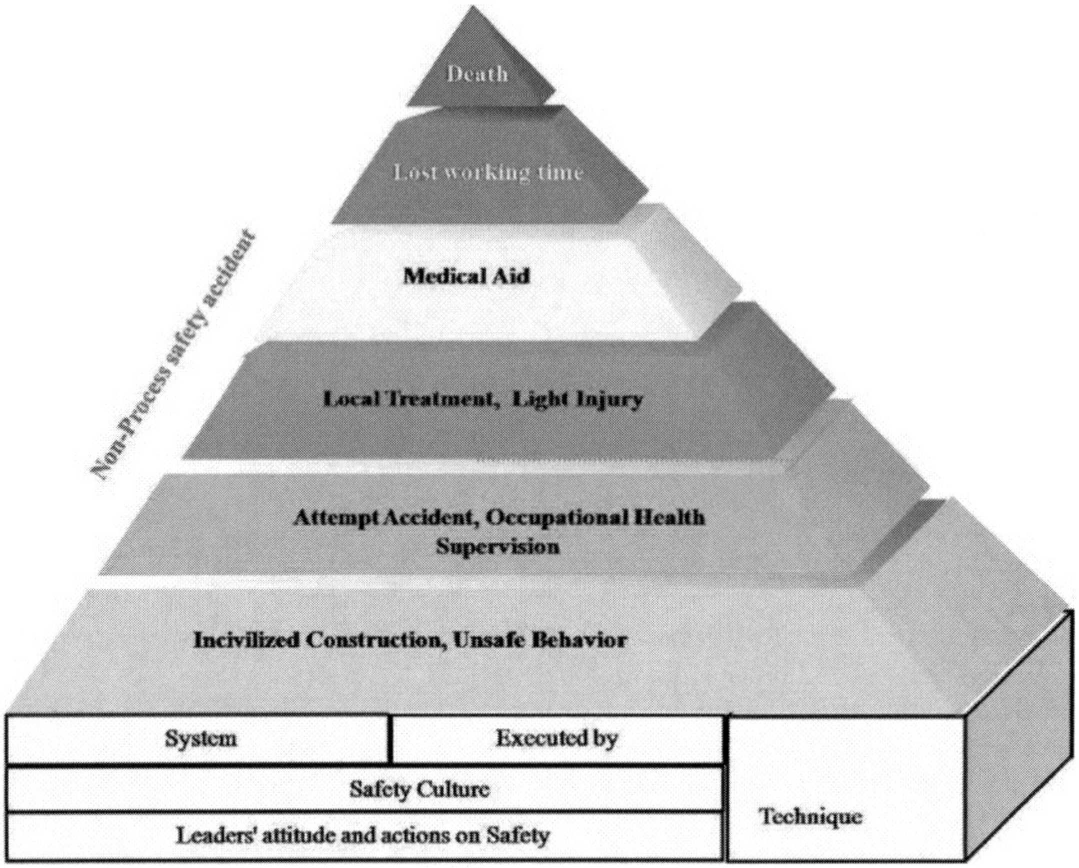

Fig.1: HSE Accident Pyramid

From Heinrich's "Accident Pyramid", we can know a rather important principle for accident prevention, that is if we want to prevent fatal and severe accidents, it is mandatory to prevent minor accidents first; if we want to prevent minor accidents, it is mandatory to prevent non-injured accidents and attempt accidents; if we want to prevent non-injured accidents and attempt accidents, it is mandatory to eliminate unsafe behaviors and insecure situations; to eliminate unsafe behaviors and insecure situations (unsafe behaviors and insecure situations are also called near-miss) mainly depends on our daily management. Anyway, our daily management, that is details management, is the most important basic work for preventing fatal and severe accidents.

2.2 Accident-pyramid-based HSE Management

According to the definition made by OHSAS 18001:1999, accident is a kind of unexpected event which will cause death, disease, injury, damage or other damages.

By the definition, it is known that accident is an unexpected situation which is not able to be predicted. According to the reasons of accident, we know that during normal operation, if there is an unexpected situation, the operation will get abnormal, which is not able to be controlled efficiently. At the same time, there is no preventive measure for the employees. In this case, accident will happen.

Based on the aforementioned analysis, we know that at the beginning of an accident, unsafe behavior and insecure situation (shown in the bottom of Accident Pyramid) will be occurred because of unexpected situations; then because there is no efficient control, attempt accident (no injury) will be resulted in, or even minor accident, severe accident and fatal accident. Therefore, we can say that accident control includes a series of actions, if there is something wrong with any action, the accident control might be failed and an accident might be caused. In our daily operation, based on the operation conditions in one company, comprehensive rules and regulations should be drawn up; the leaders should strictly follow the rules and regulations; the rules and regulations should be fully adhered to; daily safety management shall be emphasized by well managing details. In this way, unsafe behaviors and insecure situations will be lowered, by which general policy made by the company will be realized, major accidents will be prevented and employees' safety will be guaranteed.

3. PSM (Process Safety Management)

PSM targets to guarantee the safe design and secure operation of some processes used by chemical factories, oil refinery factories, natural gas processing factories and offshore drilling platforms. PSM focuses on the prevention of major process accident, like fire, explosion and leakage of toxic chemicals etc. In 1980s, a series of severe accidents happened, like leakage of toxic gas in Bhopal India. Under this background, the first law related to PSM was issued. In America, the most important PSM law is *Process Safety Management for Highly Hazardous Chemical* (29CFR1910.119) which was issued by OSHA in 1992. Even though PSM in China is not as advanced as that in America, it is believed that if PSM is able to be well carried out, a lot of accidents like blowout in Kaixian County Chongqing, Jilin Petrochemical Company and leakage in Shanghai Insecticide Factory are able to avoided.

3.1 PSE Pyramid

There is a definition in APIRP 754 for PSE, which is an unplanned or uncontrolled LOPC of any material whether or not on the UNDGL (e.g., steam, hot condensate or compressed air) from a process, or an undesired event or condition that, under slightly different circumstances, could have resulted in a LOPC of a material. From the definition, we know that PSE is an unplanned or uncontrolled LOPC of any material (whether it is listed on UNDGL, like steam, hot condensate or compressed air); or PSE is an unexpected event or condition which cause LOPC of any material. In API RP754, PSE are classified, see Fig.2.

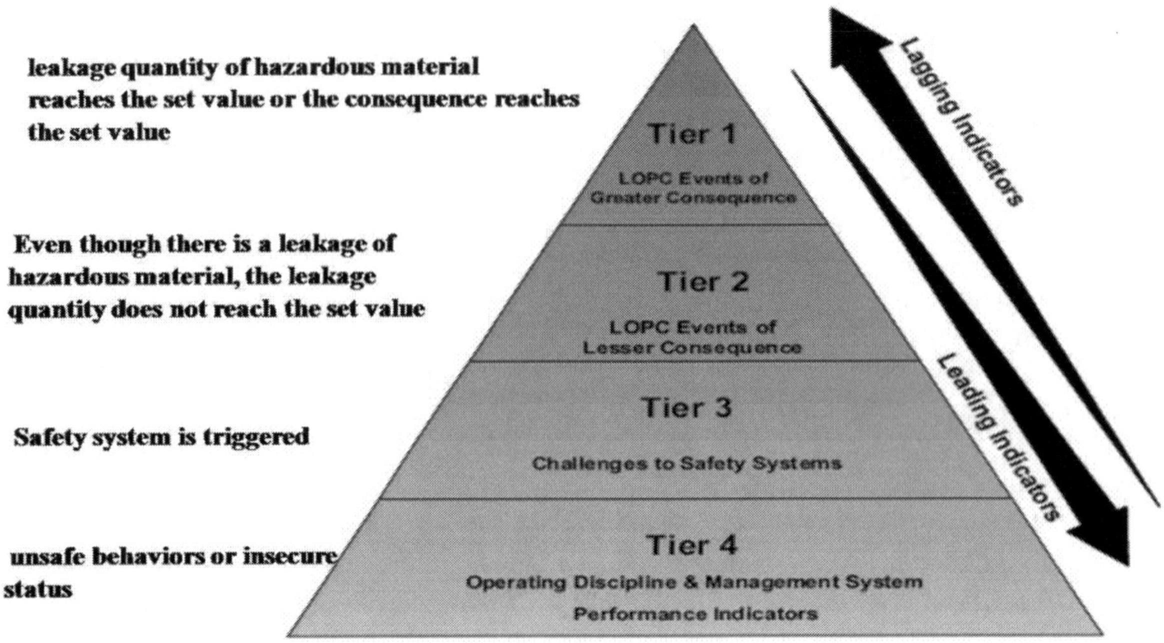

Fig.2: API RP 754 Accident Pyramid

In API RP 754, there are specific descriptions of PSE tiers, including the quantity of hazardous material's leakage, the economic loss and the personal injury. For example, if a PSE causes one injury or one death, community evacuation or community sanctuary, fire or explosion, a direct cost of $25,000 or more, an open air leakage of 500kg liquid (boiling point is less than 35 °C (95 °F) and flash point is less than 23 °C (73 °F)), or an interior leakage of 250kg liquid, it will be considered as a PSE of Tier 1. The PSE management is also based on Heinrich's theory, that is the management shall be started from the bottom of the pyramid, requiring that the small leakage, alarming, interlock trigger shall be lowered as much as possible, as well as wrong operations.

Combing HSE pyramid used in China, a PSE pyramid is able to be got. As it is shown in Fig.3, PSE are divided into accidents, attempt accident and near-miss, of which the upper four tiers are accidents, the fifth tier is attempt accident and the sixth tier is near-miss.

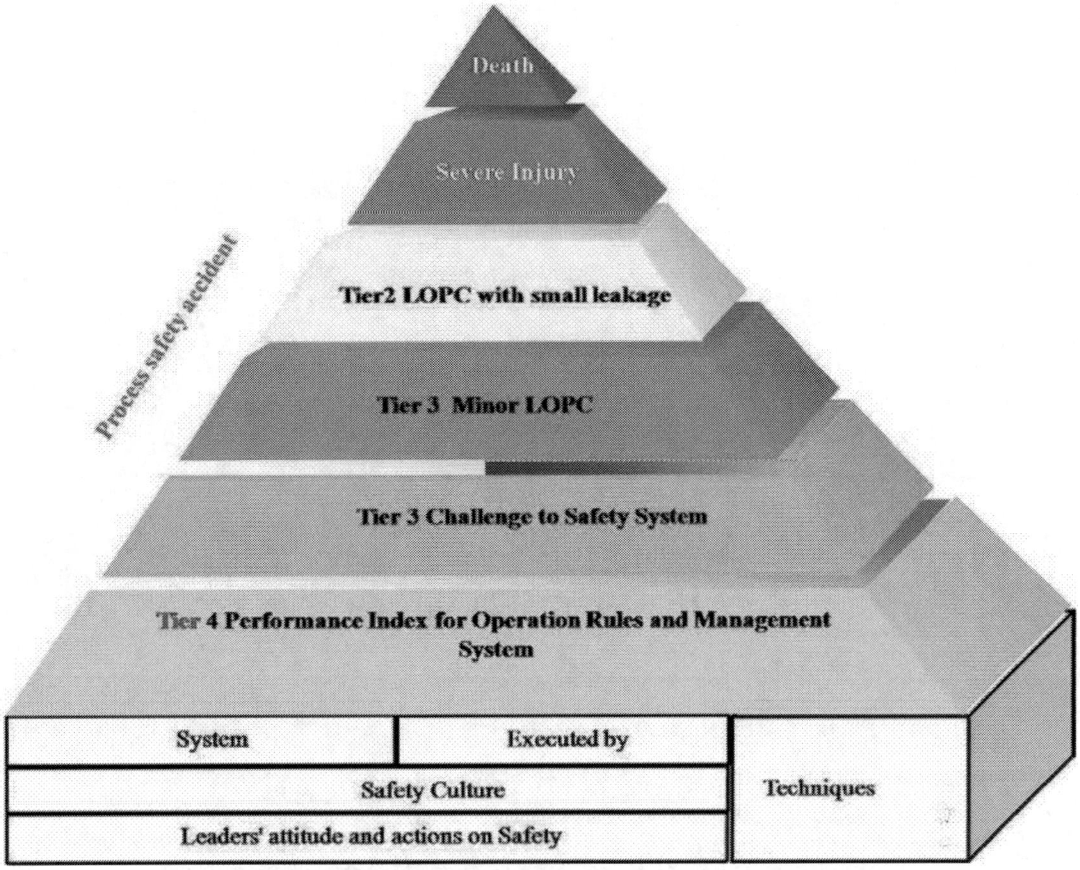

Fig.3: PSM Accident Pyramid

In order to avoid Tier 1 PSE, it is mandatory that our daily management should start with PSE in Tier 3 and Tier4. That is we should reduce the alarming in daily production and lower the trigger of safety interlock as much as possible. In addition, we should also require that all production-line workers shall strictly follow the operation regulations and SOP (Standard Operation Procedure). So in this case, near-misses will be eliminated and attempt accidents will be avoid, by which accident will not happen.

4. Combination of PSM and HSE

According to the aforementioned, both HSE and PSM, on the basis of management mechanism, safety culture, leaders' attitude on safety and leaders' actions, are used to control and manage the process risks and operation risks. By combining the HSE pyramid and PSE pyramid, we can see a new accident pyramid (refer to Fig. 4).

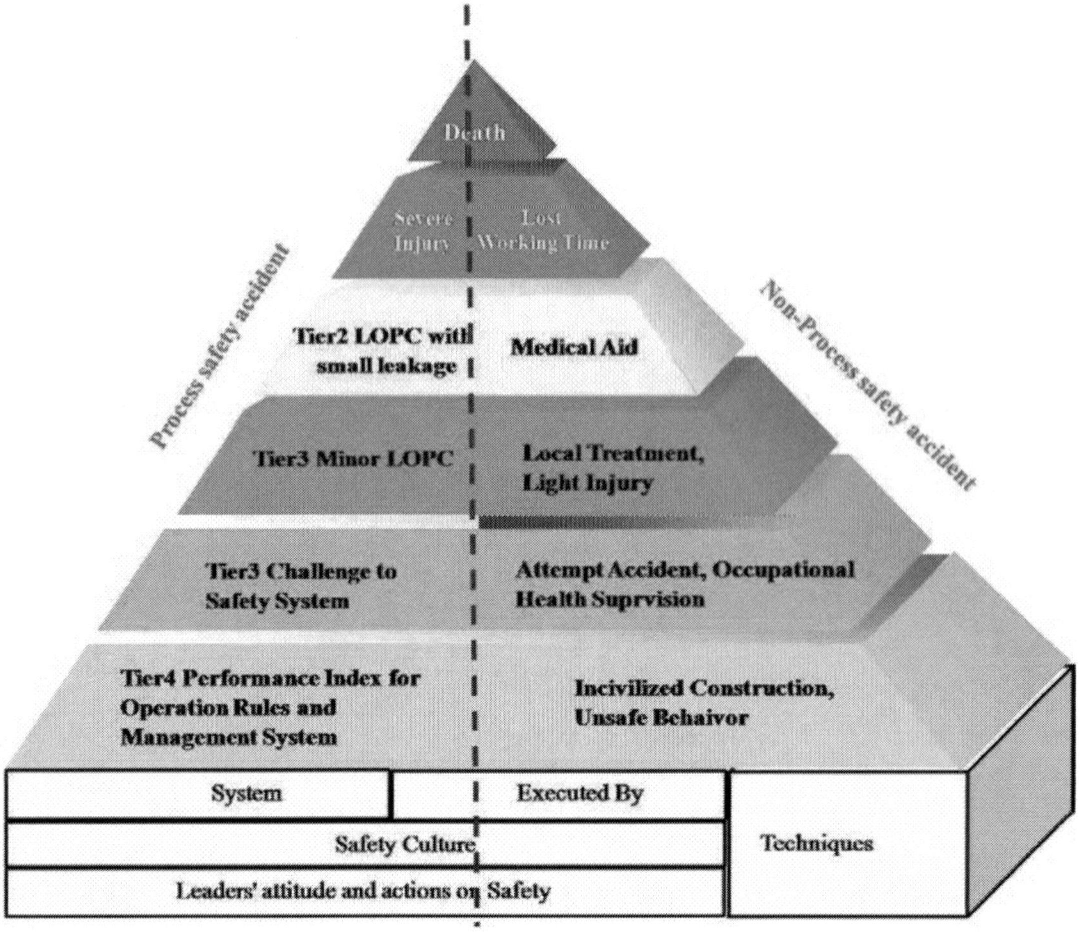

Fig.4 Accident Pyramid of HSE Combines with PSM

Based on the pyramid in Fig.4, we can make a conclusion that the combination between PSM and HSE is able to be carried out by the way of " Three Directions and Four Lines", with PSM played a core role. The final goal of PSM and HSE is to avoid accident and reach the goal of safety management.

4.1 "Three Directions"

From Fig4., we can see that risk control shall be started with three directions, which are "to avoid any accident", " to prevent any attempt accident" and "to eliminate any near-miss". First, focus on near misses. Potential near misses shall be found out by on-site inspection and hazard identification; then take relevant measures to eliminate these near misses. In this case, probability of the attempt accidents and accident might be lowered. Second, keep an eye on operations and behaviors and pay attention to the lessons. Well learn the happened attempt accidents and timely find out the potential attempt accident; propose some requirements on the systems and system execution. In this case, the probability of accidents might be reduced. Therefore, PSM shall be combined with HSE management. That is to say when it is to eliminate near misses, the relevant process safety concerns shall also be eliminated, by which attempt accidents will be prevented; still, attempt accidents linked with HSE and PSM shall be taken into account. So we can conclude that it needs to consider process safety while we are thinking about HSE. By combining HSE and PSM, the management will be more comprehensive.

4.2 "Four Lines"

From Fig.4, we can see that our management can be started with four lines --- first line, have a head-to-toe inspection to find out the accident reasons; second line, make a head-to-toe foundation; third line, eliminate all near misses; forth line, safety culture and participation by all staff.

As for the first line, have a head-to-toe inspection to find out accident reasons, it is necessary to carefully think about the process safety events or non-process safety events occurred in the our company or other companies; then make a conclusion that what can we learn from these events and why these events would happen and what consequences are caused. In addition, it is necessary to find out the accident chain (the reason and the consequence). Based on the chain, have a troubleshooting in the company; if there is any management weakness, then draw up relevant management system and strengthen the management; if there is any technological weakness, timely launch technical improvement and go on with the production by rather safe techniques.

As for the second line, make a head-to-toe foundation. From Fig.4, we can see that at the bottom of the pyramid, there are company systems, system execution, techniques used by a company, safety culture of a company, leaders' attitude and actions on safety. All these can be considered as the foundations of a company's safety management. Only the foundation is solid enough, can the accidents, attempt accidents and near misses be reduced or avoided. Therefore, if a company wants to have a good safety management, it is necessary to start with the company system, making sure that the systems are perfect and rational. And it is necessary to draw up some systems which reasonably comply with the company's real situation and cover all operations and productive actions performed in the company. When drawing up systems, it is necessary to take into account of the management on different kinds of operations, as well as the management on productive operation, material drainage, start-up and shut-down. Once perfect and rational systems are drawn up, the leaders should make themselves an example to other staff and positively take part in the safety management. In this case, all staff will emphasize safety and do their best to guarantee the safety of the company.

As for the third line, eliminate all near misses. In fact, the most severe near miss is that we do not know what the near miss is. HSE management is all for the process production. Some operations covered by HSE management, like working at heights, climbing high operations, operations in confined space, are based on the requirements of process production. From Fig.4, we can see that at first, in the view of process safety, it is necessary to eliminate near misses related to process safety, like alarming ratio and interlock trigger ratio. If the process production is in a stable and good condition, HSE operations will be lowered and the accident will be reduced accordingly. Of course, HSE near misses shall also be eliminated, for example, on the basis of troubleshooting system in HSE system, we can make a new troubleshooting system in PSM system. Additionally, near misses linked with process production should be considered, for example, if an operation in confined space (distillation column) is required, it is mandatory to carefully think about the near misses caused by combustible gas in the distillation column and it is mandatory to think about the near misses caused by entering the distillation column.

As for the forth line, safety culture and participation by all staff. The concept of safety culture was firstly defined in an accident report (INSAG-1) issued by INSAG while the definition of safety culture was firstly given on caution lights issued by CCPS in June 2006, which is how to perform when there is no supervision. PSM is not a procedure given by the management layer to the employees and contractors; instead, it is a procedure involving all staff. The key words of PSM are not only communication and management but also "participation". That is to say all stall, including administrative employees, employees and contractors, shall undertake the responsibility of PSM. In order to make PSM successfully executed, it is required that the top management shall implement and promote safety systems by themselves; the top management should take part into safety management and make themselves examples to other staff. Besides, the top management shall take part into other actions, like drawing up management systems, system execution, supervising the system execution. Since process safety plays a core role in safety management, so the 14 elements of process safety shall be performed in an efficient way. How to play a core role by process safety shall be learned by HSE management. That is to say every one shall take part; even though there is no supervision from the leaders, safety production and safety culture shall be always remembered and carried out.

5. Conclusion

It is well known that HSE and PSM have their own advantages. However, according to the current production and management situation, it is badly required that PSM should be implemented on the basis of HSE and PSM should play a core role in safety management. From Fig.4, it is concluded that on the basis of HSE, it is necessary to add some PSM actions, like management contents and management documents. What's more, strictly adhering to "Three Directions and Four Lines" (to avoid any accident, to prevent any attempt accident, to eliminate any near miss; have a head-to-toe inspection to find out accident reasons, to make a head-to-toe foundation, to eliminate all near misses, safety culture and participation by all staff), the performance of safety management in a company will be highly improved, the accidents will be avoided and safety production will be achieved.

References

[1] Process Safety–Recommended Practiceon Key Performance Indicators OGP Report No. 456, November 2011

[2] API. 2010. ANSI/API Recommended Practice 754:Process Safety Performance Indicators for the Refining and Petro- chemical Industries, (First Edition, April 2010). Available from API at: http://www.api.org/Standards/new/api-rp-754.cfm

[3] Safety production accident, near-miss management regulations February 2011

[4] Occupational health and safety management system implementation guidelines, GB/T28002—2011。

15AIChE
Spring Meeting, Austin, TX

Heat Exchanger Design Parameters Effect on Low Pressure Side Maximum Pressures during a Tube Failure

Dustin Smith
Smith & Burgess, LLC
7600 W. Tidwell Rd., Ste. 600, 77040
Dustin.Smith@smithburgess.com

John Burgess
Smith & Burgess, LLC
7600 W. Tidwell Rd., Ste. 600, 77040
John.Burgess@smithburgess.com

Zubin Kumana
Smith & Burgess, LLC
7600 W. Tidwell Rd., Ste. 600, 77040
Zubin.Kumana@smithburgess.com

Prepared for Presentation at
American Institute of Chemical Engineers
2015 Spring Meeting
11th Global Congress on Process Safety
Austin, Texas
April 27-30, 2015

UNPUBLISHED

Heat Exchanger Design Parameters Effect on Low Pressure Side Maximum Pressures during a Tube Failure

Dustin Smith
Smith & Burgess, LLC
7600 W. Tidwell Rd., Ste. 600, 77040
Dustin.Smith@smithburgess.com

John Burgess
Smith & Burgess, LLC
7600 W. Tidwell Rd., Ste. 600, 77040
John.Burgess@smithburgess.com

Zubin Kumana
Smith & Burgess, LLC
7600 W. Tidwell Rd., Ste. 600, 77040
Zubin.Kumana@smithburgess.com

Keywords: Overpressure Protection, Dynamic Tube Rupture, Cooling Water Exchanger Design

Abstract

Commonly, heat exchangers are used to heat and cool very high pressure fluids (over 1,000 psi) with an incompressible low pressure fluid. API STD 521 recommends that dynamic analysis be performed for heat exchangers in this service to determine the maximum pressure on the low pressure side in the event of a tube failure. This paper describes the effects of various design criteria that are used for construction and installation of heat exchangers on peak pressures after a tube failure. This analysis will focus on high pressure natural gas being cooled with water. Using a transient simulation of both the momentum and fluid flow properties of a high pressure vapor injection into a liquid full system, both the shock effects of the acceleration and the steady state relief is considered. The results of the modeling will be discussed and general construction and installation guidance will be given.

1. Introduction

This paper looks at a series of different installation considerations for heat exchangers with high pressure gas on the shell side and cooling water flowing on the tube side. Heat exchangers with this type of setup are being used with increasing frequency in petrochemical industries, and the pressures of the gases are increasing.

This paper helps the relief systems or plant designer understand how different design factors affect peak pressures if a heat exchanger suffers a full tube rupture. The pressures presented in this paper are based on the model presented by Sumaria et al.

This paper looks at the following variations in design parameters and how they affect the maximum pressures in various components of the system.

1. *Cooling water lateral diameter* - For each model, the exchanger was assumed to be attached to a 24" main cooling water supply and return header, and a 4", 6" and 8" diameter cooling water lateral was used to connect the exchanger to the main headers.

2. *Cooling water lateral lengths* - For each of the different lateral diameters that were investigated, lengths of 45', 75', and 150' were used to connect the exchanger to the main headers.

   Cooling water lateral flow resistance - For one of the cooling water lateral systems (6" diameter and 75' length), the effect of varying the fitting resistances (K values) was investigated.
3. *Asymmetric lateral diameters* – The effects of having different inlet and outlet lateral diameters were investigated for 75' of lateral piping.

4. *Effect of a rupture on exchangers in series* – A system with three heat exchangers in series was investigated to see the effects of a tube rupture in one exchanger on the other two.

5. *Effect of a rupture on exchangers in parallel* – A system with three heat exchangers in parallel was investigated to see the effects of a tube rupture in one exchanger on the other two.

2. Description of Model

The model is an implementation of the one presented by Sumaria et al. [1], consisting of a defined network of control volumes connected via inertial segments. The pressure within each control volume is defined by a single average value across the control volume. One or more segments are identified as the location where a tube rupture occurs.

The flow entering the system via the ruptured tube is modeled based on an orifice flow calculation using the tube internal diameter, upstream pressure and physical properties, and the pressure in the control volume in which the rupture occurs. Flow from a single orifice was introduced to the both inlet and outlet channels; thus, from an overpressure protection standpoint the system was modeled with high pressure vapor inlet flow from 2 orifices.

The differential equation governing a control volume is:

$$\left(\frac{V_l}{B_l}+\frac{V_g}{kP}+\frac{VD}{TE}(1-u^2)\right)\frac{dP}{dt}=\frac{W_i-W_o}{\rho_l}+\frac{W_g}{\rho_g} \qquad \text{Eq. (1)}$$

Where:

V_l = Liquid phase volume
V_g = Vapor phase volume
V = Liquid phase volume
B_l = Bulk modulus
k = Isentropic coefficient
D = Diameter
T = Material thickness
E = Modulus of elasticity
u = Poisson ratio
W_i = Flow into control volume
W_o = Flow out of control volume
W_g = Gas flow into control volume
P = Pressure
ρ_l = Liquid density
ρ_g = Vapor density
t = Time

The differential equation governing an inertial segment is:

$$\frac{L}{Ag_c}\frac{dW}{dt}=P_i+P_o-\Delta P_{flow} \qquad \text{Eq. (2)}$$

Where:

L = Actual length of fluid mass
A = Cross - sectional area of fluid mass
g_c = Gravitatio n conversion constant
P_0 = Pressure at time zero
P_i = Pressure at time step i

The pressure nodes, associated control volumes, and linking inertial segments were chosen to suit the layout of the system under analysis. The main cooling water supply and return headers were treated as infinite reservoirs at the stated supply and return pressures. Each segment was solved to obtain a steady state prior to the introduction of the tube rupture. The resulting system of differential equations was solved using a classical fourth-order Runge-Kutta method.

3. Results Single Exchanger Modeling

This section of the paper describes the results of the items investigated for changing the parameters for a system with a single heat exchanger. For single exchanger systems, the following parameters were used:

1. *High pressure gas* – A Natural Gas with a MW of 18.4 at 200 °F. The gas pressure upstream of the rupture was varied from 250 psig to 5,000 psig.
2. *Heat exchanger* – The heat exchanger channel was 1.5' long by 25" in diameter with 250 36' long ¾ BWG 16 u-tubes.
3. *Cooling water system* – The cooling water system consisted of a supply pressure of 75 psia and a return pressure of 60 psia. The main headers were 24" diameter.
4. *Overpressure protection* – The exchangers were modeled without any overpressure protection other than the normal cooling water flow paths to the tower (which are presumed to remain open).
5. *Cooling water flow* – The flow of the cooling water was not restricted. There were no check valves on the supply side to prevent flow reversal of the cooling water during the high pressure transient.

Some of the parameters such as resistance of fittings in the lateral and additional flows in the main header and differences in header lengths and lateral resistances were further investigated in the various sections.

3.1 Variation of cooling water lateral diameters and lengths

For all of these models, the exchangers were assumed to be attached to a 24" main cooling water supply and return header, and a 4", 6" and 8" diameter cooling water lateral was used to connect the exchanger to the main headers.

Figure 1 shows the results for the variation in inlet diameter and lengths for laterals. The 45'x4" in the legend is shorthand for a 4" cooling water lateral that is 45' on the inlet and outlet. The peak pressures shown in Figure 1 are the highest pressure predicted in the cooling water lateral and/or inlet and outlet exchanger tube side channel. Figure 2 shows the peak pressures are higher in the inlet and outlet piping than the exchanger. This was representative of a majority of the cases investigated in this study.

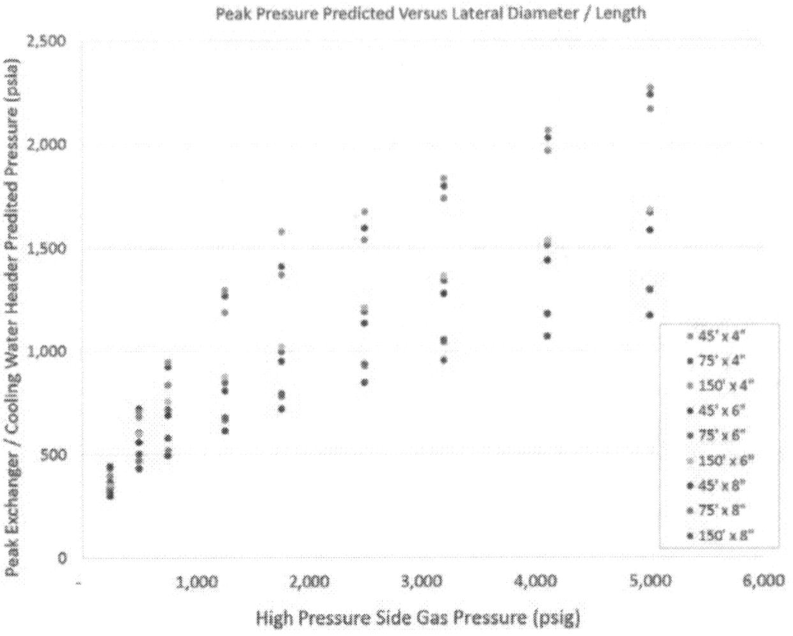

Figure 1. Comparison of Lateral Diameters and Lengths

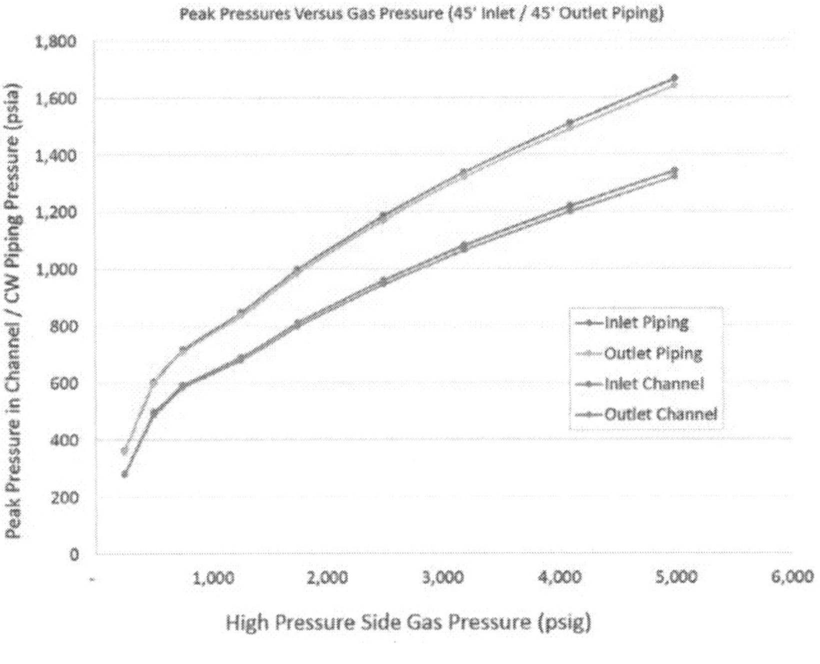

Figure 2. Comparison of Peak Pressures for a Single Exchanger Run

The results shown in the preceding charts suggest that the most important parameter for limiting pressure to the cooling water system is the diameter of the lateral piping. While it was not investigated, it could be presumed that at some point the diameter of the header piping would also be an important factor.

Figure 3 shows the maximum pressures in the systems with 75' of lateral piping for cases with asymmetric inlet and outlet lateral diameters. The results shown in Figure 3 indicate that the peak pressures predicted for asymmetric piping are between the larger and smaller lateral diameters (as contrasted to being similar to the peak pressures predicted for either the smaller or larger diameter lateral).

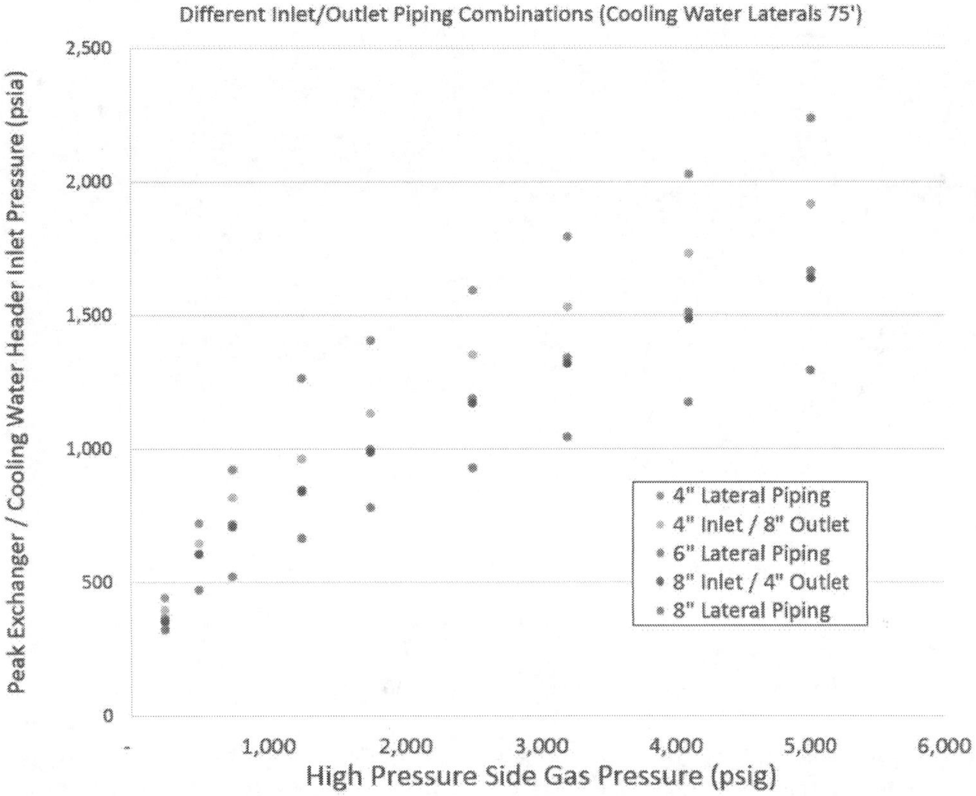

Figure 3. Comparison of Asymmetric Lateral Diameters

3.2 Variation of cooling water header lengths and lateral resistances

The length of the cooling water header supply and return piping was increased some (quadrupled) and then increased significantly (ten times) to determine the effect on the maximum predicted pressure in the exchanger and lateral piping. In addition, the flow resistance was increased (doubled) and then increased significantly (six times) to determine the effect on the maximum predicted pressure in the exchanger and lateral piping. The base case (Figure 1) that was used for comparison was the one with 75' of 6" lateral piping to the heat exchanger.

The flow resistances in the laterals were increased by first doubling and then subsequently tripling the doubled K values for the lateral piping in the model.

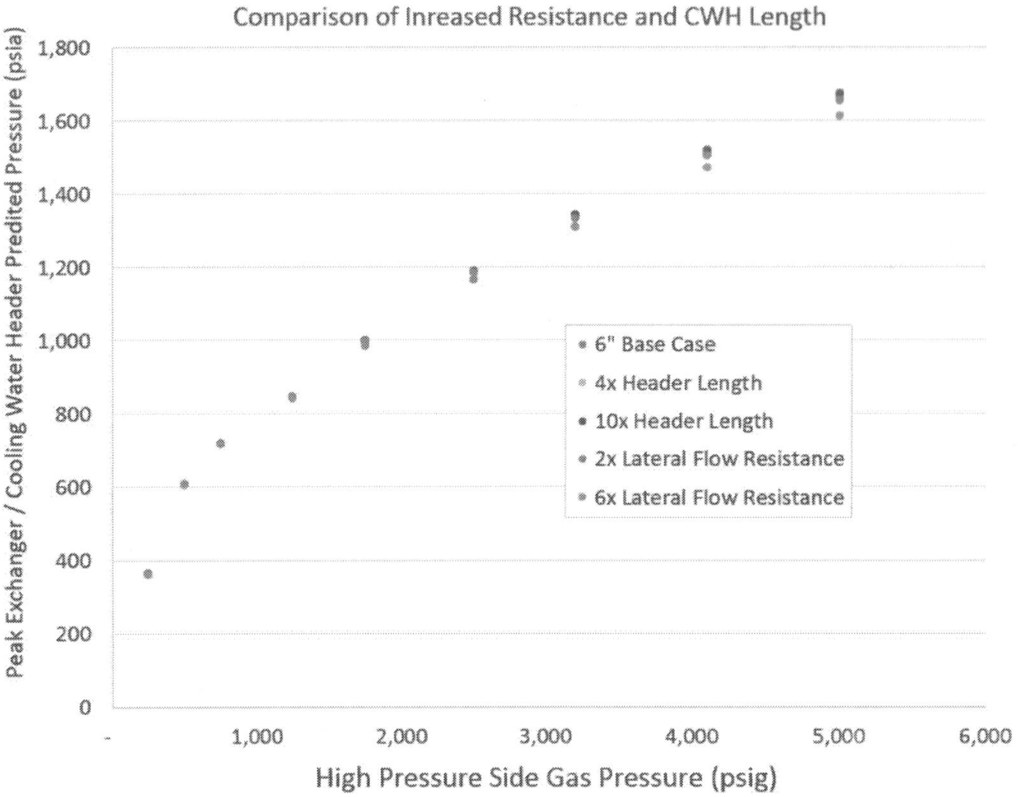

Figure 4. Comparison of Header Lengths and Lateral Flow Resistances

Since the series significantly overlap in Figure 4, it is shown that within the magnitude of the changes reviewed, increasing the header length and/or the lateral resistances do not impact the peak pressures in the system.

4. Multi-Exchanger System – 3 Exchangers in Series

This section of the paper describes the results of the consequences associated with a tube rupture in a system where there are three heat exchangers on a single lateral (as shown in Figure 5). For these runs the same parameters as for the single exchanger plus the following parameters were used:

1. *Exchanger Layout* – The inlet and outlet laterals were modeled as 45' and the cooling water (tube) side of these heat exchangers were assumed to be bolted to each other flange to flange.
2. *Rupture* – Each exchanger was assumed to rupture independently. Thus, it was not assumed that a rupture in one exchanger would cause another in series to also fail.

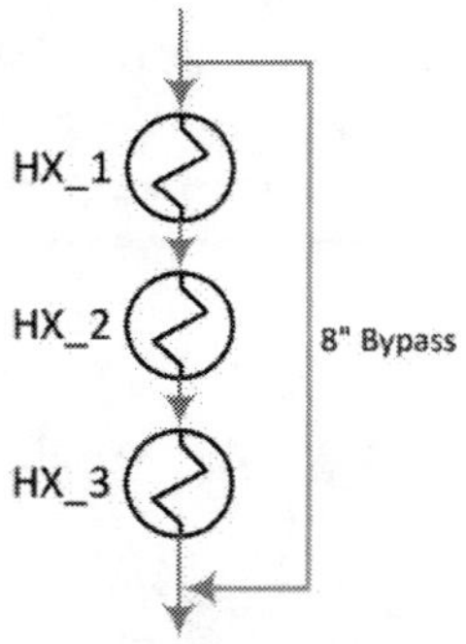

Figure 5. Arrangement of heat exchangers in series (shown with bypass variant)

As shown in Figure 6, when one of the exchangers in the series ruptures, the peak pressures in the other exchangers are within 15% difference of each other. Thus, all exchangers in the series need to be designed for the peak pressures.

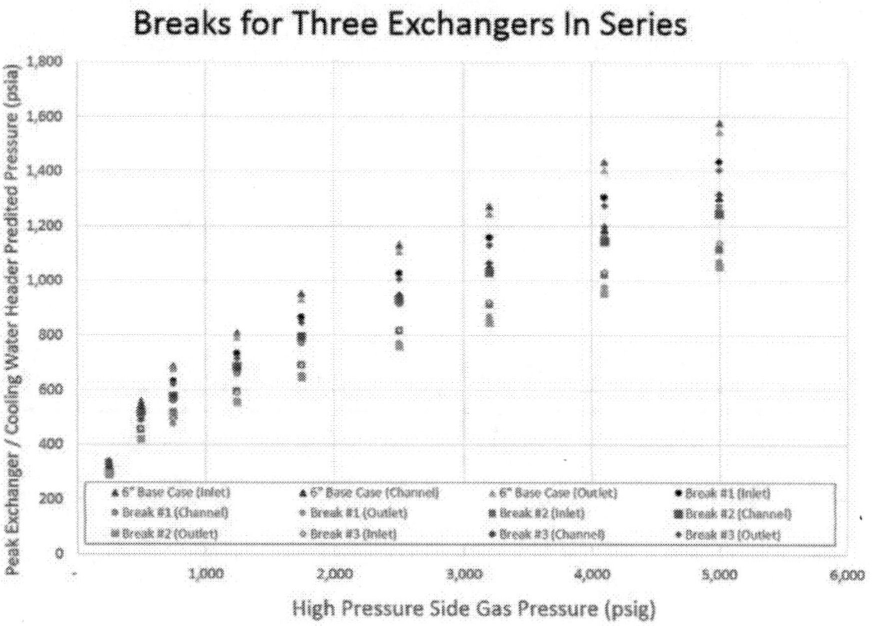

Figure 6. Comparison of Peak Pressures for Exchangers in Series

In Figure 6, Break #1 (Channel) indicates that the pressures reported are the peak pressures for all of the exchangers when there is a tube rupture in Exchanger #1. Inlet shows the pressure in the inlet lateral to the bank of exchangers and outlet shows the pressure in the outlet lateral. A run, similar to those shown in Figure 6, was performed with an 8" bypass around the exchangers connecting the cooling water supply header with the cooling water return header. This was done to estimate the effects of increased cooling water flow rates in headers on the peak pressures. Other than lower peak pressures (15% difference) in the cooling water inlet and outlet headers, the pressures were within 0.1% difference of those in Figure 6. Note that these comparisons were for a break in the middle exchanger. Breaks in the first and last exchanger were not investigated due to the lack of change in the peak pressures from the analysis performed on the middle exchanger.

5. Multi-Exchanger System – 3 Exchangers in Parallel

This section of the paper describes the results of the consequences associated with a tube rupture in a system where there are three heat exchangers connected to a common supply and header in parallel lateral (as shown in Figure 7). For these runs, the same parameters as for the single exchanger plus the following parameters were used:

1. *Exchanger Layout* – The inlet and outlet laterals were modeled as 45 feet and the cooling water (tube) side of these heat exchangers were assumed to be connected to the header 20 feet apart.
2. *Rupture* – Each exchanger was assumed to rupture independently. Thus it was not assumed that a rupture in one exchanger would cause another in series to also fail.

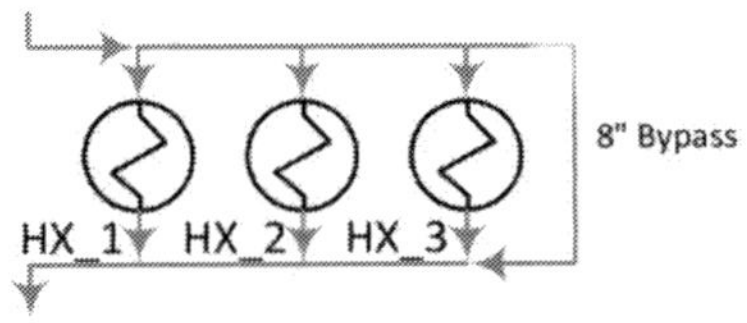

Figure 7. Arrangement of heat exchangers in parallel (shown with bypass variant)

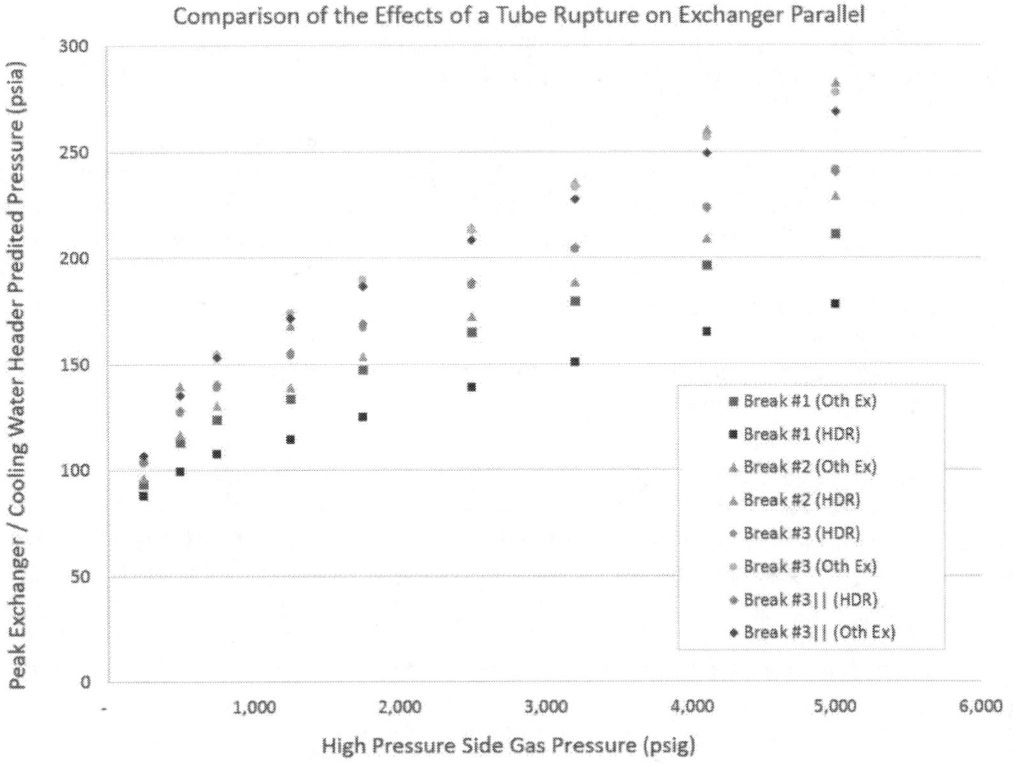

Figure 8. Comparison of Peak Pressures for Exchangers in Parallel

In Figure 8, (Oth Ex) is the peak pressure in the non-ruptured exchangers and laterals and (HDR) is the Peak pressure in the supply and/or return headers. The peak pressure of the exchanger that ruptures was not included in Figure 8 as it was similar to the other pressures presented in this paper and made it difficult to see the difference.

As shown by Figure 8, the larger system with higher header flows do not significantly affect peak pressures in the other system exchangers. This is consistent with the results found for the exchangers in series comparison.

3. Conclusion

When designing a cooling water system, a designer should consider the following parameters.

1. *Lateral Diameter* - The biggest impact as seen throughout the figures in this paper is the diameter of the lateral that leads from the supply header to the exchanger and then back to the return header.

2. *Lateral Length and/or Resistance* – The length of the laterals and the resistance (K) from fittings did not significantly affect the peak pressures observed in this model (as compared to the lateral pipe diameter).

3. *Header Lengths* – Changing the lengths of the cooling water supply and return headers did not increase the system peak pressures appreciably. However, this may not be true for a smaller diameter header (or larger diameter laterals). The smallest ratio explored in this study was a 24" diameter header with 8" laterals.

4. *Exchangers In Series* - The pressure in the exchangers in series were all significantly increased when one ruptured. A system designer should be careful that the combination of any high pressure exchangers with lower pressure exchangers on a single line may overpressure all exchangers on that lateral.

5. *Exchangers in Parallel* – When a system has multiple exchangers hydraulically close, the rupture in one can affect the peak pressures in nearby exchangers. This effect is less severe than when exchangers are in series, but could exceed the design pressure especially dependent on the high pressure side gas pressure.

6. *Peak Pressures* - As seen in many of the figures, the peak pressures seems to be a strong function of the vapor flow through the broken tube. This rate is dependent on the square root of the high side pressure.

4. References

[1] Sumaria, V.H., Rovnak, A., Heitner, I, Herbert, R.J. Model to Predict transient Consequences of a Heat Exchanger Tube Rupture, API Proceedings – Refining Department, Vol 55. 1976. Pp 631-654.

Research on Security Access Model of Enterprises in Petrochemical Ports

Haoyi Wang, China University of Petroleum (East China), Qingdao, China
Dongfeng Zhao, China University of Petroleum (East China), Qingdao, China

Abstract

Port security has initiated increasing attention, with the vigorous development of China's port undertakings. Port layout planning based on risk currently in the country is still in its infancy, is a relatively new area of research. Reasonable harbor layout planning can decrease the risk and improve the overall safety level. This article focused on research conducted for the ports enterprise access, first elaborated the principle of access to the harbor and the main process steps; second models and calculation methods for quantitative risk assessment has been described; the current quantitative calculation didn't consider personnel vulnerability, so improved algorithm is proposed; according to the characteristics of harbor this article selected the 12 secondary indicators and 30 third indicators from three aspects which constructed qualitative evaluation index system, and made interpretation of selection of indicators; thirdly this paper established a comprehensive index of qualitative evaluation model, using AHP to determine the weight, fuzzy comprehensive evaluation determining access and sort of candidate based on TOPSIS method.

Introduction

The number of China's large-scale ports have increased constantly with economic development these years. The function of ports is partly storage of hazardous chemical, therefore considering domino effect and risk for the port itself and surrounding area the optimization of plant layout is must-have during design period.

China's related law and regulation have stipulated some compulsive rules to ensure the management of hazardous chemicals such as periodical safety assessment. However Port Law did not regulate the specific steps of how to choose enterprise and keep harbor's risk under control. Most decision makers do not have more reliable and objective method to evaluate the risk level of an enterprise applying for admittance compared with petrochemical industry park, which already has established mature software to carry out risk assessment of planning scheme. Therefore, port enterprises access and related researches are not developed systematically.

There are a few studies on risk management for the related literature in port projects. Yip (2008) studied port traffic risk issues by discussing historic accidents in the Hong Kong port for ameliorating risk management. Balas et al. (2000) developed a project management model that handles the risk factors and uncertainties encountered in the design and construction of coastal projects, named the optimum risk management (ORM) model. Hassan Hashemi et al. (2013) presented a novel compromise ranking approach based on bootstrap confidence intervals according to a new weight of the strategy for the majority of risk assessment criteria and optimism level of experts in port projects. Barbara Zanuttigh at al. (2013) presented An open-source Spatial Decision Support System developed within the THESEUS Project to help decision makers to scope optimal strategies to minimize coastal risks.

From the safety viewpoint, referring to overseas ports environmental assessment standard, this paper constructs an enterprise access qualitative and quantitative model based on risk. In this work, QRA calculation has been optimized by personnel vulnerability from age structure, health condition, training level and information acceptance ability four factors; qualitative assessment focused on identifying key factors of affecting ports risk level and forming qualitative evaluation index system, using analytic hierarchy process (AHP) determining each factor's weight, fuzzy comprehensive evaluation combined with TOPSIS method to rank different candidates. Results generated from this approach will enhance safety in the conceptual design and layout stages of plant design.

Problem statement

In this paper, qualitative and quantitative risk assessment are used to be standards evaluating plant risk simultaneously. If planning scheme failed to satisfy either of them, it cannot be accepted as eligible scheme from safety assessment government department.

Quantitative Risk Assessment mainly uses individual risk curve and social risk curve this two factors as evaluation criteria, and the PHAST developed by DNV could help calculate these two curves and analyze whether it could meet risk tolerable criteria (RTC) of 40th government practice. If simulated results are above RTC, the planning scheme should be modified by reduction of risk such as layout reconstruction or adding more protective equipment, until it is up to standard.

Qualitative comprehensive assessment primarily needs to understand the whole operation and management process of port, select and identify crucial factors profoundly effecting the safety level, construct three-class evaluation system. Analytic hierarchy process (AHP) determined weight of each index, fuzzy comprehensive evaluation and TOPSIS ranking method evaluated whether enterprises could be stationed in this port yard.

Optimization of QRA based on personnel vulnerability

Staff in port enterprises, relative administrators, and people in surrounding unincorporated villages consist of the whole population in ports. Different qualities could result in different ability fighting for disaster. The following analysis focuses on influencing factors considering personnel vulnerability and how to optimize the calculation of individual risk.

3.1 analysis of several personnel vulnerability factor

3.1.1 Age organization and gender proportion

The elderly suffer body function aging while children are still immature and undeveloped physically and psychologically. In comparison with male, female usually would become nervous and flustered so it can be detected that female have higher vulnerability than male. Once accident happened, they are more likely to miss the best escape time as a result of lack of escape knowledge and weak physical quality. As a consequence, the higher proportion of this sort of people is, the casualties may be more serious in disaster.

3.1.2 Health condition

People in poor health always receive, understand, respond to emergency evacuation information slower, for instance, patients in hospital, the disabled and etc.

Qianwen Song researched residents experiencing toxic release and found that whether disable people are at home are dominant factor affecting escape time. Chuansheng Jiang found the disabled are the most vulnerable within limited escape time and space.

As can be seen the area vulnerability rises as the proportion of people in bad physical condition increases.

3.1.3 Training level

Yunfeng Deng investigated emergency evacuation of a chlorine leakage accident in Chongqing, 76.8% people with no training and experience escape with 30 minutes compared with 88.2% people with training and experience people with training and experiences are more likely to shorten escape preparation time. It can be concluded that increasing ratio of people with knowledge of self-rescue can reduce the casualty rate.

3.1.4 Information identification and reception

The reason why this factor could directly change response time after accident is the information spreading channel and management organization. Information spreading via internet and media sometimes could misguide the public. Factory, school or residential area are cohesive places which could achieve high-effective information spreading in a short period. In conclusion, distributed personnel places like ports are faced with high vulnerability.

3.2 QRA optimization of personnel vulnerability

This paper proposes a compensation value to right social risk curve (F-N curve), as shown in function 3-1

$$SP(x, y) = IR(x, y) \times \rho_p \times P_e \qquad (1)$$

Where Pe is the correction factor used for vulnerability, which has a value considering these four aspects.

$$Pe = Pa \times Ph \times Pt \times Pi \qquad (2)$$

Where Pa, Ph, Pt, Pi are respectively age structure and gender proportion, physical condition, training level and information acceptance.

As for how to determine values of these four factors, there are some acceptable methods to use.

(a) The value of Pa is related to average age. According to the bar graph of age structure of neighboring residents, value of Pa could be dependent on the different feature of graph. Average age is easy to implement but the accuracy and objectivity is not as good as the characteristic of age structure.

(b) The proportion of the disabled in China is 6.54% as benchmark (=1), proportion of weak physical condition in the evaluated area increase by 1% so that the value of Ph added to 0.01.

(c) Contingency Plan, safety training, emergency exercises recording could reflect the training level of people. So assessing these factors would help determine the value of Pt

(d) The value of Pi depends on the ways that people get reliable information and the organization structure of port yard and neighboring area.

The qualitative assessment system of port enterprises access

Besides QRA mainly concentrating on risk, there are some factors affecting the evaluation of enterprises admittance. Those are vital and useful to reflect the entire safety condition. So to select key factors and construct assessment system among complicated and impact port production are on the basis of safety systems engineering.

4.1 principles of qualitative assessment system construction

Qualitative assessment is always constituted by a group of experts, the objectivity and rationality concerns whether the results can be acceptable and guide the siting planning. Here are some principles shown as follows to help experts make decisions.

(a) Comprehensiveness:
The core of evaluation is to keep risk tolerable and locate industry reasonably, therefore the factors are selected from all aspects of port safety.

(b) Systematization
A complete system ensures each aspect are taken into account and the evaluation are scientific.

(c) Operability
Each index meaning should be described without confusion, help consulting group investigate and collect materials.

(d) Policy conformity
The determination and design of every factor must not be against current law and regulations.

(e) Emphasis
Although the comprehensiveness should be considered, design of factor should emphasize on some influential aspects.

4.2 system construction Procedure

After selection of industry material, the goal of evaluation and identification of influential factors should be finished by comparing similar factors and picking up essential ones.

Application of AHP to determine the weight of index is the first step of assessment, next input expert scores, carry out fuzzy comprehensive evaluation and draw a conclusion. Finally rank siting plans to be evaluated with TOPSIS, and find the optimal case among candidates.

Given that ports siting planning are less researched than chemical industry park, this paper referred many literature related to chemical industry park planning, attempted to establish assessment system fully understanding the current actual situation of port yard. First class factors determined are interior enterprise, admittance enterprise towards neighboring area, and neighboring area towards admittance enterprise three directions.

Complete system is show as follows

Table 1. Qualitative Assessment System of Port Safety

First class	Second class	Third class
Interior enterprise A_1	Layout of enterprise B_{11}	Rationality of equipment layout C_{111} Process risk C_{112} Facility reliability C_{113}
	Safety management B_{12}	Occupational health C_{121} Emergency rescue C_{122} Safety management system C_{123}
	Fire fighting function B_{13}	radius of protection C_{131} fire protection facilities and watering C_{132}
	Navigation safety B_{14}	Ship reliability C_{141} Personnel administration C_{142} Navigation environment C_{143}
neighboring area towards admittance enterprise A_2	Neighboring fire protection resource B_{21}	Fire-fighting resourceC_{211} Rescure time C_{212}
	Social assistance such as medical treatment B_{22}	Offshore rescue C_{221} Onshore rescue C_{222}
	Neighboring vulnerability distribution B_{23}	Neighboring C_{231} Vulnerability of Neighboring area C_{232}
	Port safety management B_{24}	Safety monitor and operation of produciton C_{241} Safety management organization and regulation C_{242} Safety monitoring technique C_{243}
	Residental Acknowledge B_{25}	Information acceptance level C_{251} Current acknowledge C_{252}
admittance enterprise towards neighboring area A3	Hazard resource B_{31}	Mobile hazards traffic capacity C_{311} Properties of mobile hazards C_{312} Fixed hazards storage C_{313} Properties of fixed hazards C_{314}
	Fire-fighting resources and emergency B_{32}	Allocation of firefighting C_{321} Integration of emergency rescue C_{322}

	Minimum safety distance and protection B_{33}	Minimum safety distance C_{331} Protection measures C_{332}

Conclusion

This paper combine QRA with qualitative comprehensive evaluation to establish the model of enterprises access to port yard. Here are some unsolved problems require further discussion.

a) Current QRA criteria does not take personnel vulnerability into account, so criteria needs some adjustment to match the new calculation.
b) Numerous factors of complicated system like port effects the safe state of enterprises. Standardization of vague factor could help describe and evaluate the whole safety condition.
c) Various types of ports acquire different standards and requirements to access, practical and standardized enterprises access schemes and procedures are in need.

Reference

1. Yip, T. L. (2008). "Port traffic risks—a study of accidents in Hong Kong waters." *Transport. Res. E Logist. Transport. Rev.*, pp. 921–931.
2. Balas, C. E., Ergin, A., Piskin, G., and Williams, A. T. (2000). "Reliability based risk assessment in project management of coastal structures." *Proc. 10th Int. Offshore and Polar Eng. Conf., the International Society of Offshore and Polar Engineers*, California, USA.
3. Hassan H., S. Meysam M., R. Tavakkoli-Moghaddam; and Y. Gholipour. (2014), "Compromise Ranking Approach with Bootstrap Confidence Intervals for Risk Assessment in Port Management Projects," *Journal of Management in Engineering*, October, pp. 334-344.
4. Barbara Z., Dario S., Stefano B., Fabio B., Luca P., Fabio Z., Simon H., and Robert J. N.(2014), "THESEUS decision support system for coastal risk management," *Coastal Engineering*, December, pp. 218–239.
5. Chen G., Zhang J., Zhang H. Yan W. Chen Q. (2006). "Study on Regional Risk Assessment Methodology," *China Safety Science Journal*, June, pp.112-117.

397904 Introduction to Safety Instrumented System Management

Abstract:

Many Automation and Instrumentation Designers are aware of Safety Instrumented Systems (SIS) as the combination of field instrumentation, logic solvers, and application code designed together according to specific technical standards to execute the safety controls, alarms and interlocks.

Are you as aware of the equally critical importance of managing the human aspects of SIS over the complete safety lifecycle?

Related industry standards and recognized publications outline lists of required operating discipline content. However, it falls to the facility and implementation projects to create the documentation and management systems which both meet these technical requirements and do so in manner which is CONSISTENT, COMPLETE, CLEAR and CORRECT so that the management system remains sustainable over time.

Covering all aspects of SIS documentation and management systems is not achievable within the limits of a single session. So in this introductory presentation, we will focus on some of the most common elements of SIS related documentation across the safety lifecycle as represented in IEC61511/ISA-84. The author will address the key concepts and content of these documents which in her experience are the most essential to facilitate successful management of the overall SIS program over the years and across the many functions involved.

397732 Safety Integrity Level Sil Assessment about Deepwater Bop System

Abstract:

As the key equipment of offshore drilling and completion work, BOPs can be used effectively to prevent the dangerous accidents, such as well kick and well blowout, and also play an important role on protecting people, equipment, marine environment and oil and gas resources from heavy damage and pollution. On the view of the important role of BOPs on offshore oil exploitation, and according to the requirements of functional safety international standards, this paper treated the BOP and its control system as a safety instrumented system, and analyzed its safety instrumented function based on the various equipment failure data. By establishing the overall safety instrumented system life cycle system, identifying risk and allocating SIL, an investigation is made on the possible failure mode and redundancy settings of each part. The analysis on the structure and function of BOP can effectively assess overall system safety integrity level and may be used to provide certain ideas and guidance for BOP quantitative evaluation.

398016 Lessons on Achieving Cost Effective ISA 84 SIS Lifecycle Sustainability

Abstract:

Compliance with ISA 84 (IEC 61511) adds significant documentation and testing requirements to already strained budgets and resources. Once the verification documentation and validation procedures are in place, maintaining them through the life of an operating facility presents a significant challenge. This paper presents two distinct case studies of PSM facilities that have struggled through the effort to execute and sustain the ISA 84 Safety Instrumented System (SIS) Lifecycle in a cost effective and efficient manner. The lessons learned will examine the use of standardization methods across an enterprise vs localized practices for execution and closing the loop on the lifecycle, data base automation vs file based documentation, and lagging vs leading key performance indicators presented by different work practices. The lessons presented in this paper will provide investment justification toward a robust program for realizing the functional safety objectives of the ISA 84 SIS Lifecycle.

The application of the bifunctional environmental deodorizer for sulfur and sour water in SINOPEC TIANJIN Company

Aijun Wang*, Jiliang Chen
Vice President
Jinan Ruidong Ltd. , 161-1 Jiefang Road, Jinan 250013, P. R. China
Email: wangaijun.jn@yeah.net

Prepared for Presentation at
American Institute of Chemical Engineers
2015 Spring Meeting
11th Global Congress on Process Safety
Austin, Texas
April 27-29, 2015

UNPUBLISHED

The application of the bifunctional environmental deodorizer for sulfur and sour water in SINOPEC TIANJIN Company

Aijun Wang*, Jiliang Chen
Vice President
Jinan Ruidong Ltd. , 161-1 Jiefang Road, Jinan 250013, P. R. China
Email: wangaijun.jn@yeah.net

Keywords: Shutdown process; Malodorous gas; FeS self-ignition; Bifunctional deodorizer; Emission

Abstract

SINOPEC TIANJIN Company Storage and Transportation Department adopted bifunctional environmental deodorizer to clean repugnant substance in the containment, including reaction tower and tank containing sulfur, ammonia and nitrogen during the shutdown process. This action has achieved good deodorization results, and purified the air, protected the environment and the personal safety and facilities, which improve the quality of maintenance and shorten the maintenance duration. After cleaning, waste emissions is allowed.

1. Introduction

Oil refining equipment including hydrofining, coking, sulfur and sour water stripping exists sour and nitrogen water with high concentration. Especially when more high-sulfur crude has been refined and the shutdown process has been prolonged, the proportion of repugnant substance has increased in sour and nitrogen water which has on obvious environmental effect [1]. However during the shutdown process, repugnant substance contained in the tank, reactor and pipeline will seriously pollute the environment, affect the residents around, and also endanger the operator and maintainer. In order to clean up the residual wastewater, we used to flush the containment with fresh water, but some repugnant substance still can adhere to the wall of container and waste disposal plant would be attacked with so much water, so the effect is limited. Another normal method is steam blowing which polluted the air seriously and still can not clean up the repugnant substance thoroughly. Nowadays more refinery plants adopted deodorizer to decrease the effects of repugnant substance. Technical section researched and decided to use bifunctional environmental deodorizer [2] produced by Jinan Ruidong company to clean repugnant substance in the containment, including more than 20 reaction tower and tank containing sulfur, ammonia and nitrogen during the shutdown process. This action has achieved good deodorization results, and purified the air, protected the environment and the personal safety and facilities, which improve the quality of maintenance and shorten the maintenance duration.

2. Cleanout Process

2.1 Cleanout Principle

Polluted water treatment agent ZCJ-961 [3] is composed of organic, inorganic chemicals and water solubility additive. It can dissolve, neutralize and complex with pollutants including hydrogen sulfide, ammonia nitrogen, sulfur alcohol and organic amine to clean up the source of pollutants, protect the environment and personal security. This deodorant is bifunctional and environmental whose main function is deodorization with passivity iron sulfide. Its performance index is in Table 1.

Table 1. Physical Characteristic of ZCJ-961

Appearance	pH Value	Density
yellowish-green transparent liquid	4~6	1.05~1.07

2.2 Main Features

This cleanout process has the following features:

(1) Fast responding speed without toxic or harmful gases emissions;
(2) The product is steady after reaction without decomposing toxic material directly discharged into sewage treatment plant;
(3) High-efficiency, nontoxic, safe and no corrosion of equipment;
(4) Convenient for packaging and transportation with low cost;
(5) Not necessary to change the existing process, or need the additional equipment and material.

2.3 The process and content of passivation cleaning

Jinan Ruidong Company is responsible for the passivation cleaning. The agent is formulated as 10% to 15% aqueous solution, which is used for system circulation, spray or immersion cleaning with of means existing processes. The plan is developed by the technology and transportation department of the Tianjin Company, Jinan Ruidong Industries. The time of its plan may be 7 day or so.

The main cleaning equipments include N1, N2, N3, N4, N5, N6, N7, N8, N9, N10, N11, N12, N13, N14, N15, N16, 0#, R1, R2, R3, R4, R82#, R11, R12, high pressure gas separation tank, absorber and other equipment.

3. The analysis and discussion of the cleanout process

After the cleanout process, there is on stench in the equipment. The system circulation only takes a little time yet with obvious effect. The immersion cleaning takes more time of 8 to 12 hours to clean depending on the amount of hydrogen sulfide, sludge and other substances in equipments.

The deodorizer for sulfur and sour water is light yellow transparent liquid. After adding into the polluted water and oil, it changes to muddy black liquid with a small amount of suspended solids.

The cleanout fluid features of sulfides, COD, pH value and suspended solids value are shown in Table 2, indicating that the passivation solution is complied with the emission standard [4].

After open human-air displacement of each equipment, the sampling analysis data of combustible gas is up to the standards.

Any phenomenon of spontaneous combustion about ferrous sulfide is not occurred after Opening the ventilation replacement equipment. The demister and metal intalox saddle in absorber appears the metal color.

The plan which is specific, clear, comprehensive is developed by the technology department of the Tianjin Company and Jinan Ruidong Industries. It actually took 4 days instead of 7 days planed.

Table 2. The cleanout fluid features

Sulfide	COD	pH value	Suspended solids
14.2	273	8.5	82

4. Conclusion

(1) During the lockout, over twenty towers and tanks within sulfur ,ammonia and nitrogen used the deodorizer which had added this bifunctional environmental water treatment agent to remove residual malodorous substances in the container equipment to keep this shutdown smooth, reduce the downtime, protect the equipment and persons, clean air, protect the environment. And it also set a foundation to improve the quality of maintenance for our workshop and shorten the repair period.

(2) This method is used for passivation cleaning of ferrous sulfide, especially ferrous sulfide content is not very high. The long duration of ventilation and the position where is full of oil and gas have better effect.

(3) The production called ZCJ-961 in this deodorant cleaning is made by Jinan Ruidong Industrial Co. The result of the sample analysis is up to environmental emission standards. The method is non-corrosive, and does impact the sewage treatment plant, which is completely different from caustic (sodium hydroxide, sodium carbonate, sodium bicarbonate) and pickling.

(4) This time, the Chemical cleaning generally achieves its goal. It not only protects the safety of equipment and persons but also buys the time for equipment maintenance and offers protection to the machine.

Reference

[1] Y. Z. Zhao, Harmfulness of malodorous gas and its treatment technology, Guangdong Chemical Industry 41 (2014), 170-171.

[2] Z. Y. Zhang, F. Yang, Analysis of heat release causes of deodorizing drum at overhead of coke quenching water tank in delayed coking unit and corrective measures, Petroleum Refinery Engineering 44 (2014), 25-28

[3] X. Tao, Application of the new type ZCJ-961 sulfur-coated waste water modifier, Environment Protection in Petrochemical Industry, 26 (2003) 38-40

[4] People's Republic of China Integrated Wastewater Discharge Standard, GB8978--1996

Dust Explosion Incident Analysis, Prevention and Controlling

Ge Anka
College of Urban Construction and Safety Engineering, Nanjing Tech University
NO.30 Puzhu Road(S), Nanjing, China
Email address: geankal@hotmail.com

Zhang Lijing
College of Urban Construction and Safety Engineering, Nanjing Tech University
NO.30 Puzhu Road(S), Nanjing, China
Email address: zhanglj@njtech.edu.cn

Prepared for Presentation at
American Institute of Chemical Engineers
2015 Spring Meeting
11th Global Congress on Process Safety
Austin, Texas
April 27-29, 2015

Dust Explosion Incident Analysis, Prevention and Control

—Explosion Incident in China

Ge Anka
College of Urban Construction and Safety Engineering, Nanjing Tech University
NO.30 Puzhu Road(S), Nanjing, China
Email address: geankal@hotmail.com

Zhang Lijing
College of Urban Construction and Safety Engineering, Nanjing Tech University
NO.30 Puzhu Road(S), Nanjing, China
Email address: zhanglj@njtech.edu.cn

Keywords: Dust Explosion   Incident Analysis   MOC   Chemical Technical Management
Combustible and Explosive Features

Abstract

16th April 2014, an explosion happened in one chemical plant in China which was caused by stearic acid dust, the incident caused 8 death and 9 injuries. On the day of the incident happened, No.1 stearic acid granulator tower is in normal operation. When welding on the cone of tower bottom, material was not discharged from the tower and the tower was in operation. A stearic acid dust explosion was triggered in the system, and caused series of explosion; the workshop was caught on fire and finally collapsed. This article reviews the process how the incident happened. The immediate cause from the analysis was: welding at the cone of tower bottom in normal operation status without material withdrawn caused the stearic acid dust explosion. Work violating procedure, MOC process not implemented, not design, constructed and installed following requirements, deficiency knowledge of stearic acid dust combustion and explosion was the indirect cause. This article raised specific prevention and control recommendations to dust explosion based on above analysis, including reinforcement of safety management during operation, MOC, technical management of chemical plants and safety management of combustible organic dust in work place.

1. Introduction

On 10:00:49, 16 April 2014, the pelletizing workshop in one chemical plant in China suffered a dust explosion incident. The incident resulted in 8 killed, 9 injured, and complete collapse of the workshop. After the incident, an incident investigation team involving relevant technical and managerial experts conducted a comprehensive investigation to the incident. The investigation team carried out a 6-day incident scene rescue, cleaning and scene investigation, collected relevant information at the company, progressed on-site sampling and analysis, simulation test, asked the witnesses and did investigation to individuals concerned, and look up related domestic and abroad information. Through full discussion and analysis to the findings identified during the investigation, the incident causes are preliminarily identified.

1.1 General information of the incident company

The chemical plant's main business is production and sales of stearic acid and glycerin. It has been engaged in trade of stearic acid since 1997, started stearic acid production and business in 1999 when the plant was built. After development for sevaral years, the plant scale was gradually expanded, now the annual production capacity was 200k tons stearic acid and hydrogenated oils. The chemical plant covers an area of 92,746 square meters, with 27,254 square meters of building area. It has 90million CNY of fixed assets. And sales amount in 2013 came to 0.41 billion dollors (including trade). The company is consists of General Manager Office, Safety and Environment Integrated Management Department, QC Department, Accounting Department, Production Department and Storage and Transportation Department, and has employees numbered 246 at the moment.

1.2 Stearic acid production process and plant information

The chemical plant currently has production plants, including hydrogen production by methanol cracking (one new and one old plant, with 800m3/h and 400m3/h hydrogen production capacity respectively), hydrogenation (an old one and an new one), hydrolysis, distillation, and pelletizing. It is a complex, and main feedstock is methanol and palm oil. H2 generated by methanol cracking directly goes to the hydrogenation plant; the hardened oil generated by catalytic hydrogenation of palm oil become stearic acid by high temperature and high pressure hydrolysis, and then the stearic acid is distilled and sent to the pelletizing workshop for pelletizing and packaging.

1.3 Basic information on the pelletizing workshop

Construction of the pelletizing workshop started in 2006, and the workshop was put into service in 2007. The building is at the northwest side of plant area of the plant, beside the wall at its north side is 5 glycerol tanks and 2 stearic acid black oil tanks; the plant's forklift workshop, power distribution room and cooling tower are 5m away from the south side; finished products warehouse, auxiliary materials and maintenance workshop is 19.5m away from the west side; close to external wall at east side is 5 stearic acid tanks, 2 fatty acid tanks and 1 glycerol tank, see plotplan below for details.

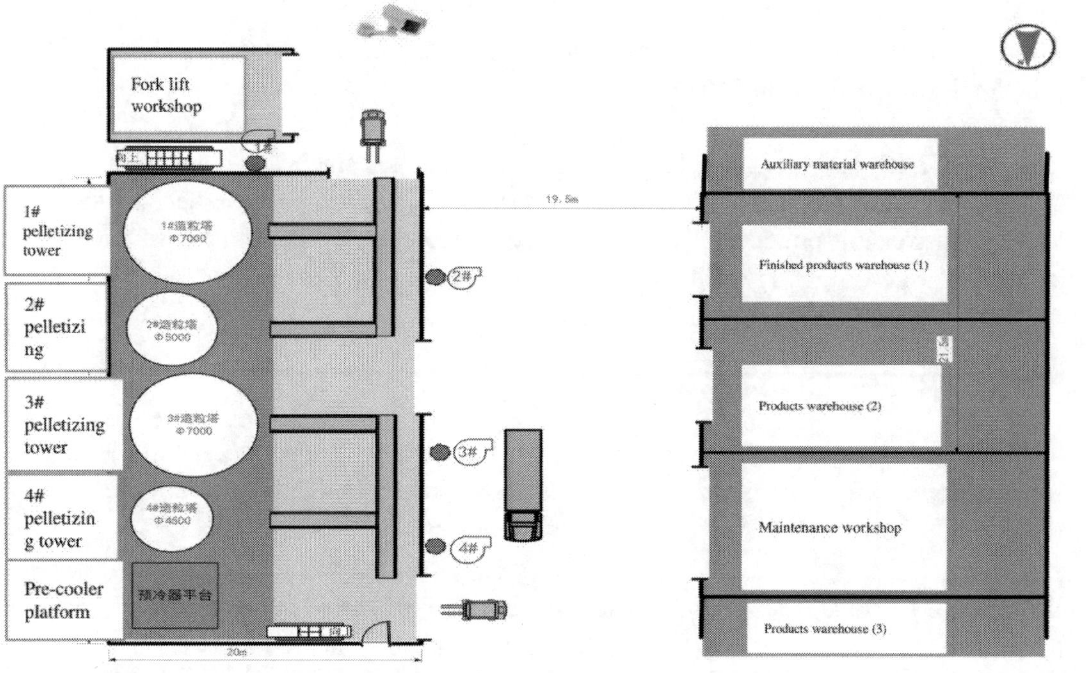

Layout plan for the pelletizing workshop in the chemical plant

Figure 1 layout plan for the pelletizing workshop

Arrangement of equipment in the pelletizing workshop: there are 4 pelletizing, packaging lines in total in the workshop, including 4 pelletizing towers, 4 condensers, 16 cyclone separators, 4 blowers, and 4 induced draft fans. At the southeast of the workshop, run from south to north, 4 pelletizing towers with 4.5-7.0m in diameter and 27m in height numbering 1-4# (diameter for 1# and 3# towers is 7m and diameter for 2# and 4# towers is 5 m and 4.5m, they are all manufactured and installed by this chemical plant itself), the pelletizing tower top elevation is 36m and bottom elevation is 9m. The discharge port of the pelletizing towers is located at the west side; the body of 1-3# towers and the fluidized bed are equipped with 6-8 mist spray inlets. Each pelletizing tower is installed with one open vibration sieve, and provided with oil dissolving tank, which equipped with around 1m3 of built-in steam coil, manually recover the big agglomerates on the vibration sieve. Cyclone separators were located at west side of platform on the second floor, each set of production line has 4 cyclone separators designed in parallel, bag is used to recover stearic acid powder on a regular basis. Each pelletizing tower with its 4 cyclone separators constitutes one independent set of pelletizing system.

The control room of workshop is at the southwest side of a 2-layer platform, and power distribution cabinet is located at the west side of the control room; blower and induced draft fan are arranged on the first floor respectively. Main building of the pelletizing workshop, covers an area of about 740m2 (37m long north-south ×20m wide west-north) , is a double-deck light steel-frame structure. The 2-layer platform is designed as staggered structure, the design and construction is done by this plant itself.

Anticorrosive paint is used for the bearing beams of steel frame and upright column steel structure for the whole workshop. Outside the workshop fire water line is installed, inside the workshop fire hydrant is provided. There are 2 fire evacuation exits on each layer of the building. Sealed lamp is used as lighting within plant area, motor protection level is IP44, and overall electrical design is non-dust-proof design.

1.4 Process flow for pelletizing plant

The pelletizing workshop only have stearic acid (fire class: class C, normal combustible solids), and no other chemicals. The pelletizing process is a pure physical process of cooling, spray pelletizing.

Main process: 70 ℃ of stearic acid in a molten state from the distillation plant is transferred to the finished product tank by pump to pelletizing workshop, and pump into top of pelletizing tower by the material pump, and is sprayed from the nozzle on the top of the pelletizing tower, the induced draft fan is used on top of the tower to suck and form a tiny negative pressure state inside the tower; the blower at the bottom of the tower is used for transferring air, the air countercurrent touch the atomized liquid drop sprayed from the tower top to make it cool down and solidify to form stearic acid particles, then go to the fluidized bed and flow to the vibration sieve from the discharge point of silo at the bottom of the tower, the on-spec products go to the scales for metering and packaging, and big agglomerates separated by the sieve are manually put into the melting oil tank and enter to the wastes tank after melting or return to distillation plant for recovery. Fine dust inside the pelletizing tower experience dust removal by the cyclone separator, and the stearic acid dust inside the cyclone separator is cleaned, melted and recovered on a regular basis.

The product goes to automatic weighing machine at the first floor for weighing , gets packed into PE bags, and then the bags get sewed by automatic bag sewing machine; after that they are conveyed by roller conveyor and get stacked up and palletized by auto-manipulator and finally delivered to warehouse by forklift.

1.5 Routine production of pelletizing workshop

The Production of the chemical plant implements 3-shift-per-day system, morning shift: 08:00-16:00, afternoon shift: 16:00-24:00, night shift: 24:00-08:00. There are 11 people for each shift, plus, the plant equips with 2 forklift drivers and 1 electrician for each shift. During normal operation, 1# and 3# pelletizing towers produce 8-9 T/H and 2# and 4# pelletizing towers produce 4-6 T/H.

2. Incident scene survey and investigation

2.1 Damages of pelletizing workshop and equipment

The major consequence of the incident was that the pelletizing workshop building integrally collapsed westwards, the tube structure of 4 pelletizing towers had no signs of explosion or crack, but they experienced mechanical deformation due to collapse force; the automatic packaging system was burnt down; all cyclone separators got deformed due to collapse force of pelletizing towers. 1cyclone separator of both 1# and 4# cyclone separation units had been blasted open respectively; the upper section housing of 4# draft fan was found at 63 meters away north, the lower section of the draft fan became bloating, the stack cap was found at 42 meters away to west by north, which smashed through the finished product warehouse roof.

Figure 2, the pelletizing and packaging workshop integrally collapsed.

2.2 Damages of the neighbour buildings and structures of the pelletizing workshop

The window glasses of the stearic acid finished product warehouse and auxiliary materials warehouse 19.5 meters away to west of the incident workshop had partially broken, the finished product warehouse roof at northwest was smashed through by the stack cap, the color steel plate at eastern door of maintenance workshop was damaged by explosion wave; the external insulation, aluminum sheet, of glycerin storage tank at north of the incident workshop was partially damaged due to burning, the window glasses of the coal-burning boiler house at north was partially damaged; partial insulation layer of stearic acid tank at east of the incident workshop was damaged and fell off due to burning; it posed no influence to other process units at east; the forklift shelter, 5 meters away to south of the incident workshop was partially burnt, 1 forklift was burnt down, the CW tower's FRP shell at southeastern corner was partially burnt down, the window glasses of chip-cutting workshop at south was partially damaged.

Figure 3, the finished product warehouse roof was smashed through

2.3 Incident scene survey

Inside the pelletizing workshop building, the electrode holder with welding rod in position was found at west of 1# pelletizing tower, the cable of the electrode holder was multiple-copper-core soft cable, which was specially used for electrode holder; there were a lot of welding rods left on the ground. The bottom cone silo outer wall of 1# pelletizing tower was about 80 cm right above the open-ended discharge hole; at west of it, a piece of web plate had been fixed and welded; a Φ 30mm reserved water spray nozzle was 8cm away from the western web plate, the nozzle was directly connected to inner side of pelletizing tower, the pipe support (170X30mm) for the spray nozzle had signs of repairing welding, according to visual appearance, it was free of corrosion, same as the welding seam of the web plat, which was clearly different from other welding seams of other supports; A piece of web plate was fixed by spot welding at south of the silo outer wall, the welding seam on one side had been completed, the inner wall surface of the fluidized bed showed a mark caused by high temperature of welding. Besides, 2 pieces of unwelded web plates (21X23cm) and 8 uninstalled pneumatic hammers were found on scene.

Figure 4, the welding rods left on the ground.

According to the survey, the body of all 4 pelletizing towers adopted reinforcing-ring design, but the appearance structure of the fluidized bed at the bottom of tower was different, a Φ30mm water spray nozzle was reserved at outer cone wall of the fluidized bed of 1-3# pelletizing towers which was built in 2006; However, the outer cone wall of the fluidized bed of 4# pelletizing tower which was built in 2008 did not have such design. The overall structure of the towers was relatively enclosed. When explosion happened, the upper section of 4# draft fan was blown 63 meters away, the lower section of the draft fan became bloating, the stack cap was blown 42 meters away.

3. The timeline and development of the incident

3.1 Field personnel distribution and production situation prior to the incident

In the early morning on Apr 16 (night shift), all 4 pelletizing tower units were under normal operation. At around 1:00am, two operators who were in charge of material control on the second floor removed the powder out of 4 cyclone separators in 4# pelletizing tower unit, more than 100 kg stearic acid powder was collected; at around 2:00am, the two operators restarted 4# pelletizing tower unit after they removed stearic acid powder agglomerates; 6:00am, one operator shut down 2# pelletizing tower unit as liquid stearic acid supply from upstream distillation unit was insufficient.

7:45a.m., the pelletizing workshop night shift operators successively handed over their work to day shift operators, 8:00a.m. shift handover was finished. In first floor of pelletizing workshop, 3 operators were in charge of 1# pelletizing tower unit and 3 operators were in charge of 3# pelletizing tower packaging unit, 1 operator was in charge of 4# pelletizing tower packaging unit,

1 operator was in charge of batch-number printing of finished product packaging bags; on the second floor of pelletizing workshop, 2 operators were in charge of material control. Another operator was arranged to another workshop after the shift handover. The plant also arranged another two forklift drivers in charge of finished product handling; there were 4 truck-loading/unloading workers and 1 truck driver beside the western gate of the workshop. 2# pelletizing tower unit remained shutdown.

8:00a.m., 1 operator in charge of 1# pelletizing tower unit in first floor started filling finished product bags at the discharge hole of 1# pelletizing tower, before 9:00am he saw welding sparks were falling to the ground at east of 1# tower from the gap between 1# tower and the second floor platform, but he did not see the welder (because the first and the second floor platforms were separated by steel floor slab); 9:00a.m., the operator handed over the work to another operator, and then he went for finished product pallets.

3.2 事故发生前的维修作业情况 Repair work prior to the incident

8:37a.m., the electrician was informed by phone call that the forklift repairman was going to use welding machine and he was told to connect power supply from the power distribution cabinet at second floor of pelletizing workshop; soon after, the electrician went to pelletizing workshop and saw the welding machine behind the first floor staircase outside the pelletizing workshop, then he brought the welding machine power cable through the gap between the staircase to second floor and plugged it in the power distribution cabinet, he did not find people around in the second floor; then he brought the secondary electrode holder cable to second floor; at about 8:45a.m., the electrician connected power supply and took off.

After 8:45a.m., 1 forklift driver, 1 forklift repairman and 2 repairmen successively arrived at second floor of pelletizing workshop, they started welding the web plates for installation of pneumatic hammers at bottom cone of 1# pelletizing tower.

10:00a.m., an explosion happened to the pelletizing workshop, and multiple explosions happened in succession within 3 minutes, causing a big fire; at 10:09, all 4 pelletizing towers successively collapsed westwards, the collapse of 4 pelletizing towers led to the steel structure of pelletizing workshop building integrally collapsed westwards.

4. Incident cause analysis

4.1 Identification of explosion time

According to video monitoring from the chemical plant: at 9:57:26, the video shows 1 forklift was driving into pelletizing workshop as normal; at 9:57:39, the video showed a large area of firelight, with fireball bursting out, this was the phenomenon when a fire was caused due to explosion, Therefore the first explosion time is identified at 9:57:39.

Please see the following video captures, the explosion spread out from south to north in succession, the explosion first happened at 1# pelletizing tower at south;

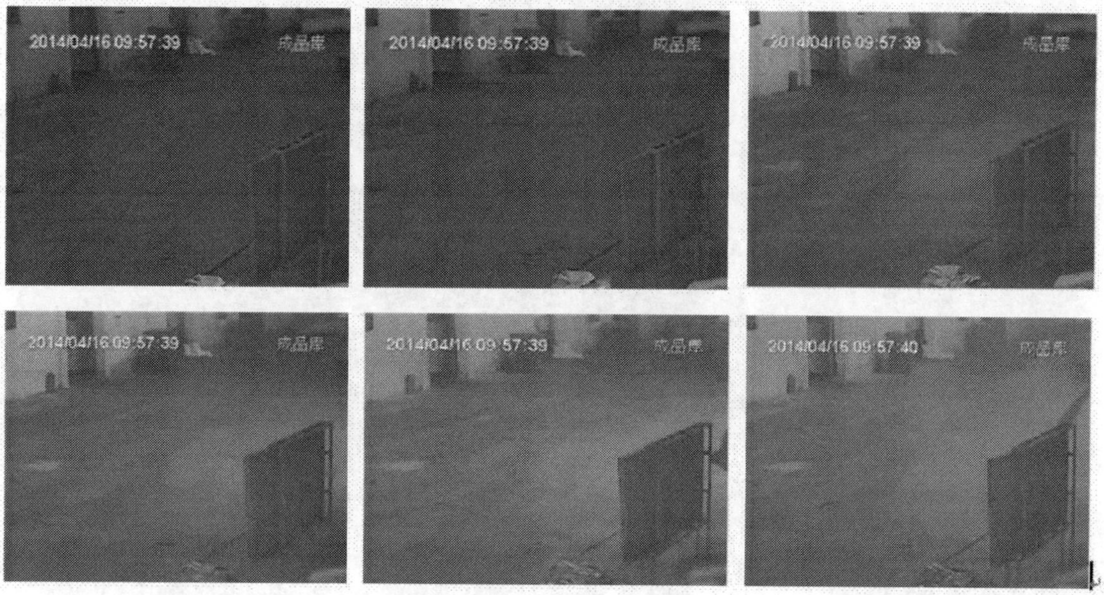

Figure 5, video captures

4.2 Identification of incident explosion point

According to incident scene survey and investigation, the stearic acid dust in 1# pelletizing tower decomposed and ignited due to high temperature welding at tower bottom cone, the heat generated led to explosion of stearic acid dust in pelletizing tower system; according to the incident scene photos (see below), we can see when the explosion incident just happened, only the location of 1# pelletizing tower caught fire. We can judge from the photos that during the early stage of the fire, the location of 1# tower had already caught fire and the fire developed both in vertical and horizontal directions, the color steel plate wall at south was damaged, but the building structure at 3# and 4# towers at north was intact and there were no signs of fire. Therefore, the explosion point of the incident is identified at 1# pelletizing system in pelletizing workshop.

Figure 6, incident scene fire

4.3 Identification of explosive material

Judgment can be drawn from the process of the pelletizing workshop and explosion incident phenomenon: this is a dust explosion. We can understand dust explosion as "combustible dust particles fully exposed to O2 in air, leading to instantaneous violent oxidation reaction under specific conditions, the massive heat released from reaction generates gas and develops dramatically increasing pressure wave" [1].

According to the plant's production process, H2 generated from methanol pyrolysis is purified through PSA (pressure: 1.5 MPa, temperature: 200℃), so we can exclude the possibility of methanol entrained in H2; the hardened oil from palm oil hydrogenation has to go through hydrolysis before it is used to produce stearic acid, hydrolysis temperature is about 245~260℃, pressure is about 5.2~5.4MPa, there is no H2 presence in liquid stearic acid. Therefore, the only chemical present at pelletizing workshop is stearic acid (liquid stearic acid, stearic acid pellets (finished product) and stearic acid dust).

According to stearic acid's MSDS provided by NRCC, stearic acid is combustible material, ignition temperature is 395℃, when stearic acid dust exposes to air, it could form explosive mixture.

After the incident, the minimum ignition energy and minimum explosion concentration of dust for the Plant's finished stearic acid product were tested, according to the test results, the minimum ignition energy and minimum explosion concentration of stearic acid dust are 1-3mJ and 30-40g/m3 respectively.

According to comparisons of the minimum ignition energy and minimum explosion concentration for following several materials, we can see: although stearic acid dust's minimum explosion concentration is close to PS's, which is far higher than aluminite powder's minimum explosion concentration, however, its minimum ignition energy is far lower than the other two materials.

Table 1, comparisons of the minimum ignition energy and minimum explosion concentration for following several materials

Name of material	Minimum ignition energy	Minimum explosion concentration
Stearic acid dust	1-3mJ	30-40g/m3
PS	120 mJ	20 g/m3
Aluminum powder	15 mJ	37-50mg/m3

4.4 Immediate cause analysis

From the development of the incident, we can see during welding of web plates for installation of pneumatic hammers at the bottom cone of 1# pelletizing tower which was under normal operation, the stearic acid dust inside the tower was decomposed and ignited due to high temperature due to fast welding heat transfer (because the temperature of the welding spot was above 1600℃, and the wall thickness of tower bottom cone was about 8mm), the heat generated led to explosion of stearic acid dust inside 1# pelletizing system. Besides, we cannot exclude the possibility that the welding sparks entered into the tower via water spray inlet while repairing the welding for water spray pipe support and caused explosion of stearic acid dust inside the tower, soon after, it evoked secondary space explosion of stearic acid dust inside the workshop, and ignited the whole workshop and led to multiple explosions in succession; 3 minutes later, it caused further explosions of 4# pelletizing system; 9 minutes later, the steel structure of the workshop integrally kept collapsing westwards. Therefore, the immediate cause of the explosion incident: as 1# pelletizing tower did not shut down for material evacuation, instead, the welding activity was directly carried out at the bottom cone of the pelletizing tower, causing explosion of stearic acid dust inside pelletizing system, and successive explosions were triggered off, leading to ignition of the whole workshop and collapse of workshop building.

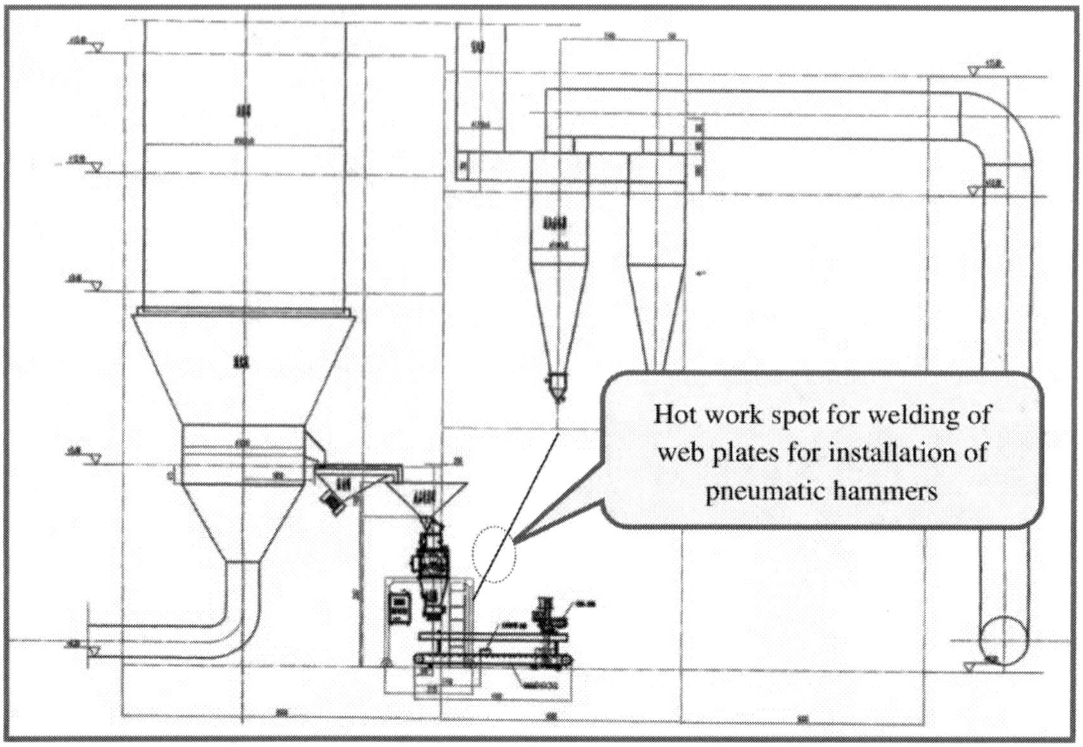

Figure 7, equipment sketch

4.5 Indirect cause

For dust explosion of production workshop (workshop building), from the micro perspective, the explosion only happens when dust concentration, ignition source and O2 concentration meet the conditions at the same time. The reason it happened is mainly because personnel unsafe behaviors, unsafe status of matters and unsafe conditions of environment make the 3 micro factors satisfied at the same time, and consequently cause dust explosion incident.

4.5.1 COW

The chemical plant neither developed maintenance work plan nor conducted safety disclosure for the maintenance work. They violated the regulations by carrying out welding activity on equipment body without shutting down the system and obtaining PTW for Hot Work; the electrician violated the regulations by connecting temporary power supply to the welding machine without obtaining PTW for Temporary Power Utilization.

4.5.2 MOC system was not implemented

Although the chemical plant developed MOC system, they did not obey the MOC system regarding the technical modification project of installing pneumatic hammers on pelletizing towers, they failed to go through technical verification and risk assessment and failed to make safety precautions and formal design.

4.5.3 insufficient technical force and personnel ability

The chemical plant wasn't equipped with dedicated technical management people on equipment, electrical, process, etc. The whole plant's equipment management, electrical management, process management is only undertaken by the director of production department who only had the education background of senior high school and didn't have the evidence to prove he accepted corresponding training, had corresponding ability; The maintenance people died in the incident didn't have the certificate of welder, didn't have the ability to do welding activity

4.5.4 violation on design, construction, and installation

The pelletizing workshop with the incident was self-designed and installed by the plant itself; the main equipment in the workshop was self-designed, fabricated, and installed by the plant itself.

4.5.5 insufficient awareness on explosion characteristics of stearic acid dust

There was less awareness on danger and hazard characteristics of stearic acid dust in the industry. There were also less data of characteristics for combustible organic dust in related Chinese standards and regulations, which cause many similar companies didn't know the explosion risk of stearic acid dust. The chemical plant didn't fully carry out the risk identification, assessment for stearic acid dust work place and also didn't put corresponding fire and explosion precaution measures in place.

5. Discussion on dust prevention measure

3P safety theory, i.e. plant (equipment and facility), procedure, people, the key is to minimize the "hole" of each line of defense, so that the incident can be minimized.

The chemical plant incident shows: in order to prevent dust explosion incident, we need to focus on chemical plant's hardware and design, development and implementation of management system and employee training and competence. Chemical plant must be subject to formal design, it must have a complete MOC system and it must be strictly implemented, hot work must be carried out as per requirement of COW, good employee competence must be effectively secured through training, etc.

According to statistic analysis of the incidents, the majority of the dust explosion incidents are caused by human unsafe behaviours. Human factor is the major factor of development of explosion incident, therefore it must draw attention. Preventive actions can be broken down as follows:

(1) Carry out comprehensive process hazards analysis.

(2) Monitor the areas with dust explosion hazards or high risks, i.e., timely monitor dust leakage, etc.

(3) Develop COW regulations and implement them in a strict manner.

(4) Strengthen safety induction and training for production employees to improve employee competence.

Matter factor determines the severity of dust explosion incident, the blast wave or shock wave generated by the dust explosion might trigger off secondary explosion of dust accumulated on outside the equipment, the damage might be escalated or the damage area might be further developed. Preventive actions can be summed up as follows:

(1) Improve the process and equipment, adopt advanced equipment and process. For example, the fine-particle-related process shall adopt wet type or inert gas protection[3].

(2) Set up safety protection and monitoring alarm device. For example, dust concentration monitoring facility and fire alarm device.

(3) The dust accumulated in production workshop shall be regularly removed and the production workshop shall be ventilated.

(4) Regularly detect and maintain the dust leakage at valves, welding seams, etc.

Research on Safety Management Assessment Technology in Oil Tank Farms

Dongfeng Zhao
China University of Petroleum(East China)
No. 66 , Changjiang West Road , Huangdao District , Qingdao , China , 266580

Mei Dong
China University of Petroleum(East China)
No. 66 , Changjiang West Road , Huangdao District , Qingdao , China , 266580
melody199012@163.com

Zhiqiang Zhao
China University of Petroleum(East China)
No. 66 , Changjiang West Road , Huangdao District , Qingdao , China , 266580
championzhaozq@163.com

Key words: Petrochemical tank farms; Safety management assessment; Analytic Hierarchy Process(AHP); Risk.

Abstract

Health, Safety and Environmental management system (HSE management system) has been vigorously promoted in the Petroleum and Petrochemical Industry in China, and it is no exception for the safety management of the tank farm. But because of the special properties of storage material, it is easy to cause the serious consequences if the accident occurred. This article concluded the safety management assessment factors of storage tank with Fishbone Diagram, and constructed elements level with AHP (Analytic Hierarchy Process), so as to give weight to elements at all levels and establish the safety management system evaluation method which is composed of 12 first class indicators, 37 second class indicators. Finally, the critical incident method is used to quantify second class indicators to complete the safety management evaluation system of storage tank. Through the application of safety management assessment system applied in oil two unit tank area of CNPC, the level of safety management system of storage tank and the problems existing in the security management of enterprises were obtained.

1. Introduction

With the rapid growth of demand for oil in China, scale of tank farm in the

Petroleum and Petrochemical Industry has been increasingly expanded. Because of the Special physical and chemical properties of the oil, it is easy to cause serious consequences if the accident occurred. As the infrastructure of the oil storage, oil tank farms play an important role in the management in the Petroleum and Petrochemical Industry. Under the circumstances, a reasonable and effective system to manage the oil tank farms is necessary in the Petroleum and Petrochemical Industry.

In the primary period of safety management, many large enterprises view good practice as the main points in the safety management system[1], e.g. MOTOROLA Six Sigma Management Method[2-4], Dow Chemical Method, 5S Management Method[5], Dupont STOP Administration Method[6-7] etc.

2. Safety Management Evaluation Analysis of Tank Farms

2.1 Safety Management of Tank Farms

The target of the safety management is to make the tank farm stay safety. Based on the risk principle, it is necessary to identify the risk in the oil tank farm. There are two meanings about the conception risk. For one is the risk which hides in the production process of oil tank farm, the other is that hides in the process of safety management system.

Safety management system includes the formulation, plan, implementation, realization review, improvements, and to maintain for the organizational structure in the corporation safety targets, goal planning, activity implementation, hold duty, convention, production process and resource guarantee are necessary etc.

2.2 Safety Management Procedure of Tank Farms

The aim to make the safety management evaluation is not to get the evaluation scores but to make the most efficient safety management with the least resources. Through building the evaluation elements system of safety management, a strict and systematic procedure of safety management will be concluded to realize optimization of resources. The procedure of safety management can be concluded in Fig.1 below.

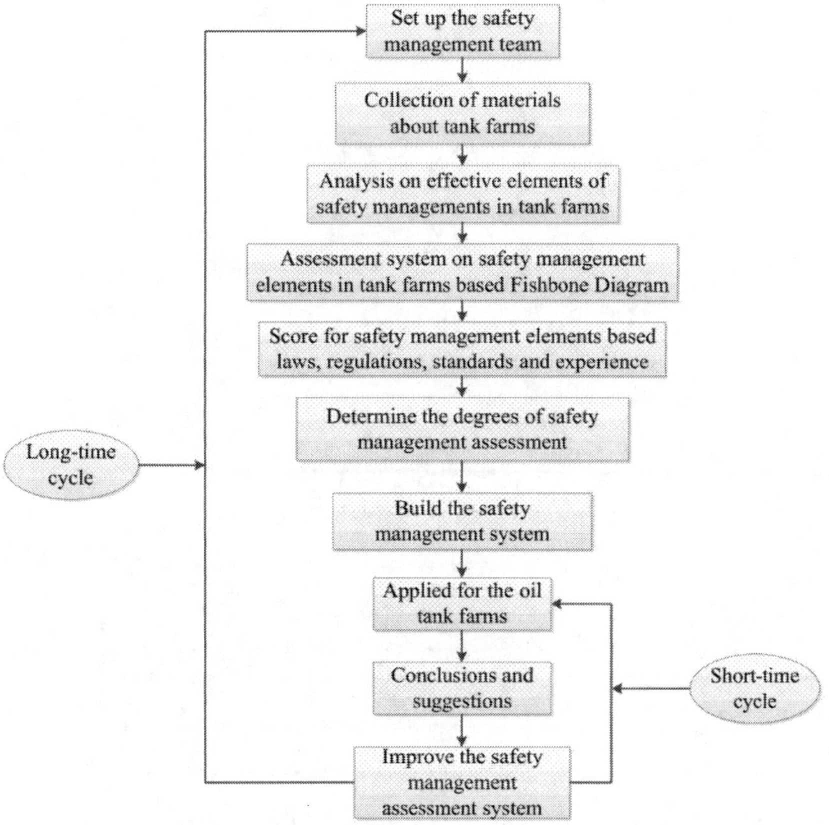

Fig.1 Procedure of safety management

In the application of the procedure in safety management, it is necessary to realize that if no major accidents/ major changes occur in tank farms in short-term, the existing safety management evaluation system can be used circularly. But oppositely, if it occurs in long-term, the system needs to be revised. And the safety management evaluation procedure of tank farms can be seen in Fig.2 below.

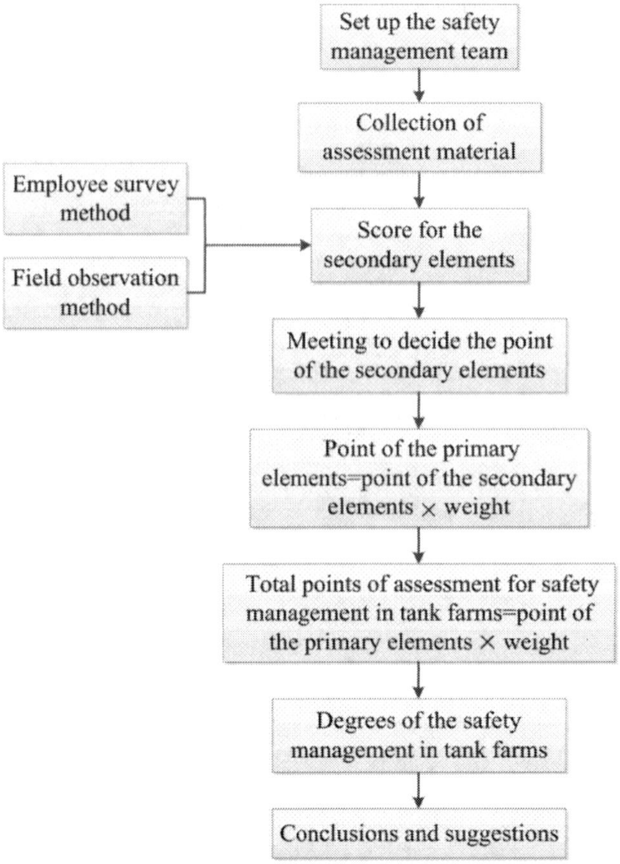

Fig.2 Safety management evaluation procedure of tank farms

3. Construction of Safety Management Evaluation Elements System

3.1 Evaluation Elements Analysis

It's difficult to clearly analyze the relationship between the evaluation elements because of the complex structure in the safety management of oil tank farms. Thus, in this paper, a new method combined with the Fishbone Diagram[8] and AHP (Analytic Hierarchy Process)[9] is proposed to solve this problem. Firstly, Fishbone Diagram is used to analyze the evaluation elements, and then build the elements hierarchy system. Finally, AHP is used to empower for the elements in each hierarchy.

According to the analysis results of influence essential factors in the safety management of oil tank farms, and refer to some advanced management technologies, e.g. NOSA Five Star Management System, Diamond Integrated Risk Management System[10], International Safety Rating System(ISRS), Management Oversight Risk Tree(MORT) Technology[11], Health Safety and Environment(HSE) Management System, evaluation elements can be concluded in detail by using Fishbone Diagram, just as the Fig.1 shows below.

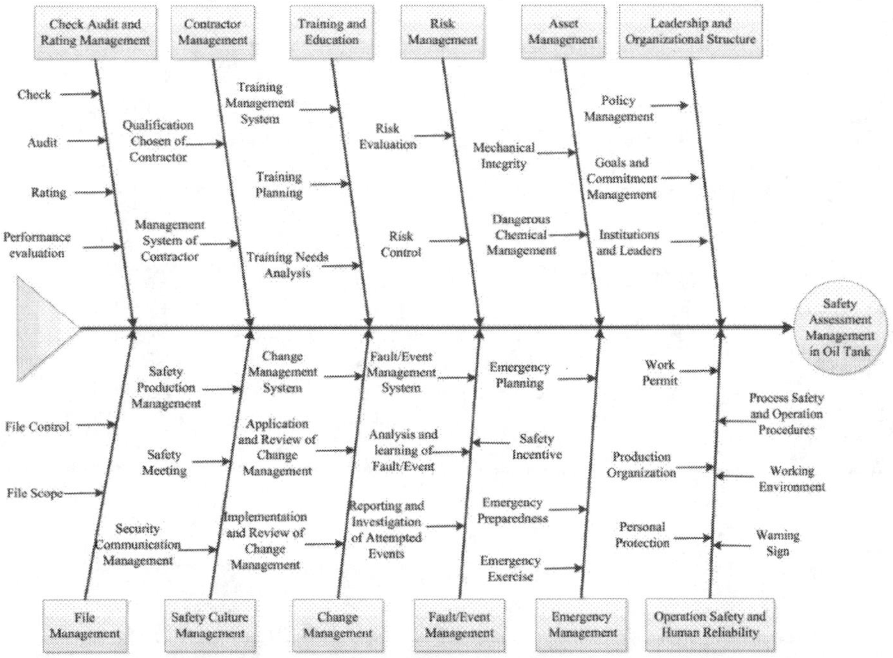

Fig.3 Evaluation elements analysis of the safety management

3.2 Weight Assignment for Elements

According to the AHP, the scale of judgment matrix is built as Tab.1.

Tab.1 Scale and definition of the judgment matrix

Scale a_{ij}	Definition	Explanation
1	Equal importance	Two factors contribute equally to the objective.
3	Somewhat more important	Experience and judgment slightly favor one over the other.
5	Much more important	Experience and judgment strongly favor one over the other.
7	Very much more important	Experience and judgment very strongly favor one over the other. Its importance is demonstrated in practice.
9	Absolutely more important	The evidence favoring one over the other is of the highest possible validity.
2 4 6 8	Intermediate values	When compromise is needed.

Based on the Tab.1, judgment matrix of safety management evaluation elements will be built as Tab.2.

Tab.2 Judgment matrix of safety management evaluation elements

No	Primary elements	Structure of Leadership and	Asset management	Risk management	Training and education	Operation security and human	Emergency management	Fault/Event management	Change management	Contractors/Suppliers management	Files management	Safety culture management	Check audit and review
1	Structure of Leadership and organization	1	1/8	1/8	1/4	1/7	1/6	1/5	1/3	1/6	1/3	2	1/2
2	Asset management	8	1	2	5	3	4	5	6	3	6	9	7

3	Risk management	8	1/2	1	4	2	3	4	5	3	5	8	7
4	Training and education	4	1/5	1/4	1	1/4	1/3	1/2	2	1/3	2	5	3
5	Operation security and human reliability	7	1/3	1/2	4	1	3	4	5	2	5	7	6
6	Emergency management	6	1/4	1/3	3	1/3	1	2	5	1/2	5	7	6
7	Fault/Event management	5	1/5	1/4	2	1/4	1/2	1	3	1/3	3	5	4
8	Change management	3	1/6	1/5	1/2	1/5	1/5	1/3	1	1/4	1	4	2
9	Contractors/ Suppliers management	6	1/3	1/3	3	1/2	2	3	4	1	5	7	5
10	Files management	3	1/6	1/5	1/2	1/5	1/5	1/3	1	1/5	1	4	2
11	Safety culture management	1/2	1/9	1/8	1/5	1/7	1/7	1/5	1/4	1/7	1/4	1	1/3
12	Check audit and review management	2	1/7	1/7	1/3	1/6	1/6	1/4	1/2	1/5	1/2	3	1

Then it is calculated that the vector approximation of elements weight based Square Root Method through Eq.1.

$$w_i' = \left(\prod_{j=1}^{n} a_{ij}\right)^{\frac{1}{n}} \quad (i=1,2,3\ldots n) \qquad \text{[Eq.1]}$$

After that, to calculate the w_i with normalization process, just as the Eq.2 shows below:

$$w_i = \frac{w_i'}{\sum_{k=1}^{n}\left(\prod_{j=1}^{n} a_{ij}\right)^{\frac{1}{n}}} \quad (i=1,2,3\ldots n) \qquad \text{[Eq.2]}$$

Then the maximum eigenvalue of judgment matrix λ_{max} and Eigenvector W can be calculated based MATLAB. Through the Eq.3 and Eq.4, the $C \cdot R$ (Consistency Ratio) can be calculated as:

$$C \cdot I = \frac{\lambda_{\max} - n}{n-1} \quad (n>1) \qquad \text{[Eq.3]}$$

$$C \cdot R = \frac{C \cdot I}{R \cdot I} \qquad \text{[Eq.4]}$$

The value of $R \cdot I$ in the Eq.3 can be referred as the Tab.3.

Tab.3 Value of $R \cdot I$

n	3	4	5	6	7	8	9	10	11	12
$R \cdot I$	0.58	0.9	1.12	1.24	1.32	1.41	1.45	1.49	1.51	1.54

Generating the corresponding value of $R \cdot I$ into the Eq.4, and then calculate the $C \cdot R$ value to judge its consistency.

Finally, judgment matrix of primary elements can be built in this method. Evaluation elements weight of safety management of tank farms in the Petroleum and Petrochemical Industry will be constructed as Tab.4.

Tab.4 Distribution Table of evaluation elements weight of safety management in tank farms

No.	Primary elements	Primary elements weights	No.	Secondary elements	Secondary elements weights
1	Structure of Leadership and organization	0.03	1.1	Policy management	0.10
			1.2	Goals and commitment management	0.26
			1.3	Institutions and leaders	0.64
2	Asset management	0.21	2.1	Mechanical integrity	0.75
			2.2	Dangerous chemical management	0.25
3	Risk management	0.17	3.1	Risk evaluation	0.75
			3.2	Risk control	0.25
4	Training and education	0.05	4.1	Training management system	0.23
			4.2	Training plan	0.12
			4.3	Training needs analysis	0.65
5	Operation security and human reliability	0.15	5.1	Production organization	0.11
			5.2	Process safety and operation procedures	0.41
			5.3	Work permit	0.25
			5.4	Personnel protection	0.11
			5.5	Warning sign	0.04
			5.6	Working environment	0.08
6	Emergency management	0.09	6.1	Emergency plan	0.25
			6.2	Emergency preparedness	0.10
			6.3	Emergency exercise	0.64
7	Fault/Event management	0.06	7.1	Fault/Event management system	0.29
			7.2	Analysis and learning of Fault/Event	0.48
			7.3	Reporting and investigation of attempted events	0.15
			7.4	Safety incentive	0.08
8	Change management	0.04	8.1	Change management system	0.65
			8.2	Application and review of change management	0.23
			8.3	Implementation and review of change management	0.12
9	Contractors/ Suppliers management	0.11	9.1	Qualification chosen of Contractors/Suppliers	0.25
			9.2	Management system of Contractors/Suppliers	0.75
10	File management	0.04	10.1	File scope	0.67
			10.2	File control	0.33
11	Safety cultural management	0.03	11.1	Safety production management	0.10
			11.2	Safety meeting	0.64
			11.3	Security communication management	0.26
12	Check, audit and	0.04	12.1	Check	0.46

review management	12.2	Audit	0.14
	12.3	Review	0.10
	12.4	Performance evaluation	0.30

It's common that safety managers quantify the secondary elements based critical incident technique, which quantifies the critical incidents by marks, and the total points of critical incident is the evaluation results of the secondary elements. Based on the study of a lot of documents specified, a strict and systematic method to mark on the secondary elements will be completed with some national standards, e.g. *General Norms for Safety Standardization of Hazardous Chemical Enterprises*[12], *Guideline of Safety Management System Implementation for Petrochemical Corporations*[13], *Provisional Rules of Dangerous Chemical Supervision and Management of Major Hazard Installation*[14], etc.

3.3 Classification criterion of evaluation system

Functions will be realized based on the safety management evaluation system[15-17], e.g. operation efficiency and weakness of the safety management system, gap between the theoretical and practical value of the safety management etc. In this paper, Program Management Maturity Model is introduced into the safety management evaluation system, so the safety management status of tank farms can be divided into 5 degrees: Original, Primary, Standard, Reinforced and Advanced. The points of each degree can be referred as Tab.5.

Tab.5 Point of each degree in tank farms

Degree	**Original**	**Primary**	**Standard**	**Reinforced**	**Advanced**
Points	0-50	51-70	71-80	81-90	91-100

4. Case Study

4.1 Introduction of the tank farms

The approach will be applied in two corporations (Corporation A and Corporation B) of China National Petroleum Corporation (CNPC). The detailed introduction of these two corporations can be seen below.

Corporation A: The atmospheric tank farm stands in the west of the chemical factory. And there are 12 storage tanks: Three 20000m^3 naphtha feedstock tanks (Inner floating roof), Three 5000m^3 Hydrogenation tail material oil tanks(Vault), Two 5000m^3 Hydrogenated gasoline tanks(Inner floating roof), Two 10000m^3 Rough pyrolysis gasoline tanks(Inner floating roof), one 1000m^3 Conditioning tank(Vault), one 1000m^3 Heavy dirty oil tank(Vault), and one 20m^3 Steam condensate tank.

Corporation B: It is mainly responsible for the receipt, storage, blending of the raw product, truck loading and delivery of finished product of fuel oil and liquefied gas in the branch company. The kinds of oil which stored in the corporation includes: gasoline, environment-friendly petrol, light diesel, military diesel fuel, 3 # jet fuel, residual oil, asphalt, 300 # liquid wax, solvent oil, isopropyl alcohol, benzene, dimethyl benzene, liquefied gas, propylene, propane, MTBE, methanol etc. And among the storage oil, benzene, dimethyl benzene, liquefied gas, propylene, propane, MTBE, methanol are toxic and harmful substances.

4.2 Evaluation results

5 safety experts in the field of safety and 2 HSE supervisors form each corporation are invited into this evaluation. Score for both tank farms of Corporation A and B based the points of each degree in Tab.5, the evaluation results can be calculated as Tab.6 below.

Tab.6 Comparison of evaluation results of Corporation A and B

Evaluation Program	Corporation A(Atmospheric tank farms)	Corporation B(Propylene tank farms)
Structure of Leadership and organization	87.8	83.8
Asset management	78.8	75.0
Risk management	86.3	67.5
Training and education	88.5	77.6
Operation security and human reliability	89.2	89.1
Emergency management	86.1	85.6
Fault/Event management	81.6	79.2
Change management	92.7	72.1
Contractors/ Suppliers management	75.0	75.0
File management	92.0	85.7
Safety cultural management	91.5	91.5
Check, audit and review management	83.9	83.9
Total points of safety management in tank farms	86.03	80.1
Safety management degrees in tank farms	Reinforced	Reinforced

Safety evaluation points of atmospheric tank farms in Corporation A is 86.03 refer to the Tab.6 above, while the points of propylene tank farms in Corporation B is 80.1. Both of their safety management evaluation degrees are Reinforced Degree. Comparison of these two corporations on the safety management evaluation points can be seen in Fig.4.

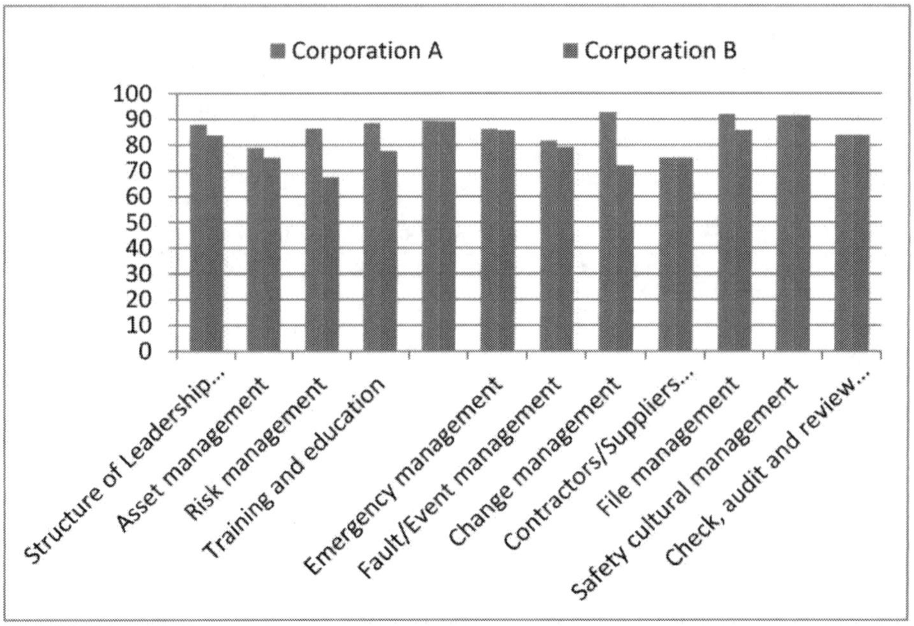

Fig.4 Comparison of two corporations on the safety management evaluation points

From Tab.6, it can be concluded that both of these two corporations have reached Reinforced Degree, and that is to say they both have a mature safety management system already.

Also it can be concluded form Fig.4 that Corporation A has done well in the Change management, File management and Safety cultural management, but some improvements are needed in the Contractors/Suppliers management and Asset management. Corporation B has done well in the Operation security and human reliability and Safety cultural management, especially in the Safety cultural management, while weakness of Corporation B is Risk management and Change management. In conclusion, Corporation A should focus on Contractors/Suppliers management and Asset management while Corporation B should concentrate on Operation security and human reliability and Safety cultural management. Especially, risk management plays an important role in the corporation safety management as it's the second important factor which affected the safety level in tank farms.

5. Conclusions

In this study, a new approach which combines Fishbone Diagram and AHP is proposed, and a procedure to make the safety management is concluded as well. The case study of two corporations is provided to illustrate the proposed approach and concerned numerical results are obtained. Through comparing the evaluations of the safety management system in tank farms of these two corporations, feasibility of this assessment system in the petrochemical enterprise can be proved well. Points of each item can effectively reflect the gap between the two corporation tank farms, and the strengths and weaknesses in safety management of each corporation.

There are two key points in the assessment process, one is the integrity in safety management, institution integrity and safety management institutions integrity; the other is implementation status of the safety management institutions. As it is hard to evaluate the human behavior in the safety management, so more attentions need to be paid to the safety management institutions. Secondary, as a whole, evaluation elements in the safety management are both mutual independence and interaction. However, the evaluation system proposed in this paper only pays more attention on the mutual independence of elements, but interaction between elements is little in the system proposed in this paper, and this is also a difficult point in the study of safety management evaluation technology.

6. Reference

[1] Wu Changcheng. Research on Quality Evaluation Method of Safety Management System[D]. China University of Mining and Technology (Beijing), 2012.

[2] Ye Yong. Research on Application of Six Sigma Management[D]. Huazhong University of Science and Technology. 2004.

[3] Samuel K. M. Ho. Management Art and Science: From 5-S to 6-σ. International Journal of Management Science and Engineering Management,1(2006),63-70.

[4] Yuan Yu-feng. Key points of 6σ management. Automobile Science and Technology,2004.

[5] Samuel K. M. Ho. 5-S Practice: The first step towards total quality management. Total Quality Management,10(1999),345-356.

[6] Yue Zeng and Jiang Nan. Research on safety personnel management mode based Dupont STOP system [J]. Safety and Environmental Engineering. 20(2013), 127-130.

[7] Huang Yin, Tian Shuicheng and Li Tingjun. Safety audit research of spotting behavior based Dupont STOP system [J]. Safety and Environmental Engineering. 20(2013), 92-96.

[8] Fishbone Diagram. Available at

http://baike.baidu.com/link?url=PAgR0JTOujVs90ADHCYMoYo7gYqBxCcAX1K55_bRXKx5l-U-ojiMtCqs3cz-yrJ39CA9fivUgJ1W8THo3vixYq

[9] Liu Lina and Yang Li. Post evaluation index weight based on AHP[J]. Cuide to Business. 4(2011), 86-87.

[10]Huang Guozhen. Introduction of Diamond Safety Health and Environment Integrated Risk Management System[J]. Electric Safety Technology. 9(2007), 29-31.

[11] Gao Ming. Safety Performance Evaluation Research based on MORT[D]. China University of Geosciences(Beijing). 2011.

[12] AQ3013-2008. *General norms for safety standardization of hazardous chemical enterprises*[S]. State Administration of Work Safety. Beijing, China. 2008.

[13] *Provisional Rules of Dangerous Chemical Supervision and Management of Major*

Hazard Installation. Chinese Regulation Code No.40 of the State Administration of Work Safety. Available at http://www.gov.cn/flfg/2011-09/13/content_1945888.htm.

[14] AQ/T3012-2008 *Guideline of Safety Management System Implementation for Petrochemical Corporation*[S]. State Administration of Work Safety. Beijing, China. 2008.

[15] Zhao Qi. Theory and Application Research of Project Management Maturity Model[D]. Nanjing: Nanjing University of Aeronautics and Astronautics. 2006.

[16] Zhan Wei. Research on Project Management Maturity Model and Its Application[J]. Journal of Beijing University of Aeronautics and Astronautics. 20(2007),18-21.

[17] Liu Bo. Employee Competence Maturity Model(SC-MM) and Research on Related Model[D]. Nanjing University of Science and Technology. 2013.

Study on Separation Distance Determination of LNG Filling Station on Water

Gujun Wan
Qingdao Safety Engineering Institute ,China Petroleum & Chemical Corporation,
218#,Yan'an 3rd road, Qingdao, Shandong, China
wangj.qday@sinopec.com

Wenyi Dang
Qingdao Safety Engineering Institute, China Petroleum & Chemical Corporation
dangwy.qday@sinopec.com
Zhifeng Wu
Qingdao Safety Engineering Institute, China Petroleum & Chemical Corporation
wuzf.qday@sinopec.com

Zhongshu Han
Qingdao Safety Engineering Institute, China Petroleum & Chemical Corporation
hanzs.qday@sinopec.com

Yongzhong Bai
Qingdao Safety Engineering Institute, China Petroleum & Chemical Corporation
baiyz.qday@sinopec.com

Prepared for Presentation at
American Institute of Chemical Engineers
2015 Spring Meeting
11th Global Congress on Process Safety
Austin, Texas
April 27-29, 2015

UNPUBLISHED

Study on Separation Distance Determination of LNG Filling Station on Water

Gujun Wan
Qingdao Safety Engineering Institute ,China Petroleum & Chemical Corporation,
218#,Yan'an 3rd road, Qingdao, Shandong, China
wangj.qday@sinopec.com

Wenyi Dang
Qingdao Safety Engineering Institute, China Petroleum & Chemical Corporation
dangwy.qday@sinopec.com

Zhifeng Wu
Qingdao Safety Engineering Institute, China Petroleum & Chemical Corporation
wuzf.qday@sinopec.com

Zhongshu Han
Qingdao Safety Engineering Institute, China Petroleum & Chemical Corporation
hanzs.qday@sinopec.com

Yongzhong Bai
Qingdao Safety Engineering Institute, China Petroleum & Chemical Corporation
baiyz.qday@sinopec.com

Keywords: LNG filling station on water, separation distance, LNG diffusion, LNG fire

Abstract

In China, more and more ships use LNG as fuel. So, more LNG filling station on water will be constructed in the future. In order to assure the safety of the community around the LNG filling station on water, the appropriate separation distance should be maintained between LNG filling station on water and community. In this paper, according to the regulation of NFPA 59A, the numerical simulation was conducted to calculate the separation distance between LNG filling station on water and buildings around it. The scenario of LNG tank and filling arm on pontoon leakage were considered. The commercial software FLACS was used to calculate the LNG diffusion behavior and independent development software LNGFHR was used to calculate the LNG fire thermal radiation. The extended distance of concentration of methane in air of 50 percent of the lower flammability limit (LFL) and LFL were obtained. Moreover, the affected distance of different heat flux, including $1.5kW/m^2$, $5kW/m^2$, $9kW/m^2$ and $13kW/m^2$, were obtained too. According to the results of LNG diffusion distance and thermal radiation affected distance, the separation distance between LNG filling station on water and buildings around it, such as important public buildings, bridge, wharf, railway, etc, were determined.

1. Introduction

There are about two hundred and thirty thousand transportation ships in China. The ships fuel mainly is diesel and fuel oil, which cause severe air and river pollution due to exhaust emission and fuel leakage. Therefore, the project of 'convert oil fueled ship to LNG fueled ship' is conducting in China. In the next three years, there are about fifty thousand ships will be refitted. But this project makes slow progress because of lackage of LNG filling station on water. Up to now, there is only one LNG filling station on water has been built in China. There are four primary bunkering ways to ship: the truck-to-ship bunkering, pontoon-to-ship bunkering, shore-to-ship bunkering and ship-to-ship bunkering. The first LNG filling station on water in China adopted the pontoon-to-ship way. In 2014, China Classification Society (CCS) issued the 'Rules for classification and construction of LNG bunkering pontoon on water' [1], which described the regulations of LNG bunkering pontoon design and construction. The separation distance of the LNG filling station didn't be considered in this standard. Fan and Zhang[2] assessed the hazard distance of pool fire and determined the layout of LNG bunkering pontoon. Thus, they didn't consider the LNG vapor cloud disperse hazard. In this paper, the separation distance of LNG filling station on water was determined according to the hazard of LNG vapor cloud disperse and LNG pool fire.

2. Determination method of separation distance

The separation distance is the distance between a hazard source and an object (human, equipment or environment), which will mitigate the effect of a likely foreseeable incident and prevent a minor incident escalating into a larger incident. There are two primary methods can be used to determine the separation distance: consequence based method and risk based method. In 2013, American National Fire Protection Association (NFPA) issued NFPA 59A: Standard for the Production, Storage, and Handling of Liquefied Natural Gas (LNG) (2013 Edition) [3]. This standard adopted the consequence based method to calculate the spacing of an LNG impoundment to the property line that can be built upon. Two consequence scenarios were considered: LNG vapor cloud dispersion and LNG pool fire. The standard specified the design LNG spill, consequence calculation model and weather conditions. A predicted concentration of methane in air of 50 percent of the lower flammability limit (LFL) shall not extend beyond the property line that can be built upon. The maximum radiant heat flux from a fire shall not exceed the limits listed in Table 1. In this paper, the consequence based method was adopted according to the NFPA 59A regulations.

Table 1. Radiant Heat Flux Limits to Property Lines and Occupancies

Radiant Heat Flux(W/m^2)	Column Title
5000	A property line at ground level that can be built upon for ignition of a design spil
5000	The nearest point located outside the owner's property line at ground level that, at the time of plant siting, is used for outdoor assembly by groups of 50 or more persons for a fire in an impounding area
9000	The nearest point on the building or structure outside the owner's property line that is in existence at the time of plant siting and used for assembly, educational, health care, detention and correction, or residential occupancies for a fire in an impounding area
30000	A property line at ground level that can be built upon for a fire over an impounding area

3. Calculation model

3.1 LNG filling station on water 3D model

LNG filling station on water 3D model was shown in figure 1. The LNG pontoon is 150m long and 18m wide. The filling station can be divided four regions: the control room, LNG tank area, LNG filling pump area and LNG filling arm area.

Figure 1. LNG filling station on water 3D model

3.2 Mathematical model

3.2.1 LNG dispersion model

The commercial software FLACS was used to calculate the LNG dispersion process [4]. The process obeys mass, momentum and energy conservation.

Conservation of mass:

$$\frac{\partial}{\partial t}(\beta_v \rho) + \frac{\partial}{\partial x_j}(\beta_j \rho u_j) = \frac{\dot{m}}{V} \qquad \text{[Eq. 1]}$$

Momentum equation:

$$\frac{\partial}{\partial t}(\beta_v \rho u_i) + \frac{\partial}{\partial x_j}(\beta_j \rho u_i u_j) = -\beta_v \frac{\partial p}{\partial x_i} + \frac{\partial}{\partial x_j}(\beta_j \sigma_{ij}) + F_{o,i} + \beta_v F_{w,i} + \beta_v(\rho - \rho_0) g_i \qquad \text{[Eq. 2]}$$

where Fw,i is flow resistance due to walls and Fo,i is flow resistance due to sub-grid obstructions:

$$F_{o,i} = -\rho \left|\frac{\partial \beta}{\partial x_i}\right| u_i |u_i| \qquad \text{[Eq. 3]}$$

The stress tensor σ_{ij} in the above equations is given by:

$$\sigma_{ij} = \mu_{eff}\left(\frac{\partial u_i}{\partial x_j} + \frac{\partial u_j}{\partial x_i}\right) - \frac{2}{3}\delta_{ij}\left(\rho k + \mu_{eff}\frac{\partial u_k}{\partial x_k}\right) \qquad \text{[Eq. 4]}$$

The effective viscosity μ_{eff} is defined as follows:

$$\mu_{eff} = \mu + \rho C_\mu \frac{k^2}{\varepsilon} \qquad \text{[Eq. 5]}$$

Transport equation for enthalpy:

$$\frac{\partial}{\partial t}(\beta_v \rho h) + \frac{\partial}{\partial x_j}(\beta_j \rho u_j h) = \frac{\partial}{\partial x_j}\left(\beta_j \frac{\mu_{eff}}{\sigma_h}\frac{\partial h}{\partial x_j}\right) + \beta_v \frac{Dp}{Dt} + \frac{\dot{Q}}{V} \qquad \text{[Eq. 6]}$$

Transport equation for turbulent kinetic energy:

$$\frac{\partial}{\partial t}(\beta_v \rho k) + \frac{\partial}{\partial x_j}(\beta_j \rho u_j k) = \frac{\partial}{\partial x_j}\left(\beta_j \frac{\mu_{eff}}{\sigma_k}\frac{\partial k}{\partial x_j}\right) + \beta_v P_k - \beta_v \rho \varepsilon \qquad \text{[Eq. 7]}$$

Transport equation for the dissipation rate of turbulent kinetic energy:

$$\frac{\partial}{\partial t}(\beta_v \rho \varepsilon) + \frac{\partial}{\partial x_j}(\beta_j \rho u_j \varepsilon) = \frac{\partial}{\partial x_j}\left(\beta_j \frac{\mu_{eff}}{\sigma_\varepsilon}\frac{\partial \varepsilon}{\partial x_j}\right) + \beta_v P_\varepsilon - C_{2\varepsilon}\beta_v \rho \frac{\varepsilon^2}{k} \qquad \text{[Eq. 8]}$$

Flow shear stresses, G_s, wall shear stresses, G_w, buoyancy, G_b, and sub-grid objects, Go contribute to the production of turbulent kinetic energy P_k:

$$P_k = G_s + G_w + G_b + G_v \qquad \text{[Eq. 9]}$$

The production rate of turbulent kinetic energy due to shear stresses appears from the derivation of the transport equation and reads:

$$G_s = \sigma_{ij} \frac{\partial u_i}{\partial x_j} \qquad \text{[Eq. 10]}$$

Production due to buoyant forces modelled by a simple gradient model:

$$G_b = -\frac{1}{\rho} \frac{\mu_{eff}}{\sigma_b} g_i \frac{\partial \rho}{\partial x_i} \qquad \text{[Eq. 11]}$$

The turbulence generation due to sub-grid obstructions is modelled by:

$$G_o = C_o \beta_v \rho |\vec{u}| u_i^2 f_i \qquad \text{[Eq. 12]}$$

where C_o is a model constant and f_i is a parameter depending on sub-grid objects.
The production of dissipation, P_ε, is modelled as follows:

$$P_\varepsilon = C_{1\varepsilon} \frac{\varepsilon}{k} P_k (1 + C_{3\varepsilon} R_f) \qquad \text{[Eq. 13]}$$

where the model for the buoyancy term is modelled as follows:

$$R_f = -\frac{G_b}{P_k} \frac{|\vec{u} \times \vec{g}|}{|\vec{u}| \times |\vec{g}|} \qquad \text{[Eq. 14]}$$

In Flacs, spreading pool (spill) is decribed by the shallow-water equations in two-dimensions. The shallow-water equations are solved on a Cartesian grid, which is equal to the xy-grid in Flacs.The equation solved for the spill height is given by:

$$\frac{\partial h}{\partial t} + \frac{\partial h u_i}{\partial x_i} = \frac{\dot{m}_L - \dot{m}_V}{\rho_l} \qquad \text{[Eq. 15]}$$

and the momentum equation is written as:

$$\frac{\partial h u_i}{\partial t} + u_j \frac{\partial h u_i}{\partial x_j} = F_{g,i} + F_{\tau,i} \qquad \text{[Eq. 16]}$$

where the gravity term is modelled as follows:

$$F_{g,i} = hg\Delta \frac{\partial (h+z)}{\partial x_i} \qquad \text{[Eq. 17]}$$

The elevation of the ground, z, has been included such that spill on sloping terrains and the effects of obstacles and embarksments can be calculated. The parameter D equals one for pools on solid surfaces and $\Delta = (1-\rho_l/\rho_w)$ for pools on water. The shear stress between the pool and the substrate is given by the general formula:

$$F_{\tau,i} = \frac{1}{8} f_f u_i |u_i| \qquad \text{[Eq. 18]}$$

3.2.2 LNG pool fire model

The LNGFIRE3 was used to calculate LNG fires. The model used in LNGFIRE3 assumes that the flame behaves as a grey cylinder emitting thermal radiation from its surface at a uniform level. This level depends on the depth of the flame. The emissive power of the flame, E_f, is given by:

$$E_f = \varepsilon E_{max} \qquad \text{[Eq. 19]}$$

where ε is the emissivity of the flame and E_{max} is the maximum flux achievable from LNG fires. This flux value was 190 kW/m^2 in LNGFIRE3.
The emissivity of flames may be expressed by:

$$\varepsilon = 1 - exp(-kD) \qquad \text{[Eq. 20]}$$

where k = an attenuation coefficient (m^{-1})
D= depth of the flame (m)

The value of k was 0.3 m^{-1} in LNGFIRE3. The depth of the flame, D, was set equal to the diameter of the circular impoundment area or its dimension perpendicular to the direction of wind for rectangular impoundments.
A target exposed to a cylindrical fire will receive radiation at a rate determined by the following equation:

$$q'' = \tau F E_f \qquad \text{[Eq. 21]}$$

where, τ= atmospheric transmissivity to thermal radiation from the fire

F = a geometric view factor which is a measure of the fraction of energy emitted by the flame that is intercepted by the target

The detailed procedure to calculate the absorptivity of water vapor is described in the SFPE Handbook of Fire Protection Engineering [5].

To determine the geometric view factor F, the flame dimensions and angle of tilt from the vertical must be estimated. For circular pools, the flame is assumed to be a tilted cylinder with a flame length that is calculated from the equation developed by Thomas [6]:

$$\frac{L_f}{D} = 42\left[\frac{m''}{\rho_a\sqrt{gD}}\right]^{0.61} \qquad \text{[Eq. 22]}$$

Where, L_f = flame length (m)

D = pool diameter (m)

m'' = mass burning rate (kg/m^{2}s)

ρ_a = ambient air density (kg/m^3)

The burning rate, m'', for circular fires is calculated from the following experimentally derived equation:

$$m'' = m_{max}[1 - exp(-0.46D)] \qquad \text{[Eq. 23]}$$

Where, m_{max}=maximum mass burning rate (kg/m^{2}s), m_{max}=0.11 kg/m^{2}s in LNGFIRE3.

LNGFIRE3 has been validated against the results of large-scale LNG fire experiments on land. But it is not suitable for LNG fire on water. Therefore, software named LNGFHR was developed to calculate the LNG fire on water according to the results of LNG fire experiments on water. Figure 2 shows the interface of LNGFHR, which can be used to calculate the LNG fire thermal radiation of circular and rectangular pool fire on land or water.

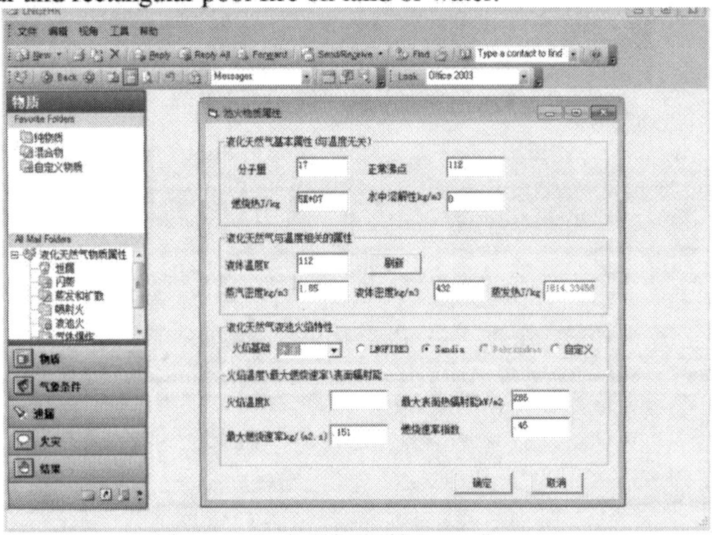

Figure 2. LNGFHR interface

3.2.3 LNG spill scenario

According to NFPA 59A regulations, LNG spill scenario was shown in Table 2.

Table 2. LNG spill scenario

No.	Leakage Position	Pipe diameter (mm)	Leakage hole diameter (mm)	duration time (min)	Leakage rate (kg/s)	Pressure (bar)	Wind velocity (m/s)	Atmospheric stability category	Wind direction
010101	tank	150	design spill[a]	1	32.11	4	2	F	Vertical to pontoon
010102	tank	150	design spill	1	32.11	4	2	F	45°
010103	tank	150	design spill	1	32.11	4	5	D	Vertical to pontoon
010104	tank	150	design spill	1	32.11	4	5	D	45°
010201	tank	150	design spill	10	32.11	4	2	F	Vertical to pontoon
010202	tank	150	design spill	10	32.11	4	2	F	45°
010203	tank	150	design spill	10	32.11	4	5	D	Vertical to pontoon
010204	tank	150	design spill	10	32.11	4	5	D	45°
020101	filling arm	100	rupture	1	4.3	7	2	F	Vertical to pontoon
020102	filling arm	100	rupture	1	4.3	7	2	F	Horizontal to pontoon
020103	filling arm	100	rupture	1	4.3	7	5	D	Vertical to pontoon
020104	filling arm	100	rupture	1	4.3	7	5	D	Horizontal to pontoon
020201	filling arm	100	rupture	10	4.3	7	2	F	Vertical to pontoon
020202	filling arm	100	rupture	10	4.3	7	2	F	Horizontal to pontoon
020203	filling arm	100	rupture	10	4.3	7	5	D	Vertical to pontoon
020204	filling arm	100	rupture	10	4.3	7	5	D	Horizontal to pontoon

a See NFPA 59A for design spill

4 Results and discussion

The LNG spill dispersion results were shown in Table 3. The methane dispersion distance of LNG tank leakage is bigger than filling arm leakage, because of tank LNG leakage rate is much bigger than filling arm leakage rate. It's more obvious when LNG leakage duration time is 10min. That's because the hindering effect of LNG tank dike is more significantly when LNG leakage duration time is 1min. For filling arm leakage, due to the pontoon blockage effect to wind, the methane dispersion distance is smaller when the wind direction is vertical to pontoon. Based on comprehensive consideration of NFPA 59A regulations and LNG filling station actual safety measures, when the weather condition is 2F and leakage duration time is 1min, the methane mean dispersion distance of different wind direction was used to determine the safety distance of LNG filling station. That is, the methane in 0.5LFL dispersion distance is 146m and 103m for tank and filling arm leakage respectively. The methane in LFL dispersion distance is 103m and 85m for tank and filling arm leakage respectively.

Table 3. LNG spill dispersion results

spill scenario	Wind direction	Concentration of Methane	0.5LFL				LFL			
		Leakage duration time	1min		10min		1min		10min	
		Weather condition	2F	5D	2F	5D	2F	5D	2F	5D
tank leakage	Vertical to Pontoon	dispersion distance	137	92	512	130	97	52	258	70
	45°		154	124	680	253	101	55	452	100
filling arm leakage	Vertical to Pontoon		56	26	66	28	50	14	52	16
	Horizontal to pontoon		150	70	180	88	120	32	138	60

The LNG pool fire calculation results which used LNGFIRE3 and LNGFHR were shown in Table 4. For the LNG tank leakage, the LNG pool fire calculation results were similar between LNGFIRE3 and LNGFHR due to the same pool fire model was used. But for the filling arm leakage, there is great difference. LNGFHR used the LNG pool fire model based on the results of LNG fire experiments on water. The value of E_{max} in Eq.19 and m_{max} in Eq.23 was 286 kW/m^2 and 0.147 kg/m^{2}s in LNGFHR respectively But in LNGFIRE3, the value were 190. kW/m^2 and 0.11 kg/m^{2}s. The results obtained from LNGFHR were used to determine the safety distance of LNG filling station.

Table 4 LNG pool fire calculation results

No.	leakage position	software	pool diameter(m)	emissive power (kW/m^2)	burning rate (kg/s)	flame length (m)	flame tilt angle	affected distance(m)			
								1.5kW/m^2	5kW/m^2	9kW/m^2	13kW/m^2
010101	tank	LNGFIRE3	20.4	189.82	0.11	39.6	0	149	83	62	51
		LNGFHR	20.4	189.82	0.11	39.6	0	152	84	61	49
010201	tank	LNGFIRE3	22.1	189.9	0.11	41.9	0	158	89	66	55
		LNGFHR	22.1	189.9	0.11	41.9	0	162	90	65	53
020101	filling arm	LNGFIRE3	7.04	175.93	0.106	18	27	70	38	30	25
		LNGFHR	7.04	264.81	0.145	35	20	105	60	45	37
020201	filling arm	LNGFIRE3	11	186.27	0.109	26	16	101	54	40	34
		LNGFHR	11	280.38	0.15	49	0	145	78	55	44

According to the requirements of threshold of methane concentration and LNG pool fire thermal radiation in NFPA 59A, different safety distance was determined between equipments in LNG filling station and buildings around it. The results were shown in Table 5. For important public buildings, railway, wharf and bridge, the safety distance was determined according to the concentration of methane in air of 50 percent of the LFL and thermal radiation intensity of 1.5kW/m^2. While for common buildings, the safety distance was determined according to the concentration of methane in LFL and thermal radiation intensity of 5kW/m^2.

Table 5 The safety distance between equipments in LNG filling station and buildings around it (m)

Buildings around LNG filling station	equipments in LNG filling station	
	LNG tank	filling arm
Important public buildings	150	120
Common buildings	105	85
railway	150	120
wharf	150	120
Bridge	150	120

5 Conclusion

In order to determine the safety distance of LNG filling station on water, according to the regulations in NFPA 59A, the commercial software FLACS was used to calculate the LNG diffusion distance. Furthermore, software named LNGFHR was developed to calculate the LNG fire on water. Compared to software of LNGFIRE3, LNGFHR pool fire model was based on the results of LNG fire experiments on water. Therefore, more accurate pool fire calculation results of LNG filling station on water could be obtained by using LNGFHR. The scenario of LNG tank and filling arm on pontoon leakage were considered. The extended distance of concentration of methane in air of 50 percent of the lower flammability limit (LFL) and LFL under different weather conditions were obtained. Moreover, the affected distance of different thermal radiation intensity, including 1.5kW/m^2, 5kW/m^2, 9kW/m^2 and 13kW/m^2, were obtained too. According to the requirements of threshold of methane concentration and LNG pool fire thermal radiation in NFPA 59A, different safety distance was determined between equipments in LNG filling station and buildings around it. For important public buildings, railway, wharf and bridge, the safety distance between them and equipments in LNG filing station was determined according to the concentration of methane in air of 50 percent of the LFL and thermal radiation intensity of 1.5kW/m^2.That is, for LNG tank, the safety distance is 150m. For filling arm, the safety distance is 120m. The safety distance between common buildings and equipments in LNG filing station was determined according to the concentration of methane in LFL and thermal radiation intensity of 5kW/m^2. That is, for LNG tank, the safety distance is 105m. For filling arm, the safety distance is 85m.

6. References

[1] CCS *Rules for classification and construction of LNG bunkering pontoon on water.* China Classification Society, Beijing, 2014

[2] H.J. Fan, H. Zhang and J.Y. Xu, Assessment of the Hazard Distance of Pool Fire for LNG Bunkering Pontoon, Shipbuilding of China 4 (2013), 186-195

[3] NFPA *Standard for the Production, Storage, and Handling of Liquefied Natural Gas (LNG).* National Fire Protection Association, Quincy, MA, 2013

[4] GexCon AS *FLACS v9.1 User's Manual.* GexCon AS, Bergen, MA, 2011

[5] K.S.Mudan and P. A. Croce, Fire Hazard Calculations for Large Open Hydrocarbon Fires, in The SFPE Handbook of Fire Protection Engineering, P. J. DiNenno, Editor-in-Chief, National Fire Protection Association, Quincy, MA, 1988

[6] P.H. Thomas, The Size of Flames from Natural Fires, 9th International Combustion Symposium, 844-859, 1963

A Case-Study of a Fire Incident of Trichlorosilane Process and Response Measures

Horng Jaojia[1] , Liao Kuangyu[2], Liou kehun[1], Yuan Paochiang[3]

1. Department of Safety, Health, and Environmental Engineering, National Yunlin University of Science and Technology, Taiwan.
2. Graduate School of Engineering Science and Technology, National Yunlin University of Science and Technology, Taiwan.
3. College of Science, Engineering and Technology, Jackson State University, Mississippi, US.

Prepared for Presentation at
American Institute of Chemical Engineers
2015 Spring Meeting
11th Global Congress on Process Safety
Austin, Texas
April 27-29, 2015

UNPUBLISHED

A Case-Study of a Fire Incident of Trichlorosilane Process and Response Measures

Horng Jaojia[1] , Liao Kuangyu[2], Lioukehun[1], Yuan Paochiang[3]

1. Department of Safety, Health, and Environmental Engineering, National Yunlin University of Science and Technology, Taiwan.
2. Graduate School of Engineering Science and Technology, National Yunlin University of Science and Technology, Taiwan.
3. College of Science, Engineering and Technology, Jackson State University, Mississippi, US.

Keywords: Tricholorsilance, fire distinguisher, liquid nitrogen

Abstract

There was a fire occurring of a overheating Trichlorosilane (TCS) reactor in a solar panel manufacture plant in 2011. Although no injury happened, the leak-out TCS caught fire and produced acid plume creating great concerns for responding governmental agencies, nearby general public and news media. Since TCS is an important raw material for petrochemical, semi-conductor, and solar panel manufactures, its hazardous properties of low flash point, easy combustible, water-reactive, and acid-producing have created many difficulties in response as its leaking out.

All incidents of chlorosilanes (CS) materials would produce large acid plume as large amount of leaking lead to burning. The fires could not be controlled by water due to their water-reactive properties. When other distinguishers such as foams were applied, the uses of water jetty were essential to cover the burning surface and to reduce the acid plume. The best practice was to continuously put foams on top until it burn out. However, this action would prolong the response and could not control the damages well. Therefore, the prevention of acid plume and the correct use of foams were essential for fire response to CS and TCS.

Upon our field tests, we found that no fire distinguishing was observed for powder, carbon dioxide, water and halon on small-scale fire (3 kg) of TCS. All continuously produced acid plume and some even reacted to form harmful byproducts. The fires would be distinguished as the application of foams and liquid nitrogen. The effective practices were that 6% mixed middle-expansion foams with 15 cm of covering height and liquid nitrogen with volume (ml) of 177+0.287*(TCS volume). However, two methods adopted totally different principles for fire extinguishing.

Our study indicated that when using liquid nitrogen to effectively distinguish the TCS fire was due to it adsorbed reaction heat of burning and stop the reaction. We further prove that no property change of TCS after our application. This application was safe and was different from applying foams that water actually reacted with TCS and producing dangerous hydrogen gas. This finding might be used in future development for cooling agency for CS fire application.

1. Introduction

The production outputs for solar panel were about 15% and 45% for Taiwan and China[1] with mostly by polycrystalline silicon manufacturing process owing to production technology and cost. Among them, the Siemens Reactor was used as reactor with Trichlorosilane (TCS, chemical formula $SiHCl_3$) as main source of silicon. TCS could easily catch fire and would produce aerosol and gas of hydrochloride acid and chlorine gas. Those byproducts of TCS fire could enter our breathing tract and proximal alveolar region in order to create harmful effects on environmental safety and personnel health rather than leaking only [2, 3].

This study studied literature and case-study of TCS related events, including references on response measures of TCS fire and leaking from Safety Data Sheet, 2012 Emergency Response Guidebook (ERG) and ASTM MNL33 Manual. A set of field tests was performed in order to find the feasible response measures on TCS fires. In addition, several researchers suggested that liquid nitrogen (LN2) could effectively distinguished pool fires of ethanol, propanol, diesel and wood blocks.[5,6] Some studies reported that 0.5 liter of LN2 could distinguish 1 meter square of pool fire [5] as well as the fire by metal sodium[6]. This study will also test on the effectiveness of the LN2 cooling agent to deal with TCS leaking and fires.

2. Literature Review

2.1 TCS Case-Studies [7] [8] [9] [13] [14] [18] [19]

In recent years, the popular solar panel production continuously increased and the usages of TCS were also increased accordingly. There were several severe incidents TCS related leaking and fire events as follows.

1. As Silicon Tetrachloride leaked out from connection line of storage tank and pipeline on April 1974, acid fog spread out to large area (400 m×450m×1600m) and more than 14,000 people were evacuated at Calumet near Chicago.
2. As a set of train derailed near Freeland, Michigan, one TCS tank leaked and caught fire for more than six days with more than 1000 people evacuated and 11 injured on July 1989.
3. As a TCS tanker's transfer pipe broke in Gainesville, Florida, more than 5000 gallons TCS leaked and created a three-mile acid cloud on July 1994
4. As a lifted TCS pipeline leaked for two hour in Midland City of Michigan on August 1997, 5000 pounds of TCS created acid cloud polluted the area for more than 6 hours.
5. As pipe broken in a SiCl4 and TCS polycrystalline process in Morse Lake City, Washington on October, 1998, two un-protected workers were injured and died as 34000 pounds of mixture leaked and spread.
6. As a glass manufacture's equipment was washed and under maintenance in HuBei, China on December 2008, TCS leaked and produced hazardous acid fog and 18 persons were send to hospitals
7. As a fluidized reactor bed over-heated in a sloar panel production plant in Pen-Tung, Taiwan on January 2011, the gaskets failed and TCS leaked out and caught fire. Fortunately, no one injured in the event.

There were several observations in the above events.

1. As large amounts of leakages caught fires, the broader effects of acid cloud created many problems. The public evacuation and using water fog to disperse and to wash out acid gases were the main tasks among response measures.
2. Owing to the water reactive substances, such as TCS, could not be controlled by water spreading, the dry power extinguishers might also be of limited effects because of the in-effective coverage and possible turbulence during spreading.
3. Using foam to extinguish TCS fires might be in-effective and unable to reduce acid cloud if the covering area could not be controlled. In addition, the water portion in fire-fighting foam would accelerate the reaction with TCS and create large acid smoke until extinguished[10].
4. As the defensive response enacted until burning subsided, the response time would be extended as the acid clouds spreading out to broader areas for evacuation in a larger scale.

2.2 Fire Extinguishing Measures for TCS

Using different extinguishers on TCS fires was studied in field tests [11] with several observations below.

1. The application of dry powder could not extinguish TCS fires but to enlarge through spreading. The observation by the Infrared Thermography indicated that the temperature of outer flame was higher and could quickly raise back to 500 Celsius as the application of dry powder stopped.
2. No segregation or asphyxiation effects were observed as applying CO2 extinguisher on TCS fires.
3. The application of halon extinguisher produced colored smokes which might be halon de-composted into highly reactive and toxic fluorine or chlorine in TCS fires.
4. Foaming AFFF could extinguish TCS fires as the foams accumulated to 15 cm high. However, the water in inflated AFFF could hydrolyze TCS to produce hydrogen gas as small explosion observed above foams.

Upon consulting Safety Data Sheet, 2012 Emergency response Guidebook, ASTM MNL33 Manual and relevant references for water-reactive TCS fires, the main points were to reduce the possible hazards by the by-products during response. Thus, the suggestions included the use of dry power and CO2 extinguishers or using copious amount of water to accelerate reactions for small TCS fires. At the large TCS fires, the suggestions were to use low or medium inflated Aqueous Film Forming Foam (AFFF) for fire extinguishing [10]. Therefore, the extinguishing methods for TCS fires include the follows.

- Dry chemical powders that containing salts would inhibit chemical and ignition reactions.
- Fire extinguisher with carbon dioxide could reduce the concentration of comburant such as oxygen and would smother fire burning.
- Medium inflated AFFF would inhibit the byproducts such as acid gases and reduce surface reaction through covering surface.
- Alkaline lime and soda ash could neutralize acid and react with chlorosilane to produce flammable and explosive gases.
- Copious amount of water could hydrolyze with chlorosilane, shorten the reaction time and extinguish fire eventually.
- Covering with sands and soils would reduce the dispersion of acid gas into the air and asphyxiate the fires.

2.3 TCS Fire Field Test

A set of fire extinguishing test for TCS fires was executed at a remote fire training field on July 2014 at Ksoshiung City, Taiwan [16]. The TCS fires were initiated in with constant weight of TCS in iron plated and then ignited until stable burning at about 2 minutes. Then, the test agents of extinguishers were applied to observe the effectiveness. Several thermal imaging cameras were used for observations of temperatures of fire sources, acid gases and smokes. All personnel wear protective masks, gloves and clothes to avoid the contacts with those hazardous byproducts. Several observations and results were achieved as described elsewhere [16].

- The tests estimated that the TCS's combusting mass flux was about 1.98kg/m^2 • min which was 10 times higher than the evaporation rate. TCS showed a great combustion potential.
- TCS fires with byproducts of thick acid smoke could impact and hazard environment and public significantly.
- The byproducts of TCS fires would emit acid aerosol of $HCl_{(p)}$ and acid gases of $HCl_{(g)}$ and $Cl_{2(g)}$. Thus, full-faced respirators/masks, gloves and clothes were all needed for personal breathing, eyes, and skin protection during the responses to TCS fires or leakage.
- TCS fires were difficult to extinguish because of high reaction rate and energy output. Although medium inflated AFFF was suggested to use for extinguishing fires, it seemed that some techniques were needed.

2.4 Liquid nitrogen as Extinguisher

LN2 was used to suppress pool fires of metal sodium [6]. It was thought to be the effects of cooling and asphyxiation. LN2 does not react with metal sodium, does not burn or assist burning, and does not produce toxic derivatives. In addition, LN2 has high heat-absorbing capacity, forms asphyxiation above fire, and is relatively low cost and stable. LN2 has been considered as a good extinguisher for water-reactive materials, especially TCS[12].

3. Results and Discussion

3.1 Extinguishing TCS Fire in Field Tests [12]

As TCS burning, white smoke was coming of the liquid surface accompanying with heat generation and fuming. The extensive white smoke was reported to be the mixture acid aerosols HCl, SiO_2 and acid gases of HCL and Cl_2 releasing from the TCS burning reaction (Fig.1a and b). Observing by infrared thermal image and measuring the temperature distribution on TCS pool burning, we observed highest flame temperature above 510℃ (detection limit of thermal imager) as in white color (Fig.1a). A linear relationship between burning time (kg/min) and superficial area of TCS was observed and it was not related to the initial TCS weight (Fig. 2 & 3). From those data, the TCS mass burning rate was 2.41 kg/m^{2}min. The value was between of Methanol 1.02 kg/m2-min and gasoline 3.3 kg/m2-min. Since TCS burning often produce many acid aerosol and gases, a medium burning rate indicated that smoke spread quickly as TCS combusting.

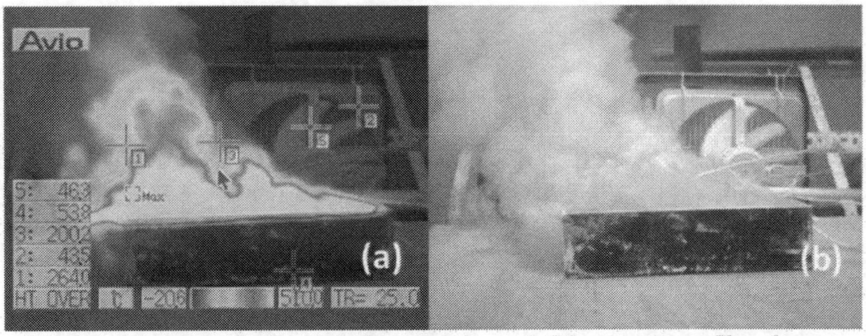

Fig.1 (a) thermal infrared observations ; (b) corresponding image

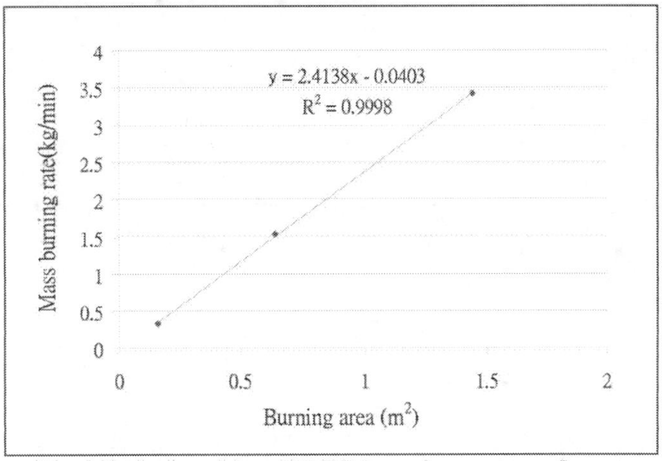

Fig.2 Relationship of TCS burning rate and area

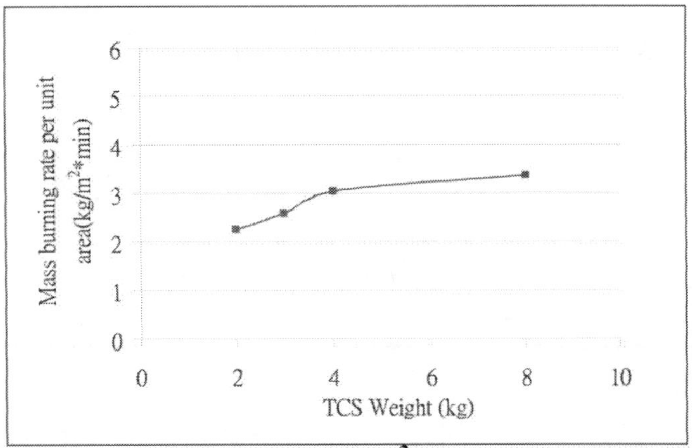

Fig.3 Relationship between burning rate (kg/m^2s) and initial TCS weight with area of 40cm by 40cm.

3.2 LN2 for TCS fire suppression

As LN2 was adding into the TCS pool fire (usually decanting at 1-1.2 sec), both cooling ans asphyxiation effects worked on extinguishing TCS burning simultaneously. When the pool fire burn at the area of 40 cm by 40 cm, the addition of LN2 would first absorb TCS burning heat and asphyxiate the fire. It took 30 sec to 1 min to extinguish fire. As showed in Fig, 5, a 500ml LN2 was pouring into 2kg TCS fire (a). The flame would first extend (a) and then temperature would drop (b). As LN2 quickly evaporation (c), TCS fire would extinguish (d). The weight of TCS pool fire correlated to decanting LN2 volume regressed as formula 1.

Volume LN2 (ml) = 177 (ml) + 221 * TCS weight (kg) [Eq. 1]

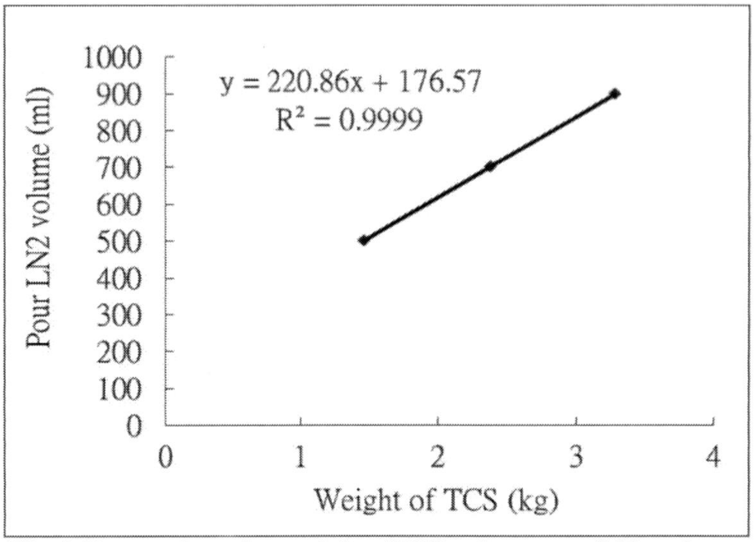

Fig.4 Regression of remaining TCS volume v.s. LN_2 volume

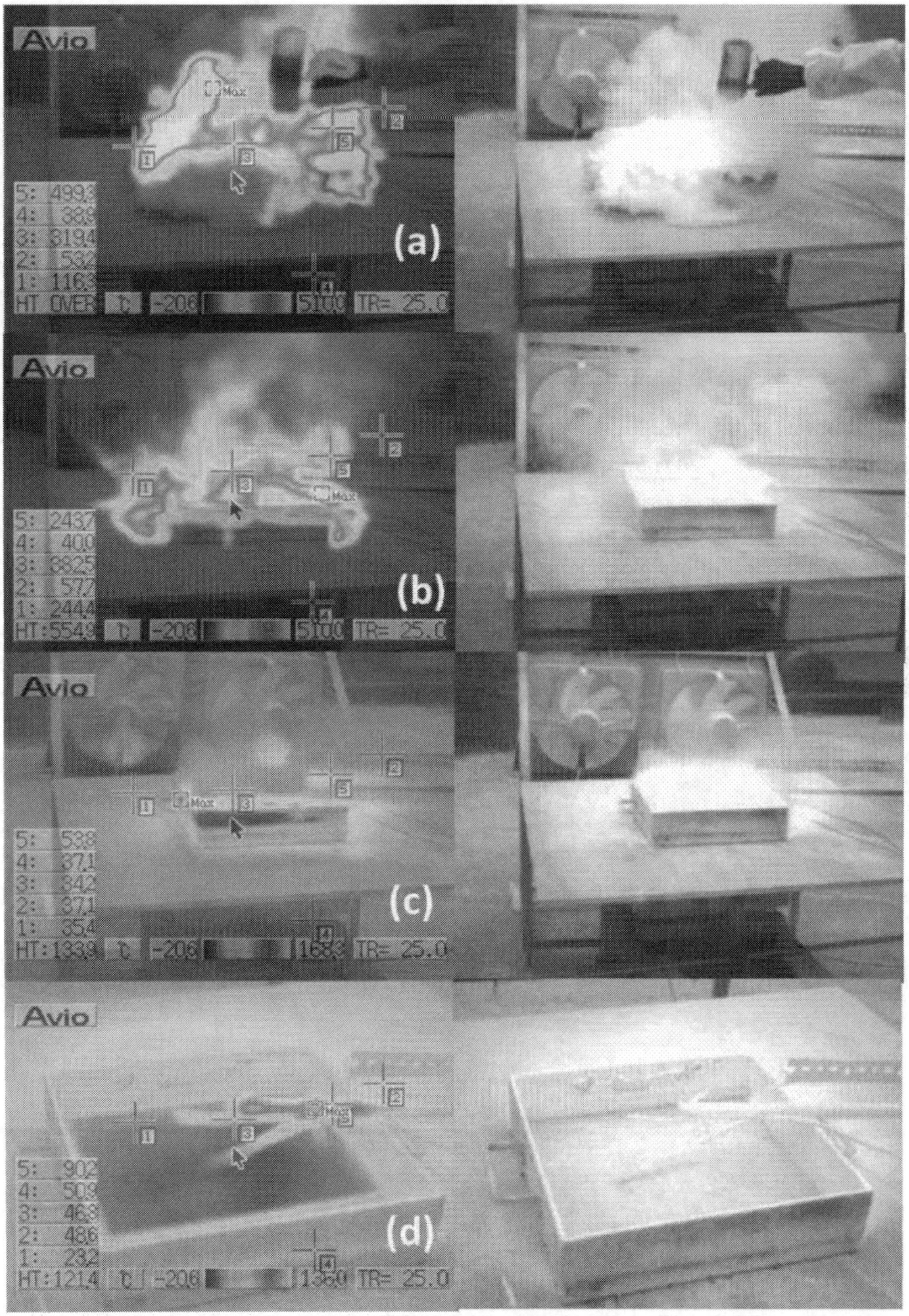

Fig.5 Infrared thermal image, (a) as 500ml LN_2 pouring into 2kg TCS where first flame extending, (b)then temperature dropping, (c)LN_2 quickly evaporation, (d) TCS fire extinguished.

Moreover, we also observed the following phenomena.

(1) The efficiency of adding LN2 would correlated to its flow velocity which more rapid spreading LN2 on TCS pool yielded more effective towards TCS fire extinguishment.

(2) Our observations also showed that an effective LN_2 usage need to form a nitrogen gas blanket above TCS burning pool in order to produce suffocative effect and to avoid re-ignition.

(3) Using LN2 seemed to bring about partial cryocoagulation to limit the flowing of TCS liquid and to slow the spread of flammable smoke.

(4) Using LN2 in an open or half-open space for TCS fire, the cooling effect that lower the self-ignition tempeture of TCS was thought to be the main mechanism of fire extinguishment rather than sphyxiation.

3.3 TCS re-ignition property after LN2 application

After TCS fire extinguished by LN2, the TCS pool was frozen (Fig. 6). The frozen TCS remains were then tested by thawing and igniting with fire. Those TCS remains could still ignited into fire (Fig. 7) and burning to produce those acid aerosols and gases. Thus, the application of LN2 could extinguish the TCS pool fire and freeze TCS but it still held the water-reactive and flammable properties. We could conclude that applying LN2 on TCS pool firing was merely a physical reaction of cooling and asphyxiation to put out TCS burning reaction. If TCS fire occurred, the addition of LN2 not only could extinguish the fire, reduce the harmful byproducts, and freeze the TCS pool for easy handling. However, the TCS hazards still existed after the thawing.

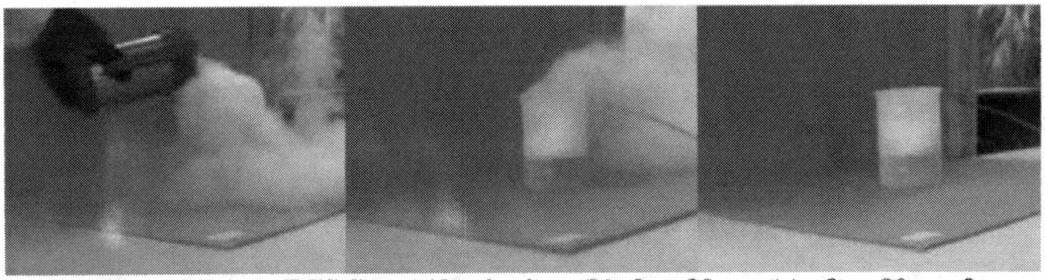

Fig.6 adding LN2 into TCS fire, (a)beginning, (b)after 30sec, (c) after 90sec, frozen.

Fig.7 TCS re-ignition tests

3.4 Dry ice application on TCS fire

The application of dry ice could also extinguish small TCS fire with some delaying as shown in Fig. 8. Dry ice of sublimation latent heat 571.3 kJ/kg is higher than LN_2 evaporation latent heat 199.1 kJ/kg. When applying LN_2 on TCS pool fire the contact area would be uniform and even than dry ice, the TCS fire should be extinguished faster with LN2. Our tests showed that the dry ice could not stop the TCS fire until 90 seconds after the addition. And, TCS pool was not frozen after the dry ice application although the sublimation heat of $CO_{2(S)}$ is larger than the TCS. The temperature differences by adding dry ice into TCS pool fire were smaller than those of LN2. Since dry ice were solid and LN2 was liquid, lesser heat transfer by solid dry ice should be obvious and reduce the effects on fire extinguishment. In addition, $CO_{2(g)}$ gas density 1.98 kg/m^3 is higher than $N_{2(g)}$ 1.25 kg/m^3, the suffocation effect should be better for $CO2_{(g)}$ and might show more effective extinguishing ability in semi-open or confined spaces where asphyxiation was important.

Fig. 8 Dry Ice application on TCS fire (a)burning initated, (b) adding dry ice, (c) after 10 sec. , (d) 30 sec., (e)90 sec., (f)120 sec.

4. Conclusions and Suggestions

TCS is a water-reactive and hazardous chemical to use. Unless the production process could be changed, TCS will remain as the main silicon sources for solar panel production. Therefore, proper prevention and mitigation measures were still needed for production operation as well as leakage or fire response. Our study hope to contribute toward this direction.

4.1 Conclusions

After the field tests on extinguishing TCS fires, there were several conclusions could be drawn as the follows.

(1) This study estimated that the TCS mass burning rate was 2.41 kg/m^2-min. The value was between of Methanol 1.02 kg/m^2-min and gasoline 3.3 kg/m^2-min. Since TCS burning often produce many acid aerosol and gases, a medium burning rate indicated that smoke spread quickly as TCS combusting.

(2) Field tests indicated that LN2 could effectively extinguish fire with the main mechanism of the cooling which LN2 evaporation could quickly absorb the burning heat and stop the burning reactions as observed by infrared thermography on flame temperature below TCS self-ignition temperature. In addition, the suffocation of TCS fires through asphyxiation as nitrogen gas blanketed the burning surface was observed.

(3) As a 40cm by 40cm pool TCS fire occurred, pouring LN2 at volume of 177 (ml) + 221(ml) * TCS weight (kg) all at once could extinguish the TCS effectively. The required LN2 volumes might be related to the cooling and asphyxiation simultaneously. However, the efficiency of cooling could not be calculated due to the rapid dispersion of nitrogen gas above TCS pool.

(4) Another cooling agent of dry ice was also used for extinguishing TCS fires. Since the dry ice would sublimate from solid to gas with less heat absorbed, it indicated that the asphyxiation might be the main mechanism of TCS fire stop as denser CO2(g) would

strengthen the suffocation effect to put out TCS burning. Hence, dry ice might show more effective extinguishing ability in semi-open or confined spaces where asphyxiation was important.

4.2 Suggestions

Due to the popularity of solar panel application, the manufacture use of hazardous TCS has increased in recent years. Since TCS is water-reactive and easily react with humidity, its by-products or derivatives of TCS fires were also harmful to environment and human health. Therefore, the mitigation and prevention measures against TCS should be studied and regulated in order to respond to possible spills or fires. Our studies showed that several extinguishers were not effective again TCS fires and its byproducts. However, a promising LN2 application could absorb burning heat and asphyxiate TCS fires to lessen the impacts to the environment and public safety. Although frozen TCS could still be hazardous after its thawing, solid form of TCS would be easy to handle than liquid or gas types of products. However, several research directions still need to be fulfilled in order to eliminate the TCS hazards, such as the application equipment for LN2, application techniques, emergency response procedure, possible asphyxiation hazards, and personal protection needed, etc. It is still many challenges needed to develop the required technology for using LN2 as extinguisher for water-reactive chemicals besides TCS. We hope our studies could contribute toward those directions somehow.

5. References

[1] Jäger-Waldau A., Chapter IIC-2 – The Photovoltaic Industry in **Solar Cells** (Second Edition), (2013), 565-583.

[2] Soo J.-C., S.-R. Li, J.-R. Chen, C.-P. Chang, Y.-F. Ho, T.-N. Wu, and P.-J. Tsai. "Acid Gas, Acid Aerosol and Chlorine Emissions from Trichlorosilane Burning Processes". Aerosol and Air Quality Research, 11, (2011), 323–330.

[3] Chen, C.-J., P.-C. Tsai, and C.-W. Li, Trichlorosilane of Emergency Operating Procedures and Implementation of Field Testing Program, Southern Taiwan Science Park Administration , (2008).

[4] Woditsch P., & W. Koch, "Solar grade silicon feedstock supply for PV industry. Original Research Article Solar Energy Materials and Solar Cells", Volume 72, Issues 1–4, (2002), 11–26.

[5] Levendis Y., A. Ergut, and M. Delichatsios, "Cryogenic extinguishment of liquid pool fires." Process Safety Progress, Volume 29, Issue 1, (2010), 79–86.

[6] An D., P. B. Sunderland, and D. P. Lathrop, "Suppression of sodium fires with liquid nitrogen," Fire Safety Journal, vol. 58,(2013), 204–207.

[7] Illinois Legislative Investigation Commission, Chemical Leak at the Bulk Terminals Tank Farm, A Report to the Illinois General Assembly, by the Illinois Legislative Investigation Commission, June 1975.

[8] National Transportation Safety Board, "Derailment of CSX Transportation Inc Freight Train and Hazardous Materials Release Near Freeland, Michigan on July 22, 1989", NTSB Report RAR-9 1/04, 1991.

[9] Reed, L. A., Safety Recommendation In Reply Refer To: R-91-57 and -58, National Transportation Safety Board, October 9, 1991

[10] Higgins, J. T., W. E. Hubbs, R. A. Kayser, P. Kremer, S. W. McMahan, L. Patrick, and M. Strong (ed.) , Manual on Chlorosilane Emergency Response Guidelines, ASTM MNL 33, 1998.
[11] Liao, K.Y ,J.-J. Horng, K. Liou, Y. Liao, and J. Chen, Investigating the efficiency of different extinguisher means on Trichlorosilane fires. CCPS, First Asia-Pacific Conference on Process Safety, 2013.
[12] Liao, K.Y, J.-J. Horng, K. Liou, Y. Liao, and J. Shen, Extinguishing efficiency of liquid nitrogen against trichlorosilane fires. CCPS, 2th Asia-Pacific Conference on Process Safety, 2014.
[13] Mores, T. F., New Aspects of Chemical Vapor Deposition, Trichlorosilane, Literature Survey and Combustion Experiments. Fundamental Studies of Laser Interaction in Materials Preparation, DTIC, 1984.
[14] U.S. Nuclear Regulatory Commission, Chapter 3 estimating burning characteristics of liquid pool fire, heat release rate, burning duration, and flame height, Fire Dynamics Tools, 2004.
[15] Dow Corning, Responsible Care Report, Available at http://www.dowcorning.com/content/about/aboutehs/PDF/Responsible_Care_Report_99.pdf , 1999.
[16] Violations alleged in fatal Moses Lake silicon plant explosion, Available at http://www.lni.wa.gov/news/1999/pr990414a.asp, April 14, 1999.
[17] Chen, T.-L., Y.-C. Ye, Y.-L. Chen and S.-S. Kao, "Public hazardous substances places using the fire-fighting equipment design of trichlorosilane as an example to technology plant," WuFeng Journal, 19, (2011), 15–29.
[18] Xin hua net, Hubei Yichang CSG company spill occurred trichlorosilane, Available at http://news.qq.com/a/20081210/002881.htm, December 10, 2009.

Case Analysis of Oil and Gas Pipeline deflagration accidents

Presenter/Principal Author: Shi Li
Affiliation: Department of Environmental and Safety Engineering, China University of Petroleum (Este China)
Address: No. 66, Changjiang West Road, Huangdao District, Qingdao, China
Email address:lishi19785460@163.com

Other Author(s): Hong Shen, Dongfeng Zhao
Affiliation(s): China University of Petroleum (Este China)

Keywords: Pipeline leakage; Deflagration accident; Case analysis; Risk control

Abstract

Pipeline transportation plays an important role in the petrochemical industry. As a main form of the current oil transport, with a series of advantages of large volume, low cost, and low energy consumption, etc. On the other hand, frequent occurrence of oil and gas pipeline safety accidents in recent years make people recognize the safety and environmental risk of pipeline transport in some new angles. In this paper, the reasons and consequences of typical deflagration accident of oil and gas pipelines are analyzed, and related experience and lessons are summarized. On this basis, the future safety and environmental risk prevention measures and suggestions are put forward.

1. Overview of the present situation of pipeline applications

1.1 Pipeline construction has entered the peak period

Pipeline is an important way for oil and gas transport in China[1]. Up to Oct. 2013, the total length of China's oil and gas pipeline was about 106000 km, which included 60000 km of natural gas pipelines, 26000 km of crude oil pipelines and 20000 km of products pipeline. These pipelines have covered almost all provinces and the special administrative region. In the period of "Twelfth Five-Year Plan", China will focus on the construction of pipeline in China and Kazakhstan, China and Myanmar, Central Asia natural gas, West-East natural gas transmission project third and fourth lines. According to the "Twelfth Five-Year Plan", by the end of 2015 the total length of oil and gas pipelines will reach 150000 km or so[2]. The rapid increase of oil and gas pipelines will promote a substantial increase in the supply of the scale and optimize the energy consumption structure.

1.2 Pipeline accident happened frequently

With the increase of the length, the growth of the age, the pipeline accidents occurred frequently in China. According to related statistics, the pipeline accident average rate was 3 times/1000 km • year in China, far more than 0.5 times/1000 km • year in the United States and 0.25 times/1000 km • year in Europe[3-4].

1.2.1 Complex working environment

In China, most of the pipeline buried underground. The long distance transportation lead to the pipeline face complex natural environment which makes the pipeline easy to corrosion, even leakage.

1.2.2 Pipeline aging

With the increase of using time, a considerable number of pipelines had entered the aging period. About 20000 km of pipeline was worked over 20 years in China, Pipeline aging and badly corrosion conditions made the leakage accidents happen occasionally [5].

1.2.3 Pipelines tie up issues

With the rapid development of China's urbanization, some old pipelines were surrounded by residential areas, plants, and even tied by building (structure)[6]. There is the cross-mixed phenomenon of oil and gas pipelines and municipal pipelines.

2. Typical pipeline leakage and explosion accidents analysis

Since 2010, there were some serious pipeline accidents happened in China, such as Xin-Gang oil pipeline explosion accident in Dalian, Lan-Cheng-Yu oil pipeline leakage accident, Dong-Huang oil pipeline explosion accident, etc. These accidents not only caused serious casualties and heavy economic losses but also had badly effect on society and politics[7-9].

2.1 Qingdao"11·22" leakage and explosion accident of Dong-huang oil pipeline

Dong-huang oil pipeline was put in use in 1986, which was used to transport oil from Huangdao oil depot to DongYing City. At the beginning of the establishment of Huangdao oil depot, surrounding residents were rare. With the increasing urbanization, the living area, school and malls were gradually established. This made Dong-huang oil pipeline very closed to the roads in urban. Especially, the urban covered channels run perpendicular to Dong-huang oil pipeline and they met at the T-junction.

On November 22nd, 2013, because of the corrosion, the pipeline cracked and leaked. Leaked crude oil crushed out of the road and ran to the south on the street. Finally, the crude oil ran into the urban discharge culvert by the gully. Volatile oil and gas extended along upstream and downstream of the airtight discharge culvert, and extended to estuary.. Before the explosion, the tide level was at the highest level, seawater flowed back gradually to discharge culverts, which made the leaked oil and the mixture of oil and gas spread and diffuse to the upstream.

After the leakage accident, related personnel rushed to the scene to repair. To deal with the damaged pipelines, on-scene commanders decided to open the prefabricated panels, use excavators and hammer carrying drilling operations. When drilling operations were underway, explosion accident happened. The explosion caused 62 people died, 136 people injured, and 7517.2 million RMB of direct economic losses.

2.2 Dalian "6.30" Xinda oil pipeline leakage& fire accident

In June 30, 2014, a CNPC Xinda oil pipeline leaked and fire occurred at a distance of 6 km away. The accident caused no casualties, and 2,019,150 RMB of direct economic loss.

When horizontal directional drilling machine was operating drilling, the drill hit oil pipeline wall and cutting formed voids on it which makes the crude oil spilled. The leakage of crude oil and the mixture of oil and gas spread and diffuse under bridge which was comparatively hermetic space. The oil and gas air mixture was ignited easily, finally caused a fire accident. The accident caused no casualties, and 2,019,150 RMB of direct economic loss.

2.3 Comparison of "11.22" and "6.30" accident

Study of both "11•22" accident and "6•30" accidents, the characteristics of the two accidents were summarized to learn the lessons.

2.3.1 <u>The similarity of the two accidents</u>

The leaking pipes are buried long-distance and mismanaged;

Pipeline and municipal pipeline are crossing at sensitive areas;

Both accidents occurred at urban living area, which is not easy to evacuate;

The oil leaking from pipeline flowed into the municipal pipeline and and oil spread to sensitive areas after leakage;

All spilling oil contained more light component——similar properties of pipeline media.

2.3.2 The differences of the two accidents

2.3.2.1 Different leakage causes

In "11.22"accident, the main reason is the corrosion of pipeline, however in "6.30" accident, it is attributed to improper operation which was cutting form voids on the wall of pipeline.

2.3.2.2 Amount of spilled oil

Comparison these two accidents, it could be found that both the length and the diameter of the pipeline are longer, and the duration of leakage is longer in "11 • 22" accident, which means large quantity of spilled oil.

2.3.2.3 The wayof oil flowed into municipal pipeline

In "11.22" accident, the oil flowed into municipal pipeline by over ground and underground way. However, in "6. 30" accident, the surroundings of underground pipeline are filled with soil and stone, the spilled oil flowed into municipal pipeline only by over ground way.

2.3.2.4 Difference of municipal pipeline layout

In the "11.22" accident, the culvert led to the sea, which was easily affected by the tide. Before the accident the sea was at high tide, with the increase of seawater level, seawater flowed back to the culvert, making a greater concentration of oil and gas in limited spaces. On the other hand, in "6.30" accident, pipeline was led to river and Sewage Treatment Plant.

2.3.2.5 The different air tight features of fire and explosion area

In "11.22" accident, the ignition point was at culverts with higher airtight level and near the leak point. But in "6.30" accident, ignition point was at out of the estuary of rainwater pipe networks and the confined space was under Light Rail bridge.

3. Pipeline leak and Explosion risk analysis

In addition to these two typical pipeline leak and explosion accidents, there were a number of similar accidents happened in the last decade. For example, in 2006, inlet pipelines explosion in Fujia gas station, 10 dead, 3 seriously injured, 47 slightly injured. In 2010, California natural gas transmission pipelines explosion, 7 dead and several injured. August 2014, Taiwan propylene chain explosion, 26 dead and 280 injured. Through the study of these accidents, it can be inferred that leakage is the primary reason to induce subsequent deflagration and a series of serious safety and environmental problems.

3.1 Risk factors of pipeline leakage explosion

3.1.1 Medium in pipelines

Internal pipeline corrosion caused by high-sulfur crude oil and increased the risk of pipeline leakage[10]. The more light component content in oil, the more volatile flammable vapors. Once this kind of vapors release into the limited space and reach explosion limits, fire and explosion will occur.

3.1.2 Pipeline Corrosion

Pipeline faced complex natural environment, because the long distance transportation has to cross wide areas. This can cause pipeline corrosion, resulting in oil pipeline leak .

3.1.3 External forces

The destruction of human activities on the pipeline is the main cause of pipeline leak. In traversing pipeline near human activities, such as construction activity, production will have an impact on pipeline safety, such as 6.30 accident in Dalian.

Environmental forces is one of the factors that influence the stability of the pipeline, such as pressure from surrounding soil, stress caused by the temperature difference between transport medium and the environment, stress generated in the pipe pressure, bending stresses produced by differential settlement of soil. All these will cause the pipeline stress concentration, instability and the strength damage in the pipeline, increasing the risk of pipe leakage.

3.1.4 Pipes crossing the sensitive area properties

Through the experience of the accidents, it could know that municipal pipeline layout is an influence factor. Municipal pipeline type, overlapping of municipal pipeline and oil and gas pipelines have important influence on the further expansion of oil and gas pipeline leakage accident. The surroundings of the pipeline crossing area is another factor. For example, whether the regions pipeline crossing exist densely populated areas, and soil environment, airtight properties have important implications for the development of pipeline leakage accident.

The ignition source which may exist around the pipeline was always a directly factor caused the accident. Hot work, sparks, open flames, static electricity, lightning, arson play an important role on the evolution of fire and explosion.

3.2 Pipeline leakage and explosion risk types

The typical consequence of pipeline leakage accident is fire. When leaking, oil gas and air mixed to form a combustible gas mixture, in the event of fire or other sources of ignition case, would cause a fire. More serious consequence would be explosion, when the concentration reached the explosion limit in limited space and exposed to an ignition source, explosion will happen. For high-sulfur oil and gas leaks, hydrogen sulfide and other substances have serious toxic effects and may cause death. Pollution problem is normal, if oil and gas leakage happens in important ecological protection area or goes into municipal water pipe network. Oil leak can cause irreversible effects on the surrounding natural environment, biological, urban drinking water and so on.

4. Outlook of oil and gas pipeline risk control

4.1 Urban layout planning and management

Strengthen the planning and management of urban utilization and adjusted the underground pipe network layout timely. Related departments should set up factory relocation plan, and make the safe distance meet the requirements of the relevant laws and regulations to ensure that the density of population in high-risk areas will reduce as far as possible. The government should check the hidden risk and implemented corrective measures, for those pipeline with higher level risk should be stopped using. Supervising and urging enterprises to establish a system for removing hidden risk, and interpreted the responsibility of hidden risk management definitely. For major hazard installations, figure out the layout of the underground pipes and examine the degree of the risk. Otherwise, building automatic control systems and surveillance systems should be used to strength the management of plumbing facilities and safety, especially the underground pipeline which safety distance is not compliance, and currently do not having the relocation conditions.

4.2 Technical improvement

Through the modern pipeline leak detection and location technology development, we could accurately detected the location of leak, such as real-time detection method, negative pressure wave method, flow or pressure mutation method and so on[11].

In order to avoid the leakage accidents in maximum, digital technology should be expanded the range of application, such as pipeline leakage detection system, pump operating parameters monitoring technology [12].

4.3 Fundamental research

Although the scientists have already studied on different aspects of oil evaporation since several decades ago, it is still not formed maturational theory system. To reduce the possibility of pipeline deflagration accidents, study of the oil evaporation regulation is indispensable and focused on the relationship between the evaporation regulation and all kinds of experimental factors like temperature, space condition, time and so on[13,14]. Enhanced links between oil properties and regularities of oil evaporation, meanwhile, established the equation of evaporation rate[15].

5 Conclusion

(1) The aging pipeline, corrosion and incorrect manipulation are major reasons of leakage;

(2) With the development of the city, oil pipeline and municipal pipeline was more easily staggered than before. Furthermore, the spilled oil will have more opportunity to flow into municipal pipeline which might expand the range of leakage.

(3) After the spilled oil flow into the municipal pipelines, oil and gas are accumulated in confined or semi-closed confined space which could cause fire or explosion. Accident characteristics are closely related to the tightness of the space.

In order to reduce the risk of the accident, it should strength the planning and management of the city, improve the detection and location technology, enhance the scientific research on oil evaporation.

6. References

[1] Z.Z Wu and S.Z. Zhu, et al, Zhang, A method for long-distance oil and gas transmission pipeline route selection based on potential loss of life, CSSJ 24 (2014), 71-76.

[2] L.K. Wang and H.C. Wang, et al, Analysis and proposal on leak detection of long-distance oil pipeline, OGST 33(2014), 1198-1201.

[3] Y. Kang, Oil and gas pipeline engineering, China Petrochemical Press, Beijing, 2008.

[4] Y. Di and J. Shuai, et al, Study on Methods for Classifying Oil & Gas Pipeline Incidents, CSSJ 23(2013), 109-115.

[5] M. Pu, Current status and future prospects for oil & gas pipeline development in China, International oil economy 17(2009), 40-47.

[6] F.H. Shu, Advocating Pipeline Protection Culture and Enhancing Soft Power so as to Safeguard Oil Pipelines, Petroleum & Petrochemical Today 21(2013), 42-46.

[7] Q.B. Zhang and Y. Jia, et al, Hazard Analysis of Long distance Oil and Gas Pipelines, CSSJ 18(2008), 134-138.

[8] R. Liang and C.Y. Zhang, et al, Comparative Analysis of Evaluation Models for Explosion Consequence Caused by Leakage of Natural Gas Pipelines, CCSJ 17(2007), 131-135.

[9] Y.B. Xu and X.M. Qian, et al, Quantitative Risk Analysis on the Leakage of Compressed Natural Gas Pipeline, CCSJ 18(2008), 146-149.

[10] K.M. Wu and K.Z. Wang, et al, Corrosivity of Sour Crude on Oil Pipelines, OSGT 29(2010), 616-619.

[11] H.S. Wang and H. Ye, et al, Oil Pipeline Leakage Detection Based On Wavelet Algorithm, Information and Control 31(2002), 456-460.

[12] J. Feng and H.G Zhang, et al, Development on Leak Detection and Location System for Oil Pipeline, Journal of Northeastern University(Natural Science) 24(2003), 731-734.

[13] M. Fingas, Modeling evaporation using models that are not boundary-layer regulated, J Hazard Mater 107(2004), 27-36.

[14] L. Zhu and J.Q. Chen, et al, Experimental analysis of the evaporation process for gasoline, J Loss Prevent Proc 25(2012), 916-922.

[15] K. Okamoto and M. Hiramatsu, et al, Evaporation and diffusion behavior of fuel mixtures of gasoline and kerosene, Fire Safety J 49(2012), 47-61.

Research of Process Safety Management Platform Architecture Based on Internet of Things

Presenter Author: Yang Guoliang
China Academy of Safety Science and Technology
Chaoyang District Beijing China, 100012
yangguoliang@outlook.com
Principal Author: Wu Zongzhi
China Academy of Safety Science and Technology
Chaoyang District Beijing China, 100012
wuzongzhi@vip.sina.com;
Wang Rujun, Duo Yingquan
China Academy of Safety Science and Technology

Keywords: Process safety management, Internet of Things, platform architecture, information fusion, information decision

Abstract

With the rapid development of Chinese chemical technologies, there is a rapid growth in quantity and scale of petrochemical enterprises. Meanwhile, the relatively weak link - chemical process safety management, may easily cause frequent occurrence of production safety accidents. In order to make the chemical process proceed with programmed, standard-based safety management, a conception of setting up the safety management information platform of Internet of Things (IOT)-based chemical process is proposed in this paper. A hierarchical model for the safety management information platform of chemical process is set up, which including perception layer, network layer, basic support layer, application service layer, by combination of the IOT architecture. Main functions and features of a software architecture of the information platform is introduced, which is designed based on twelve elements in chemical process safety management. This platform could providing a means to promote the information technology level of chemical process safety management as well as the meticulous safety supervision.

1. Introduction

With the rapid development of Chinese chemical technologies and high-speed growth of economy, there is a skyrocketing growth in quantity and scale of petrochemical enterprises. Up to 2013, chemical enterprises above designated size have reached over 24,000[1]. In this background, the relatively weak link - chemical process safety management in chemical industry may cause safety production risk due to its inadequacy. According to statistics, in 2012, the number of accidents above ordinary level and the death toll in the chemical industry of China amounted to 2.2 and 4.87 times the same in USA respectively [2]. The small chemical workshops without safety guarantee amount to over 80% of the total [3].

In order to improve the fundamental work in safe production of chemical enterprises, a series of laws, regulations and standard specifications have been issued in our country to intensify the corporate process safety management and the overall process supervision by government. With some progresses achieved in this regard, problems still remain concerning the intensification and implementation of enterprise's principal responsibility, as well as the overall process supervision and trace management by government. This is caused by diversities in mode and level of management of numerous enterprises. However, IOT technology, characterized by overall perception, reliable transmission and smart treatment, matches with the requirements for check, application, approval and implementation in the above safety management. The chemical process safety management platform integrated with IOT technology provides a technological support to implement corporate responsibility, staff involvement, government supervision, self-regulation and social supervision.

This paper delivers a solution to build the chemical process safety management platform by presenting an IOT-based platform software architecture and discussing the functions and features of main modules.

2. Demand analysis of IOT-based platform

2.1 IOT concept and technical feature analysis

IOT refers to a giant network combining information sensing devices (RFID device, infrared sensor, GPS, laser scanner, etc.) with internet [4]. It aims to connect all things with internet for automatic, real-time identifying, positioning, tracking, supervising and activating [5].

Such IOT technology integrating internet and physical things has the features below [6]:

(1) Fusion of network technology and information technology in information field and control technology and automation technology in physical system.

(2) Reliable, real-time information acquisition and interaction.

(3) Conversion of discrete events into continuous process to enable fusion management of time and space parameters, by network-aided acquisition of time & space related information in physical world.

2.2 Chemical process safety management and trend analysis

Intensifying the chemical process safety management is an internationally advanced method to prevent and control major industrial accidents, and an important fundamental job to eliminate potential safety hazard, prevent accidents and to build a long-term mechanism for safe production. To improve the fundamental work in safe production of chemical enterprises, as well as to strengthen the process safety management and overall process supervision by government, our government issued regulations and standard specifications, including Measures for the Safety Supervision and Administration of Hazardous Chemical Construction Projects (Decree No.45 issued by State Administration of Work Safety), Instruction on Intensifying Chemical Process Safety Management by State Administration of Work Safety (A.J.Z.G.S [2013]No.88) and Guidelines for Process Safety Management of Chemical Corporations (AQ/T3034). With some progresses achieved in this regard, problems still remain with "two opposing extremes": on one hand, large chemical enterprises, featured by large-scale equipment, complex process, intensive industry and capital intensification, can effectively reduce the occurrence of major/extra serious accidents due to their perfect process safety management; on the other hand, small and medium-scale enterprises, featured by backward process, poor professional quality and low equipped level, are prone to frequent accidents due to their poor process safety management.

At present, IOT technology applications in chemical industry prove to be fruitful. Let's take for example a set of emergency rescue and response information system set up in a chemical industry park. The system uses IOT technology to make a real-time acquisition of dangerous chemical information and sensing data like ambient temperature, pollutants, wind direction and wind speed, uses the accident consequence model to instruct the emergency rescue by real-time computing of the scope of influence in related accidents, and send users corresponding evacuation and rescue messages[7].

IOT technology is used to effectively improve the corporate process safety management information level and implement overall process supervision by government. As twelve elements are involved in the chemical process safety management and the job for each of them is subject to tremendous acquisition and storage of information, the whole process shall be featured by clear management. In need of deployment of numerous manpower and material resources to ensure effective implementation of the management, enterprises and governments can use the IOT-based platform to enable all-round, programmed enforcement of the process safety management by enterprises, and to enable the trace supervision by supervisory departments.

3. Hierarchical Model for the Safety Management Information Platform of Chemical Process

3.1 Safety management information platform of IOT-based chemical process

The platform, an intelligent one backed by IOT technology, is used for process implementation of the elements in corporate process safety management, information management of relevant systems, operation specification and process equipment data, and for the trace management of examination of the safety facilities, license application and change in construction project by safety supervision department. In this way, the standard process safety management can be expected and the safe production level promoted for enterprises.

3.2 Hierarchical model for the safety management information platform of IOT-based chemical process

According to the features of IOT, the safety management information platform of IOT-based chemical process is divided into four hierarchies: perception layer, network layer, support layer and application layer, which are shown in figure 1.

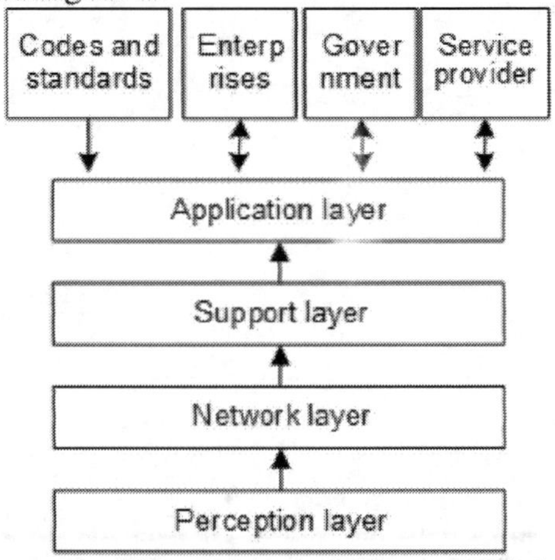

Figure 1. Hierarchical platform model

From the bottom up, the hierarchical model is composed of the perception layer, network layer, support layer, application layer and the service provider as well as service objects (government, enterprises). Such layers are defined as below: in the perception layer, the information perception (sensing) devices are used to make acquisition and real-time tracking for all safe production information related with chemical hazard, technology and device in production process; in the network layer, network (private network, wireless network) technologies are used to enable interconnection and interworking by transmission of the acquired messages to the support layer and the application layer; in the support layer, universal and fundamental IOT platform services are offered to enable fusion and unification of messages, support automatic management and operation of platform, and provide the application layer development with platform environment; in the application layer backed by the other layers, relevant functions are designed and developed to serve the service objects in the platform, based on Measures for the Safety Supervision and Administration of Hazardous Chemical Construction Projects (Decree No.45 issued by State Administration of Work Safety), Instruction on Intensifying Chemical Process Safety Management by State Administration of Work Safety (A.J.Z.G.S [2013]No.88) and Guidelines for Process Safety Management of Chemical Corporations (AQ/T3034).

4. Safety Management Information Platform Architecture for IOT-based Chemical Process

4.1 Overall platform architecture

In view of the major business requirements for the chemical process safety management and the overall process supervision by government, as well as based on the above hierarchical platform model, the design of the safety management information platform architecture for IOT-based chemical process is shown in figure 2.

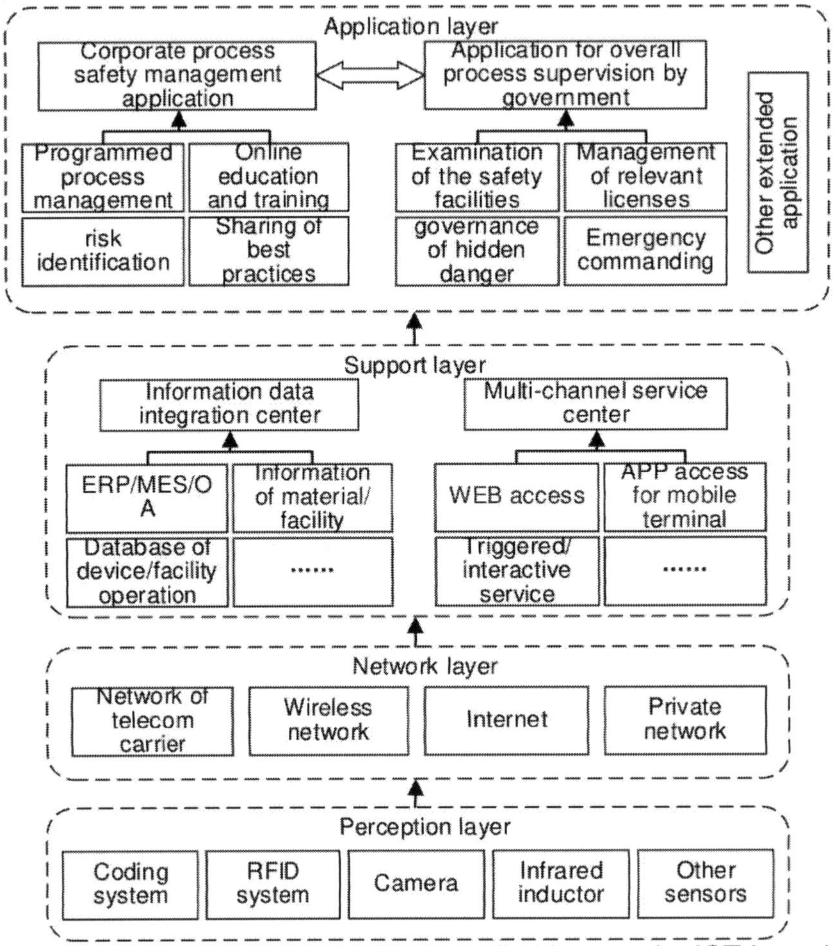

Figure 2. Safety Management Information Platform Architecture for IOT-based Chemical Process

4.2 *Perception layer and network layer*

In the perception layer, the original data required for analysis is offered for the other parts in the platform. The operation data of devices are imported via front-end sensors such as coding system, RFID system, GPS and video acquisition system to enable real-time acquisition and supervision. Overall safe production information on chemical hazard, process and device in production is acquired via data entry in system and file uploading.

In the network layer, based on the protocol of flexible information acquisition and forwarding, massive front end perception information is transferred to the support layer and the application layer in IOT platform via uniform internet, private network and mobile communication network, enabling the access of such information.

4.3 *Support layer*

The support layer composed of an information data integration center and a multi-channel service center, universal and fundamental IOT platform services are offered to support automatic management and operation of the platform, providing the application layer with information source and service approach.

The information data integration center, composed of information integration and fusion service and operating environment, provides the platform with powerful computing function and information storage space and provides application with the support environment for virtual data center operation. The information integration and fusion service refers to fusion and sorting as well as normalized processing of the information required in chemical process safety management. Such information mainly comprises those obtained via RFID and image recording by the perception devices, those containing dynamic in-service and storage volume of hazardous chemicals, those of devices, materials and safety facilities, and those on operation of devices and facilities. The operating environment, featured by cloud services, comprises five parts such as dynamic operation building module, dynamic service management module, co-scheduling module, batch job scheduling module and operation fault-tolerance and migration module[5], providing the whole platform with application support

The multi-channel service center offers an approach to implement the functions of platform application layer, depending on different users and functions, such as web access based on daily management application, APP access based on field hidden danger check and remedy, activating response service for devices and facilities in emergency, and push service for important information issuance.

4.4 *Application layer*

As a vital part of applying IOT technology to enterprises and government, the application service layer, which dominates the design and layout of the perception layer, network layer and support layer, shall be planned and designed to meet the real needs of enterprises and government as per relevant laws and regulations and standards. The application layer mainly concerns three parts such as corporate process safety management application, the application for the overall process supervision by government and other extensive application.

4.4.1 Corporate process safety management application

To meet the demand of enterprises for the chemical process safety management, the programmed process safety management module, online education and training platform module and risk identification aid decision making module. In addition, to shorten the gaps between enterprises in the chemical process safety management, a best practice and experience sharing module is configured to provide backward enterprises with guidance.

The programmed process safety management module is used to serve the daily corporate management, build a programmed process safety management regulation, guide relevant corporate personnel to implement the elements of the process safety management as required, and trace the operation of each step for analysis, gradually upgrading the management level. The module architecture is shown in figure 3.

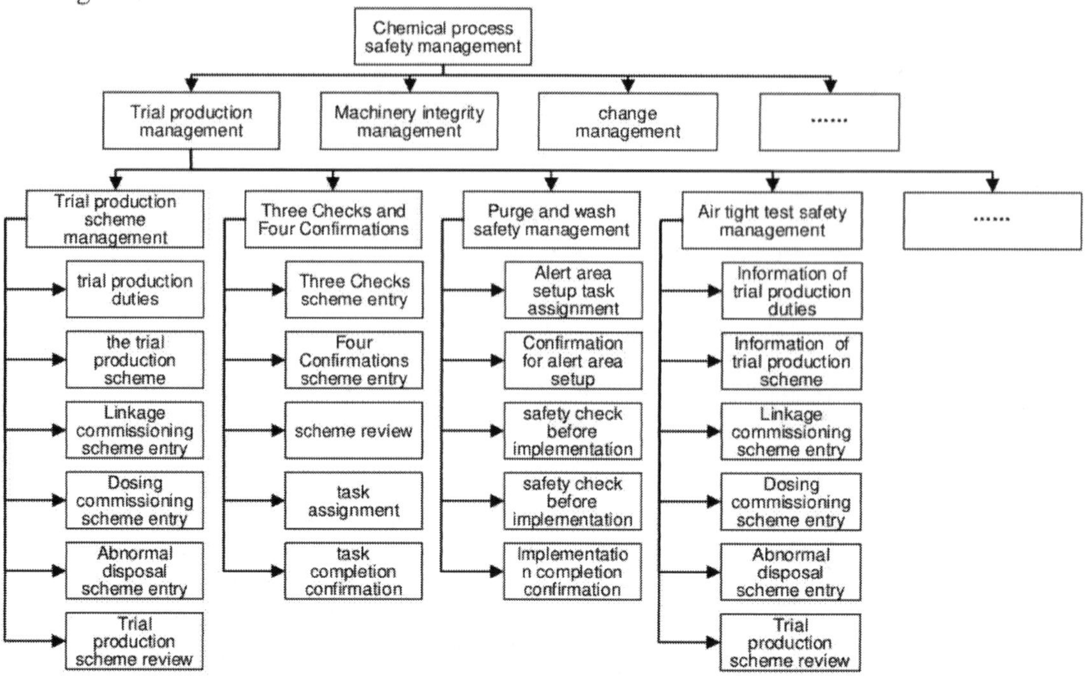

Figure 3. Architecture of the process safety management module

In figure 3, Three Checks means checks before trial production, include omission in design, project quality and hidden danger, and uncompleted quantities; and Four Confirmations means confirm task, personnel, time and measures for issues founded in Three Checks.

The online education module is mainly used to provide corporate staffs with job education and training. Relevant training plan is prepared and implemented according to the job feature and skills of staffs. In the platform, specialized teaching video and training materials as well as teaching resources recorded and uploaded by enterprises are available to ensure training and teaching quality. Meanwhile, the training effect evaluation function is integrated into the module for improvement and optimization of the training scheme.

Before a unit is put into operation, the process hazard analysis and the efficacy thereof can be ensured by an intermediary to offer specialized risk identification report. However, as for the risk identification in case of temporary change during daily management or before special operations like hot work and work in confined space, there is a higher demand for field operators. Thus, the risk identification aid decision making module can be a good solution to cater for such demand. The module, which fuses identification methods such as what-if, checklist, FTA and HAZOP [8,9], can be enabled to enhance the risk identification reliability by helping staffs finish the risk identification as per steps, based on the progress of the process safety management.

4.4.2 Overall process supervision application for safety supervision departments

Overall process supervision application for safety supervision departments meets the demand of safety production supervision departments for their daily work[10], by helping them implement informative, transparent process supervision during the whole life cycle of the hazardous chemical construction project (including project approval, design, construction, startup, operation, maintenance, and shutdown). According to the domestic laws and regulations system, the overall process supervision application mainly includes the safety supervision and management module for the construction project, application, change and extension of licenses (safe production license, safe use license and operating license of hazardous chemicals), hidden danger check and governance information management module, and emergency command module. The overall process supervision application module for supervision departments has the functional architecture shown in figure 4.

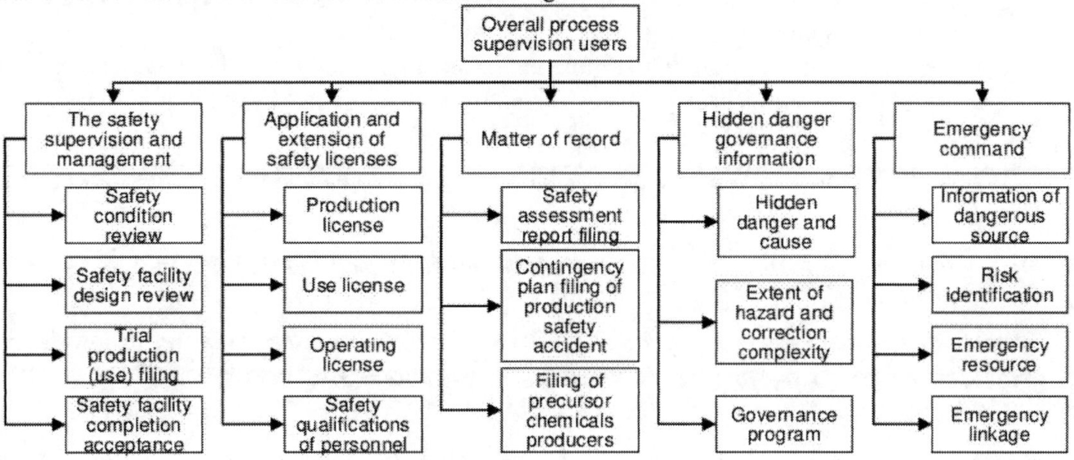

Figure 4. Functional architecture of overall process supervision application module for supervision departments

4.3.3 Other extensive applications

As open platforms subject to utility upgrade, other extensive applications are used to provide users with tailored services as required by them and with information service such as supply & demand information issuance of materials, products, information exchange of safety and technology, analysis of typical cases and yellow pages of safety and technology service organizations.

5 Conclusion

By introduction of IOT technology into the chemical process safety management field, the paper presents the hierarchical model and the functional architecture, especially the architecture applicable to the application modules related with enterprises and safety supervision departments. Backed by the safety management information platform for IOT-based chemical process, a mechanism of facilitating corporate responsibility, staff involvement, government supervision, self-regulation and social supervision can be set up, so as to promote safety production standardization, shrink the gap in safety management level among enterprises, and provide governments with efficient means to enable process supervision and trace management, for the purpose of carrying out policies and regulations related with the process safety management.

6 Acknowledgment

This research is sponsored by the National Science and Technology Support Program (project no. 2012BAK13B01), and by China Academy of Safety Science & Technology Basic Research Program (Grant No: 2014JBKY07).

7 References

1. 2013 China Statistical Yearbook. Available at http://www.stats.gov.cn/tjsj/ndsj/2014/indexch.htm Accessed on January 8, 2014.
2. Du Hongyan, Wang Yanping, Lu Junchen. Statistical analysis on petroleum and chemical accidents at home and abroad during the period of 2012[J]. Journal of Safety Science and Technology, (9)2013, 184-188.
3. 2014 Safety Production Work Report. Available at http://news.xinhuanet.com/talking/2014-12/24/c_1113757840.htm Accessed on December 24, 2014.
4. Wang Baoyun. Research Overview for IOT Technology, Journal of Electronic Measurement and Instrument, (23)2009.1-7.
5. Wang Jing, Quan Chunlai, Zhou Xiang. Research of Public Security Platform Software Architecture based on Internet of Things [J]. Computer Engineering and Design,(10)2011, 3374-3378.
6. Shen Subin, Fan Quli, Zong Ping, etc. Study on the Architecture and Associated Technologies for Internet of Things, Journal of Nanjing University of Posts and Telecommunications (Natural Science), (6) 2009, 1-7.
7. Preliminary development of IOT in the chemical industry. Available at http://www.cnscn.com.cn/news/show-htm-itemid-1085.html. Accessed on July 20 2012.
8. State Administration of Work Safety, Instruction on Intensifying Chemical Process Safety Management by State Administration of Work Safety[Z]. 2013-7-29.
9. AQ/T3034-2010, Guidelines for Process Safety Management of Chemical Corporations [S]. State Administration of Work Safety, 2010.
10. State Administration of Work Safety, Measures for the Safety Supervision and Administration of Hazardous Chemical Construction Projects[Z].2012-1-30.

Application of Quantitative RBI on the Hydrocracking Unit

LIANG Shuang
CNPC Research Institute of Safety & Environment Technology
A, CNPC Innovation Base, Xishatun BridgeWest, Shahe Town, Changping District, Beijing, China
anita25ps@163.com
LI Xiang
Chinese Special Equipment Inspection Institute
Building 2, Xiyuan, Heping Street, Chaoyang District, Beijing, China
stevenurban@163.com
LUO Fangwei
CNPC Research Institute of Safety & Environment Technology
A, CNPC Innovation Base, Xishatun BridgeWest, ShaHe Town, Changping District, Beijing, China
luofangwei@cnpc.com.cn

Keywords: Risk-Based Inspection (RBI), Hydrocracking Unit, Inspection Program, Safety Assessment

Abstract

The loss of accident increases while the influence of social deteriorates because of the maximization and complication of equipment in petrochemical industry. Hydrocracking, as one of the most important catalytic processes of petrochemical industry, is also a frequent accident process. Developing optimization risk management in hydrocracking unit has function of significance in guaranteeing process smoothly of exploration and production. Under these demand, the Risk-Based inspection (RBI) was presented and widely prevailed in petrochemical enterprises of Europe and America.

Quantitative RBI on equipment of a hydrocracking unit starts at inspection data collection which is followed by identification and assessment of damage mechanism and failure modes. Equipment risk is assigned with the help of Orbit-onshore (DNV) software. According to quantitative RBI results, risk level for each piece of equipment item is established and prioritized by criticality matrix so that equipment rank helps defying critical equipment which would be inspected thoroughly. Eventually, an appropriate inspection program for this unit is designed. Meantime, numbers of proper countermeasures are proposed through quantitative RBI assessment to control the frequency of shutdown and to extend inspection interval by assigning resources of inspection and maintenance.

1. Introduction

As an indispensable approach of heavy oil processing, hydrocracking technology has become priority of relieving insufficient of oil and gas, increasing production of high quality transportation fuel and satisfying the supply of ethylene and PTA. While, hydrocracking unit conducts under high temperature, high pressure, hydrogen service, flammable and toxic condition, full with hydrogen and poisonous material.

In order to reduce the risk and accident, the traditional method which was based on single inspection frequency and measurement had to be modified to manage safety, environmental and business-interruption risk in an integrated, cost-effective manner. RBI, in accordance with inspection supporting code and technical supporting code, can identify risk of equipment and reduce them by offering proper inspection activities which allow more effective use of the inspection resources and finally achieve reliability of equipment using optimum inspection costs. It's not only the requirement of hydrocracking unit for increasing operating time, but also the demand of refinery for improving economic returns.

Based on RBI program, a refinery applies Orbit-onshore software from DNV to analyze and identify quantitatively the hydrocracking unit. After prioritizing the equipment and piping on damage mechanism and measured risk, an optimum inspection program is set up and countermeasures against risk are suggested among equipment in this unit to put control measures into effect.

2. Theory of RBI

2.1 status of RBI study method

RBI is systematic methodology by means of screening for risk with considering both safety and cost-effective, estimating risk value associated with the operation of all equipment item, identifying areas of potential concern, and developing a prioritized list for more in-depth inspection so as to achieve the optimum combination of inspection methods and frequencies [1].

The American Petroleum Institute (API) RBI Project was initiated in May 1993 by an industry sponsored group to develop practical methods for RBI in petrochemical enterprises [2]. After 3 years, draft API Base Resource Document (BRD) 581 was presented, followed by national standard API 581 and API 580 separately in May 2000 and May 2002. The research of RBI in China started in the end of last century. In cooperation with BV, Mouming petrochemical company (SINOPEC) and Hefei General Machinery Research Institute applied RBI in ethylene cracking unit as Huabei petrochemical company (CNPC) and SK did in catalytic cracking unit and diesel hydrogenation unit [3]. General Administration of Quality Supervision, Inspection and Quarantine enacted a regulation about the promotion of pilot project on RBI technology, which supported the extension of this technology in 2006 [4]. In 2009, *Pressure Pipe Safety Technology Supervision Regulation for Industrial Pressure Pipe* was published and the result from RBI was permitted to make inspection program. So far, RBI, used widely both in China and abroad, has been demonstrated the advanced management technology for enhancing reliability of the unit and optimizing inspection resource.

Quantitative RBI starts at capture of inspection data including information of design, manufacture, process parameters and history. After division of process flow, damage mechanism and failure modes, such as thinning, corrosion, operational error and process upset are identified for every subsection, likelihood and consequence of failure, are identified and assessed. Utilizing professional software, the risk multiplied by likelihood and consequence is performed for each scenario of corresponding hole size. Assessment result audit is followed by modified inspection program as ending of the procedure.

2.2 potential damage mechanism

The analysis about the data of material balance and actual material sampling accounts for the corrosive impurities in this hydrocracking unit: active sulfur(H_2S,element sulfur and etc.), H_2, NH_3, amine so and so on, which derive main potential damage mechanism: thinning(general thinning and localized thinning are included), stress corrosion cracking(SCC), external damage and hydrogen attack[5]. The main damage mechanism in this unit is portrayed in the following content.

2.2.1 thinning

High temperature H_2S/H_2 corrosion, naphthenic acid corrosion, ammonium chloride corrosion, ammonium hydrogen sulphide corrosion, amine corrosion and high temperature sulfidic corrosion are involved in this unit.

2.2.2 SCC

Wet H2S cracking and amine cracking may happen in this unit. Former generally exists in place with carbon steel and low alloy steel where hydrogen blistering, HIC, SOHIC, SSCC probably appears. Amine cracking is related to the type, density, temperature and stress of amine.

2.2.3 external damage

External damage can be divided into atmosphere corrosion and corrosion under isolation (CUI). Former is only sensitive to carbon steel, low alloy steel and copper aluminum alloy. And later is also sensitive to 300 stainless steel and duplex phase stainless.

2.2.4 hydrogen attack

Temperature, hydrogen partial pressure and time combined stress decide hydrogen attack. This situation happened concentrates in hydrofining reactor, hydrocracking reactor and their relative main piping.

3. Illustration of Hydrocracking Unit

3.1 modeling of safety assessment

The in-service pressure vessels and piping in hydrocracking unit are all covered for risk analysis in this paper. The amounts of them are listed below.

Table 1. Statistic Data of Equipment in Hydrocracking Unit

Equipment	Number	Analyzed Element

pressure vessel	129	228
piping	459	459

There are 16 material loops and 10 corrosion loops, sum up 687 analyzed elements totally.

3.2 identifying of risk source

The main composition of material in hydrocracking unit contains hydrogen, acid gas, LNG, naphtha, fuel gas, aviation oil, diesel, mixture raw oil, reactor effluent, fresh hydrogen, recycle hydrogen, barren liquor, rich liquor, water(vapor included), waste water, ammonia and air so and so on. First six of them are flammable and combustible, and the equipment stored hydrogen sulfide in other system is hazard area.

4. Result of Quantitative RBI on the Hydrocracking Unit

4.1 risk ranking

Defining Jun 30, 2014 as risk assessment point, data collected are inputted to Orbit-onshore software to make quantitative risk analysis calculation for this Hydrocracking Unit. The risk matrix for integral safety is presented in Fig.1 where risk is composed of high risk, medium (medium and medium high) risk and low risk. Matrixes for pressure piping and vessel are addressed separately in Fig.2 and Fig.3.

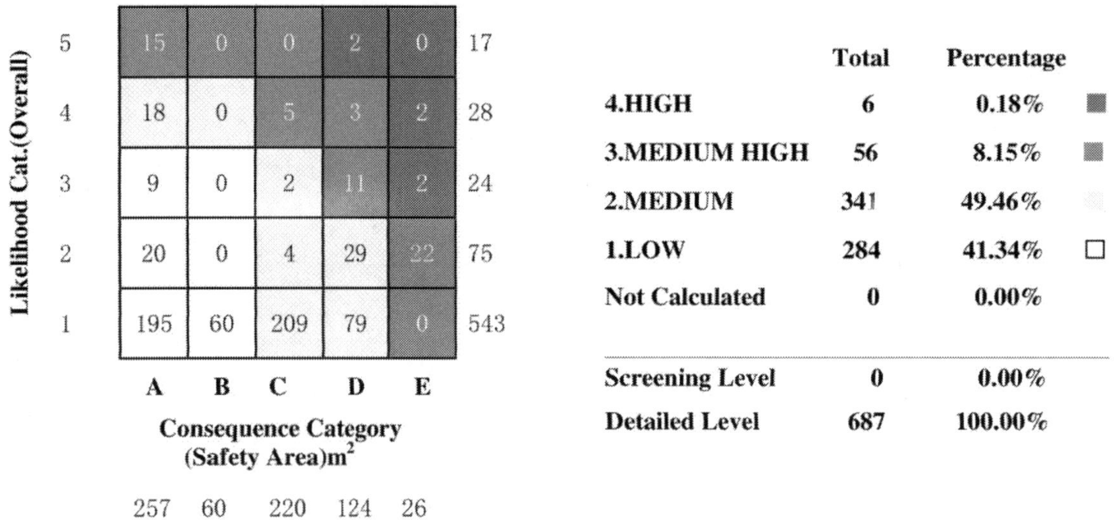

Fig.1 Risk Matrix for Hydrocracking Unit

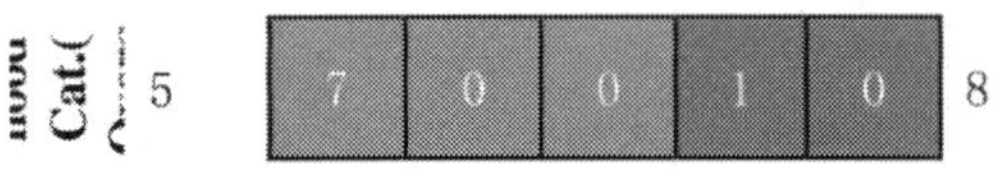

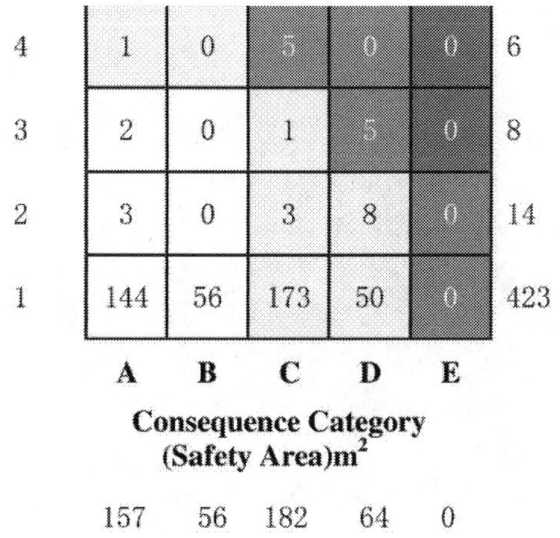

	Total	Percentage
4.HIGH	1	0.22%
3.MEDIUM HIGH	17	3.70%
2.MEDIUM	236	51.42%
1.LOW	205	44.66%
Not Calculated	0	0.00%

Fig.2 Risk Matrix for Pressure Piping

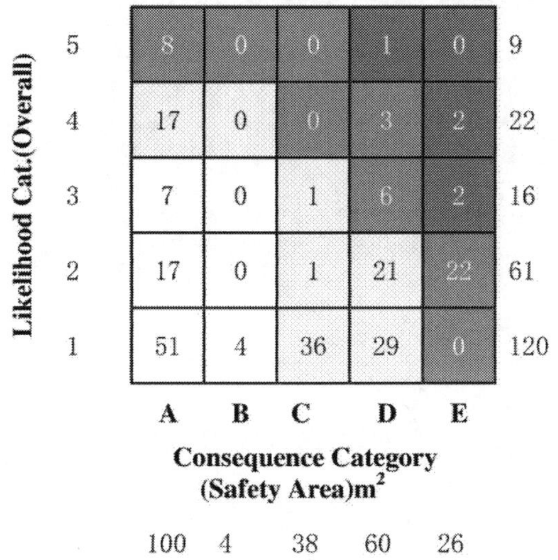

	Total	Percentage
4.HIGH	5	2.19%
3.MEDIUM HIGH	39	17.11%
2.MEDIUM	105	46.05%
1.LOW	79	34.65%
Not Calculated	0	0.00%

Fig.3 Risk Matrix for Pressure Vessel

Based on the risk evaluation, 5 vessels and 1 pipeline locate high-risk level, 39 vessels and 17 pipelines are medium high-risk, medium-risk units are made up of 105 vessels and 236 pipelines, and 79 vessels and 205 pipelines are plotted on low-risk.

Make a general survey of Fig.4, the average risk rank from high to low separately is vessels, columns, heat exchangers and pressure piping. In the other hand, the accumulated risk rank from high to low is pressure vessels and piping, heat exchangers and columns in turn. Vessels take up a main situation in average risk rank the same as pressure piping in accumulated risk rank.

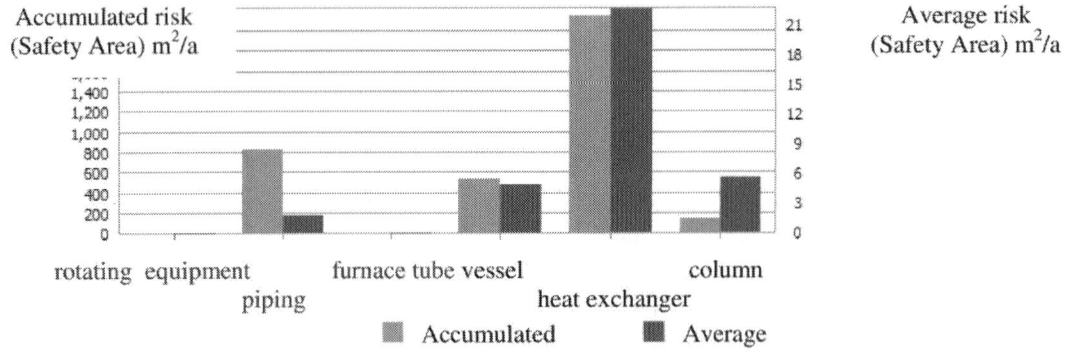

Fig.4 Risk Distribution Diagram of Whole Unit

A large percent of the total unit risk is concentrated in a relatively small percent of the equipment items, which is illustrated in Fig.5.The 10% equipment items who contributes more than 90% risk requires greater attention though a revised inspection plan to reduce risk of integral unit. The cost of the increased inspection effort can sometimes be offset by reducing excessive inspection effort in the area identified as having lower-risk.

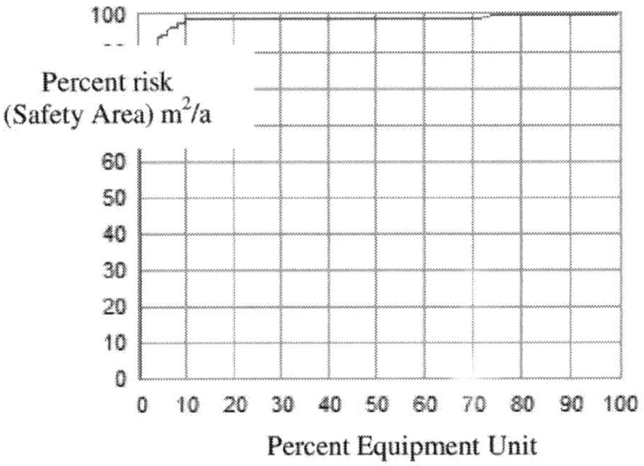

Fig.5 Distribution Diagram about Percent Risk of Whole Unit

4.2 influential factors and risk

Table 2. Top 4 of Equipment Item in Risk Rank

ID	Equipment Item	likelihood	consequence	Risk (safety area) m2/a	proportion
V1002-top	Knock-out Pots and Dryers	4	E	7.02E+02	19.3%
V1007-top	Knock-out Pots and Dryers	4	E	6.79 E+02	18.7%
P1021	Pipeline	5	D	3.4 E+02	9.3%
E1001-cha	Heat Exchangers Shell	5	D	3.02E+02	8.3%

Asset Integrity Management defines risks as the product of two separate terms-the likelihood that a failure will occur and the consequence of a failure. Table 2 describes the 4 equipment items that owns larger proportion of whole risk. For integral unit, likelihood beyond 4 covers 45 equipment items and their main damage mechanism are thinning, SCC and external damage. For equipment item whose likelihood down to the high degree, bring specific anticorrosion treatment, monitoring and inspection techniques into force by analyzing damage mechanism can be conductive to risk mitigation. For example, weight loss test is used for measuring equipment items with high corrosion rate, and NDT is applied for those with high likelihood of SCC to limit risk on line and updated texture if necessary.

4.3 optimization of inspection program

Based on the technical regulations, guidance and optimization inspection is the destination of risk assessment taking operation condition and risk level of equipment into account. Excess, lack and invalid inspection, as traditional methods, need to be avoided.

- Selection of inspect method complies with feasibility and specific of the inspection program for equipment items. For vessel, inner inspection is priority when damage occurs at interior and external test is also necessary when higher risk exists at the same time.
- Pay more attention to high-risk equipment items and reduce relatively inspection proportion in low-risk items. Beside annual survey, extra test should be increased to identify distribution and quality of risk in case of doubt.
- Refer to API, effective method for corrosive thinning is visual check combined with wall thickness measurement, and key points fall down on vulnerable part as reducing, feed and outlet piping, elbow, concentric reducer and tee etc. macroscopic check applies with magnetic particle testing or penetrant testing aim to SCC, and welding joint as an example of stress concentration place deserves to be put high value.

5. Conclusion

- By June 30 2014, the hydrocracking unit is found that 6 analyzed elements are defined as high-risk, 56 analyzed elements fall down to medium high-risk and rest of pressure vessels and piping are plotted on medium-risk and low-risk region. The result spells out less than 10% of equipment items concentrated in approximately 90% unit risk. Thus, greater attention paid to small percent of the equipment items may effective control the large percent of the total unit risk, which could make effective use of inspection resources and reduce risk to a great extent.
- Potential damage mechanisms in hydrocracking unit include: thinning(general thinning and localized thinning), stress corrosion cracking(SCC) and external damage (CUI is covered). Combine potential damage mechanisms with actual inspection result to determine damage mode existed in equipment item.
- For those equipment item with high-risk or medium high-risk and whose likelihood is 4 or more, they are suggested to be more concerned in the next overhaul.
- An appropriate inspection program for press vessels and piping of this unit has established according to the result of risk analysis and is planned to conduct during overhaul for comparing result and improving database. Finally, regulation of risk evaluation and inspection for hydrocracking unit would be made on this basis.

6. References

[1] *Risk-Based Inspection Base Resource Document API 581,* American Petroleum Institute, New York, 2008.

[2] HE Lei, CHEN Guohua, Application of RBI in petrochemical industry, Petroleum Refinery Engineering,33(2003), 21-24.

[3] Geng Xuefeng, Zuo Yantian and Xue Xiaolong, Research on the application of RBI technology in China, Chemical Equipment Technology, 33(2012), 17-21.

[4] Dong Shaoping, Application of RBI in synthetic ammonia unit, Petro-Chemical Equipment Technology, 01(2008), 59-64

[5] T.L.Willke, U.S. risk management can reduce regulation, enhance safety, Oil & Gas Journal, 95(1997), 37-46.

Fire from the Cascading Failure of an Oxygen Supply System

Delmar "Trey" Morrison, Ph.D., P.E.
Exponent, Inc.
4580 Weaver Parkway, Warrenville, IL 60555 USA
tmorrison@exponent.com

Sean J. Dee, Ph.D., P.E., and Brenton Cox, Ph.D., P.E.
Exponent, Inc.

Vijay Kumar
EQUATE Petrochemical Company
PO Box 9717, Ahmadi, 61008, Kuwait
kumarv@equate.com

Mohammad Al-Shamary, Abdulqader Al-Qabandi
EQUATE Petrochemical Company

Prepared for Presentation at
American Institute of Chemical Engineers
2015 Spring Meeting
11th Global Congress on Process Safety
Austin, Texas
April 27-29, 2015

UNPUBLISHED

Fire from the Cascading Failure of an Oxygen Supply System

Delmar "Trey" Morrison, Ph.D., P.E.
Exponent, Inc.
4580 Weaver Parkway, Warrenville, IL 60555 USA
tmorrison@exponent.com

Sean J. Dee, Ph.D., P.E., and Brenton Cox, Ph.D., P.E.
Exponent, Inc.

Vijay Kumar
EQUATE Petrochemical Company
PO Box 9717, Ahmadi, 61008, Kuwait
kumarv@equate.com

Mohammad Al-Shamary, Abdulqader Al-Qabandi
EQUATE Petrochemical Company

Keywords: fires, explosions, oxygen enriched atmosphere, case study

Abstract

This paper discusses the investigation, findings, and root cause analysis of a fire due to the loss of oxygen containment in an ethylene oxide process. This is one of many types of processes that may contain a high pressure oxygen supply system. In such systems, the pressure and velocity of the oxygen stream are related to the potential for ignition of system components. Thus, measures are taken in design, operation, and maintenance to utilize ignition resistant materials of construction.

While more ignition resistant than carbon steel, stainless steel is still ignitable under certain process conditions in oxygen enriched atmospheres. In this incident, unanticipated conditions led to ignition of internal parts and stainless steel components in the high purity oxygen system. The resulting loss of oxygen containment caused a fire that destroyed the oxygen injection system and allowed a significant release of a combustible process gas stream. While automated safety instrumented safeguards shut down the oxygen supply to the system, the ensuing fire involving the combustible process gas stream continued. Consumption of both steel and ignition-resistant alloy components in the system during the oxygen enriched fire masked the underlying causes of the system failure. No injuries were sustained in the incident, but the process unit sustained major damage and required a lengthy outage for repair.

Given the level of damage to the oxygen injection system, the failure analysis required rigorous and systematic investigation by a multidisciplinary team. Through evaluation of the process data, metallurgical analysis of the damaged components, and understanding of the process dynamics, detailed hypotheses were developed, tested and refuted leaving a few most likely hypotheses explaining the cause of the incident. The lessons learned from the incident were used to develop a system redesign and maintenance guidelines for the rebuilt unit to mitigate the likelihood of future failures.

1 Introduction

A fire occurred in the oxygen supply system in an Ethylene Glycol Unit of the Shuaiba, Kuwait facility of EQUATE Petrochemical Company (EQUATE), which led to significant damages, unit shutdown, but no injuries. Since the fire involved an oxygen system, it escalated rapidly, but emergency shutdown systems quickly isolated oxygen flow. The piping system was breached at two locations: a carbon steel piping flange upstream of an automated block valve and a manual stainless steel gate valve downstream of check valve in the oxygen piping. The remainder of the process gas feeds was isolated at the unit's battery limits during the emergency shutdown, but the inventory of flammable process gas partially blew down through the breached oxygen system. This contributed to a large local fire that heavily damaged the oxygen system, process unit structure, and surrounding equipment.

Given the destructive nature of the fire, a detailed investigation was undertaken to identify the cause of the initial loss of containment in the oxygen system. The oxygen system failure analysis required rigorous and systematic investigation by a multidisciplinary team. Through evaluation of the process data, metallurgical analysis of the damaged components, computational fluid dynamics (CFD) analysis, and understanding of the process dynamics, detailed hypotheses were developed, tested and refuted leaving a few most likely hypotheses explaining the cause of the incident. The lessons learned from the incident were used to develop a system redesign and maintenance guidelines for the rebuilt unit to mitigate the likelihood of future failures.

2 Background on Pure Oxygen Systems

Pressurized oxygen systems pose significant hazards due to the acceleration of combustion of materials when exposed to an oxygen-enriched environment. Industrial experience has shown that oxygen enhanced fires rapidly accelerate in intensity since even apparently non-combustible materials can be consumed [4]. The ignition of both metallic and nonmetallic components is thus a concern in O_2 enriched atmospheres (i.e., greater than 23.5% O_2). Components that would not normally burn in ambient air (i.e., 21% O_2) may burn vigorously in an O_2 enriched environment. This can lead to loss of containment of high purity oxygen and rapid escalation of fire and explosion incidents. High oxygen concentration combined with a fuel and ignition source can result in a breach in oxygen containment systems, rapid loss of containment of pressurized oxygen, rapid fire growth, and rapid fire extension to other areas.

Purity and concentration of oxygen in the system directly affect the dominant design considerations. Oxygen enriched systems are defined as containing at least 23.5% O_2 by volume with the remaining volume being inert [1]. Oxygen-enriched systems can be classified into a number of categories depending on purity, including [1]:

- Low purity oxygen – contains less than 35% O_2 by volume
- Standard purity oxygen - equal or greater than 99.5% O_2 by volume
- Ultra high purity oxygen - greater than 99.999% O_2 by volume

Industrial standards and the underlying research efforts by industrial oxygen users have led to the development of safety and design guidelines and standards including ASTM G94-05, ASTM G63-99, ASTM G88-05, CGA G-4.4, NFPA 50, and NFPA 53.

2.1 Understanding Potential Modes of Ignition

Initial ignition of a component or particle within an oxygen system may lead to a kindling chain mechanism of fire growth where the energy released from an initial, smaller, easily ignitable fuel source causes ignition and combustion of a more ignition-resistant material [Reference [2], Section 5.2.14]. The kindling chain may quickly lead to fire growth within a system and overcome (i.e., ignition and continued combustion) the combustion resistance of the components in a piping system leading to loss of O_2 containment. Ignition and combustion resistance of materials are related to the material composition, thickness, configuration, and environment (e.g., temperature, pressure, and velocity). Several potential ignition mechanisms for metal alloys have been identified in the literature [1]:

- Particle impact
- Mechanical friction[1]
- Adiabatic compression
- Promoted ignition
- Electrical arcing (e.g., static electric discharge, lightning strike, electrical system fault)

The collision of high velocity particles with stationary surfaces can generate localized heating of the particles due to the conversion of kinetic energy to heat at the impact point. If the heat generated is sufficient, ignition and combustion of the particle may occur which could then ignite the impacted material or other nearby fuels (e.g., polymeric gaskets). Both impinging (impact face-on in the direction of flow), and non-impinging (impact obliquely or parallel to the direction of flow) impact with internal piping system surfaces may result in ignition of either the particle or the surface material [Reference [2], Section 5.2.6]. Particles may be introduced from the attrition of surface scale deposits from corrosion in piping, mechanical wear of system internals, or other sources of particle contamination. The nature of the particle along with the velocity and other system conditions will affect the likelihood of ignition of the particle. For example, silica particles are non-ignitable, but rust particles are ignitable.

Mechanical frictional heating may result from metal-to-metal rubbing, which can be experienced in moving valve internals. This rubbing mechanism can wear protective coatings from surfaces as well as generate small particles [Reference [2], Section 5.2.8.1]. Adiabatic compression of oxygen can occur when a valve is opened to a dead end space in a pipe with O_2 at a lower pressure. The resulting mechanical compression of the gas in the dead space as it is pressurized to the pipeline pressure can result in heating sufficient to ignite non-metallic valve seats or gaskets [Reference [2], Section 5.2.7]. These could, in turn, ignite metallic components leading to a system fire.

---

[1] While frictional heating may occur from gaseous flow against a pipe wall or through a small opening, flow frictional heating causing ignition of metals is generally only supported by "unverifiable anecdotes" in high pressure oxygen systems (e.g., greater than 500 psi) [ASTM G88-05, 5.2.12 – 5.2.12.1]. These pressures are significantly higher than those in the system discussed in this manuscript. Hence, frictional heating is more appropriately discussed from a standpoint of surface friction induced by internal moving components in valves, pumps, or other devices with moving parts.

Several design steps can be taken to limit the potential for ignition, or to stop the kindling chain before the system is breached. The use of filters can reduce the presence of particles, but it is infeasible to remove all particles for an oxygen system [Reference [2], Section 5.2.6.1]. The likelihood of ignition from both frictional heating and particle impact can be limited by selecting combustion-resistant materials of construction [Reference [2], Sections 7.6.3 and 7.8.3]. The likelihood of ignition by adiabatic compression can be limited by system design and throttling controls, for example to limit the rate of valve opening/closure or ignition resistance of materials in the adiabatic compression zone.

2.2 *Metal Alloys in Oxygen Service*

Metal alloys for use in O_2 service are classified as either combustion-resistant or exempt, based upon the alloy, configuration, temperature, pressure, and gas flow velocity. A combustion resistant material at the given conditions may undergo ignition, but the combustion process will quench with limited propagation away from the point of ignition. An exempt material has no thickness or pressure limitation for use in O_2 systems, based upon the standards [1]. Although the alloys listed are referred to as "exempt" from velocity limitations, this does not mean that the metals are immune to oxygen-enhanced combustion. Sufficient evidence exists in the literature and from practical experience to indicate that damage can occur to these materials from repeated combustion events without causing a propagating combustion front.

The kindling chain may quickly lead to fire growth within a system and overcome the burn resistance of the components in a piping system leading to loss of O_2 containment. Ignition and burn (i.e., combustion) resistance of materials is related to the material composition, thickness, configuration, and environment (e.g., temperature, pressure, and velocity). For example, carbon steel alloys are unsuitable for use in oxygen systems at concentrations greater than 35% by volume due to their reduced resistance to combustion [Reference [1], Section 4.2.3.2].

Metals are more difficult to ignite than non-metals due to their higher ignition temperature, higher thermal conductivity, and ability to generate ignition-resistant oxide coatings [Reference [3], Section 5.4.2]. If sufficient energy can be added to initiate combustion (i.e., ignition), then a given material may start to burn. However, if the oxygen concentration, density, and velocity are not within certain bounds, the combustion event will extinguish. In practical terms, repeated particle impact may be able to eventually cause a hole in a non-exempt material part. Various modeling approaches have been discussed that can be used to estimate the rate of consumption of a metal containment system [4],[6],[7],[8]. Estimates of the propagation rate of the combustion front can be used to design containment systems (e.g., wall thickness) and the timing for automated shutdown systems.

Examples of metallic alloys used in O_2 distribution systems are provided in Reference [1], Appendix C and D. The table lists some common stainless steel alloys and nickel alloys. Carbon steel alloys are typically unsuitable for use in oxygen systems at concentrations greater than 35% by volume due to their reduced resistance to combustion [Reference [1], Section 4.2.3.2]. The exemption status of a given metal is determined based upon a maximum observed consumed length of 30 mm in some standardized tests, or maximum consumed lengths of 10 cm to 15 cm, in others [Reference [1], Appendix B]. For a material in a thickness less than that tested, the standards indicate that it should be treated as flammable at pressure and velocity above the exemption curves as displayed below.

Table 1. Example O_2 oxygen system alloy exemption pressures and minimum thicknesses.

Material	Nominal Composition	Min. Thickness	Exemption Pressure
MONEL® 400	67Ni, 32Cu, 1Fe	None Specified	21 MPa (3000 psig)
MONEL® K-500	66.5Ni, 30Cu, 3(Al, Ti)	None Specified	21 MPa (3000 psig)
Hastelloy C-276	56Ni, 12Cr, 13.5Mo, 4Fe, 3W, 2.5Co	None Specified	5.3 MPa (750 psig)
INCONEL® X-750	74Ni, 15.5Cr, 7Fe, 2.5Ti, 1Al	None Specified	6.9 MPa (1000 psig)
Stainless Steel, cast, CF-8M	19Cr, 11Ni, 3Mo, 1.5Si, 1.5Mn, Bal. Fe	3.18 mm (0.125 in) 6.35 mm (0.250 in)	1.4 MPa (200 psig) 2.0 MPa (290 psig)
Stainless Steel, wrought, 316	17Cr, 12Ni, 2Mn, 3Mo, 1Si, Bal. Fe	3.18 mm (0.125 in) 6.35 mm (0.250 in)	1.4 MPa (200 psig) 2.0 MPa (290 psig)
Stainless Steel 317 [5]	19 Cr, 13 Ni, 2 Mn, 3.5 Mo, 0.75 Si, ~Bal. Fe	Not listed	Not listed
Gray Cast Iron	3C, 2Si, 0.8Mn, Bal. Fe	3.18 mm (0.125 in)	0.27 MPa (25 psig)

For non-exempt materials (i.e., combustion-resistant), velocity limitations for impingement and non-impingement oxygen flow are shown in Figure 1 and Figure 2 respectively [Reference [1], Sections 4.4.2 and 4.4.3]. For each graph, systems containing oxygen-contact stainless steel alloy components should be designed to keep the pressure and velocity below and to the left of the curve to limit particle kinetic energy [Reference [2], Section 7.6.2]. The velocity limit as a function of pressure is lower for the impingement curve than the non-impingement curve because a theoretical particle impact face-on to a surface will generate more energy than an oblique impact.

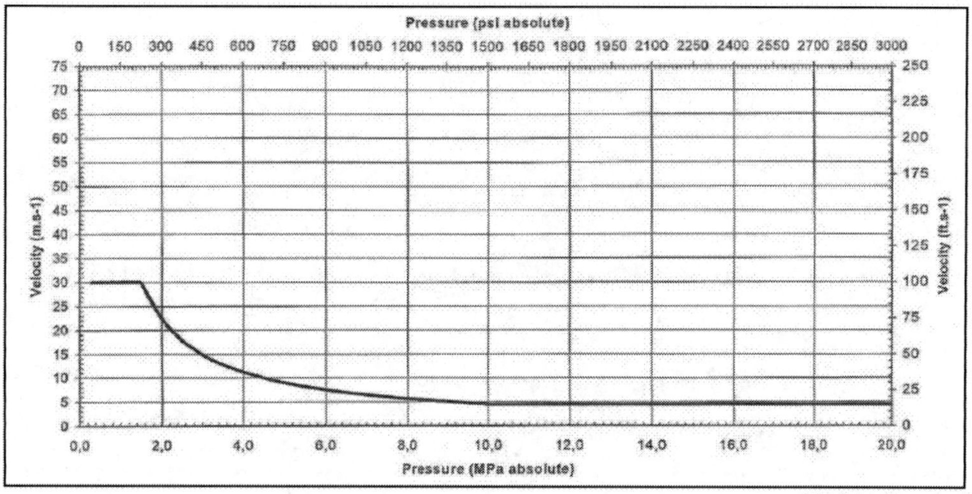

CGA G-4.4; Figure 1, p12

Figure 1. Impingement velocity curve.

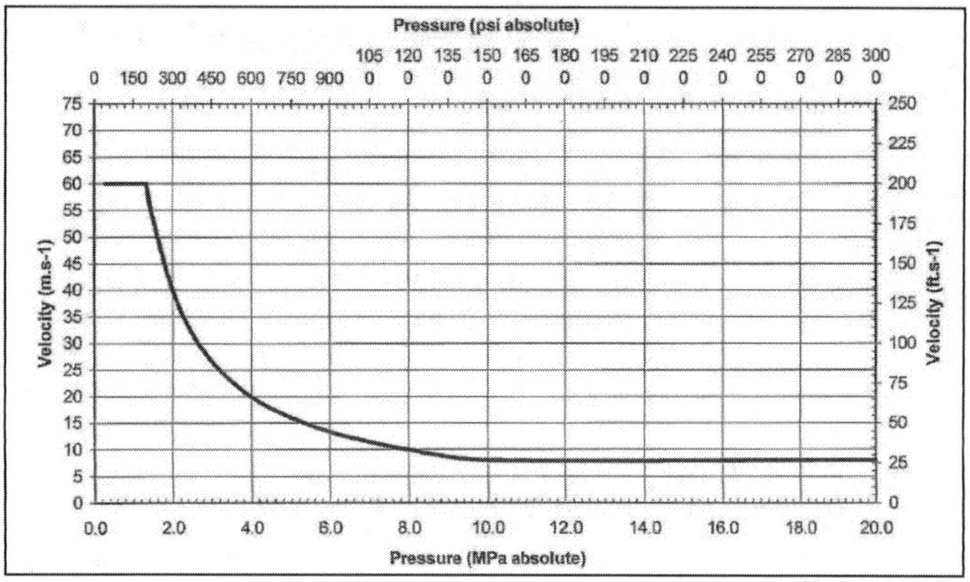

CGA G-4.4; Figure 2, p13

Figure 2. Non-impingement velocity curve.

3 Failure Analysis of Oxygen System

In undertaking an investigation, the scientific method of analysis is an essential framework for completing a thorough and objective analysis. Figure 3 provides a flowchart that outlines the approach. This is a hypothesis-driven approach where scientifically defensible answers are sought. The key objectives of the investigation are to gather the facts, perform an objective and systematic analysis, and determine the most likely causes of the incident. After these most likely causes have been identified, evaluated, and understood, the investigation can reliably move toward providing useful lessons learned and recommendations for preventing a similar incident from recurring.

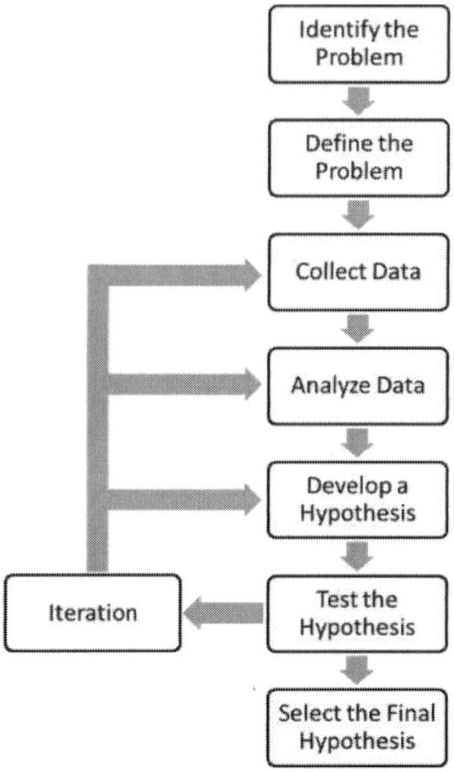

Figure 3. Flow chart for the scientific method of investigation.

3.1 Overview of the Oxygen System

The oxygen system was installed in approximately 2000, and the piping diameter was increased from 4-inches to 6-inches sometime after the initial installation to meet operational requirements. The system was equipped with an emergency shutdown system that tripped based on temperature or pressure deviations outside the acceptable operating envelope. In approximately 2005, the emergency system actuated. The root cause investigation determined that the most likely causes involved fugitive particles in the system. The system operated without incident for several years until the loss of containment incident in 2012.

The system supplied oxygen at approximately 27 bar (400 psig) and 13 kg/s (29 lb/s) to the process. The oxygen feed was filtered before passing into the piping control system. A piping isometric describing the system is provided in Figure 4. Components made from ignition resistant alloys are colored green in Figure 4, while stainless steel alloys are red. In the oxygen handling system, high purity oxygen was passed through two sets of filters before it was fed to the O_2 Isolation Section. In the O_2 Isolation Section, oxygen flow to the process was modulated through a MONEL® flow control valve (6). Two automated stainless steel block isolation valves (5 and 7) were tied to the emergency shutdown system to isolate and stop oxygen flow in the event of an emergency. Oxygen then flowed through a 6-inch tee connection (8) to the N_2 Purge Section. Oxygen reverse flow back into the N_2 Purge Section was restricted by a stainless steel swinging check valve (10). MONEL® piping connected the O_2 Isolation Section to the O_2 Process Section through a tee at this location. In the O_2 Process Section, oxygen passed through a MONEL® dual disk check valve (3) with internal components made of either MONEL® or INCONEL® alloys. Directly downstream of the dual disk check valve was a stainless steel manual gate valve (2) that connected to additional MONEL® piping before the oxygen was fed to the process unit. Within the system, stub ends were comprised of MONEL® with non-ignition resistant flanges. Additionally, gaskets within the system were found to have stainless steel inner rings and MONEL® coils. Thus, in the oxygen system, the oxygen contact surfaces were found to be composed of ignition resistant materials, with the exception of any exposed surface of the gaskets' inner ring.

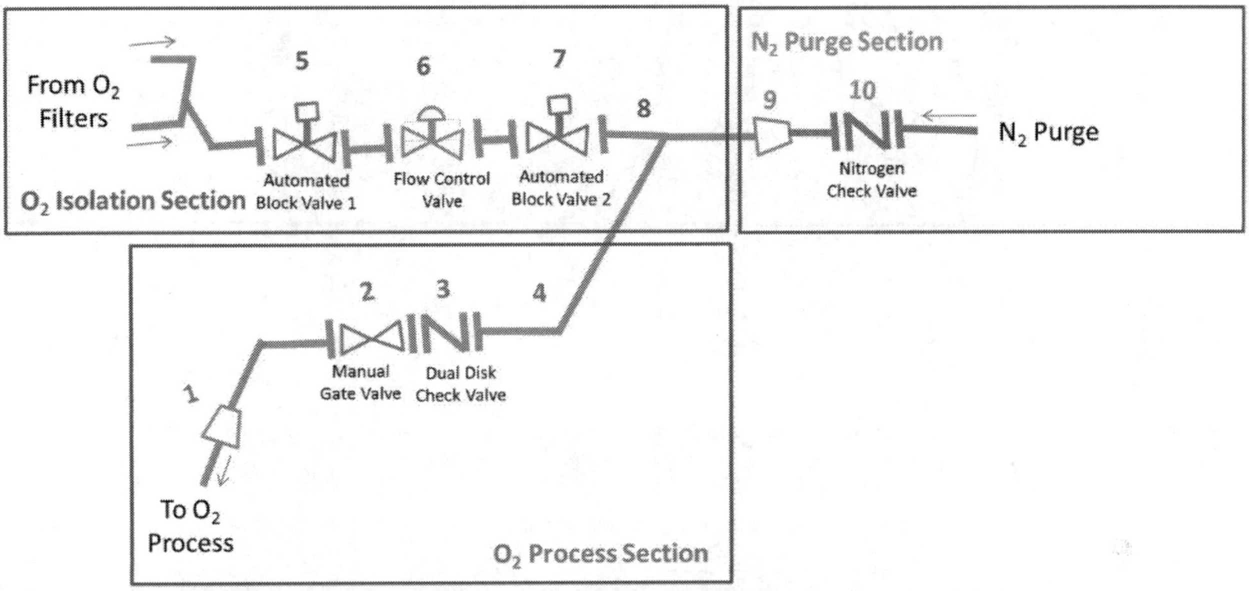

Figure 4. Piping isometric diagram of the O_2 system.

A pre-incident image of the O_2 system is provided in Figure 5. The O_2 filters are located out of the image to the left with flow from the left to right matching the isometric figure. The blue actuators correspond to the automated isolation block valves (5 and 6). The green actuator corresponds to the flow control valve (6). The hand wheel shows that the manual gate valve (2) is in the open position. A post-fire photograph of the O_2 system is depicted in Figure 6 indicating that most of the piping system components partially survived the fire.

Figure 5. An aerial view of the O_2 system prior to the incident.

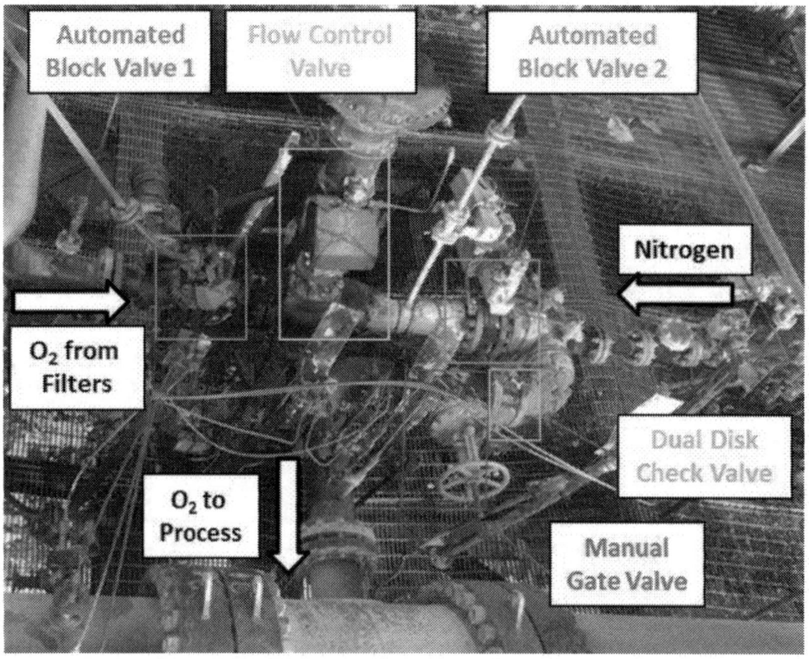

Figure 6. Overhead view of piping from oxygensystem fire.

3.2 Determining the Initiating Event

Two potential areas were identified that could correspond to the initial loss of containment incident: the carbon steel flange upstream of automated block valve (5) and the stainless steel manual gate valve (2). Both components were found to contain significant material loss due to oxygen-enhanced combustion. The upstream flange was refuted as a potential source of the initial LOC due to the evidence of automated isolation system action and a delayed pressure blowdown observed in the recorded process data. Thus, the investigation focused on determining the cause of the breach at the manual gate valve. The most likely potential initial ignition modes for this high purity oxygen system were determined to be frictional heating and particle impact. To characterize their relative likelihood as the potential cause, the average operating conditions based on process data for the oxygen system leading up to the incident are summarized in Table 2.

Table 2. Average operating conditions of O_2 system prior to incident.

Pressure bar (psi)	Temperature (°C)	Mass Flow Rate $kg\ s^{-1}$ ($lb\ s^{-1}$)	Calculated Volumetric Flow Rate $m^3\ s^{-1}$ ($ft^3\ min^{-1}$)	Calculated Nominal Velocity $m\ s^{-1}$ ($ft\ min^{-1}$) 6-inch Diam.
27.9 (396)	22.5	13 (29)	0.4 (850)	20 (3900)

An important aspect of designing systems and processes that utilize high purity oxygen is conducting an oxygen compatibility assessment (OCA) [9]. In a typical OCA, oxygen-wetted materials are evaluated under their potential worst case operating environment to assess their potential flammability. If a material is deemed to be flammable in the operating environment, then the potential ignition mechanisms are evaluated and scored according to their likelihood to promote ignition of the flammable material. Finally, the potential for a kindling chain or reaction effect is evaluated with respect to other nearby materials.

In this particular incident, OCA was used as a tool to assist in evaluation of potential origin and cause hypotheses. Within the oxygen system, five metals were evaluated for oxygen compatibility at 400 psi at ambient temperature: MONEL® 400, MONEL® K-500, INCONEL® X-750, 316 Stainless Steel, and 317 Stainless Steel. The MONEL® alloys are considered ignition resistant at 400 psi and ambient temperature because published literature has demonstrated that these alloys can withstand pressures up to 10,000 psi without failing a promoted ignition test. 300 Series stainless steel has been shown to burn at pressures as low as 111 psi, and fail promoted ignition tests at 200 psi. Therefore, the stainless steel alloys in the oxygen system would be considered flammable at 400 psi. INCONEL® X-750 has not been as extensively tested in the literature as MONEL® or steel alloys. The available data has only demonstrated that INCONEL® X-750 fails the burn length criteria at 1,000 psi, and has not been tested at lower pressures. At 400 psi, the flammability of INCONEL® X-750 is undetermined.

The classification of materials as "non-flammable", as opposed to "ignition resistant", must be approached with caution from a fire and explosion hazard assessment. Traditional classification of a material as "non-flammable" is based on the material exhibiting less than 1 inch of combustion consumption of a 0.125 inch diameter rod in a promoted ignition test [Reference [1], Appendix B]. Thus, materials classified as "non-flammable" can still exhibit metal combustion and consumption but "pass" the promoted ignition test. Promoted ignition and combustion of these "non-flammable" materials may still occur. For example, in the oxygen system under investigation, MONEL® may be considered "non-flammable" in an OCA, but particle impact, frictional heating from metal/metal rubbing, or kindling chain reactions may still result in combustion of the alloy, albeit in a prolonged fashion.

The limited available data for special alloys, can also lead to confusion in material selection related to OCA. In the example above, INCONEL® X-750 is listed as being exempt up to 1,000 psi, but the available data demonstrate that at 1,000 psi INCONEL® X-750 failed the promoted ignition tests. Further testing data was not available at lower pressures. It is of critical important that consideration of material exemption pressures as a safety limit or threshold be supported with test data.

4 Evaluating the LOC Hypothesis

The most severe damage and metal consumption occurred on the underside of the stainless steel manual gate valve and on the flange between the check valve and gate valve. Once ignited in the O_2 environment inside the piping system, combustion of the stainless steel components would be anticipated to occur rapidly leading to a breach in the valve body and release of O_2 pressure from the piping system. There are several distinct features of the damage that were used to identify the most likely point of origin for the fire inside the assembly.

4.1 Explaining Physical Damages (Metal Consumption)

Based on the literature, a representative rate of consumption for 316 stainless steel rods of 1 to 1.2 cm/sec was estimated for the valve body at 2.76 MPa (400 psi) [6]. Although this rate cannot be directly correlated to the more complex geometries involved, it does give a frame of reference of a few seconds to reach the degree of metal consumption observed [4].

The stainless steel gate valve body was consumed internally on the bottom side of the valve, but did not contain internal wall thinning on the top inside of the valve. The valve body consumption was focused around two holes on the bottom side located on either side of the valve seat. Based upon the differential consumption and metal wall thickness, it appears that these two holes were a consequence of the thinner wall thicknesses at these locations in the valve body casting when compared to the thicker cross section near the valve seat. The high pressure O_2 release caused the outer valve body to be consumed in jagged, irregular patterns. A few bolts in the valve flanges and the valve bonnet area were also partially or totally consumed during the O_2 release.

Given the nature of enhanced O_2 combustion of metal, the greatest amount of consumption can be related to the duration of combustion at that location. For a thin wall thickness specimen with continued O_2 flow across the surface, the hole will grow quickly in diameter with the perimeter of non-consumed metal moving quickly away from the point of origin. After O_2 flow through the hole stops (i.e., O_2 system isolation), consumption of the metal will likely stop immediately since ambient O_2 concentration is insufficient to support the metal combustion and a process gas-fueled fire is insufficient to cause stainless steel consumption in an ambient oxygen environment.

The damage to the gate valve and adjacent check valve is most severe at the connection of the two components as evidenced by the thickness of the original flange, which was completely burned through, and is located at one of the bottom bolts connecting the gate valve flange to the check valve body (see Figure 7). Not only is the gate valve-to-check valve flange consumed through its full thickness, but the adjacent gate valve bonnet flange is also externally consumed near this location. The damage patterns, both internally and externally, point back to this location as the origin of the initial high pressure leak in the piping system. The damage, and thus the combustion of the valve, extends away from this point of origin.

Figure 7. Localized consumption of stainless steel gate valve flange adjacent to check valve. (Installed such that the circled region was facing downward.)

The section of valve flange shown in Figure 7 was unique. No other flange thickness was completely compromised. The degree of consumption is consistent with this being the point of the initial loss of containment; thus, this it was most likely near the origin of the internal fire in the stainless steel valve body. Examination of the spiral wound gasket between the valve body and check valve indicated that the inner ring, which was in oxygen contact, was composed of stainless steel. The inner ring, spiral winding, and outer ring were consumed on the downward-facing edge, consistent with the area of greatest damage to the valve body and gate valve flange.

4.2 Cause of Ignition

Given the high consumption rate for combustion of the stainless steel components, the point of ignition must have been very close to the origin of the damage in the valve assemblies (see Figure 7). There are two likely modes of ignition that could cause ignition in the oxygen system either individually or when combined: frictional heating and particle impact.

The flow control valve was not being stroked at the time of the incident, and the seat was largely intact. However, evidence of particle creation and particle impact upstream in the O_2 system was noted during the investigation (see Figure 8). During disassembly, that upstream gasket of the flow control valve was found to be sloughing off layers of corrosion products that could be carried in the O_2 stream.

Figure 8. An apparent particle impact point with non-propagating combustion of the stainless steel check valve component was observed on the O_2 side of the N_2 check valve.

Two types of damage were found to the MONEL®/INCONEL® check valve internals just upstream of the gate valve: mechanical wear and melting. The mechanical wear is due to relative motion of the internal parts. The melting is either due to frictional heating or due to heat from nearby non-propagating combustion of the components. Movement of the parts is required for either potential mechanism. Given the high oxygen flow velocity and evidence of molten metal droplets being carried downstream in the piping, the check valve damage was not likely caused by the metal fire moving upstream from the gate valve to the check valve against the flow of O_2.

The wear and melt-over-wear damage patterns to the components as shown in Figure 9 were consistent with prolonged oscillation of the valve disks. The dual disk check valve opened and closed each disk with independently operating torsion springs. O_2 flow pushed on the disks to open the valve, but in the event of a loss of O_2 flow, the torsion springs were intended to shut the disks and prevent reverse flow. After one or both springs were compromised, the disks were no longer restrained, and the oscillations could increase in frequency, thereby accelerating the mechanical wear process and resulting in potentially significant frictional heating. The washers displayed evidence of wear on their internal circumference and some washers actually cut through the hinge pin.

Figure 9. View of the check valve internals. The two worn and melted gouges on the stop pin (top pin) are consistent with the location of the torsion spring restraints (see Figure 11). The stop pin was determined to be a MONEL® alloy. The grooves on the surface of the center of the hinge pin (bottom pin) are consistent with mechanical wear from the torsion spring.

The check valve internal components were combustion-resistant MONEL® and INCONEL® alloys, thus a single particle impact could not be wholly responsible for causing the observed damages to the check valve. Particle impact may have damaged one or both INCONEL® alloy torsion springs, which may have accelerated mechanical wear within the check valve due to a loss of the restraining force for the disks. However, mechanical wear over time due to disk oscillation without particle impact may also be sufficient to create the observed damage. Thus, mechanical wear alone could not be refuted as the root cause of the failure.

A view of the internal components prior to disassembly is depicted in Figure 10. The internal damage on the base of the check valve flows downstream and connects with damage to the gasket and gate valve body. Based on the flow dynamics, the configuration of the components, and the localized fire damage, the most likely cause of the internal fire damage at the MONEL® check valve is due to frictional heating at the bottom hinge-to-check valve body interface. Additional sprays of molten (black) metal alloy are visible on the disks along with erosion of the stop pin.

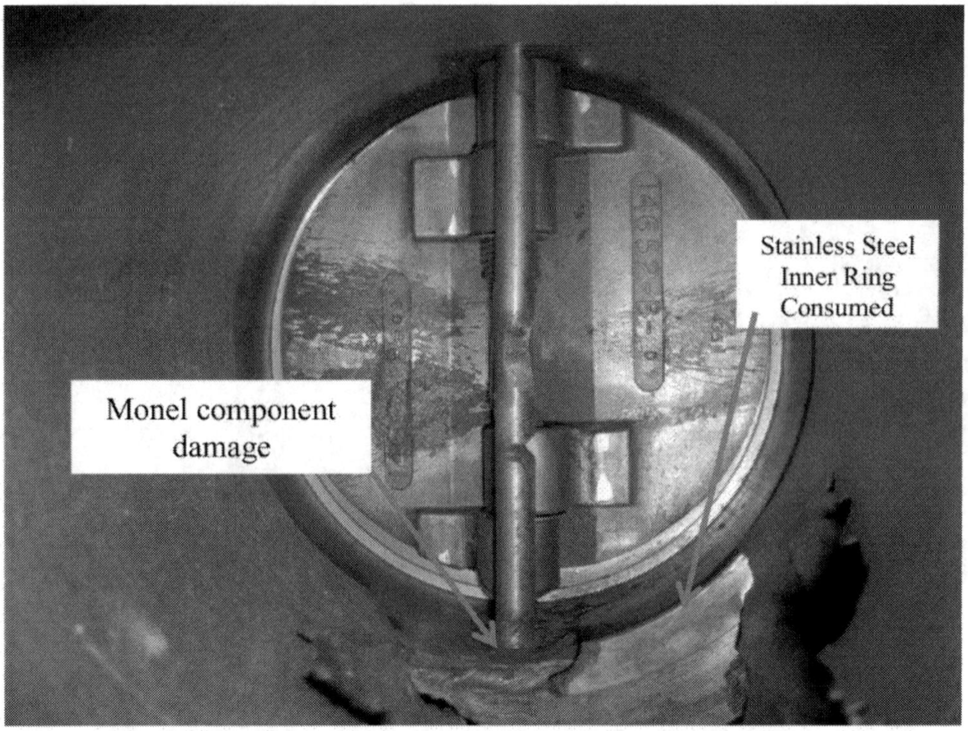

Figure 10. View through the gate valve at the downstream side of check valve. The area at the bottom of the valve that includes the stop pin and hinge contains surface melting and MONEL® consumption up to the mating surface of the partially consumed stainless steel inner ring for the spiral wound gasket.

Although M-400 is a velocity exempt material and is considered to be combustion resistant, there was evidence of non-propagating, localized combustion and melting on many of the internal components. The area of damage at the bottom of the check valve body around the contact area for the bottom hinge included the stainless steel inner ring of the spiral wound gasket. The thermal damage to the bottom of the MONEL® valve body appears to have been sufficient to cause ignition of the stainless steel inner ring of the gasket. Combustion of the inner ring led to combustion of the adjacent stainless steel gate valve body, which then propagated through the gate valve further downstream. Failure of the thin gasket likely led to the high-pressure release and noise identified by witnesses prior to the louder sound of the deflagration. The gasket failure, combined with a high-velocity oxygen release, also led to consumption of the local full thickness of the gate valve flange and external damage to the nearby valve bonnet flange.

4.3 Review of Preventative Maintenance

The check valve was initially thought to have been in service since approximately 2001. However, the investigation discovered that the check valve was scheduled to be replaced in 2010, but a defect in the replacement valve body led to just the internals being replaced. No detailed report was included in the maintenance file assessing the state of the original internals.

A few photos were provided from the 2005 emergency shutdown investigation, which depicted the check valve's internal components after five years of service. This is the same check valve body that was in service during the incident.

In hindsight, two photos indicated the progressive nature of the mechanical wear damage to the valve internal components. Figure 11 shows the downstream side of the valve with the disks held closed by the pair of torsion springs. In the figure, indications of wear-related failure are apparent for the bottom two washers. There is no washer between the bottom two hinges, which indicates one of three possibilities: (1) no washer was originally installed, (2) the washer has been worn through and its remains passed downstream, or (3) the washer has severed the shaft and passed downstream. Given the fact that the top hinge pair contains a washer, possibility (1) is unlikely. No evidence of heating or ignition of the MONEL® or INCONEL® alloy components is apparent in the photo.

In the slightly different view of the image in the right side of the figure, additional mechanical wear damage is visible. The center washer is missing, and a groove is worn in the hinge pin where the washer was originally located. This indicates that the washer was worn through and passed downstream. Additionally, the top spring in the photo has surface damage on the coil immediately adjacent to the missing washer. Although the 2005 investigation was searching for particles within the O_2 piping system, they did not find the two missing washers. In the current incident, three washers were unaccounted for, but no remains were found. The reason for the disappearance of washers could not be explained, but it is possible that they were consumed in the fire, or overlooked or lost during equipment disassembly.

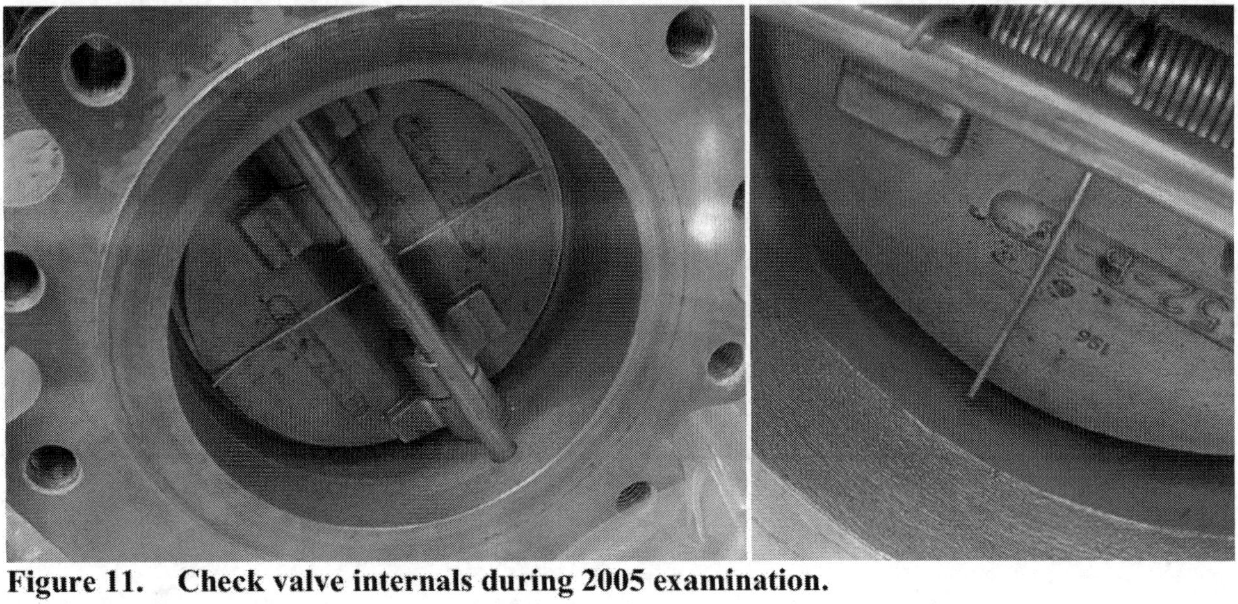

Figure 11. Check valve internals during 2005 examination.

4.4 Oxygen Flow Dynamics

Design calculations for the O_2 system indicated that the bulk average flow velocity of the system was below the limiting velocity curve. However, the actual piping geometry including the 6-inch dual disk check valve, which varied from a 5.56-inch inlet diameter to a 6.75-inch outlet diameter, caused local regions of higher velocity flow. The check valve disks should open when oxygen flow starts and close when it stops. If the flow rate through the valve is lower than the minimum velocity required to keep the valve open, then the valve disks may oscillate ("flutter") with perturbations in the local gas flow. The available documentation for the valve did not provide a minimum velocity to reduce or eliminate disk oscillation [10]. The minimum velocity to pin the disks in the open position was estimated as 35.6 m s^{-1} based on calculations derived from the listed "cracking" pressure to cause the disks to start to open. The process data indicated that the mass flow rate supplied to the O_2 process during the hours prior to the incident fluctuated from 12.4 kg s^{-1} to 13.1 kg s^{-1}, corresponding to a nominal velocity of 19.1 m s^{-1} to 20.2 m s^{-1}. This velocity range was therefore about half that estimated for pinning the disks in the open position to prevent oscillation.

A computational fluid dynamics (CFD) model demonstrated that the local geometry had a dramatic impact on the flow velocity through the dual disk check valve. The check valve geometry created streamline velocities as high as 40 m s^{-1} on the outside port of the valve and much lower on the inside port (relative to the upstream elbow). The unbalanced streamline velocities at the check valve face (from ~20 m $s^{-1}$to 40 m s^{-1}) likely also contributed to oscillation of both disks (i.e., fluttering) when coupled with the force required to open the disks. The installation manual from the manufacturer of the dual disk check valve recommended placement of the valve at least 5 pipe diameters downstream of any piping elbow or pump discharge "to reduce the likelihood of turbulent flow through the valve, which could shorten valve life due to component vibration."[10] Additionally, 40 m s^{-1} is significantly higher than the velocity limitations suggested for both impingement (16.3 m s^{-1}) and non-impingement (29 m s^{-1})[2] surfaces in oxygen service as shown in Figure 12. Thus, the flow rate acceleration caused by the internal components of the check valve led to a velocity increase that not only increased the wear of the internal components (potentially creating particles), but also left the stainless steel gate valve susceptible to particle impact ignition when compared against the velocity limitations for the stainless steel components.

Oscillation of the disks in the check valve likely provided the mechanism for mechanical failure of the valve. This mechanical failure caused parts of the valve such as the washers to pass downstream into the stainless steel gate valve. It also appears to have caused non-propagating local ignition of the combustion-resistant MONEL® and INCONEL® alloys. Although the damage to these alloy components was limited, the adjacent stainless steel components were not exempt from velocity limitations in high pressure O_2 systems.

---

2 At an assumed operating pressure of 400 psi.

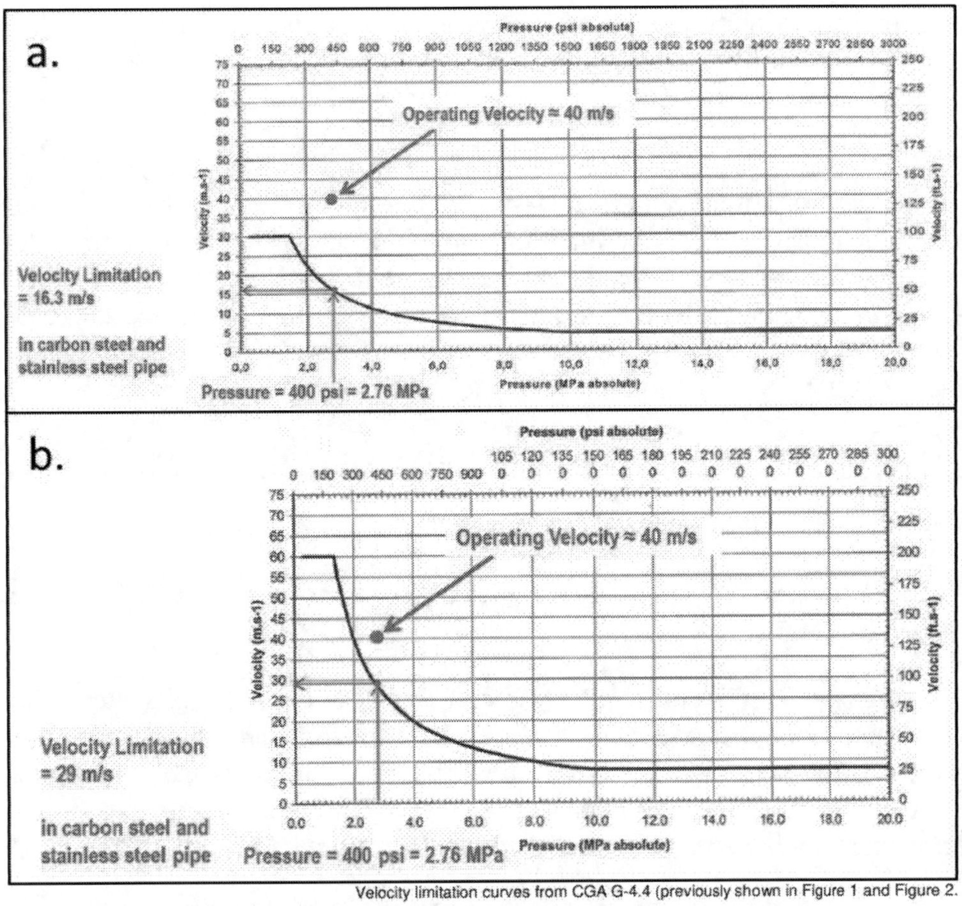

Velocity limitation curves from CGA G-4.4 (previously shown in Figure 1 and Figure 2.

Figure 12. Velocity limitations for impingement (a) and non-impingement (b) for oxygen flow at 400 psi. Note: Suggested velocity limitation for O_2 system pressure (400 psi) is indicated on the plots.

5 Conclusions and Recommendations

Based on the result of the investigation, several lessons were learned. The investigation identified complexities of oxygen flow and dual disk check valve performance that were not apparent to the original designers. Although the nominal bulk velocity for O_2 through the system fell below the recommended threshold for particle impact ignition, the complex internal geometry of valves drastically increased the local velocity. Additionally, the necessary velocity for maintaining the spring-loaded dual disk check valve pinned in the open position exceeded the maximum velocity limitation for stainless steel in oxygen service.

The system contained a combination of ignition resistant alloy components and stainless steel components. Even though ignition resistant alloys were used, the high velocity oxygen flow still facilitated non-propagating ignition of these alloys. Unanticipated and undocumented mechanical wear in the dual disk check valve components also provided the most likely source of ignition energy for the incident. Frictional heating at the bottom hinge-to-valve body interface and particle generation in the check valve are the most likely ignition sources to have caused the fire. Excessive wear to the internal dual disk assembly in the oxygen check valve was likely caused by fluctuations in flow velocity through the valve and operation below a minimum velocity required to pin the disks in the open position. Ignition of the stainless steel appeared to be a direct result of non-propagating ignition of adjacent MONEL® or INCONEL® alloy components.

As a result of the investigation findings, the preventative maintenance for the oxygen system was modified to include more frequent examination and replacement of internal components that were prone to wear. Technicians were also trained to identify and address signs of particle impaction inside the oxygen system during maintenance activities. More detailed analysis of the effects of piping and valve geometry on the oxygen flow dynamics was also implemented as part of the Management of Change program.

The following specific lessons were implemented to prevent future recurrence:

- Replace the SS316 gate valve with an INCONEL® gate valve.
- Replace all gaskets in oxygen service with gaskets that have a MONEL® inner ring for oxygen contact.
- Careful management, identification and control of materials of construction in critical services such as oxygen are vital. It is important to manage the materials originally selected for use, verify installation of those materials, and to manage later changes during subsequent maintenance activities. The special materials and their usage in oxygen systems must be documented and shared during training and certification programs for maintenance staff.
- Management of change, understanding what constitutes a change and the use of this system to prevent unwanted material, geometry, and operating changes is critical. Whenever a change is done it needs to be evaluated how this change will impact the system according to the MOC process. For example, in this case even small change such as allowing the purchase of a stainless inner ring gasket had significant impacts long after the change occurred.
- Careful documentation is required when working on, inspecting or repairing critical oxygen systems. The equipment owner must review and approve what was done (repair/ change/ no finding). This includes clear documentation of what was found, what repairs were made, what specifications were used, and verification of materials of construction. This will help ensure incorrect materials of construction/ specifications are not installed.

We would like to acknowledge the dedicated staff of EQUATE Petrochemical Co. and Exponent for conducting the detailed root cause investigation and sharing the learnings with the industrial fraternity for improvement of process safety and operations.

6 References

[1] CGA G-4.4, "Oxygen Pipeline and Piping Systems," Compressed Gas Association, 2012.

[2] ASTM G88-05, "Standard Guide for Designing Systems for Oxygen Service," ASTM International, November 2005.

[3] ASTM G94-05, "Standard Guide for Evaluating Metals for Oxygen Service," ASTM International, October 2005.

[4] Cox BL, Dee SJ, Hart RJ, Morrison DR. Development of a steel component combustion modelfor fires involving pure oxygen systems. Process Safety Progress 2014; 33(3):299–304.

[5] Sandmeyer Steel Company Specification for 316, 316L, 317L Stainless Steel. http://www.sandmeyersteel.com/images/316-316L-317L-Spec-Sheet.pdf. Note: Specification for 317L. Averages were calculated where ranges were reported in the specification.

[6] Benz, F. J., Shaw, R. C., and Homa, J. M., "Burn Propagation Rates of Metals and Alloys in Gaseous Oxygen," *Flammability and Sensitivity of Materials in Oxygen-Enriched Atmospheres: Second Volume. ASTM STP 910.* M. A. Bennings, ED., American Society of Testing and Materials, Philadelphia, 1086, pp. 135-152. Higher pressure consumption rates are listed in ASTM G94-05, Table X1.1, p 15.

[7] Slockers, M. J., and Robles-Culbreth, R, "Ignition of Metals at High Temperatures in Oxygen," Journal of ASTM International, Volume 3, NO. 9, 2006.

[8] Sato, J. and Hirano, T., "Fire Spread Limits Along Metal Pieces in Oxygen," Flammability and Sensitivity of Materials in Oxygen-Enriched Atmospheres: Third Volume, ASTM STP 986, D. W. SchroU, Ed., American Society for Testing and Materials, Philadelphia, 1988, pp. 158-171.

[9] Beeson H. D., Smith, S.R., Steward, W.F., Safe Use of Oxygen and Oxygen Systems: Handbook for Design, Operation, and Maintenance. ASTM International, 2nd edition, Baltimore, MD 2007.

[10] Crane Brochure for Duo-Chek II High Performance Check Valves, 2002

Lessons Learned from an Incident at a Cryogenic Gas Processing Facility

Adrian Pierorazio
BakerRisk
304-5515 North Service Road,
Burlington, Ontario, Canada
AdrianP@BakerRisk.com

Prepared for Presentation at
American Institute of Chemical Engineers
2015 Spring Meeting
11th Global Congress on Process Safety
Austin, Texas
April 27-29, 2015

UNPUBLISHED

Lessons Learned from an Incident at a Cryogenic Gas Processing Facility

Adrian Pierorazio
BakerRisk
304-5515 North Service Road,
Burlington, Ontario, Canada
AdrianP@BakerRisk.com

Keywords: Incident investigation, cryogenic, hot work, welding, coke oven gas

Abstract

In September 2009, a fatal incident occurred in a cryogenic gas processing plant. The investigation of the incident indicated a number of potential issues that may have contributed to or caused the event. These issues include hot work procedures, electrostatic discharge, electrical conduit sealing, convective "breathing" due to multiple vents, equipment age, maintenance, and worker training. This paper provides an overview of the process and facility, a timeline of events, a summary of the investigative process, and a discussion of the lessons learned from this event.

1 Introduction

Incidents that occur in industry often provide key learnings for other operators. While many of the specifics may change from site to site, it is valuable to consider how the lessons learned at one facility may apply to others, even if the processes are quite different. The incident being considered here is part of a coke oven system but illustrates a number of issues that have wider application. In this case, the process is a sublimation process to extract the fuel gases from the toxic and inert components of coke oven gas. The cryogenic and sublimation aspects are interesting, but should not detract from the key lessons learned. This paper starts with an overview of the process and then presents the facts about the activities underway at the time of the incident before moving on to a narrative of the event and an analysis of the technical causes of the incident. Some of the interpretations of the events may differ between the experts who investigated this incident, but the purpose here is to promulgate the lessons learned.

2 Process Description

The system was designed, built, and commissioned in the 1960s to convert coke oven gas to fuel gas for other parts of the complex. The system uses a cryogenic process to convert coke oven gas into three streams:

- Downriver gas
- Underfire gas
- Sub gas

The downriver and underfire gases are used as fuels in other areas of the complex. One part of the facility is downriver, giving rise to the "downriver" name. The underfire gas is a lower pressure fuel and is used as part of the fuel for the coke ovens. The sub gas contains the components that have been taken from the coke oven gas to produce the two product fuel streams. The sub gas is processed further in another area to remove more fuel gas and further concentrate the heavier components out of the stream.

The processing of the gas is performed using cryogenic separation. The coke oven gas is flowed over a bed of cold stones in a series of vessels called "regenerators." The stones are cold enough (as low as -258°F) to cause the gases to liquefy or freeze. The inlet flow is then turned off and the pressure is dropped. The control of pressure and the use of sweep gas cause the components to sublimate, allowing the capture of the three product streams by switching the outlet at the appropriate time.

The regenerators are clustered into groups of four, called "quads." Each quad has its own insulated "cold box" enclosing the bottom of the regenerators and the valves and piping there. These quad cold boxes are constructed with an insulated double wall with a void for the pipes and valves. The outer wall is made of steel plate that is seal-welded to hold it in place. The inner skin is made of thin metal sheets that are clipped in place. Mineral wool insulation is installed between these skins and along the floor and roof to help maintain the cryogenic process temperatures. Each quad is numbered, with a total of seven quads being installed in the main regenerator area.

The piping and valves that connect the quads together are housed in cold boxes at the end(s) of each quad. There are three such cold boxes and these are designated using letters A through C. These lettered cold boxes differ from the quad cold boxes in that they are packed with insulation rather than having an insulated double wall around a hollow space.

The lettered cold boxes (A through C) are connected through the quads and also through a "cross-tie" box that permits gas to bypass the quads and flow between the A and B Cold Boxes. In this report, the cross-tie box is referred to using the letters of the cold boxes at the ends of the cross-tie (i.e., "AB").

The cold boxes are designed to keep the space under the Quad 3 regenerators cold. This is done by attempting to seal the cold box to prevent air from being introduced into the cold box and to prevent gas from inside the cold box from leaking out.

The numbered cold boxes each have a nitrogen supply and backpressure regulator system. The quad boxes each have a single nitrogen injection point consisting of a 2-inch nominal diameter pipe penetrating the double-wall of the box. There are two 10-inch nominal diameter vents on the quad cold boxes, each equipped with a flapper-style relief valve to relieve overpressure and to allow air to be drawn in to relieve vacuum to prevent the cold box from collapsing in the event of a substantial vacuum forming in the cold box. One of the vent lines on each quad box is also fitted with a backpressure regulator with a separate open-ended vent line that is designed to maintain a positive pressure in the cold box. By contrast, the lettered cold boxes receive nitrogen from a sparger pipe and have a single vent pipe that runs up to roof level.

3 Summary of Facts

The following sections provide a summary of the key facts obtained during the investigation.

3.1 Activities at the Time of the Incident

Extensive maintenance activities had been underway since 2005 and were ongoing at the time of the incident. Quads 1, 2, and 3 and Cold Box A were in operation and work was progressing on the other equipment. Cold Box B was in the process of being rehabilitated, with the east half (i.e., the half connected to the south end of Quad 3) having been completed and returned to service weeks before the incident. The west half of the Cold Box B was being stripped of insulation at the time of the incident, in preparation for repair, re-insulating, and return to service. The insulation was being removed by two vacuum trucks that were parked outside of the main regenerator area with hoses running into the cold box.

Work was being performed in the main regenerator area on the day of the incident. Rock wool insulation was being vacuumed from the west side of Cold Box B. Grinding and welding repairs were being made at the south end of the AB Cross-tie Box adjacent to the north face of Cold Box B, on the south face of Cold Box B, and on top of the Quad 5 Cold Box. Hot work was being performed in the control room 2 building around the time of the incident. This included the insulating work in Cold Box B as well as welding and grinding work that was being performed to repair the wall plates of the cold boxes (Cross-Tie AB, Cold Box B and Quad 5 being worked on the day of the incident). There were numerous portable gas meters in use in the west half of the B Cold Box, under the AB Cross-tie Box, and on top of Quad 5 at the time of the incident.

Two work crews were in the area at the south end of the AB Cross-tie Box (i.e., on the north wall of the B Cold Box near the junction of B Cold Box and Quad 3). A welding crew was welding a plate at the intersection of Cold Box B and the AB Cross-tie Box, within a few feet of Quad 3's cold box. Another crew was working at ground level, preparing to mix concrete to refurbish a footing for Cold Box B at the time of the incident. It was reported that the welding crew on the north side of Cold Box B (approximately 50 feet from their return lead) had just struck an arc on the AB Cross-tie.

Quads 1, 2, and 3 were in operation, along with Cold box A and parts of the Cross-tie AB box and Cold Box B. At the time of the incident, one of the control valves in Quad 3 was switching.

3.2 Equipment Condition Prior to the Incident

The main regenerator equipment was being repaired to allow it to return to service and some cold boxes had been out of service for some time.

During the refurbishment project, the flapper valves at the top of the vent stacks were being inspected. Some of the flapper valves were repaired and/or replaced. Following the incident, some of the flapper valves were found to have eroded and the seals were missing.

The welding crew on the north side of Cold Box B was using one of the group of four individual Witherup welding machines that were stacked and/or bolted together near the road on the west side of the Control Room 2 area and shared a piece of angle welded to the opposite side of Cold Box B for their return lead location. The crew had welded a support to Cold Box B and had used a bottle jack to lift a member on the bottom of Cross-tie Box AB in order to restore a support clip for it back to the Cold Box B structure. The original connections had failed and new connections were being made as part of the refurbishment project.

During the investigation, one expert opined that parts of the system were beyond their lifespan. Whether this was true or not was subject to debate regarding the assumptions in the lifespan model. In the end, the company had implemented an inspection program to find and address cracks and any other flaws in the equipment and repair these when found. Many chemical processing facilities are of a similar vintage, or older, and rely on such inspection/repair programs to extend the life of their assets, so this is a common condition.

3.3 Safety Procedures

Standard operating procedures in the regenerator area required workers to use personal gas monitors to monitor for flammable gas, oxygen, and carbon monoxide as well as area monitors and periodic checks by operations with a hand-held monitor. A review of the logs and the results of testing of the personal monitors collected after the incident confirms that none of these gas detectors sensed alertable levels of flammable gas or carbon monoxide or an oxygen deficiency in the area prior to the incident.

There was a specific requirement for patching the live cold boxes to be performed as "cold work" with a gas sampling port and a nitrogen purge port to allow the atmosphere behind the plate to be sampled and then purged before hot work (i.e., welding) could commence under a separate hot work permit. The welding crew was performing hot work at the junction of the High Box and the Cross-tie Box, specifically on the Cross-tie Box, using hot work, with no separate permit, no purge, and no continuous monitoring of the sample port.

3.4 Damage Sustained

On September 3, 2009, an "explosion" was reported in Quad 3, resulting in one fatality and one worker being sent to the hospital for treatment. The fires that followed the event burned for a number of hours and were concentrated in the area around and under the Quad 3 regenerators.

Inspections after the event showed that the Quad 3 Cold Box floor had been deformed outward to the point that it impacted the concrete below, and the walls of the Quad 3 Cold Box had become detached and displaced/deformed outwards on both sides of the cold box to the point where they were no longer enclosing the space underneath.

A double wythe brick infill wall to the east of Quad 3 failed and fell into the axi compressor building, away from the Quad 3 Cold Box.

The damage was generally confined to the immediate vicinity of Quad 3. The only significant structural damage observed in Cold Boxes A and B were to the walls that separate them from Quad 3. This damage resulted in wall movement that moved the pipes running between these lettered boxes and Quad 3, resulting in some mechanical damage to the valves and piping in the lettered boxes. The walls were deflected away from the Quad 3 Cold Box.

The damage patterns are consistent with an "explosion" having occurred inside the Quad 3 Cold Box.

4 Interpretation of the Event

The investigation concluded that the incident that occurred on September 3, 2009 was a confined vapour cloud explosion (VCE). The limited extent of the damage indicates that the energy involved was similarly limited. The damage is consistent with a relatively small vapour cloud, and the intensity of the overpressure also indicates that the VCE was a deflagration rather than a detonation.

A vapour cloud explosion is a combustion event and, as such, requires three elements to be present in sufficient quantity:

- Fuel
- Oxidizer
- Ignition

Each of these elements must exist in balance with the others. The fuel must exist in a concentration between the lower and upper flammable limits, the oxidizer (oxygen from the air in this case), must exist at above the minimum oxygen concentration (MOC) required to support combustion, and the ignition source must be in the correct location and of sufficient strength to produce a self-sustaining combustion. Each of these elements is discussed in the sections below.

4.1 Fuel Source

The underfire, downriver, and sub gas lines run through the Quad 3 Cold Box. These lines run at cryogenic temperatures. When these lines are serviced and the quad is brought on-line, the pipes are cooled to below -250 °F. The stresses due to cryogenic processes can cause flanges and other connections to leak. When these leaks occur, they are generally small. Even small leaks in the relatively sealed space of the cold box can produce accumulations of gas and the concentration can increase over time. After the incident, the piping was purged with nitrogen and maintained at a slight positive pressure of nitrogen. As part of their investigation, handheld quad gas meters were used to detect locations with low oxygen readings. These low oxygen areas were identified as areas where nitrogen was being dispersed from the piping. Several such areas were identified and it was concluded that, as expected, there were small leaks in the piping that could allow gas to escape. While flow rates were not measured, it was concluded that these were a credible source of fuel for the event.

4.2 Oxygen

As designed (and as in operation on the day of the incident), the nitrogen is introduced at a single point from the cross-tie alley (i.e., on the west of the odd numbered quads and on the east of the even numbered quads) at the south end of each quad cold box. With the erosion of the vent flapper valves, the nitrogen supply was unable to provide a significant positive pressure in the cold box. The leak in the flapper valves could result in the venting of the incoming nitrogen through the nearest vent (located just south of the C regenerator) and the nitrogen would not sweep through the cold box to purge the air.

With hydrogen-rich gas (i.e., underfire or downriver gas) leaking from the process piping, the gas mixture would have a lower density (i.e., be "lighter") than normal air. The result of this was that the gas would rise out of the vents creating a convection loop in which the light gas rises out of the cold box and denser oxygen-laden air entered the cold box to replace it and providing the oxygen for combustion to take place. Calculations show that this mechanism is a credible source of oxygen to support combustion in the Quad 3 cold box, if sufficient fuel is present along with a competent ignition source.

4.3 Ignition Source

With fuel and oxygen present in the cold box the only remaining required element is ignition. A number of potential ignition sources were considered including the following:

- Electrical equipment or wiring inside the quad 3 cold box
- Electrical equipment or wiring outside the quad 3 cold box
- Static electric sparks from the vacuum truck operations
- Static electric sparks from within the quad 3 cold box
- Ignition due to hot surfaces in the quad 3 cold box
- Ignition due to deformation/failure of metal components in the system
- Ignition due to stray currents induced by process equipment in the control room 2 area
- Ignition due to stray currents induced by the welding in operation at the time of the incident

The following sections provide a summary of how each potential ignition source was evaluated.

4.3.1 Electrical equipment or wiring inside the quad 3 cold box

One of the first hypotheses considered was that there could be a short or malfunction of electrical equipment or wiring in the quad 3 cold box. The only wiring that was found to run inside the cold box itself was thermocouple wiring to measure the process temperatures. The voltage expected from a thermocouple under normal operating conditions in the cold box would be less than 10 mV and would be a static voltage; this is well below what would be required to ignite a flammable gas cloud.

4.3.2 Electrical equipment or wiring outside the quad 3 cold box

As a corollary to the scenario above, there is an amount of electrical equipment outside of the quad 3 cold box that does operate at voltages of 120 volts or more. Since the wiring and equipment are in an electrically classified area, the wiring is routed inside of conduit and junction boxes intended to prevent potentially flammable gas mixtures from being present where significant voltages and/or currents were present. To investigate this, the junction boxes near the cold box were opened and inspected. It was observed that the seals between the quad 3 cold box were of an elastomeric caulking-like material, rather than the cast solid material that would normally be used in classified conduit. It was further revealed that some of the locations where cast or poured seals would be expected (i.e., where conduit is connected to a junction box) were missing seals or had gaps in the seals.

The conduit was analyzed to determine the minimum gas concentration that would produce a flame that could propagate through the conduit that connected to the thermocouple junction box. The pressure produced by the deflagration of this minimum gas concentration was then predicted and compared to the pressure capacity of the conduit and junction boxes. The analysis shows that any flammable gas concentration that can propagate through the conduit would cause failure of the junction boxes and/or conduit. Since these components did not fail due to overpressure and since there were no other signs of combustion inside the conduit and junction boxes and since there were signs that there were functional seals between the cold box and the electrical system outside the cold box, it was concluded that this was not a credible ignition source.

4.3.3 Static electric sparks from the vacuum truck operations

Work crews were working on west side (quad 4 side) of Cold Box B at the time of the incident. As part of this work, they were using a vacuum truck to suction the old insulation away to clear the area for rehabilitation. The vacuum hose ran east under quads 5 and 6, along the east side of quad 5, then under cold box B to the work location. The hose was not conductive hose and so there is a potential for static buildup and discharge from the hose. Static discharge is a high voltage/low current phenomenon which occurs locally where a static source with high voltage potential creates an arc to ground. In this case, the potential ignition locations would be inside the pipe (charged particle to hose) or along the path of the hose. Since there was no report of fire or ignition at the work site or vacuum truck, it can be surmised that there was no deflagration propagating in the hose. Since the hose was generally lying on the ground or near quad 4/cold box b, any local discharge from hose to structure would be unable to ignite a flammable gas mixture in quad 3. As a result, ignition from static discharge due to vacuum operations was ruled out.

4.3.4 Static electric sparks from within the quad 3 cold box

Potential sources of static electric discharge inside the quad 3 cold box were sought. These included moving parts that were well insulated from any metallic structure, and any particles or gases that can flow in such as was as to develop an electrical charge. None of these potential sources were found and so this source of ignition was ruled out.

4.3.5 Ignition due to deformation/failure of metal components in the system

It has been postulated that the ignition may have occurred as a result of deformation or failure of a component. The only mechanism for such an event to occur inside the quad 3 cold box is a gross failure of the process equipment. When the process equipment was inspected, the damage was consistent with mechanical failure due to the rupture of the cold box under positive pressure. The only evidence of any component that may have failed prior to combustion is the bellows at the bottom of regenerator 3a. However, this failure appears to be inwards and not outwards, indicating that it is more likely due to impact during the failure of the cold box, than due to rupture prior to the incident. In any event, the deformations in this area are relatively small and not credible as being sufficient to heat the metal close to the autoignition temperature of the fuel.

4.3.6 Ignition due to stray currents induced by process equipment in the control room 2 area

With ignition due to electrical, chemical, and mechanical systems inside the quad 3 cold box ruled out, attention turned to how operations outside the cold box could produce ignition inside the cold box. The most logical of these would be for ignition to occur as a result of electrical current flowing through the cold box from external sources.

Elementary electrical theory holds that the current between parallel paths will divide as the reciprocal of the resistances of the various paths. Stated more simply: current will flow through all available paths from a source to a sink (often ground) with low resistance paths getting a higher portion of the flow than high resistance paths. The current that flows through alternative paths such as along structural elements and process piping is referred to as "stray currents".

The cold boxes run at cryogenic temperatures, making the resistance of paths though the steel of the structure and piping very low. This makes paths through the cold boxes the preferred path for currents flowing through the structure. A search was undertaken to find electrical equipment that either did not have a viable local ground connection (to cause currents to need to go through the structure), or was capable of providing high enough currents to the system that even a small fraction of the current would be significant enough to cause ignition.

In the process area, there are a number of compressors which use high voltages and currents. These are all a fair distance from the cold boxes. The compressors were tested found not to be shorted and had reasonable connections to ground to account for stray currents produced locally. These were then ruled out.

The lighting and electrical switches in the regenerator area are much closer to the cold boxes (implying lower resistance to the cold boxes), but none were found to have shorted to the structure or otherwise lack a proper electrical return path. As a result, these were ruled out as likely ignition sources.

4.3.7 Ignition due to stray currents induced by the welding in operation at the time of the incident

In evaluating electrical sources outside of the cold boxes themselves, the welding activity being conducted at the time of the incident was also considered as a potential source of appreciable stray currents. In welding operations, currents as high as 100 amperes, or more, flow through the components being welded together. The current then flows through the metal to the return lead to complete the circuit. At the time of the incident, a crew was welding on the north side of the B cold box at the west side of the intersection with the crosstie box (i.e., approximately two stories up, very close to quad 3). The return lead (colloquially, if inaccurately, referred to as the "ground clip"), was located near the middle of the south wall of the B cold box near ground level. This means that the full welding current was being passed through a large part of the cold box structure as a result of the way the return clip was located. The combination of high current source, low conductivity path through the quad 3 cold box, and the long path to the return lead, indicates that there is a high potential for substantial currents to be flowing throughout the quad 3, crosstie box, and cold box b areas.

Witnesses reported that the ignition occurred at virtually the same time that a welding arc was struck on the AB Cross-tie Box. Electrical testing post-incident revealed that welding being performed in the plant at that time was producing stray electrical currents that were strong enough to interfere with electrical testing being performed. This indicates that the welding practices in place at that time were transmitting energy to areas away from the work location.

Locating the ground return lead on the far side of the B Cold Box approximately 50 feet from the welding location, increased the energy sent to other areas of the system. Since it was shown that stray current paths existed in the cold boxes, that the welding was being performed in a way that would encourage the creation of significant stray currents along parallel paths in the cold boxes, that there was a correlation in time between the start of the welding near Quad 3 and the incident, and since all other ignition mechanisms considered were ruled out, it is concluded that the welding is the most probable cause of ignition.

5 Summary and Lessons Learned

While the processes described in this paper are not typical of wider chemical industry, there are some lessons learned that can be applicable to other processes. As with any incident, the root cause is often a failure of the PSM system and this incident was no exception.

5.1 Training and Contractors

Despite documentation and training to the contrary, the welders (contractors) did not seem to be aware of the hazard of welding on an enclosure that could contain flammable gas. Their personal gas monitors would not be able to detect a flammable gas mixture inside the cold box from their location outside of the box and the welding current was able to penetrate the skin of the cold box easily. The safety personnel were responsible for monitoring the gases in the area and did not note this discrepancy in the sample location.

5.2 Hot Work

This incident appears to be a failure of the hot work permitting system in that the sampling and welding practices did not prevent ignition. Safe procedures were in place, but not followed.

5.3 Mechanical Integrity

As with many aging facilities, erosion, wear, and fatigue were present in this facility. While active testing and repair programs corrected many issues, the erosion of the flapper valves was critical to the safety of the operation.

Post-incident inspections and testing demonstrated that the grounding was disconnected in some places. While this was not likely a cause of the incident, it is valuable to note that this electrical issue may occur in other process areas and may go undetected unless specifically sought out. While it is not specifically mechanical, the observation that the seals in the classified electrical conduit and junction boxes were incomplete, missing, or sealed in a manner that is not consistent with good practices (i.e.,. using an elastomer rather than a castable seal) is a valuable lesson as well. The conduit and junction boxes had been installed many years prior and, it seems, had not been inspected to determine if all seals were in place. This may have been found if the seals were inspected.

5.4 Human Factors

The flapper valves were located at the top of exhaust stacks, well above the roof of the process building. To access these valves required a crane and basket and so the equipment was not regularly inspected or maintained. As such, personnel were unaware of the erosion of the valves, which led to the presence of air in the cold box.

5.5 Hazard Identification

The design, as received from the technology licensor included two large vent stacks. The presence of two stacks, rather than one, allowed a convection loop to form. The designer and operator failed to identify this hazard and modify the design appropriately.

The nitrogen purge system, coming in at a point rather than being distributed by a sparger, allowed pockets of air to remain. This hazard was not identified prior to the incident.

The welders were not aware of the hazards of welding on live cryogenic equipment which may contain flammable gases. There was a belief that their flammable gas monitors would have alarmed if flammable gas was present in the area.

The welders were unaware of the path taken by their welding current. While there was general understanding that the current would follow the path of least resistance, this was interpreted as the shortest distance through the steel without considering parallel paths or the decreased resistance of steel at low temperatures.

5.6 Incident Investigation

The cold box had a much smaller event four years prior to this incident. Investigators at that time concluded that it was caused by lightning strike, and so recommended lightning rods be installed to prevent a second strike. While there is debate about the veracity of this cause of ignition, the earlier investigation could have identified the fuel and air sources and recommended ways to control them. Had the original investigation found the eroded flapper valves and identified convection as a means of introducing air into the cold box, the valves could have been replaced or the vent piping re-configured to prevent this from occurring and so could have prevented the event.

11TH GLOBAL CONGRESS ON PROCESS SAFETY

1996 Amines Plant Explosion

Jerry Forest
Celanese
1601 W. LBJ Freeway, Dallas, TX 75234
Jerry.Forest@Celanese.com

Prepared for Presentation at
American Institute of Chemical Engineers
2015 Spring Meeting
11th Global Congress on Process Safety
Austin, Texas
April 27-29, 2015

UNPUBLISHED

1996 Amines Plant Explosion

Jerry Forest
Celanese
1601 W. LBJ Freeway, Dallas, TX 75234
Jerry.Forest@Celanese.com

Keywords: Incident, PSE, explosion, Lessons Learned

Abstract

Process safety management systems should be in place to ensure repeatability in operations. A good system consists of higher level procedures that describe what to do. At the plant level, the management system describes who does the activity and how it is done. Following procedures leads to predictability in operations. In 1996, Celanese had established a rigorous management system that consisted of 21 elements of process safety. However, failure to rigorously apply several elements led to an explosion in our Cangrejera facility resulting in 2 fatalities. This paper discusses the management system failures of conduct of operations elements, safe operating limits, standard operating procedures and training, and critical safety equipment management and how each failed due to a common weakness of management systems.

1. Introduction

In 1987, Celanese had an event that fundamentally changed the way we manage process safety. An explosion at our Pampa facility resulted in three fatalities and 39 injuries. Not only were the lives of those directly involved severely impacted, the company made a substantive shift in how we define and practice process safety.

Corrective action from this incident led to the development of a comprehensive 21 element process safety model; similar to the one that the CCPS later and independently developed with 20 elements. A strong driver for Celanese in the development and implementation was the recognition that leaders drive change and create the culture.

Despite having implemented such an aggressive approach to process safety, an explosion occurred at our Cangrejera Mexico facility nine years later that resulted in two fatalities. While the management systems were in place, they were not rigorously followed. As with many process safety incidents there were several opportunities for existing systems and controls to prevent the explosion. However, each of them failed. There was a mode failure of our process safety culture, as it allowed partial bypasses in the process safety management system.

2. The Incident

Among other chemicals, Celanese produces amines at our facility in Cangrejera Mexico. This unit uses ammonia and methanol as raw materials. The reaction that produces the amines is endothermic and relies on an electrical heat exchanger to provide the necessary heat required to initiate and maintain the reaction.

On June 7, 1996, the facility experienced a brief electrical power outage. The temporary interruption caused a loss of cooling water and a loss of power to the electrical heat exchanger that provided the necessary heat for reaction for the amine unit. During the course of the outage a pressure relief device on a nearby distillation column lifted and failed to reseat. The amines unit therefore had to be shut down rather than start-up from warm, in order to re-seat the pressure relief device.

Standard operating procedures were in place to cut power to the reaction section electrical heat exchanger during a shutdown. However, due to a communication error between the outside operator and the board operator, this did not occur. During the course of the outage, power and therefore heat, continued to the unit exceeding the maximum allowable working temperature of the heat exchanger resulting in a rupture in the shell wall.

Reactor contents consisting of ammonia and methanol were released. The resulting vapor cloud found an ignition source from a nearby welding operation. The ensuing explosion fatally injured the welder and fire watch. Ignition of the reactor contents propagated back to the heat exchanger source resulting in a secondary jet fire.

3.0 Primary Causes

As is typical with many process safety incidents that have severe consequences, there were several layers of protection that failed allowing this incident to occur. They include a failure of conduct of operation elements; failure to recognize a safe operating limit challenge; failure to follow standard operating procedures and training on those procedures; and failure to manage critical safety equipment.

3.1 *Conduct of Operations*

Conduct of operations (COO) concerns how work is performed in a manufacturing unit with the overall objective of producing consistent, repeated and desired results. COO reduces the likelihood of human error. The term operating discipline is often used to describe how to bring a consistent approach to various operations to minimize errors. [1]

Proper communication between and among operators, crafts, and supervisors is inherent to a good deal of how work is performed in a unit. Communications should be crafted in a standardized format that includes positive verification and acknowledgement. [2] The Celanese COO management system includes requirements for closed loop communications and defines the standard format that leads to proper communication for operators.

At the time of the unit shutdown described above, a communication error occurred between the outside and inside operators regarding shutting off power to the electric heat exchanger. The COO closed loop communication requirement in our management system failed. Had the communication taken place as intended, this incident would not have occurred.

3.2 ***Safe Operating Limits***

A second level of protection from this type of incident is found in defining true safe operating limits (SOL) that allow for pre-determined action by the operator to prevent an incident. Associated with the SOL are alarms and operator training on the consequence if exceeded and the actions that should be taken to prevent the consequence from occurring.

Since power was on the electrical heat exchanger during the course of the outage, the temperature in the process exceeded the maximum allowable working temperature of the heat exchanger. Failure to set a proper SOL with an associated alarm, and failure to train operators on the consequences of deviation and predetermined actions voided this layer of protection from preventing this incident.

3.3 ***Standard Operating Procedures***

The third level of protection is standard operating procedures (SOPs). SOPs should be followed as written each time an activity is performed. At the time of the Cangrejera incident, the unit shutdown SOP did define that power should be turned off to the electrical heat exchanger to prevent the consequence of vessel rupture by exceeding the maximum allowable working temperature. However, this step in shutdown did not occur.

3.4 ***Management of Critical Safety Equipment***

Communications, SOLs, and SOP layers of protection are human interactions that prevent process safety incidents. Engineered controls are also in place to prevent them.

Specifically, a typical design for this type of heat exchanger includes both a low flow and high temperature interlock. Either interlock would remove power from the electrical heat exchanger in this shutdown situation and prevented the incident. However, in this instance the temperature interlock was removed due to poor reliability without a management of change. At the time of the shutdown, the low flow interlock was bypassed – again, without a management of change.

3.5 ***Management System Failure***

How can so many levels of protection fail at the same time? As is seen with so many major process safety incidents, the answer lies in failure of the overall process safety management system. In this case, we had the process safety systems in place: COO, SOLs, SOPs, and properly designed interlocks. The common failure mode was lack of a process safety culture that insured two-way communication was always followed, safe operating limits were set and challenges investigated, safe operating procedures were always followed, and engineered controls were operable.

4.0 Response to the Incident

As a result of this incident Celanese took corrective action to update and reinforce our COO program. In particular, COO now includes a process safety officer (PSO). Each site identifies process safety critical activities that need special attention. Examples include critical startups and shutdowns where we might be more vulnerable to a process safety incident. A PSO is assigned and has the following duties:

- Present during safety critical startup, shutdowns, and other operating modes identified in the PSO reference manual
- An in-depth knowledge of the hazards of the process
- A working knowledge of unit operations
- An understanding of all applicable SOPs
- **Have no other duties that distract from their PSO responsibilities**

As applied to the amines plant explosion, the PSO would both observe two-way communication to ensure proper transmission and understanding and verify that the communicated activities are properly executed. The PSO would also verify the SOP was followed, monitor the SOLs, and understand the effect of any safety critical equipment bypasses and alternate mitigations.

Another key action taken is that all SOPs are risk ranked using our risk scoring tool. The ranking identifies how frequently tasks are performed in an SOP and the possible process safety consequences that could result if the tasks are done incorrectly or not followed. SOPs are then classified as critical, reference, and guideline.

Critical and reference SOPs must be followed as written. Critical SOPs are written in a checklist format, must be present and in hand during execution and each step is signed off by the operator. When complete, a responsible second party such as PSO or front line supervisor independently validates that all activities are complete.

Finally, MOCs are required for critical equipment bypassing, and these bypasses must be posted on a bypass board in our control rooms. In-service by-pass is not allowed in Celanese unless there is another variable that can be monitored with the same intent of the one being bypassed. More recently, Celanese has recognized that knowledge of past incidents has a definite life. If we are to prevent the mistakes of the past, we must somehow ingrain both knowledge of the incident and the corrective actions that were put in place. Thus, our process safety lessons learned program has been put in place to reinforce our process safety culture.

5.0 About Celanese Process Safety Lessons Learned Program

The fundamental premise of this program is to go beyond incident communication and deeply embed and institutionalize knowledge of incidents and the mitigations that resulted from them in the organization. This is done by a constant reinforcement of the message or to state another way, by "retelling our story". This concept of defining and strengthening culture comes from two sources. First, Trevor Kletz is often quoted that "organizations don't have memory, people do." [3] That combined with reference materials from Loyola University on how group cultures are formed and strengthened indicates that in order to define and build a culture, the story must be defined, presented and repeated. This therefore is the basis of our Lessons Learned program.

In order to achieve maximum effectiveness, our program is divided into five parts.

1. Global Lessons Learned – The major incidents in Celanese that have defined how and why we do process safety

2. PS Rules (Process Safety Rules) – Technology specific process safety rules that define the critical few incidents that could occur and what mitigations are in place to prevent
3. Event Sharing – A comprehensive catalog of all Tier 1 and Tier 2 process safety incidents that have occurred at all Celanese sites since 2010.
4. Industry Learnings – A catalog of resources with learnings from industry incidents that are applicable to Celanese
5. PS BITS (Process Safety Basic Information to Share) – short informal videos submitted by employees that encourage employee participation in the learning process.

5.1 *Global Lessons Learned*

The intention of the Global Lessons Learned is to define for each manufacturing employee why we do process safety by telling the story of the major incidents that have occurred at our sites from our past. We do this by producing high quality videos (similar to CSB) that are shown to each manufacturing employee each year. Our intention is to limit this portion of the program to a few videos so that we do not reduce the impact of each. To date, we have produced:

1. Pampa Plant Explosion, 1987 – The result of this incident was the development of our comprehensive 21 element process safety program.
2. Cangrejera Amines Plant Explosion, 1996 – Reinforces the need for strong process safety management systems to prevent incidents
3. Hot Work, Shelby 1999, and Ocotlan 2014 – Reinforces fundamentals of life critical hot work procedures and hazard recognition
4. Organic Peroxide Explosion, Edmonton – Reactive hazards (in production)

Each video has subtitles in the languages for the various countries in which we operate. Some are overdubbed in the native language depending on those countries wishes to do so. The videos use animation, historic video clips, pictures, and interviews to convey the story of the incident, the causes, the key learnings and follow-up actions from each.

5.2 *PS Rules (Process Safety Rules)*

Celanese has several groups of similar technologies that in general, have similar risks and mitigations of risks. The PS Rules Lessons Learned is an attempt to target operators with the key hazards of the process by developing a short, memorable "rule." In order to be effective and follow the premise of Lessons Learned, the PS Rules have background training material that describes the hazard and the engineered and administrative controls in place to prevent. To ensure that operator can commit the PS Rules to memory:

- There are no more than 5 rules per technology
- They are written in such a way so that they are memorable using "Always do x", or "Never do y".

An example rule for our ethylene vinyl acetate technology is "Never mix vinyl acetate with initiators". The training materials then include the possible hazard created by this mixing and what controls are in place to prevent an incident. The PS Rules become part of refresher training for operators.

5.3 *Event Sharing*

This portion of the Lessons Learned program follows industry's traditional incident sharing program. For each Tier 1, Tier 2, and serious near miss, a single Power Point incident flash is produced and distributed to each site. The purpose of the flash is to raise awareness that an incident has occurred and provide an opportunity for each site to determine if action should be taken at their site. The corporate process safety department determines if a global teleconference call is needed to broadly communicate the learnings of the incident.

What might separate our event sharing program from our peers is the way we catalog the incident flashes on our Lessons Learned portal. They are listed by technology and site and have several uses. A primary use of the portal is to prepare for PHA by identifying and analyzing incidents of interest to the PHA team. We also encourage use of the event sharing in our New Leader Process Safety Certification Program which requires a review of historic incidents for certification. Other programs where this portal is regularly accessed include the pre-work requirements for our front line supervisor (FLS) certification program, MOC hazard reviews, supplemental operator training, and toolbox meeting topics for FLSs.

5.4 *Industry Learnings*

As stated above, our PHA process requires review of incidents that are related to the processes that we operate, both internal and external to the company. Furthermore, each site is encouraged to stay current with industry incidents in order to determine applicability for our sites and lessons that can be applied to prevent similar incidents.

This portion of Lessons Learned is a convenient repository of reference materials, pictures, videos, and investigation reports for external incidents that might have applicability for our manufacturing sites. Examples include Bhopal and Texas City.

In addition to those incidents that have helped define process safety, we publish a quarterly event sharing newsletter. The newsletter highlights similar incidents that have recently occurred in industry and compares and references them to similar incidents that have occurred at Celanese. A major source of industry event information is the AFPM Advancing Process Safety Event Sharing Database. Each manufacturing site is expected to consider applicability of learnings at their site.

5.5 *PS BITS*

PS BITS, or process safety basic information to share, is the newest addition to our overall Lessons Learned program. PS BITS engages employees by having them produce short video segments on process safety topics. These clips are less than 5 minutes, can be recorded with a smart phone and edited with common tools such as Microsoft Movie Maker.

The key to this portion of Lessons Learned is that employees are telling the story in their own words for other employees.

6.0 Conclusion

Good process safety management systems are in place to understand and manage the risks of the process. By understanding the risks, we define the layers of protection to prevent consequences.

Managing the risks includes assurance that both the human controls, and engineered controls are in place and working.

The common element to understanding and managing risks is the commitment to process safety. This commitment takes the form of a leadership driven culture of process safety and ensuring that process safety management systems are in place and are not compromised. Feedback to the commitment comes in the form of learning from process safety incidents and acting to improve.

The explosion at the Celanese amines plant illustrates how these four pillars of process safety introduced by CCPS must be in place to prevent incidents. [4]

Reference

[1] J. Forest, Management Discipline, Process Saf Prog XX (2014)

[2] Center for Chemical Process Safety, Conduct of Operations and Operational Discipline, American Institute of Chemical Engineers, Newark, NJ, 2011.

[3] T. Kletz, Learning from Accidents in Industry, Butterworths, London, 1988

[4] Center for Chemical Process Safety, Guidelines for Risk Based Process Safety, American Institute of Chemical Engineers, Hoboken, NJ, 2007

11TH GLOBAL CONGRESS ON PROCESS SAFETY

U.S. Chemical Safety Board's Final Investigation Report on the Chevron Richmond Refinery Pipe Rupture and Fire

Mark Wingard
Chemical Safety Board
Denver, CO
mark.wingard@csb.gov

Steve Cutchen
Chemical Safety Board
Denver, CO
Steve.Cutchen@csb.gov

Lauren Grim, Dan Tillema, Amanda Johnson
Chemical Safety Board
Denver, CO

REPORT NO. 2012-03-I-CA
JANUARY 2015

Presentation at
American Institute of Chemical Engineers
2015 Spring Meeting
11th Global Congress on Process Safety
Austin, Texas
April 27-29, 2015

U.S. Chemical Safety Board's Final Investigation Report on the Chevron Richmond Refinery Pipe Rupture and Fire

Mark Wingard
Chemical Safety Board
Denver, CO
mark.wingard@csb.gov

Steve Cutchen
Chemical Safety Board
Denver, CO
Steve.Cutchen@csb.gov

Lauren Grim, Dan Tillema, Amanda Johnson
Chemical Safety Board
Denver, CO

Keywords: enter a few keywords here to aid searching

Abstract

On August 6, 2012, the Chevron U.S.A. Inc. Refinery in Richmond, California (Chevron Richmond Refinery) experienced a catastrophic pipe rupture in the #4 Crude Unit. The incident occurred from piping referred to as the "4-sidecut" stream, one of several process streams exiting the refinery's C-1100 Crude Unit Atmospheric Column. The pipe rupture occurred on a 52-inch long component of the 8-inch 4-sidecut line. The rupture of the 4-sidecut piping was the result of the 52-inch component being extremely thin due to unidentified, accelerated sulfidation corrosion in the low-silicon piping component.

The ruptured pipe released a flammable, 640°F light gas oil, which then partially vaporized into a large, opaque vapor cloud that engulfed 19 Chevron employees. Approximately two minutes after the release, the flammable portion of the vapor cloud ignited. Eighteen of the employees safely escaped from the cloud just before ignition; one employee, a Chevron refinery firefighter, was inside a fire engine that was caught within the fireball when the vapor cloud ignited. Because he was wearing full body fire-fighting protective equipment, he was able to make his way through the flames to safety. Six Chevron employees suffered minor injuries during the incident and subsequent emergency response efforts. The leak resulted in a large plume of vapor which traveled across the Richmond, California, area. The ignition and subsequent burning of the hydrocarbon process fluid created a large cloud of particulates which also swept across the Richmond area. In the weeks following the incident, approximately 15,000 people from the surrounding communities sought medical treatment at nearby medical facilities for ailments including breathing problems, chest pain, shortness of breath, sore throat, and headaches. Approximately 20 of these people were admitted to local hospitals as inpatients for treatment.

This report details the CSB's technical findings on sulfidation corrosion and inspection techniques, as well as delves into the Chevron organizational and Chevron emergency response factors that contributed to the incident.

http://www.csb.gov/assets/1/19/Chevron_Final_Investigation_Report_2015-01-28.pdf

11TH GLOBAL CONGRESS ON PROCESS SAFETY

Bhopal 30 Years Later

Ronald J. Willey, PE, PhD
Dept. of Chemical Engineering, Northeastern University
313 SN, 360 Huntington Ave., Boston, MA 02115
r.willey@neu.ecu

Daniel A. Crowl
Dept. of Chemical Engineering, Michigan Technological University
1400 Townsend Drive, 203 Chem Sci & Eng. Bld. Houghton, MI 49931
crowl@mtu.edu

Prepared for Presentation at
American Institute of Chemical Engineers
2015 Spring Meeting
11th Global Congress on Process Safety
Austin, Texas
April 27-29, 2015

UNPUBLISHED

in papers or printed in its publications

Bhopal 30 Years Later

Ronald J. Willey, PE, PhD
Dept. of Chemical Engineering, Northeastern University
313 SN, 360 Huntington Ave., Boston, MA 02115
r.willey@neu.ecu

Daniel A. Crowl, PhD
Dept. of Chemical Engineering, Michigan Technological University
1400 Townsend Drive, Houghton, MI 49931
crowl@mtu.edu

Keywords: Bhopal, Layers of Protection, Process Safety Management, Management of Change

Abstract

On the early morning of December 3rd 1984 a large release of methyl isocyanate (MIC) gas occurred in a pesticide production plant located in Bhopal, India. The plant was partially owned by Union Carbide (USA) and its affiliate Union Carbide India Limited (UCIL). Thousands of people died and approximately 200,000 were exposed to the MIC gas in that early morning.

This paper reviews the prevailing accident theories, the chronology of events that occurred after the accident, the industry and regulatory response after the accident, and the current state of the Bhopal site.

1. Introduction

For those us practicing chemical engineering in 1984, it is hard to believe that 30 years has passed. Worldwide, every practicing chemical engineer was influenced by the process safety incident that unfolded in Bhopal, India on December 2 and 3, 1984. Over 2,500 lives were lost due to inhalation and exposure to methyl isocyanate vapor released from a Union Carbide India Limited pesticide plant. 200,000 people suffered various levels of exposure that ranged from death, to blindness, to nausea [1]. As the horrifying photographs of the chaos reached the news media in the United States, the US public was drawn in. Our profession was challenged. Are we really the premier engineers that our salaries command? Or, are we another group that puts production above safety? The incident woke up our political system. Can it happen here? What regulations are in place to prevent a Bhopal from happening again and especially within the USA? The outcome: the establishment of the Center for Chemical Process Safety within the AIChE [2], the enactment of OSHA 29 CFR 1910.119 [3] Process Safety Management (PSM) of Highly Hazardous Chemicals, and EPA 40 CFR Part 68 [4] Sub Part G Risk Management Plan (RMP). Specific governmental requirements were enacted on chemical plants to demonstrate to the public that they can operate safely. The U.S. Chemical Safety Board [5] (CSB) an independent governmental agency was formed. Its mission is to investigate major chemical industry incidents, and share the outcomes such that future incidents are prevented. There is not one chemical engineer reading this manuscript who has not been affected by the incident in Bhopal and the resultant actions taken to prevent such again. This manuscript reviews the technical aspects of the accident and discusses what has transpired over 30 years. The intended audience is the safety and health practitioner who joined the profession after 1985.

2. The Accident

2.1.Exothermic Reactions in Closed Vessels

Explanation of this accident begins with chemistry, the storage of a very reactive chemical inside a 15,000 gallon, 40 foot long, 304 SS storage tank, and contamination by water. When chemical reactions form more stable molecules from more energetic molecules the result is the release of bond energy as heat. The energetic molecule involved: methyl isocyanate, or MIC as it has become abbreviated in many articles. Methyl isocyanate has two adjacent double bonds as shown in Figure 1 [6].

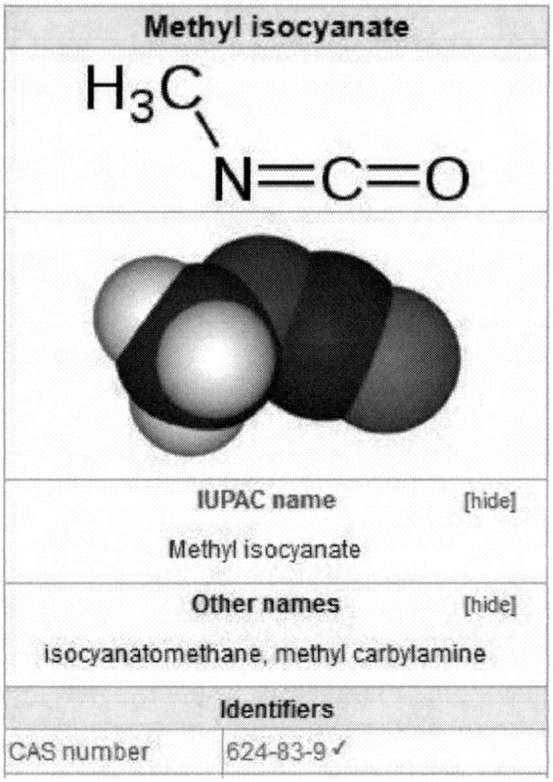

Figure 1. Molecular formula of MIC (ref. 6)

This concentration of electrons make it very reactive. It is a combination used in isocyanate chemistries to form higher molecular weight nitrogen containing organic chemicals that range from insecticides to insulation. It is an organic chemist's answer to a prayer. It is a chemical engineer's answer to a nightmare. The reason is the heat released when this chemical reacts with nearly any compound. If we had one chance to define a runaway reaction, this event is the example. When the organic chemist works on the bench scale, he is using test tubes, with high surface areas, creating plenty of cooling surface. When this reaches the application stage at 1,000's of gallons, heat transfer is much more challenging. Specifically, the combination of MIC with water releases up to 1,360 kJ/kg (585 BTU/lb) of MIC, or about 8,600 kJ/kg (3,700 BTU/lb) of water [7]. Recall that the heat of vaporization of water is approximately 2,200 kJ/kg (1,000 BTU/lb). There is enough energy available to raise the temperature of a mixture above the boiling point of the mixture and generate significant vapor pressure within a closed container. The reaction also forms carbon dioxide as a gas and this adds pressure to a closed system such as a storage tank. It was the entry of an estimated 1,000 to 2,000 lbs of water that initiated the runaway [8]. Abnormal amounts of chloroform, 1,500 to 3,000 lbs, was also present in Tank 610 creating a source for a chloride ion that attacked the stainless steel tank walls[8].

Several exothermic reactions happened in parallel and series. Three key reactions are shown in Figure 2 [9,10]. They are ordered based on the analysis of core samples collected after the accident and the initiator reaction, MIC reaction with water. The major product found was a trimer formed from 3 molecules of MIC (Rxn-1). This reaction is accelerated in the presence of a catalyst. In this case the catalyst was iron. Reported to be 0.18 to 0.26% or approximately 20 to 40 lbs within the tank [8].

Main residue, Trimethyl isocyanurate (TMI) ~40 to 55%, formed by a trimerization reaction:

$$3\ CH_3N{=}C{=}O \rightarrow \text{Trimethyl isocyanurate} \qquad -\Delta H = 36\ \text{kJ/mol}$$

MIC — trimer (a ring structure) — Rxn-1

1,3-Dimethylisocyanurate (DMI) ~ 13 to 20%

$$2\ CH_3N{=}C{=}O + HNCO \rightarrow \text{Dimethylisocyanurate} \qquad -\Delta H = ??\ \text{kJ/mol}$$

MIC — Isocyanic Acid — DMI — Rxn-2

Initiating reaction

$$CH_3N{=}C{=}O + H_2O \rightarrow CH_3NH_2 + CO_2 \qquad -\Delta H = 38\text{kJ/mol}$$

MIC + Water to Methyl Amine + Carbon Dioxide — Rxn-3

Figure 2. The three major exothermic reactions involved in forming the residual in Tank 610

Note: Heat of reactions estimated via Aspen Ver. 8.6 with existing database for chemicals.

The nature of a runaway exothermic reaction is explained by the Arrhenius effect. The Arrhenius effect models the observation that as temperature rises, reaction rates increase. An activation barrier exists between reactants and products. This barrier lowers as temperature increases. A common rule of thumb is that the reaction rates doubles every 10°C. On an absolute temperature scale, that is a doubling of a reaction rate for every 2 to 3% rise in absolute temperature. Once things are initiated, if the heat is not removed, the mixture will accelerate into high temperature regions until the reactant is fully diminished. Secondary exothermic reactions can occur and be even more severe in consequence such as the T2 incident. [11,12].

Another chemical engineering challenge with MIC is its physical properties. One specific property is its volatility at ambient temperatures. At 20°C its vapor pressure is 348 mm Hg [9]. Its boiling point is 39.1°C, not far from the ambient temperature in central India in the summer. As the exothermic reaction generates heat, and if that heat is not removed, then the contents warm up, and the vapor pressures of all components increase. In a closed vessel, this means increased pressure.

In Bhopal, once the runaway reactions gained momentum, the pressure increased from a reported 2 psig at 22:20 on Dec 2, 1984 to 10 psig 40 minutes later [13]. One hour and 15 minutes later the pressure had exceeded 55 psig and the relief system was initiated. Temperature within the tank was reported above 100°C at the time.

Runaway reaction are well known to chemical engineers designing reaction and storage vessels. The effects of a MIC runaway was on the minds of the design engineers involved in the Bhopal pesticide plant. To prevent tank explosions, relief systems are used. The Bhopal storage tanks 610, 611, and 619 had such a relief system. In this instance a 3" diameter rupture disk followed by a spring relief valve with a telltale pressure in between existed on the storage tank. The telltale gauge is required practice now; however, in the early 80's it was optional. Why is a telltale good practice? Rupture disks can corrode over time and form small pin holes. Thus, the telltale detects this leak beforehand. This relief system was the only safety system that worked – the tank did not explode. When the plant was commissioned in February 1980 many safety systems were in place and functional.

2.2. What failed?

In one sentence: The local plant management neglected their safety systems. In detail, the plant was under extreme financial constraints. It was only able to sell about 1/3 of its design capacity to the Indian market. Management was looking for possibilities to make the plant profitable. This included making the cooling system non-operational on the MIC storage tank in June 1984 [8] . The purpose was to keep the contents within the tank below 0°C. The reaction rate of MIC with itself is low at cold temperatures, and thus, the release of heat of reaction could be managed with a cooling system. This also kept the vapor pressure of MIC below 200 mm Hg [**Error! Bookmark not defined.**]. Further, at lower temperatures, the Arrhenius effect works backwards. In other words, very low reaction rates and heat generation, keeping things under control. Finally, the tank was supposed to be water free. It was designed to hold only the intermediate MIC produced within the plant and then bled off to formulate the pesticide.

Further management decisions followed. Management did not react urgently to a necessary pipe repair needed on the flare system. The flare system was a last resort safety layer intended to burn off CO and related organic vapors before release to the surroundings. The flare was taken out of service several weeks earlier due to a corroded pipeline [14].

In between the flare and the relief system was a NaOH scrubber system. Although the capacity of this system was not intended for the neutralization of 40 tonnes of MIC, if operational, then a portion of the event would have been mitigated. The scrubber system was placed on standby with the reasoning being that it served as a cost savings. Another safety layer removed.

Cost savings reached down to the operating personnel and personnel cut backs. As personnel were let go, the replacements that moved into various open process positions were not trained for their job at the level done when the original crew hired in. This created a disgruntled work force. Disgruntled to the point that one or more may have committed a sabotage act (discussed below).

The local politicians share responsibility in this accident. Initially the plant was built on undeveloped agricultural fields outside the city limits on land leased from the state government of Madhya Pradesh [9]. The central population of Bhopal city was located 1 to 2 km southward. All of the surrounding lands were zoned for industrial use [9]. However, with the siting of a major chemical manufacturing plant at the city edge a revenue source appeared for many economic endeavors. Several shanty towns were built up around the plant near the fence line or across a dividing road. These shanty towns had over 30,000 inhabitants, and were the locations of the many of the deaths and the severest of injuries. Zoning laws were in place to prevent such shanty towns. However, local politicians, in order to reach favor with their newly found constituents, looked the other way regarding the enforcement of the zoning laws.

We see above several layers of protection compromised over time. Specifically, surrounding open grounds were inhabited, the relief system, in its entirety was compromise, and training non-existent. Terrible decision making.

2.3.How was the MIC contamined by water?

In one sentence: It does not matter [15]! For the benefit of a novice audience, two explanations evolved. Remember major safety systems were in place. They were compromised well before the accident.

2.3.1. The Water Washing Theory

In this scenario a worker on site was assigned the task of connecting a hose to a collection point for flushing. The normal procedure required disconnecting the RVVH line from the filter line beforehand and placing a blank near valve labeled 19 in Figure 3 [16]. This was not done. Further, complications occurred when one of the four nozzles did not show any water exiting during the wash cycle. This washing continued for several hours. It is contended that water backed up into the relief valve vent header RVVH line and passed through a jumper line that connected to the a process vent header and that was connected to a leaking valve attached to the MIC storage tank 610. Figure 4 shows the entire pathway. This theory was dismissed by Union Carbide Investigators based on hydrodynamic arguments and the lack of liquid water being found in any low points of the PVH.

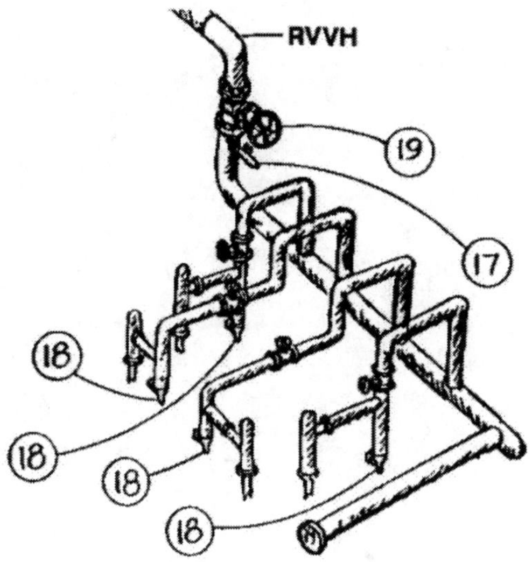

Figure 3. Filter system flushed. From [16]

2.3.2. Sabbatoge Theory

The Union Carbide investigation team concluded that water was intentionally added via a hose nearby the tank by a disgruntled worker. A missing gauge (labeled 10 in Figure 5) and a nearby water hose at tank 610 plus the testimony from an instrument supervisor lead to this conclusion. Kalelkar [16] presents a full discussion of this theory.

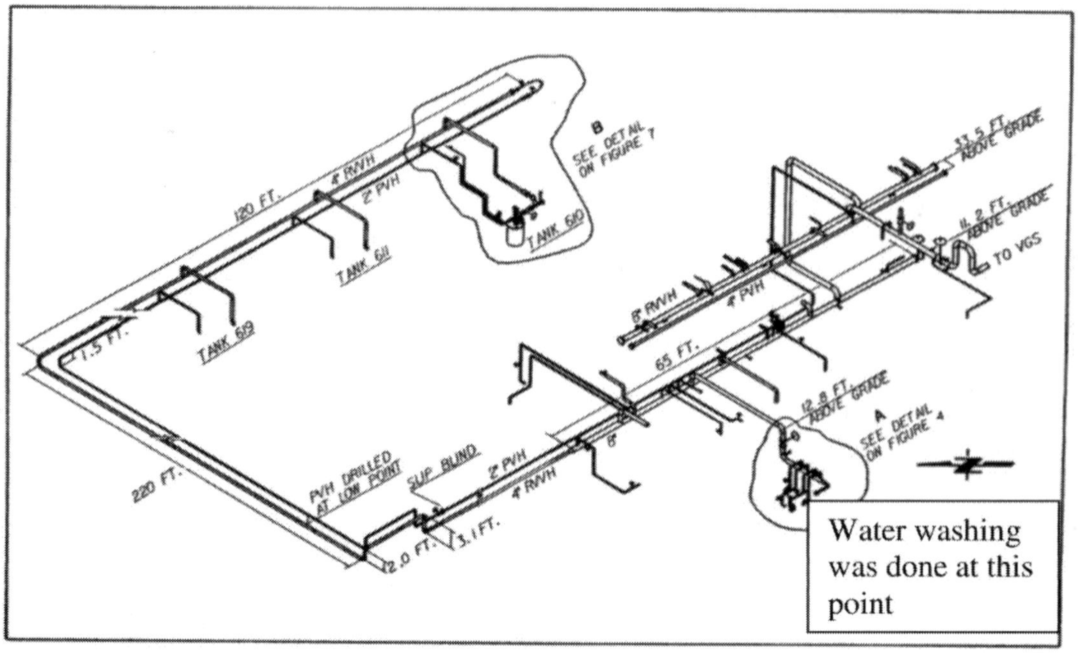

Figure 4. The layout of the PVH and RVVH. Significant distance and head exists from potential water source to storage tank. (adapted from 16)

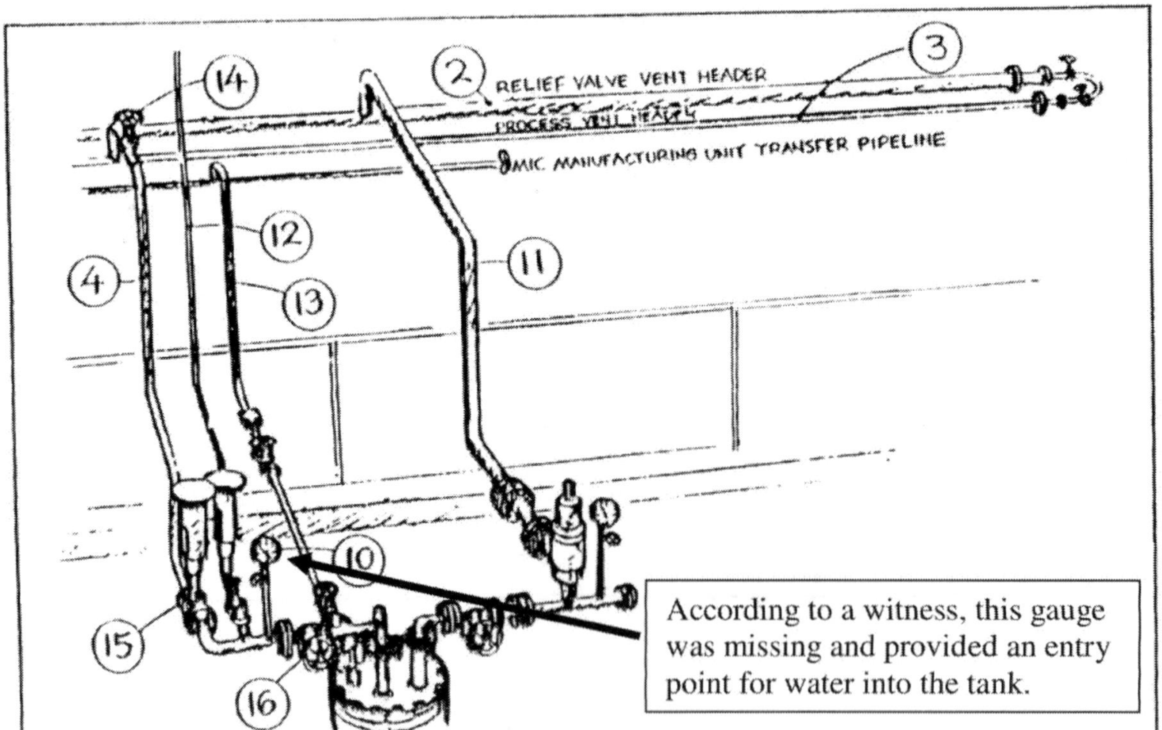

Figure 5. The process piping at the top of tank 610. (adapted from [16]

3. Impact of Bhopal on the Practice of Process Safety

It is interesting to note that the Bhopal accident sensitized the American community to chemical plant accidents. Previous accidents, such as Flixborough in 1974, Seveso, Italy in 1976 and a number of U.S. chemical plant accidents did not achieve this.

A large historical compendium could be written on what occurred after the accident. To complicate things, this history is filled with mis-information and frequent mis-statement of facts. The authors have done their best to substantiate the facts as best as possible. Only important and interesting items identified by the authors are shown below.

3.1. Historical Time Line

a) December 2 – 3, 1984, Bhopal, India accident occurs.

b) At 1 PM on December 3, Union Carbide held its first press conference on the accident at the Danbury Hilton Hotel. The company acknowledged that the accident had occurred at a plant owned by Union Carbide India Limited, with Union Carbide Corporation owning a 50.9 percent share. They announced plans to send medical assistance and technical experts to investigate the cause of the accident. They announced plans to halt production of MIC at the Institute, WV plant [17].

c) Several days after the incident, the then governor of Michigan, James Blanchard, announced that no MIC was used in Michigan. Several days later it was discovered that MIC was being shipped by tank truck from Institute, WV, to a chemical plant in western Michigan.

d) Union Carbide forms a technical team which assembles in India on December 6. The team spent 24 days at the site. However at the site, they were forbidden by the CBI (Central Bureau of Investigation, India) to inspect the tank, to interview plant personnel, and to examine plant records. Their initial focus was the safe removal of remaining MIC by using it in pesticide production. After about 17 days the team was allowed to examine Tank 610 and to take core samples from the remaining material in the tank [9]. The samples were taken back to the U.S. and an extensive experimental study was undertaken. Several years later the Indian government releases some 70,000 pages of documentation as a result of the legal discovery process [17]. The Union Carbide team believes the evidence supports the sabotage theory. The Indian Government believes the evidence supports the water washing theory.

e) On December 7, Warren M. Anderson, chairman and chief executive of Union Carbide, flew to India. He was immediately arrested by Indian authorities. After posting bail, he left the country never to return. Mr. Anderson died on September 29, 2014 at the age of 92 [18].

f) The accident marked the beginning of the end for Union Carbide. Prior to the accident, Union Carbide stock traded between \$50 and \$58. In early 1985 the stock traded at \$32 to \$40. Since the company was now worth more than its stock, it became the target of a hostile takeover by GAF Corporation. To respond to this takeover, Union Carbide sold its consumer products division, its most profitable division. In 1986 Union Carbide sold assets worth \$3.3 billion and repurchased 38.8 million shares to protect the company from further takeovers. Union Carbide retained its commodity chemicals division, which other chemical companies were divesting (UN University archives, 2015). The downward spiral continued until the remaining Union Carbide assets were purchased by Dow Chemical in 1999.

g) In February, 1989, the Supreme Court of India mediates payment of $470 million from Union Carbide and Union Carbide India and dismisses the charges against Warren Anderson [19]. Union Carbide pays the settlement within 10 days of the order [17].

h) On the 20th anniversary of the accident, a person claiming to be a representative from Dow Chemical is interviewed by BBC World. The representative states that Dow Chemical has agreed to a multibillion compensation package for the Bhopal accident. Within 23 minutes, Dow shares in Frankfurt fell 4.24%, wiping $2 billion off of its market value. The BBC eventually discovers the representative to be a hoax, and the stock shares rebounded [20].

i) In September 2006, the Indian Welfare Commission reported that all claims had been adjudicated with no further cases pending. All settlement money had been distributed [[21]. However, lawsuits in both India and the U.S. continue to this day.

3.2.U.S. Industrial Response

Much of the industrial response in the area of process safety after the accident was directed towards actions to prevent another Bhopal accident.

a) In March, 1985, the American Institute of Chemical Engineers (AICHE), responding to industry concern about chemical plant safety and the Bhopal accident, forms the Center for Chemical Process Safety (CCPS) "dedicated to improving the ability of engineers to deal with process hazards." [22]. AICHE provides $1 million in seed money and establishes CCPS as an independent organization. CCPS focusses on "conducting research and providing objective, technical information on issues related to safety in the manufacture, handling and storage of toxic and reactive chemicals"[23]. Thomas Carmody, former product safety director at Union Carbide, is selected as its first director.

b) In 1985 CCPS forms the Undergraduate Education Committee to improve process safety education in undergraduate chemical engineering curricula. The committee decides to publish a book of problems that can be used throughout the curriculum. Reid Welker and Charles Springer of the University of Arkansas are identified as authors and contractors to write this book. The book is published in 1990 and free copies are distributed to all undergraduate chemical engineering students in the U.S. An instructor's book is also published providing solutions to all the problems and supplemental material- these materials are distributed free to all chemical engineering faculty in the U.S. at that time.

c) In 1988, the Chemical Manufacturer's Association (now the American Chemistry Council) establishes an industry practice called Responsible Care®. All companies who are members of the American Chemistry Council (ACC) are required to follow the principles of Responsible Care®. Today, this program includes process safety, product safety and security features. More information on this can be found at the ACC website (responsiblecare.americanchemistry.com).

d) In 1990, the American Petroleum Institute publishes API Recommended Practice 750, *Management of Process Hazards*. This recommended practice was the predecessor for the OSHA process safety management regulation to follow.

e) In 1991 the CCPS Undergraduate Education Committee forms Safety and Chemical Engineering Education (SACHE). SACHE provides many instructional modules in process safety. In the last ten years many of these materials are converted to web-based resources. Today, over 30,000 undergraduate students annually view these resources and receive certificates of completion.

3.3. U.S. Regulatory Response

a) In 1985 California passes the Hazardous Materials Management law.

b) In January of 1986, New Jersey passes the Toxic Catastrophe Prevention Act.

c) On October 17, 1986, Congress passes the Superfund Amendments and Reauthorization ACT (SARA) to amend the Comprehensive Environmental Response, Compensation, and Liability Act (CERCLA) of 1980. This included the Emergency Planning and Community Right to Know Act (EPCRA). The EPCRA act provides requirements for local and state emergency planning with respect to hazardous chemicals, public access to chemical hazards information in their communities, and reporting requirements for facilities handling hazardous chemicals. The act also established Local Emergency Planning Committees (LEPC) [24].

d) In September of 1989 Delaware passes a regulation for the *Management of Extremely Hazardous Substances.*

e) In 1990 Congress passes the Clean Air Act. This act contains Section 112(r) which eventually results in the EPA promulgating the final Risk Management Plan Rule on June 20, 1996. This regulation requires all facilities that use extremely hazardous substances to submit information on potential releases that may affect off-site plant areas. This includes submission of total extremely hazardous chemical inventory amounts, a detailed hazard assessment estimating the consequences of the release, a prevention program and emergency response.

f) In 1992 OSHA promulgates the regulation entitled 29 CFR 1910.119: *Process Safety Management of Highly Hazardous Substances.* This rule is based primarily on the API 750 consensus standard and has fourteen elements.

4. The plant site 30 years later

4.1.Report of suffering

The tragedy still lives on. The media continues to report of people affected by the Bhopal incident. About 100,000 people who were exposed to gas continue to suffer illnesses such as cancer, and have visual, respiratory, immune and neurological disorders [25]. Birth defects such as twisted limbs, brain damage, and musculoskeletal disorders was reported to Reuters from Bhopal Medical Appeal activist Sanitath Sarangi [25].

4.2. The plant still stands, rusting away

Figure 6 shows a photograph from Reuters news photographer Danish Siddiqui [26]. The view is from the MIC plant towards the flare tower. The pipes and pipe rack are still standing. Insulation surrounding the pipes is beginning to erode away. What is shown in this manuscript are evidence of chemicals left behind. Recent photographs shows bottles of chemicals sitting, collecting cobwebs, awaiting proper removal.

Figure 7 is a recent Goggle Earth image of the former Bhopal UCIL Plant. It still stands. What is amazing is the dense build up around the plant (a portion of the build up can be seen in the upper right hand corner) and that the city has expanded from 1 million to 1.8 million. We have often wonder why the plant remains. This statement was found from an on-line article [26]:

> Rather than seeing the factory demolished, [Rachna] Dhingra said the ICJB [International Campaign for Justice in Bhopal] would like to see the site preserved as a memorial. 'It has a symbolic importance. We want it to be restored and to set a precedent in terms of environmental rehabilitation and medical rehabilitation. To the people of Bhopal these are not just structures; it's memory – memory of pain, but also, one day, memory of justice.'

This may explain why the plant is being left behind.

Figure 6. A photograph of the plant, 2014, by Danish Siddiqui, Reuters [25]

Figure 7. Satellite Image from Goggle Maps. Accessed 4 Jan 2015 [27]

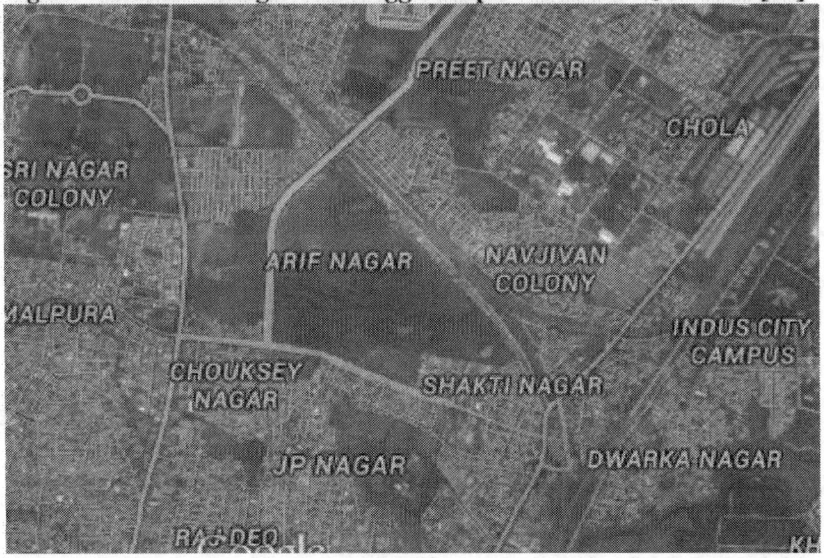

Figure 8. The surrounding communities now abutting the former UCIL Bhopal plant [27]

5. Conclusions

The authors conclude the following

1. After the Bhopal accident the United States was sensitized to chemical plant accidents.
2. Since a thorough investigation was not completed by a credible independent organization and no definitive report issued, speculation continues to run rampant on what actually occurred.

3. Union Carbide completed an extensive investigation of the incident. However, due to legal and liability issues most of this information was not published.
4. The accident was the beginning of the end for Union Carbide.
5. The accident resulted in considerable industry initiatives and government regulations related to process safety. This changed the practice of chemical engineering forever.

6. References

1. *Chem. Eng. News 62 (Dec 6, 1984) 6 and special C&E News Bhopal issue 63 (Feb 11, 1985) 14-65.*
2. http://www.aiche.org/ccps accessed 29 Dec 2014.
3. OSHA overview PowerPoint file https://www.osha.gov/pls/oshaweb/owadisp.show_document%3Fp_table%3DSTANDARDS%26p_id%3D9760 accessed 29 Dec 2014.
4. EPA, http://www.ecfr.gov/cgi-bin/text-idx?tpl=/ecfrbrowse/Title40/40cfr68_main_02.tpl accessed 29 December 2014.
5. http://www.csb.gov/ accessed 29 Dec 2014.
6. http://en.wikipedia.org/wiki/Methyl_isocyanate accessed 30 December 2014.
7. W. Worthy, "Methyl Isocyanate: The Chemistry of a Hazard," *Chem. Eng. News* 63 (6), (1985), 27-33, DOI: 10.1021/cen-v063n006.p027.
8. *Bhopal Methyl Isocyanate Incident Investigation Team Report*, Union Carbide Corporation, Danbury, Connecticut (March 1985).
9. T. D'Silva, *The Black Box of Bhopal, A Closer Look at the World's Deadliest Industrial Disaster*, Trafford Publishing, Canada, 2006.
10. S. Varadarajan, et. al *Report on Scientific Studies on the Factors Related to Bhopal Toxic Gas Leakage*, New Delhi, India (December 1985).
11. *T2 Laboratories Inc. Reactive Chemical Explosion* U.S. Chemical Safety Board,Available at http://www.csb.gov/investigations/completed-investigations/?F_InvestigationId=30. Accessed on 1 Jan 2015.
12. R. J. Willey, H. S. Fogler and M. B. Cutlip, "The integration of process safety into a chemical reaction engineering course: Kinetic modeling of the T2 incident," *Process Safety Progress* 30 (1), (2011), 39-44, DOI: 10.1002/prs.10431.
13. I. Eckerman, *The Bhopal Saga: Causes and Consequences of the World's Largest Industrial Disaster*, Universities Press (India) Private Limited, Hyderguda, Hyderabad, 2005.
14. S. Diamond,The Bhopal Disaster: How it Happened,The New York Times, A1, A6-A7, January 28, 1985
15. Personal quote from Dennis Hendershot.
16. A. S. Kalelkar, "Investigation of Large-Magnitude Incidents: Bhopal as a Case Study," The Institution of Chemical Engineers, I.Chem.E. Symposium Series No. 110, London, England, (1988), 561.
17. Browning, Jackson B., *Union Carbide: Disaster at Bhopal*, available at www.bhopal.com.
18. Martin, Douglas, "Warren Anderson, 92, Dies: Faced India Plant Disaster," *The New York Times*, October 20, 2014.

19. Endgame, *Chronology of the Union Carbide Corporation*, http://www.endgame.org/carbide-history.html, accessed 12 Jan 2015.
20 . CNN, "Bhopal hoax sends Dow stock down," 3 Dec 2004. http://www.cnn.com/2004/WORLD/europe/12/03/bhopal.hoax/index.html, accessed 12 Jan 2015.
21. Bhopal, *Union Carbide's Response Efforts to the Tragedy and the Settlement*, http://www.bhopal.com/UCC-Response-Efforts-to-Tragedy, accessed 14 Jan 2015.
22. Chemical Week, "AICHE points the way to chemical process safety," July 30, 1986.
23. CCPS Annual Report, September 1986.
24. EPA, http://www.epa.gov/superfund/contacts/infocenter/epcra.htm, accessed 12 Jan 2015.
25. http://www.buzzfeed.com/franciswhittaker/moving-images-showing-the-impact-of-the-1984-bhopal-disas#.iiXPQxxlz, Accessed 4 Jan 2015.
26. http://www.telegraph.co.uk/news/worldnews/asia/india/5978266/Bhopal-gas-disaster-the-legacy-lives-on.html, Accessed 4 Jan 2015.
27. www.google.com/maps Accessed 4 Jan 2015.
23. UN University Archives, http://archive.unu.edu/unupress/unupbooks/uu21le/uu21le0c.htm, Accessed January 12, 2015.

AUTHOR INDEX

AUTHOR INDEX